FIELD GUIDE TO THE
Birds of the Atlantic Islands

FIELD GUIDE TO THE

Birds of the Atlantic Islands

Tony Clarke

Illustrated by Chris Orgill and Tony Disley

CHRISTOPHER HELM
LONDON

First published in 2006, by Christopher Helm,
an imprint of Bloomsbury Publishing Plc

Reprinted in 2010, 2014

Copyright © 2006 text by Tony Clarke
Copyright © 2006 illustrations by Chris Orgill and Tony Disley

The moral right of this author has been asserted.

No part of this publication may be reproduced or used in any manner whatsoever without written permission from the Publisher except in the case of brief quotations embodied in critical articles or reviews.

Every reasonable effort has been made to trace copyright holders of material reproduced in this book, but if any have been inadvertently overlooked the publishers would be glad to hear from them.

Bloomsbury Publishing Plc
50 Bedford Square
London
WC1B 3DP

www.bloomsbury.com

BLOOMSBURY, CHRISTOPHER HELM and the Helm logo are trademarks of Bloomsbury Publishing Pl

A CIP catalogue record for this book is available from the British Library

Typeset and designed by D&N Publishing.

ISBN (print) 978-0-7136-6023-4

10 9 8 7 6 5

Printed in China by C&C Offset Printing Co Ltd.

MIX
Paper from responsible sources
FSC® C008047

The publishers and author would like to thank the following for permission to use their photographs:
Gaspar Avila Photography, photograsphersdirect.com: 16 (bottom); John Jackson 15 (bottom), 16 (top).

Front cover: Laurel Pigeon

Back cover (in descending order): Canary Islands Chat, Zino's Petrel, Azores Bullfinch, Raso Lark

Title page: Zino's Petrel (top), Fea's Petrel (bottom)

Contents

Preface	7
Acknowledgements	8
Introduction	
Geography of the islands	9
Climate of the islands	14
Island habitats	15
Ornithological history of the islands	18
Birdwatching areas	24
How to use this book	30
Colour plates	34
Species accounts	173
Appendices	
A Update for 2005 and 2006	331
B Distribution of species on the Atlantic Islands	
i. Canary Islands	332
ii. Madeira and the Salvage Islands	340
iii. The Azores	346
iv. Cape Verde Islands	351
C Distribution of endemic taxa on the Atlantic Islands	356
Bibliography	359
Index	363

Dedicated to the memory of Luis Monteiro and Alec Zino, two men who furthered the ornithological knowledge of the region immensely. Their loss has left a great void in the Azores and Madeira respectively, but their work will forever be their legacy in the region.

Also dedicated to the memory of Brian Johnson, with whom I spent many hours in the field on the Canary Islands; he was a personal friend and will be greatly missed.

PREFACE

THE IDEA FOR THIS GUIDE first stirred in my mind in 1988 when I visited the Canary Islands for the first time. My initial stay on Tenerife quickly made me aware that the available literature in English was somewhat limited and rather outdated. I was surprised, to say the least, that the first species of pipit I recorded on the island was actually Red-throated Pipit *Anthus cervinus* and not the near-endemic Berthelot's Pipit *Anthus berthelotii* as one would have expected. After my visit I was unable to find a single reference to Red-throated Pipit having occurred in the Canary Islands. What was even more amazing was that I had seen a group of about fifteen individuals. Obviously, a publication updating the status of all species within the Canary Islands was clearly needed to assist the visiting birdwatcher. However, it was not until a few years later, when I was approached by the publishers, that the idea came to fruition and work began in earnest.

It was decided from the beginning that a wider region should be covered, and it was agreed to base the book on the pioneering four-volume work *Birds of the Atlantic Islands* by David Bannerman, which was published between 1963 and 1968. This meant that the book would include not only the Canary Islands but the Azores, Madeira and the Cape Verde islands as well. Since the start of work on this guide, the number of observers visiting the various archipelagos has increased dramatically, particularly in the Azores and Cape Verde, both of which were severely underwatched in the past. A direct result of this has been the regular addition of species to the lists of the various island groups, and in particular to those that have been the least visited in the past. I have tried to keep the text as up-to-date as possible by incorporating these annual additions, including many records that have never reached the official rarities committees, but I have endeavoured to maintain an even treatment of each archipelago and its birds, and to obtain photographic evidence for all species not submitted to the authorities. Inevitably, there had to be a cut-off date after which no further information could be added, and this was set at the end of 2004. Despite this, the amazing autumn of 2005 on the Azores could not be ignored – a summary of the new records for 2005 is given in Appendix A.

The purpose of this book is to give visiting birdwatchers a better idea of the status of the various species within each archipelago, as well as providing a means to identify them. I have tried to cover the identification of each species in as much detail as space has allowed, with particular emphasis on the endemic taxa of the region. However, for more detail, this guide should be used alongside one of several authoritative Western Palearctic works. We have referred to the region as the Atlantic Islands, but the name by which these four archipelagos are often known is Macaronesia (not to be confused with any island groups in the South Pacific). I hope that this book will add to our collective knowledge of the birds of the region, and that it will be regarded as an invaluable addition to many people's hand luggage.

ACKNOWLEDGEMENTS

FIRST, I WOULD LIKE TO thank all those observers, too numerous to mention, who have contributed to the ever increasing knowledge of the birds of the region by submitting tour reports and/or species lists from their visits to the Canary Islands. Many of these have been sent to me direct but others have been made available to a wider audience via the internet. Specialised help for the various archipelagos has come from the late Luis Monteiro, Fátima Melo, Mark Bolton and Gerald Le Grand for the Azores; the late Alec Zino, Frank Zino, James Siddle and Niklas Holmström for Madeira; Keith Emmerson, Aurelio Martín and Toño Lorenzo for the Canary Islands; and Kees Hazevoet for the Cape Verde Islands.

Others whose help has been most appreciated include Derek Turner for his time spent on correcting the first draft of the text and the use of his library, Tony Blake also for the use of his library, Tommy Frandsen for his company on my first visit to the Azores, Phil Hansbro for encouragement and company on my first visit to the Cape Verde Islands and Helder Costa for his assistance with the Portuguese common names. Since my first visit in 1998, the Azores have been watched by an increasing number of visiting birders, some of whom have made many journeys to the islands. This has resulted in a continual flow of records (particularly in autumn) over the last few years, culminating in the astonishing autumn of 2005, which occurred too late for all the new species to be dealt with in full within the text of this book.

For all their excellent data, special thanks to Peter Alfrey, Joël Bried, Bosse Carlsson, Gonçalo Elias, Tim and Carol Inskipp, Jordi Marti, Stefan Pfützke, Staffan Rodebrand and Kris de Rouck, as well as Svante Åberg, Mashuq Ahmad, Pete Akers, Richard Allison, Michael Averland, Theo Bakker, Felipe Barata, Patrik Bergdahl, Hans Bister, Phil Bloor, Leo Boon, Simon Boswell, Simon Buckell, P. Catry, Dirk Colin, Andrew Collins, Tim Collins, Pierre-André Crochet, Erik Didner, Philippe Dubois, Hugues Dufourny, Guus van Duin, Euan Dunn, Tim Earl, Anne-Marie and Erik Ehmsen, Kari Eischer, Ian Enlander, Michael Fricke, Bob Furness, Gunner Grandström, Martin Gray, Mags Grindle, Ricard Guerreiro, Seppo Haavisto, Megan Hall, Phil Heath, Jens and Heidi Hering, Peter van Horssen, Pertti Hotta, Hannu Huhtinen, Jukka Ihanus, Kjetil Jensen, Frédéric Jiguet, H. Lauruschkus, H. Legge, Phil Lillywhite, Rami Lindroos, Steve Lister, Rasmus Maki, Saana and Jouni Meski, Dominic Mitchell, Colm Moore, Lars Mortensen, Lars Olausson, Klaus Malling Olsen, Martin Oomen, Gert Ottens, Lars Petersen, Tina Petersen, Markku Rantala, Ilkka Sahi, K. Schmitz, Michael Schmitz, Joachim Seitz, Bill Smith, Kirsi Sorvali, Rainer Sottorf, Jonas Starck, W. Steenge, Ingvar Torsson, William Velmala, Keijo Wahlroos, A. Welz and Maxime Zucca. Others who have provided additional help in one way or another include Ruben Barone, Bernard Couronne, Peter Crocker, Paul Donald, Stefan Fischer, Nico Geiregat, the late Brian Johnson, Manfred and Heidi Koch, Barry Lancaster, Cesar Palacios, Rob Payne, Marcel Peters, Dave Sargeant, Domingo Trujillo and the staff of the Natural History Museum at Tring.

I am indebted to Chris Orgill and Tony Disley, who painstakingly painted the superb plates that accompany this book. I would also like to thank Nigel Redman and Jim Martin at Christopher Helm Publishers for their encouragement and support during the long gestation of the book and for their considerable efforts in producing a very handsome title. Particular thanks are due to the copy editor, Ernest Garcia, who also helped greatly with the introductory and caption texts.

Finally, I would like to thank my wife Luisa, and daughters Emma and Chelsey, who have had to put up with long periods without me but have always shown considerable support for a project that has taken so much of their time. This book has certainly been a labour of love and without theirs it may never have appeared at all.

Tony Clarke

I AM PARTICULARLY GRATEFUL to the following for their help and guidance during the production of the plates: Don Sharp at Wollaton Hall Natural History Museum, Nigel Killips at Leicester Museum, the staff at The Natural History Museum, Tring, Nigel Slater, Mike and Tess Falgate, Martin Butler and Keith Corbett, Jean Viggers for her calming influence, Ren Hathway for his encouraging chats when the mountain looked very high, Nigel Redman and Jim Martin at A&C Black, and Tony Disley for his excellent contribution … and of course my wife Karen and son Benjamin. Without their support I would never have been able to contemplate the project at all.

Chris Orgill

GEOGRAPHY OF THE ISLANDS

Four archipelagos together make up Macaronesia, a name derived from the Greek *makaros* (happy) and *nesos* (island). The 'happy islands' are the tips of a series of undersea volcanoes, some of them still active. In terms of their natural attributes, they are indeed fortunate on account of their generally benign climate, varied landscapes and distinctive flora and fauna, these last including a number of endemic species. All four island groups are included within the Palearctic faunal region, the Azores being the westernmost territory of the Palearctic and the Cape Verdes a tropical outpost. The Azores and Madeira are Portuguese territories and the Canary Islands are Spanish, but Cape Verde is a republic, having gained independence from Portugal in 1975.

CANARY ISLANDS

The Canary Islands are situated between 27°37' and 29°24'N and 13°23' and 18°08'W. They consist of seven main islands, various islets and numerous rocky outcrops, with a total surface area of 7,490 km².

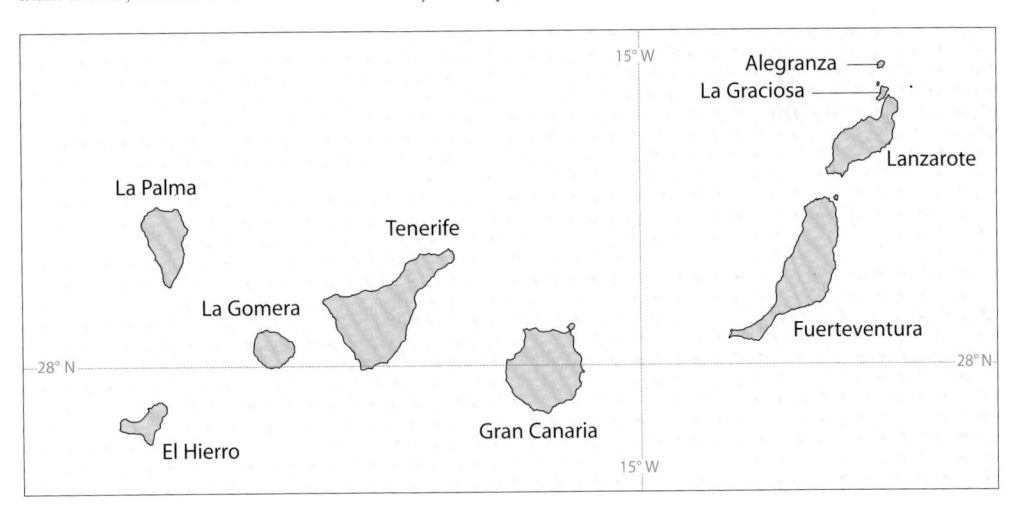

El Hierro The smallest and westernmost of the Canaries, El Hierro was once regarded as the end of the world; the zero meridian formerly passed through the Punta de Orchila, prior to Greenwich obtaining this distinction in 1911. The island covers 269 km², rises to 1,500 m at Malpaso and has a population of around 7,600. The uninhabited islets of Roques de Salamor off the north coast are known for their seabird colonies.

La Palma The largest of the three western islands, La Palma has a surface area of 708 km² and is the second-highest island in the archipelago, reaching 2,426 m at El Roque de los Muchachos. The island's centre is dominated by a huge volcanic crater, the Caldera de Taburiente, north and east of which are many spectacular ravines (barrancos) leading to the coast. The human population is 78,200, spread irregularly throughout the island. La Palma is the site of the last volcanic eruption on the Canaries where, in 1971, the Volcán de Teneguía erupted in the south of the island.

LA GOMERA The second smallest of the chain with a surface area of 370 km^2, La Gomera is roughly circular with many deep barrancos and deep valleys radiating from the island's centre, the 1,484-m Garajonay. The population is c. 16,800 and the capital is San Sebastián.

TENERIFE The largest and highest of the Canary Islands, Tenerife has a surface area of 2,034 km^2 and a maximum height of 3,718 m at the volcano Pico del Teide, which is just one of a chain of mountains that divides the island into the wetter north and drier south. The caldera in Cañadas National Park is immense, being 17 km long and approximately 10 km wide. The population numbers around 678,000 with the largest concentration centred on Santa Cruz de Tenerife and neighbouring La Laguna, which are the current and former capitals of the island respectively.

GRAN CANARIA The third largest of the archipelago with a surface area of 1,560 km^2, and the third highest, rising to 1,950 m at Pico de las Nieves. Gran Canaria is roughly circular, cut into sections by numerous deep barrancos and with an area of extensive sand dunes on the south coast. The population is around 730,000, making it the most densely populated of the archipelago. The capital, Las Palmas de Gran Canaria, is in the north-east of the island and a large percentage of the inhabitants live there.

FUERTEVENTURA In surface area, 1,660 km^2, this is the second largest of the Canaries and it is reported to be the oldest, having formed some 17–20 million years ago. The highest point is Pico de la Zarza at 807 m, on the Peninsula de Jandía, in the south of the island. The population is just in excess of 49,000, the majority within the municipality of the capital, Puerto del Rosario. Just off the north coast lies the small islet of Lobos. This island is the nearest to continental Africa, the distance being in the region of 100 km.

LANZAROTE The easternmost of the main islands, Lanzarote covers 846 km^2 but rises only to 671 m, at Peñas del Chache amongst the Famara cliffs. Most of the island comprises stony and sandy plains. The number of inhabitants is c. 130,000, most in the municipality of the capital, Arrecife. North of Lanzarote, less than 1 km offshore, is the small island of La Graciosa. which occupies only 27 km^2 and has a population of around 650 people. This island, and the islets of Alegranza, Montaña Clara, Roque del Oeste and Roque del Este, form what is often referred to as the Chinijo Archipelago.

MADEIRA

The islands of the Madeiran archipelago are situated between 30°01' and 33°80'N and 15°51' and 17°16'W, to the north of the Canary Islands.

MADEIRA The largest of the islands, although not the first to be discovered. The nearest mainland to Madeira is Morocco, 608 km away. The surface area is 741 km^2 and the island rises to 1,862 m at the summit of Pico Ruvio. The neighbouring Pico de Arieiro, famous as the only known breeding location for Zino's Petrel, is only slightly lower at 1,810 m. The population numbers around 280,000, of whom 86% live in the warmer south, and approximately 100,000 in the capital Funchal.

PORTO SANTO Some 37 km north-east of Madeira is the only other inhabited island in the archipelago, Porto Santo, with a population of c. 5,000 inhabitants. The island covers c. 50 km^2 and the highest point is Pico do Facho at 517 m. There are many offshore islets including Ilhéu de Baixo, Ilhéu de Cima, Ilhéu de Ferro and Ilhéu das Cenouras.

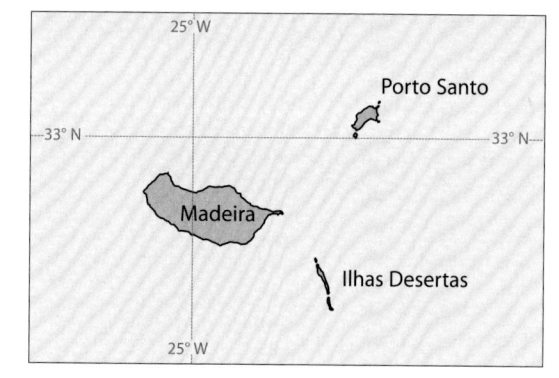

ILHAS DESERTAS This uninhabited group famous for its seabird colonies lies 16 km south-east of Madeira. Three islands comprise the group – Ilhéu Chão, Deserta Grande and Bugio, with the highest being Deserta Grande at 479 m.

ILHAS SELVAGENS (SALVAGE ISLANDS) Although much closer to the neighbouring Canaries, these small, windswept islands are officially part of the Madeiran archipelago. They lie 216 km south of Madeira and are well known for their seabirds, particularly the large colony of White-faced Storm-petrels.

THE AZORES

Situated in the mid Atlantic between 36°55' and 39°43'N and 25°01' and 31°07'W, the Azores comprise nine main islands in three distinct groups: eastern, central and western. Their total surface area is c. 2,250 km² and the islands stretch for over 480 km from east to west. The nearest continental land is Cabo Roca in Portugal, 1,408 km east of Santa Maria.

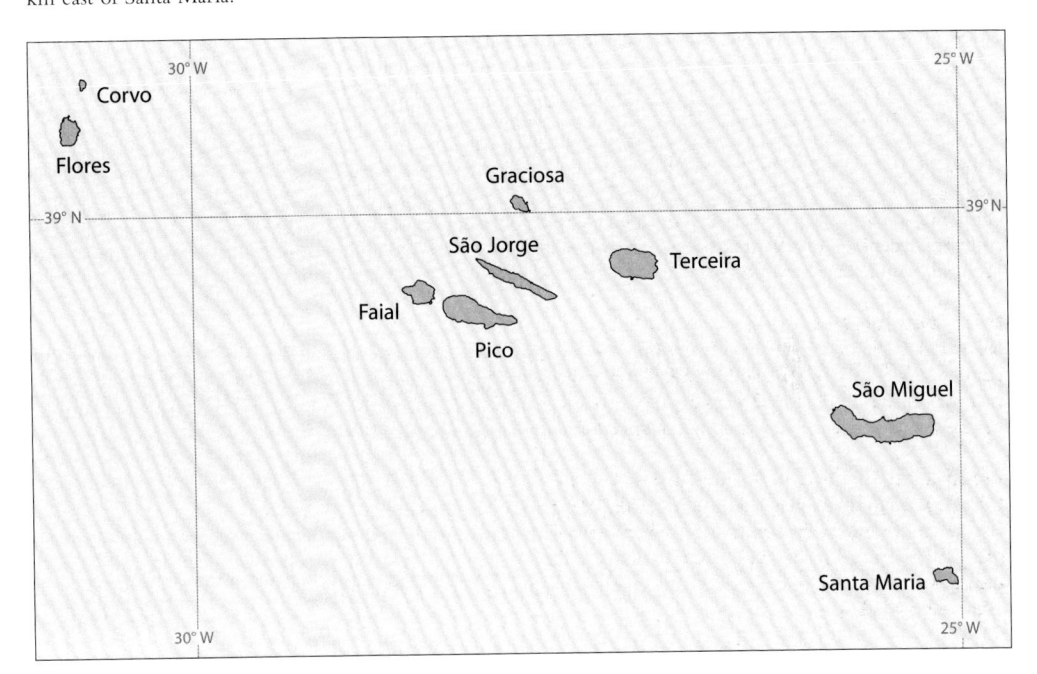

Corvo The smallest of the islands in the archipelago, Corvo covers a mere 17.5 km² and has just 400 inhabitants, primarily in the island's capital, Vila Nova do Corvo. The island is dominated by Monte Grosso, which rises to 777 m and has two crater lakes, c. 360 m below the rim and dotted with tiny islets.

Flores The westernmost island of the archipelago, with an area of 143 km², Flores is separated from Corvo by a strait of c. 24 km; together, they form the western group of islands. The population of Flores is c. 4,000, mostly in the municipalities of Santa Cruz and Lajes. The highest point is Morro Grande at 941 m and the island is almost completely surrounded by high sea cliffs. The south-central region comprises a large area of moorland that is very swampy and the location of the island's seven main lakes. Climatically, Flores receives the most rainfall of any island in the group.

Faial Separated from the western islands by over 190 km, Faial is the westernmost of the central group, with an area of c. 173 km² and rising to a maximum height of 1,043 m at Cabeço Gordo. The caldera on this island is c. 2 km in diameter, 400 m deep and, as usual, has a lake in its base. The population of Faial is c.15,000 and the capital is Horta.

Pico Separated from Faial by a mere 7 km this island is named after, and dominated by, the highest peak within the entire archipelago, the imposing, 2,351 m, Pico. The second-largest island of the group, Pico has a total surface area of 447 km² and a population of some 15,000 inhabitants.

São Jorge Situated just 17.6 km off Pico, this rather long, thin island covers 246 km² and has a central montane ridge which rises to 1,053 m at Pico da Esperança. The main human centres are the municipalities of Velas and Calheta, and the island has a total population of c. 9,500. Off the eastern end lies the small islet of Topo which holds one of the most important seabird colonies in the archipelago,

Graciosa Some 34.6 km north of São Jorge lies the smallest island of the central group with an area of c. 61 km². Graciosa is also the least populated of this group, with just 4,500 inhabitants centred on the two main towns of Santa Cruz and Lajes. It is the lowest lying of the Azores, the highest point being just 402 m above sea level on the southern rim of the caldera. Offshore are two small islets, Praia and Baleia, both of which are important for colonies of breeding seabirds.

TERCEIRA About 52 km east of São Jorge and the easternmost of the central group is Terceira. Only slightly smaller than Pico, this is the third largest of the archipelago with a surface area of 382 km^2. It is elliptical in shape and, with a population of $c.$ 58,000 inhabitants, is the second-most populous island in the chain. The highest point is the Serra de Santa Barbara, which is the youngest of the islands' volcanos, and reaches 1,021 m. Terceira's municipal capital is Angra do Heroísmo and Praia da Vitória is the other important town. Off the south coast lie a small group of islets known as the Ilhéus das Cabras (Goat Islands) which are a breeding area for seabirds. Terceira is also home to the well-known quarry at Cabo da Praia, famous for the number of American shorebirds that occur in autumn.

SÃO MIGUEL The westernmost island of the eastern group and some 145 km east of Terceira, São Miguel is the largest of the Azores, covering 759 km^2, and is also the most populated, with around 140,000 inhabitants. The highest point is Pico da Vara which rises to 1,105 m at the island's eastern end. The main municipal centre is Ponta Delgada, but other reasonable-sized townships are Ribeira Grande, Vila Franca do Campo and Povoação. São Miguel also boasts the largest crater in the Azores, at 12 km wide, which contains the two lakes at Sete Cidades, Lagoa Azul and Lagoa Verde. Apart from these crater lakes in the west of the island there is also one in the centre, Lagoa do Fogo, and one further east, Lagoa das Furnas. Offshore islets include Ilhéu de Vila Franca but this does not have any important seabird colonies.

SANTA MARIA The easternmost of the entire chain, the nearest to mainland Europe and 88 km south of São Miguel, Santa Maria is the third-smallest island in the archipelago with a surface area of 97 km^2. It rises to 587 m at the summit of Pico Alto. The population is $c.$ 5,500, most of whom live in the capital, Vila do Porto. Off the south coast lies Ilhéu do Vila, another small islet known for its seabird colonies.

CAPE VERDE ISLANDS (CABO VERDE)
The Cape Verdes are situated between 14°48' and 17°12'N and 22°40' and 25°22'W. They comprise ten main islands, of which one is uninhabited, in two distinct groups, the Windward Islands (Barlovento) and the Leeward Islands (Sotavento). They have a combined surface area of 4,033 km^2 and lie $c.$ 620 km west of Senegal.

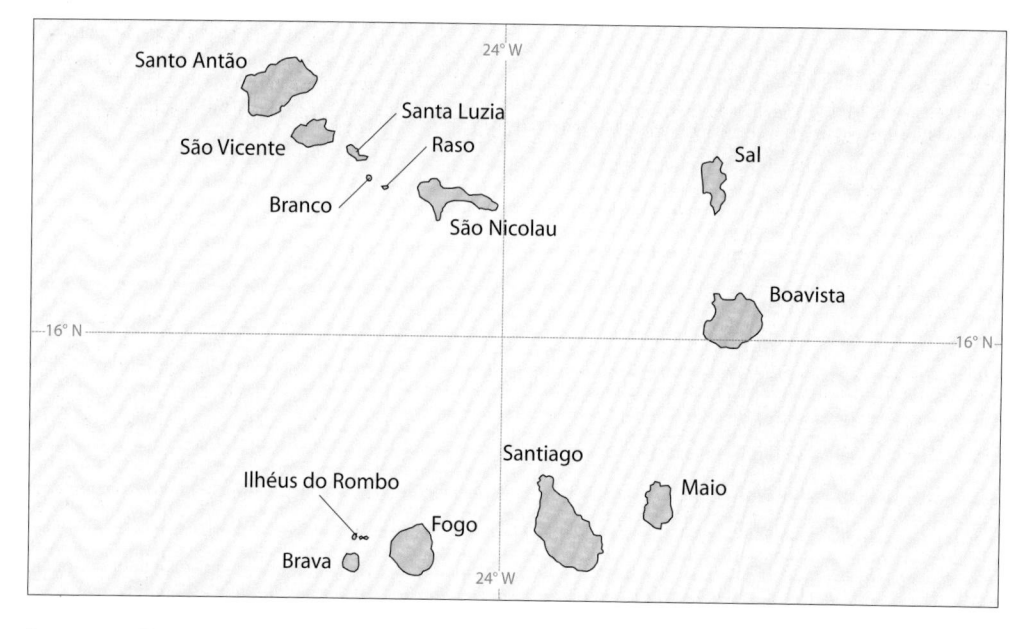

ILHAS DO BARLOVENTO
SANTO ANTÃO The northernmost and westernmost island of the archipelago, Santo Antão is the second largest, with a surface area of 779 km^2, and rises to 1,979 m at Tope de Coroa. A rugged montane ridge runs north-east to south-west, with several peaks above 1,500 m, making Santo Antão among the most spectacular of all the Cape Verdes. It has a population of $c.$ 45,000 with the main settlement at Vila da Ribeira Grande, on the coast at the junction of two spectacular, deep, well-cultivated valleys, Ribeira Grande and Ribeira de Torre. The island is also the only one of the archipelago with permanent streams, which are features of the Ribeira da Paúl and Ribeira Janela.

São Vicente South-east of Santo Antão lies the smaller but more densely populated São Vicente, which covers just 227 km^2 and reaches a height, at Monte Verde, of a mere 774 m. The lack of any substantial mountains makes this island very dry, and the landscape is dominated by barren hillsides amongst stony plains and coastal dunes. The population numbers around 55,000, with an estimated 40,000 in the capital Mindelo, the second-largest town in the islands. São Vicente is currently the most commercial island of the archipelago.

Santa Luzia, Branco and Raso These three barren, uninhabited islets lie between São Vicente and São Nicolau. By far the largest is Santa Luzia with a surface area of 35 km^2 and a high point of 395 m. Formerly known for its seabird colonies, these seem to be now deserted. Next is Raso, which covers 7 km^2. This island rises only to 164 m; it has extensive areas of relatively flat land and patches of sparse vegetation inhabited by the endemic Raso Lark. Raso also holds a number of large seabird colonies. The third island, Branco, lies just a few kilometres to the west of Raso. It is nothing more than a large rock, with a surface area of 3 km^2. Reaching a height of 327 m, Branco boasts one of the most important seabird colonies in the Cape Verdes.

São Nicolau Formerly important for agriculture, recent climatic changes have had a devastating effect on this island and very little commercial agriculture remains. The vast majority of the 20,000 or so inhabitants live in the west, which receives more rainfall than the drier eastern 'arm'. The capital, Vila da Ribeira Brava, is located inland in a valley, and the surface area is 343 km^2, rising to a height of 1,312 m at Monte Gordo.

Sal Apart from uninhabited Santa Luzia, Sal is the most barren of the archipelago. It is currently profiting a little from a newly developed tourist industry, but otherwise the island comprises stony plains and sand dunes. Sal rises to 406 m at Monte Grande in the north-east and has a surface area of 216 km^2. The population is c. 10,000, most of whom live in Espargos, the island's capital, attracted by work at the international airport and the new hotels. There is very little agriculture and in the distant past the economy was reliant on saltworks at Pedra de Lume and near Santa Maria, but the latter is now virtually unworked.

Boavista The easternmost and southernmost of the Ilhas do Barlovento and the easternmost of the entire archipelago, Boavista is the third largest of the group with a surface area of 620 km^2, but is rather flat with the high point, Monte Estância, just 387 m above sea level. The population numbers just 4,000, most of these in Sal Rei and Rabil. Boavista is very dry and mostly comprises mobile dunes and stony plains, interspersed in the interior with date palm-fringed oases. The lagoon at Rabil and the seasonal pool at Curral Velho have attracted many species new to the archipelago. Offshore are some small islets noted for their seabirds, such as Ilhéu dos Pássaros, Ilhéu de Baluarte and Ilhéu de Curral Velho. The last of these has the distinction of being the last location in the Western Palearctic where the Magnificent Frigatebird clings on as a breeder.

Ilhas do Sotavento

Maio The island of Maio covers 269 km^2 and its highest point is Monte Penoso at 436 m. The population is small at c. 4,500 with most in the capital, Vila do Maio, near which is a saltworks. Off the north coast is Ilhéu Laje Branco with an important seabird colony.

Santiago The largest of all the islands covering 991 km^2, Santiago rises to 1,392 m at Pico da Antónia and is the most densely populated with c. 160,000 inhabitants, of which around 50,000 are in the capital, Praia. The island is dominated by two mountain ranges, Serra do Pico da Antóia and, further north, the Serra da Malagueta. Santiago hosts varied habitats, ranging from lush valleys around São Jorge dos Órgãos, forested mountain slopes, and the agricultural zone, through to stony plains in the south of the island. Santiago also has a permanent, but often very small, lagoon just south of Pedra Badejo on the east coast.

Fogo Basically just a huge volcano, Fogo has a surface area of 476 km^2, emerging from the sea to reach the highest point in the archipelago at Pico Novo, a staggering 2,829 m above sea level. The population numbers around 40,000, the majority in São Filipe. Fogo has a live volcano which last erupted as recently as 1995.

Brava The smallest of the Ilhas do Sotavento, Brava has a population of only around 10,000 people. Brava covers around 64 km^2 but includes a dry central tableland and more mountainous regions, the highest part being Fontainhas at 976 m.

Ilhéus do Rombo North-northeast of Brava lie the uninhabited Ilhéus do Rombo, a small group with two main islets—Ilhéu Grande covering 2 km^2 and reaching 96 m at Monte Grande and Ilhéu de Cima with 1.5 km^2 and a high point of just 77 m—and three even smaller rocky outcrops, Ilhéu Luiz Carneiro, Ilhéu Sapado and Ilhéu do Rei. The only vegetation is some scattered grassy areas on the larger islets. Only Ilhéu de Cima is noted for its current seabird colonies, but large colonies seemingly existed in the past on Ilhéu Grande.

Climate of the Islands

The atlantic islands are generally warm, but there are wide variations in rainfall between the archipelagos, and even between individual islands. These climatic variations determine the dominant vegetation and habitats on the islands, and therefore strongly influence the avifauna.

CANARY ISLANDS

The Canary Islands have a Mediterranean climate, with warm, dry summers and cooler, wetter winters. Air temperatures are moderated in summer by the influence of the cool, south-flowing Canaries current, but sea temperatures inshore typically range from about 18°–24°C. Average sea-level temperatures on land range from 14°–26°C with hotter periods in summer, when the dry easterly 'calima' blows from the Sahara, laden with dust. The western islands receive much more rainfall than the easternmost ones, partly because of their higher relief, and also due to their greater exposure to the Atlantic. For example, La Palma has an annual average rainfall of 731 mm whereas Fuerteventura averages just 111 mm. The heaviest rainfall is associated with the passage of Atlantic depressions in winter, often accompanied by strong or gale-force westerly winds. The predominant northeasterly trade winds also bring winter rainfall, especially to northern parts of the islands. The climate varies considerably locally, depending on exposure and relief. In particular, there is a great difference between the climate at sea level and at high altitudes; the summit of Pico de Teide on Tenerife is snow-capped for some of the year, for example.

MADEIRA

Madeira enjoys a pleasant Mediterranean-type climate year-round, with temperatures moderated by the surrounding sea. Mean surface temperatures have an annual range of 18°–23°C. Average temperatures on land range from 13°–24°C with a mean annual air temperature of 19.5°C. The winter months can be quite wet; the monthly rainfall between October and March ranges from 64–89 mm, but the summer is dry. The climate is influenced by the northeasterly trade winds, which occur year-round but predominate in summer. Most rain is associated with westerlies, especially the short-lived but strong westerly and southwesterly gales that occur in winter. The 'leste', a very dry easterly wind laden with fine Saharan dust, blows occasionally, bringing hot spells in summer. The high altitude of the upper reaches of the island of Madeira gives the mountains a cooler, wetter climate than elsewhere in the archipelago.

AZORES

The Azores have a temperate maritime climate, benefiting from the warming influence of a branch of the Gulf Stream. This arrives from the north-west at Flores and Corvo and moves east-south-east past Santa Maria, keeping the mean monthly sea and air temperatures within the ranges of 14°–23°C and 13°–23°C respectively. Summer days are often pleasant but the Azores weather is characterised by its changeability, especially between October and April, when sunny periods are interrupted by frequent showers, giving what the locals call the 'days of the four seasons'. Rainfall occurs year-round but the wettest months between October and March average between 14 and 21 wet days and see up to 150 mm of rain. The driest months are June to August, each with nine to 11 wet days and 43–49 mm of rain. Fog occurs frequently and often disrupts inter-island communication. The northern mid-Atlantic anticyclone gives fine weather when it is centred over the Azores but may shift to the south, especially in winter; this brings Atlantic depressions from the west, sometimes accompanied by gale-force winds. Conversely, shifts in the Azores anticyclone to the north, mainly in summer, bring warm winds from the east or south-east.

CAPE VERDE ISLANDS

The Cape Verdes are the hottest and driest of the four archipelagos. They have a tropical climate with two seasons; a dry season from December to July and a humid season between August and November. Temperatures are relatively high, ranging from 20°–35°C, but 25°C–29°C on average. Most rainfall is associated with the southwest monsoon from August to October, but the more mountainous islands usually see much more rainfall than the lower, flatter ones. The wettest month is September, when rainfall may reach 114 mm in the hills, but prolonged droughts are frequent and there is usually little or no rainfall for nine months of the year. The rather dry north-east trade winds are strong early in the year, from December to April. A dry hot wind, the harmattan, blows intermittently between October and June, bringing Saharan dust and producing very hazy conditions.

ISLAND HABITATS

The four archipelagos between them offer a broad diversity of habitats, from deserts to moist cloud forests, all distinguished by a high degree of endemism among their flora and fauna. For example, more than a quarter of the vascular plants and nearly half of the arthropods of the Canary Islands are endemic. All the island groups have been inhabited for centuries and human activities, including the impacts of introduced animals and plants, have had profound effects on the local environments; deforestation has led to desertification at lower levels, particularly in the Cape Verdes and Canaries. The remaining patches of native habitat are of the greatest importance and some of them are now protected.

CANARY ISLANDS

The more mountainous western islands, La Palma, El Hierro, La Gomera, Tenerife and Gran Canaria, were originally forested above 500–1,000 m, and large wooded areas remain. The coastal regions are characterised by Canarian Palm *Phoenix canariensis* and a diversity of *Euphorbia* and other plants of semi-arid habitats, many of them endemic. The lower slopes are heavily affected by agriculture but originally supported extensive scrub. The Dragon Tree *Dracaena draco* is one of the most striking endemic plants of this region. The moist and shady laurel forests, the 'laurisilva', which include four members of the Lauraceae and a diversity of other shrubs, extend between 600 m and 1,500 m. Similar formations occur in Madeira and the Azores. Areas of heathland, the 'fayal-brezal', are typical of poorer soils from 500 m to 1,700 m, between the laurisilva and the pine forests higher up. Forests of the endemic Canarian Pine *Pinus canariensis* are found at low elevations in southern areas, but in the northern parts of the islands are found only from 1,200 m to 2,400 m. The highest mountains have areas of xerophytic vegetation, above 2,000 m on La Palma and Tenerife. Many of the islands have sea cliffs. Natural wetlands, however, are very limited and comprise of little more than a few coastal lagoons, but reservoirs and salt pans are important for waterbirds on most of the islands (see p. 24). The eastern islands, Fuerteventura and Lanzarote, are lower and much drier than the western group, and are covered with semi-desert and xerophytic scrub with a few patches of agricultural land. There are clumps of palms but no proper woodland. The north of Lanzarote has notable sea cliffs.

Embalse de las Peñitas is one of the few freshwater lakes on the arid island of Fuerteventura (Tony Clarke).

A regenerating native pine forest on Mount Teide, Tenerife. The pines are important for Blue Chaffinch and Tenerife Goldcrest, and for the canariensis *subspecies of Great Spotted Woodpecker (John Jackson).*

MADEIRA

Undoubtedly the most important natural habitat of the mountainous island of Madeira is the laurisilva, which covers more than 15% of the island at altitudes of 300–1,300 m and is protected within the Madeira Natural Park. As the name suggests, the principal components are members of the Lauraceae, comprising species endemic to Macaronesia. There are two main types of laurisilva: dry laurisilva on south-facing slopes, characterised by *Apollonias barbujana*, *Visnea mocarena* and *Picconia excelsa*; and moist laurisilva, with a canopy reaching some 25 m, on north-facing slopes and gorges, where *Laurus azorica*, *Ocotea foetans* and *Persea indica* predominate. Both types boast a diverse understorey. The forest, a remnant of a habitat that once covered most of the island, now largely occurs in deep gorges and in inaccessible valleys. Much of the island is now terraced and given over to agriculture, producing wine grapes, bananas and subtropical fruits as well as subsistence crops. Reforestation with non-native species, such as the Maritime Pine *Pinus pinaster* and Blue Gum *Eucalyptus globulus*, is widespread at mid-altitudes. The fields are irrigated by a network of channels, the levadas, which also provide walkways into the mountains. The coastline is rugged, with steep sea cliffs and some pebbly beaches. Rivers and streams are steep and fast-flowing.

Laurel forest on the slopes of Pico de Arieiro in central Madeira. This woodland is the haunt of Trocaz Pigeon and Madeira Firecrest (John Jackson).

Porto Santo is best known for its sandy beaches, which are almost entirely lacking in Madeira itself. Porto Santo is thinly vegetated with xerophytic *Euphorbia* scrub and other hardy coastal species. The peaks inland have lost much of their original vegetation and are the subject of botanical restoration projects. The Desertas are barren islets, their rocky coastlines enclosing grassy inland plateaux. The Salvage Islands are even more barren rocks; Selvagem Grande has a central stony plateau surrounded by steep cliffs about 90 m high.

AZORES

The Azores are distinguished by their rugged coastlines and mountainous or hilly interiors. Many of the volcanic craters are filled by picturesque lakes. Expanses of boulder-strewn lava flows are also characteristic. Most of the original laurel forest has long been cleared for agriculture but important remnants remain above 500 m, especially on Terceira, although such cloud forests are often on steep, inaccessible slopes. The gentler gradients have vineyards and orchards among the arable land, much of which is devoted to growing grass for the

Santiago lake on São Miguel, with Sete Cidades in the distance. These caldera lakes are magnets for wildfowl (Gaspar Avila).

important dairy industry. The fields are characteristically divided by stone walls. The Azores have suffered from the introduction of invasive and aggressive exotic plants, such as the Kahili Ginger *Hedychium gardnerianum* and Giant Rhubarb *Gunnera tinctoria*, which have displaced large expanses of native vegetation on some islands. There are also major plantations of exotic conifers, eucalyptus and other trees on some islands. The coastlines are interrupted by numerous bays and rocky inlets.

CAPE VERDE ISLANDS

The Cape Verde Islands were once covered with vegetation, most of which has long been destroyed by human activity, with the land having been converted to agriculture or overgrazed by goats and other introduced animals. Afforestation with pines, eucalyptus and other trees has also displaced natural habitats, notably in the mountains of Santo Antão, in São Vicente and Maio. The alien shrub *Lantana* is also widespread. As a result, the native vegetation is now highly fragmented and largely confined to the mountain peaks and steep slopes. Erosion is a problem. The lower and drier islands were covered with a steppe-like vegetation and areas of dry scrub, expanses of which still remain. The higher and wetter islands also had dry monsoon forest, of which only small remnants now survive. The three eastern islands (Sal, Boavista, Maio) have large areas of sand dunes and these also occur locally elsewhere. Boavista has oases of Date Palms *Phoenix dactylifera*. Temporary streams flow during the wet season, especially on the higher islands, but permanent water is largely lacking. Only Santo Antão has permanent streams. Tamarisks and the Giant Reed *Arundo donax*, the latter an important habitat for the endemic Cape Verde Warbler, are associated with watercourses. The three eastern islands have saltpans and lagoons.

Inland Raso. This bleak and barren place is home to the endemic Raso Lark (Tony Clarke).

The cliffs of many of the Cape Verde islands hold important seabird colonies. These are on the island of Raso. Note the Red-billed Tropicbird swooping above the shore (Tony Clarke).

Ornithological History of the Islands

CANARY ISLANDS

It is possible to find ornithological references in more general works by the various explorers who visited the Canaries following their discovery by European voyagers in the 14th century. Some of this information is of value but much must be treated with caution. French writers in the 15th century wrote of El Hierro '...and there are falcons, sparrowhawks, larks, a large quantity of quail and a type of bird that has feathers of a pheasant, the size of a parrot and it has a crest on its head like a peacock, and flies very little'. With respect to Fuerteventura, 'The countryside is very rich in small birds, in herons, in bustards, in big pigeons with the tail tipped with white, in doves so common you would not believe, but the falcons destroy everything; in quail, in larks, in other birds in uncountable numbers; and one species of the other birds that are white and as big as a goose'. In the late 16th century Torriani also made a few comments concerning birds; speaking of La Graciosa he mentions 'There are a large quantity of birds called *Pardelas* (Shearwaters) named from their dull coloration, they are almost as big as pigeons and good to eat when roasted ... these birds put their eggs in the sand, in certain burrows, like the rabbits'. In 1779 Urtusaustegui indicated that on El Hierro 'you can find large numbers of pigeons, rabbits, shearwaters and Tahoces, a beautiful bird, bigger than a blackbird, with very black plumage and the breast from the head was white, with the small feet like all waterbirds with webs; this species is tasty and if it was more common it would compensate some palates for the lack of partridges'. In 1796 André-Pierre Ledrú visited Tenerife and included in his publication of 1810 a list of birds from the island. In 1803, Bory de Saint-Vincent remarked when speaking of Tenerife that 'the ravines are full of two species of pigeons, one is probably unknown to the ornithologists'. At the beginning of the 19th century Viera & Clavijo included numerous bird names in their *Diccionario de Historia Natural de las Islas Canarias*.

The first truly scientific publication on Canarian avifauna, by Philip Barker Webb and Sabin Berthelot, appeared between 1835 and 1850 in their *Histoire naturelle des Îles Canaries*. This monumental publication was completed after a tough period of field work in 1828–30. Included in this work with the assistance of Horace Bénédic Alfred Moquin-Tandon was *Ornithologie Canarienne*, which was published in 1842. This listed 108 species of breeding species and migrants, with supporting data on distribution and plumage. These authors were responsible for the first description of Blue Chaffinch on Tenerife.

Dr Carl August Bolle visited the islands in 1852 and 1856 for a total of about two years, publishing his interesting observations in *Journal für Ornithologie*. In honour of his friend Berthelot, he named the pipit resident in the islands *Anthus berthelotii*. His arrival heralded a period of intense interest in the Canarian avifauna, characterised by much collecting of specimens and eggs, and giving rise to the description of many species and endemic subspecies.

In 1871 one of the world's most famous collectors, Frederick Du Cane Godman, made a brief visit of around a month to Tenerife, Gran Canaria and La Palma. He described the dark-tailed pigeon of the laurel forests as *Columba bollii* in recognition of the work of Carl Bolle, particularly as he was first to mention that the islands' laurel forests had two different types of pigeons. In 1879–82, D. Leandro Sera and Fernández de Moratín, from Granadilla but resident in Santa Cruz de Tenerife, published various articles about ornithology in the *Revista de Canarias*. Following this Savile Reid focused his attention on the island of Tenerife for three months in 1887.

A contemporary of Reid in the second half of the 19th and the early 20th centuries was D. Ramon Gómez, a chemist and taxidermist from Puerto de la Cruz on Tenerife. He supplied many skins and eggs to visiting naturalists and museums. Furthermore, authors of this period regularly mention personal communications from Gómez, or information obtained from his collection.

The ornithological knowledge of the islands advanced considerably with the studies of Edmund Gustavus Bloomfield Meade-Waldo. He remained in the Canaries for a period of three years and eight months between 1887 and 1890, during which he visited all of the islands. Perhaps most importantly, he described the Canary Islands Stonechat, although he also discovered several endemic subspecies. His name remains inextricably associated with an endemic species that became extinct by the mid-20th century, the Canary Islands Oystercatcher, *Haematopus meadewaldoi*.

The Reverend Henry Baker Tristram, famous for his massive collection, overlapped with Meade-Waldo in the Canaries. He published a couple of papers on the birds of La Palma and Gran Canaria, and was the only ornithologist to see Bolle's Pigeon on Gran Canaria before the Doramas forest was completely destroyed. Also, in the first half of 1889 Dr Alexander Koenig undertook an intense study of the avifauna, which was published in 1890. In 1893, D. Anatael Cabrera, a farmer in La Laguna perhaps better known as an entomologist, published a catalogue of the islands' birds based on his own observations and collecting over a period of four years. The subspecies of Blackbird in the Canaries and Madeira, *Turdus merula cabrerae*, was named after him by Hartert in 1901. His collection of skins, nests and eggs, some of which still exist, are of great interest and were mentioned by many ornithologists, one of the best known being Hartert, who himself described a few subspecies of Canarian birds and was responsible for finally giving the Laurel Pigeon its scientific name, *Columba junoniae*. In 1901, Henry E. Harris published the first photographs of Canarian birds, with many nests, along with some picturesque comments. The photographs were of a remarkable quality, and especially worthy of mention are those of the nests of Houbara Bustard, Cream-coloured Courser, Egyptian Vulture and Common Raven.

Between 1902 and 1904 Johann Polatzek stayed on the eastern islands, from where he visited the entire archipelago. His observations were published in 1908–09. His finest discovery was the Blue Chaffinch on Gran Canaria, and his name was given to this endemic subspecies, *Fringilla teydea polatzeki*, when described by Hartert in 1905. In the early 20th century Rudolf van Thanner established himself at Vilaflor on Tenerife, where he resided for the next 17 years. He published several papers and journeyed to the various islands of the archipelago. However, he is best remembered for provisioning various museums with skins, and for his occasional over-enthusiasm for collecting (he took 122 Blue Chaffinches on Tenerife and a further 76 on Gran Canaria).

In 1904, en route to the West Indies, David Armitage Bannerman made his first visit to the islands, although the majority of his studies were undertaken in 1908–13. In 1911 he married his first wife, the daughter of the collector T. P. Morgan from Las Palmas de Gran Canaria. Bannerman returned to the islands in 1920, 1927 and 1959. He published many papers in 1911–20, a book about his expeditions in 1922 and finally the *pièce de resistance*, *A History of the Birds of the Canary Islands and of the Salvages*, the inaugural volume of *Birds of the Atlantic Islands*, in 1963. Criticised more recently for his collecting, particularly for obtaining the last known example of Canary Islands Oystercatcher in 1913, his book continues to make compulsive reading and brings together the works of all the ornithologists who studied the avifauna of this remarkable archipelago.

Since Bannerman's era, G. H. Gurney spent a few weeks on Tenerife and Gran Canaria in 1927, and, following a stay of a few months in 1947, Dr Helge Volsøe produced some interesting works on both the migrant and resident species, as well as on the origin and evolution of Canarian avifauna, which was published in 1949–55. At around the same time (1948), David Lack and H. N. Southern made a short visit to Tenerife, whilst M. Cullen *et al.* worked on La Palma and La Gomera in the summer of 1949. Between 1956 and 1960 researchers on the islands that stand out include R. D. Etchécopar and F. Hüe, as well as A. M. Hemmingsen, T. Hooker and S. Knecht. Also, Captain J. H. McNeile, a renowned egg collector, made various visits to the islands. His collection and diaries are held in the Royal Scottish Museum in Edinburgh. McNeile's notes on breeding were used by Bannerman in his 1963 book. In the 1960s and 1970s the works of E. A. R. & D. Ennion, J. Cuyás Robinson, R. Lovegrove, P. R. Grant, J. Heinze and N. Krott deserve to be mentioned.

The end of the 1960s heralded the awakening of local ornithologists, the pioneering work being *Las Aves de Canarias* by Francisco Pérez Padrón, first published in 1971 and reprinted on numerous occasions since. From 1975 until the present day the work of Dr Juan José Bacallado and Keith Emmerson has permitted the formation of a group of people interested in birds, most of whom studied at the Departamento de Biología Animal at the Universidad de La Laguna or the Museo de Ciencias Naturales de Tenerife. Among this group are Aurelio Martín and Juan Antonio Lorenzo, who recently published the monumental *Aves del Archipiélago Canario*, Fernando Domínguez, Guillermo Delgado, Manuel Nogales, Vicente Quilis, José Carrillo, Miguel Angel Hernández, Rosa Maria Alonso Quecuty, Efrain Hernández and José Manuel Moreno. Working on other islands, Domingo Concepción on Lanzarote, and Octavio Trujillo, Gorgonio Díaz, Felipe Rodríguez, Francisco del Campo, Ángel Moreno, Martín Moreno, Antonio Cardona and Pedro Martín on Gran Canaria also deserve recognition. Observations by others have increased our knowledge of avian distributions, particularly the migrant species, in the archipelago. These include Domingo Trujillo, Rubén Barone, Felipe Siverio, Manuel Siverio, Juan José Ramos, César-Javier Palacios and many others. The formation of a Canarian branch of the Sociedad Española de Ornithología in 1993, directed by Cristina González, has developed conservation awareness in the archipelago. Finally, the birdlife of these islands has sparked much interest in ornithologists both from mainland Spain and from elsewhere, resulting in the publication of numerous papers by workers such as L. M. Carrascal, A. Lynch, A. Baker and B. Schottler, to name but a few.

MADEIRA

As with the other archipelagos, the earliest references to the birds of Madeira were made by early navigators such as Diogo Gomes and Alvise da Cadamosto. The first reference to specific species was made by Gaspar Fructuoso in his *Saudades da Terra*, written prior to his death in 1591, although the second volume (which refers to Madeira) was unpublished until 1873! The first work on Madeira by a British naturalist came in 1707, when Sir Hans Sloane published a travelogue describing his journey to Madeira, which he visited first in 1687. He devotes two pages to the birds he encountered and makes reference to the commercial use of Canaries for their beautiful song. Nearly 70 years later, in 1772, Georg Forster visited Madeira aboard HMS *Resolution* with Captain James Cook. Five years later the account of his visit was published in *A Voyage Round the World in His Britannic Majesty's Sloop 'Resolution'*.

In the late 1820s and early 1830s publications by the German ornithologist, C. Heineken and by Sir William Jardine followed, but it was not until 1851 that a complete scientific list of the birds of Madeira, both residents and migrants, was published. Compiled by Edward V. Harcourt, it was published in his *A Sketch of Madeira* and refers to 30 breeders and 65 visitors. Harcourt released more of his work in the *Annals of Natural History*, culminating in a 'Final list of residents and migrants combined' that was published in 1855. Harcourt is also well known for being the first to describe the Madeira Firecrest, which was then treated as a full species and, following demotion to subspecific status for many years, is recognised as such again today.

Following Harcourt there were brief visits by the German Carl Bolle (better known for his work on the Canaries), and the Englishmen Alfred Newton and Frederick Du Cane Godman. Godman's visit in spring 1871 almost ended in disaster when the vessel he chartered to visit the Desertas was inundated with water, resulting in the loss of most of the specimens and eggs that had been collected. He also made a small collection of birds on Madeira itself which included Trocaz Pigeon. Thereafter, in 1885, the *Handbook of Madeira* by J. Yate Johnson, which contained a section devoted to birds, was published.

The following year the first of many papers written by W. Hartwig, 'Die Vögel Madeiras', was published in *Journal für Ornithologie*. It was apparently based on notes made in his personal diaries and devoted much attention to the breeding season and eggs. The rest of Hartwig's work appeared in 1891–94 in various journals.

1890 was notable for visits by two ornithologists associated with other islands within the region: A. Koenig who also worked in the Canaries, and W. R. Ogilvie-Grant, who subsequently led an expedition to the Azores. Ogilvie-Grant obtained a good collection from the archipelago, from which Bowdler Sharpe described Madeira Sparrowhawk, *Accipiter granti* (now considered a subspecies of Eurasian Sparrowhawk). Also in 1890, J. J. Dalgleish wrote on the petrels of Madeira, as well as a note concerning a specimen of Soft-plumaged Petrel, which he had received from the Porto Santo group.

From his first paper in 1893 until the last in 1910, Padre Ernst Schmitz was the mainstay of Madeiran ornithology, and to this day remains one of its most influential characters. Schmitz arrived in Madeira from Germany in 1845 and finally departed the archipelago on 7 July 1908. One of his most important publications appeared in 1899 – a complete list of the migrants and accidental species brought to his attention. Unfortunately, the vast majority of his specimens have been lost, as they were allowed to fall into decay relatively soon after his departure from the island. This is a huge loss, as many were of historical value and are irreplaceable.

During Schmitz's time on Madeira another very influential ornithologist was active in the islands. Adolfo César de Noronha was born in Madeira in 1873; his work on Porto Santo in 1900–03 forms the basis of our knowledge of that island's avifauna, as well as the islets of Baixo and Cima. His diaries were published by Padre Schmitz in their entirety, translated into German, in *Ornithologisches Jahrbuch* between 1902 and 1904. Subsequently, Noronha founded the Museu Municipal do Funchal, which to this day holds a representative collection of the archipelago's birds. Also in the early 1900s, papers were published on Madeiran birds by Victor Ritter von Tschusi zu Schmidthoffen, mostly in the *Ornithologisches Jahrbuch*, and one by M. J. Nicoll in *Ibis*.

In 1914 a legend in the study of birds in the Atlantic Islands produced his detailed paper 'Distribution and nidification of the Tubinares in the North Atlantic Islands', which appeared in *Ibis*. The legend was, of course, David Armitage Bannerman, who was to be an inspiration for many who followed in his footsteps, particularly after the publication of the seminal *Birds of the Atlantic Islands* (see below). In 1924, another legendary figure, the infamous Richard Meinertzhagen and his wife Annie, collected on Madeira and the Desertas. The results were published the following year in *Ibis*. Subsequently, in 1934, Gregory Mathews was the first to describe the two distinct races of Soft-plumaged Petrel breeding in the Madeiran archipelago; *Pterodroma mollis madeira* in the mountains of the main island, and the larger *P. m. feae* on Bugio in the Ilhas Desertas (these birds are now treated as separate species, distinct from Soft-plumaged Petrel, of course – Fea's Petrel, *P. feae* and Zino's Petrel, *P. madeira*). 1934 was also noteworthy for the appearance of the first of a set of monographs by Alberto Artur Sarmento dealing with the vertebrate fauna of Madeira. That dealing with birds appeared two years later, in 1936.

In the summer of 1939 R. M. Lockley visited the islands to study the petrels. He published his observations three years later in *Shearwaters*, but it was not until 1952 that a full account of his expedition appeared in *Ibis*. The next major advance came in 1948 when the individual volumes of Sarmento were combined under the title *Vertebrados da Madeira*. Closing this volume is a useful annotated list of the birds, compiled by the museum curator G. E. Maul. It follows the list given by Sarmento but includes recent acquisitions of the museum and additions to the Madeiran list since 1936. Unfortunately, Sarmento included some rather unsatisfactory records without comment, but these were subsequently removed from the Madeiran list by Bannerman.

The following year, Swedish naturalist John Bernström visited Madeira. The culmination of his studies was the publication in 1951, of a 'Checklist of the Breeding Birds of the Archipelago of Madeira' in the *Boletim do Museu Municipal do Funchal*. This work included a review of status, a table of the breeding seasons and an excellent bibliography, all of which were referenced by future researchers. Ten years after Bernström a British naturalist, John Buxton, visited the islands. He was only present for a short period but his observations appeared in 1959 in *Bocagiana* and in 1960 in *Ibis*.

On 27 March 1960 David Bannerman and his wife Mary commenced their work on the birds of Madeira. The results of this and subsequent visits formed the nucleus of what is probably the most important work ever written on the birds of Madeira, *A History of the birds of Madeira, the Desertas and Porto Santo islands*; this was published in 1965, forming volume two of *Birds of the Atlantic Islands*. Soon after, the arrival on Madeira of Alec

Zino witnessed a new phase in ornithology on Madeira. From then on, less emphasis was placed on collecting and more on observations and conservation. From the late 1960s to the present there have been many important papers published by visiting ornithologists, e.g. Jensen (1981), Swash (1986) and Jepson & Zonfrillo (1988) to name but a few. In 1995, 'Birds of the Archipelago of Madeira and the Selvagens: New Records and Checklist' by Zino et al. was published in the Boletim do Museu Municipal do Funchal. This is the most recent study to include all of the migrants and vagrants recorded in the islands. Frank Zino has continued in his late father's footsteps, while other names in recent Madeiran ornithology include Paulo Oliveira, who discovered a further colony of the critically endangered Zino's Petrel.

In the last 30 years there has been an increase in awareness of the important avifauna of the Salvage Islands, where the first Swinhoe's Storm-petrel in the Western Palearctic was recorded. A list of migrant species, 'Les oiseaux visiteurs des Îles Selvagens' was published in 1987 (Mougin et al. 1987).

THE AZORES

The first person to mention the avifauna of these islands was Gaspar Fructuoso, who was born in the Azores in 1522. He published *Saudades da Terra* towards the close of the 16th century. Although not an ornithologist, Fructuoso mentions the following species by their local names, without unfortunately providing any description: woodpigeon, rock dove, buzzard, partridge, quail, little woodpecker, bullfinch, canary, chaffinch, blackbird, wagtail, blackcap, goldcrest, a shearwater and some migrants. His mentioning the bullfinch is significant but the reference to a 'little woodpecker' is even more intriguing. The partridge was presumably Red-legged, which was introduced but is now extinct on the islands.

The next two works on the fauna of the Azores were produced in France where, in 1860, A. Morelet published *Notice sur l'Histoire Naturelle des Açores* in Paris. The following year, H. Drouet published *Eléments de la Fauna Acorénne*, also in Paris. On 21 March 1865 the British ornithologist Frederick Du Cane Godman landed for the first time on the Azores. Accompanied by his brother, they were joined a few days later by his collector, Brewer, who, although better known in entomological circles, served as taxidermist to the expedition. During the next three months Godman visited all of the islands except Santa Maria. His specimens served as a reference for his paper 'On the birds of the Azores', published in 1866, and for his book *The Natural History of the Azores*, published in 1870. Notably, Godman was the first to describe the endemic Priôlo, or Azores Bullfinch, in 1866. His specimens are currently housed in The Natural History Museum (Tring).

Thereafter, in 1890, there is written evidence from José Maria Raposo de Amaral concerning the recent importation of Goldfinches into São Miguel; in the same letter he also mentions the earlier introduction of Greenfinch. In another letter, dated 18 April 1906, the same writer refers to the Azores Bullfinch and lists all of the places where it could be found. His evidence suggests that it was then a common bird within its limited range.

Born in 1880, Padre Ernesto Ferreira, resident of Vila Franco on São Miguel, was most active from the early 1900s until his death in 1943. His specimens were passed on to the Carlos Machado Museum in Ponta Delgada, but virtually none of them possess details concerning date or location. However, with a few exceptions that are labelled, it is known that all were collected on São Miguel.

In 1903 a British Musuem expedition led by W. R. Ogilvie-Grant visited the Azores, where they were greatly assisted by Colonel Francisco Afonso de Chaves, director of the museum in Ponta Delgada since 1884. The expedition arrived at Santa Maria on 26 February and remained in the archipelago until the end of May, during which time all of the islands were visited and a large collection amassed. On returning to England, the collection was split between Lord Rothschild, the expedition's benefactor, and the British Museum. The report was written jointly by Ogilvie-Grant, who recounted the field work, and the curator of Lord Rothschild's museum, Dr Ernst Hartert, who compiled the systematic part. This work and that of Godman served as the foundations for ornithology within the Azores, although many later authors have not afforded them the credit they deserve.

José G. Correia was born on Faial but emigrated to the United States. He made two collecting expeditions to his native islands, one to Faial and Pico in 1921–22, and the second, a major undertaking, to the islands of Terceira and São Miguel as well as those visited on the first expedition. On the first trip he collected several hundred specimens including the type of *atlantis* Yellow-legged Gull, and his collection was subsequently worked on by the American, Robert Cushman Murphy. The second trip was in 1927–28, and during this period he obtained, albeit with great difficulty, five specimens of Azores Bullfinch, all near the chapel on the shore of Lagoa das Furnas which, until the species' rediscovery in the mid-1960s, were the most recent evidence of its continued existence. Also during this period, he collected the only specimen of Snowy Owl known from the archipelago, on Faial. Murphy, together with fellow American James Chapin, published details of Correia's collection in *American Museum of Natural History Novitates* in 1929.

Dr António da Silveira Vicente was appointed to head the zoological section of the Museu Carlos Machado on 2 May 1930, whereupon he set about rearranging the bird collection and compiling a complete list of the specimens. In 1931 two other native Azoreans, Manuel Dionisio and Pacheco de Castro, collected skins and eggs on behalf of the Frenchmen Noël Mayaud and J. de Chavigny. The results were published in *Alauda* in 1932, with the egg descriptions being the most significant work. For many years this remained the most complete study of the avifauna of the Azores. José Maria Álvares Cabral was another Azorean associated with ornithology through his position in the Museu Carlos Machado. In 1959 he turned his attention to migrants and he was the first to compile a complete list of those represented in the collection at the Museu Carlos Machado.

The final great advance in our knowledge of the birds of the Azores came in 1966 when David and Mary Bannerman published *A History of the Birds of the Azores*, the third volume in their monumental *Birds of the Atlantic Islands*, which became the most comprehensive work on the avifauna of the Azorean archipelago and included full details of breeding birds, migrants and vagrants. Álvares Cabral was in charge of the Museu Carlos Machado at the time of Bannerman's visit and assisted the work in many ways. Another Azorean ornithologist who provided much help was Colonel José Agostinho. His position in the meteorological office in Ponta Delgada permitted him to travel extensively throughout the archipelago and to pursue his great interest in birds. He retired to Terceira where he has made a number of interesting ornithological discoveries.

CAPE VERDE ISLANDS

Although few details were given, some early navigators made passing references to the birds of the Cape Verdes. These included accounts by Cadamosto in 1456, Diogo Gomes in *c.* 1460, Christopher Columbus in the 1490s (who refers to the Magnificent Frigatebird), William Towerson in August 1558, and Olfert Dapper in 1668, who lists 'Herons, Turtle Doves, Guinea fowl, Quail and Flamingos' for the Cape Verdes. In the late 17th century the islands were visited twice by Dampier, who in September 1683 collected a number of flamingos on Sal, the only proof this species has occurred on the island. In October 1690 François Leguat visited Sal from where he reported various seabirds but failed to find flamingos.

The first detailed reference to birds in the Cape Verdes was made by Johann Forster and his son Georg, who travelled with Captain Cook aboard HMS *Resolution*, visiting Praia on 13–15 August 1772. Although not necessarily identified correctly, his notes make reference to species such as Brown Booby, Grey-headed Kingfisher, Iago Sparrow, Blackcap and Common Quail. Next, in 1783, João da Silva Feijó was commissioned to obtain information on the natural history and resources of the Cape Verdes. He collected some birds during his stay, which were lodged in the Royal Museum and Botanical Garden of Ajuda. Most of his collection was transferred to Paris in 1808, and by the late 19th century no Feijó specimens could be located in Portugal.

In the early 19th century A. Delalande visited the islands and obtained some specimens, including two new taxa, the Grey-headed Kingfisher and Brown-necked Raven. These specimens were sent to Paris and described by Lesson. Arguably the most famous early visitor to the islands was Charles Darwin, who stayed on Praia for three weeks, in January–February 1832, whilst travelling on HMS *Beagle*. On his return journey, in early September 1836, he found himself again at Praia. Specimens were obtained during both sojourns and these included new subspecies of Bar-tailed Lark and Black-crowned Sparrow-lark, as well as the first specimens of a new species, the Cape Verde Swift. The specimens were originally housed at the Zoological Society of London, but most were subsequently transferred to the Natural History Museum.

In the early 1840s the first two bird lists for the Cape Verdes were published, one by J. C. C. de Chelmicki and F. A. de Varnhagen in 1841, and the other by J. J. Lopes de Lima in 1844. The German naturalist Carl Bolle, although probably better known for his work on the Canaries, stayed in the Cape Verdes from July 1851 until December 1852. During this time he visited most, if not all, of the archipelago, but the vast majority of his time was spent on São Nicolau. In his 1856 publication 'Die Vogelwelt auf der Inseln des grünen Vorgebirges' he refers to only around 20 species within the archipelago, but he was the first to mention migrant shorebirds, and also made reference to the abundant seabirds off Rombo, Branco and Raso. A final contribution made by Bolle was his observations on São Nicolau of 'a rufous Sylvid, larger than *Sylvia hortensis* [Orphean Warbler]', which is almost certainly the first reference to the Cape Verde Warbler, on which he collected much information on breeding, habits and habitats.

On 24 July 1852 John Macgillivray called in at Porto Grande, São Vicente, aboard HMS *Herald*. Of most interest is his mention of frigatebirds and cormorants at this location. The first major collecting expedition to the islands commenced in December 1864 when Heinrich Dohrn from Germany and Johan Gerard Keulemans from Holland arrived at Porto Grande. They stayed until March 1865 and, as well as São Vicente, they obtained specimens from Santiago, Santo Antão and São Nicolau. Their most important contribution to Cape Verdes ornithology was the collection and subsequent description of the Cape Verde Warbler, which they named *Calamodyta brevipennis* (now called *Acrocephalus brevipennis*). One of the specimens collected by Dohrn and Keulemans on São Nicolau is currently held at Tring, as is the holotype of the *neglectus* race of Common Kestrel (described as a separate species at the time), which was also collected by this expedition. Apart from those skins in Tring, New York, Amsterdam and Basle, the whereabouts of the remainder of the Dohrn collection are unknown. Also active during this period was the Frenchman A. Bouvier, who visited the islands in 1867–69. Most of his specimens were sent to Paris, but some 20 are housed in Tring, where the few specimens collected by Henry Moseley aboard HMS *Challenger*, which visited the islands of São Vicente and Santiago in July-August 1873, are also retained. Ten years later, on 20–30 July 1883, the Talisman expedition visited the seas around the Cape Verdes. During this time they visited Branco, where they collected the type of Cape Verde Shearwater.

One of the early authorities on the birds of the Cape Verdes was the Portuguese naturalist and director of the Lisbon Museum, José Vicente Barbosa du Bocage (1823–1907). His series of 24 papers, published in 1867–83, included a list of all specimens from the archipelago housed in Lisbon, and was the first major work on the islands' avifauna. In 1901 Francisco Newton, a Portuguese collector, landed in the Cape Verdes. He was presumably there to increase the size of the Lisbon collection, but when he departed in 1902 he had only 19 skins of 15 species, although these included the first examples of Bulwer's Petrel and three migrant visitors, including the only ever record of Grasshopper Warbler.

The first of the large-scale collectors to visit the islands was Boyd Alexander, who led two expeditions to the islands in 1897, from 10 February until 21 May and from early October until 16 December. During these trips he collected many specimens, most of which are now in Tring, although some are also housed in New York. Undoubtedly his most important discovery was the unusual Raso Lark, which he named *Spizocorys razae* but is now known as *Alauda razae*. Just six days after the departure of Boyd Alexander, the Italian collector Leonardo Fea arrived on Boavista on 22 December 1897 where he remained until mid-February 1898. He subsequently made lengthy visits to Santiago, Fogo, Brava, Ilhéus do Rombo, São Nicolau and Raso, before finally leaving the islands on 15 December 1898. He collected 308 specimens of 47 species, including one new to science that was named *Oestrelata feae* in his honour; this is now known as *Pterodroma feae* or Fea's Petrel. Some of Fea's other interesting finds included the first record of Bald Ibis and the only record of Ferruginous Duck on the archipelago. This important collection is currently held in the Genova Museum in Italy.

In the early 20th century a few British ornithologists made small contributions to the picture of the Cape Verde avifauna, including M. J. Nicoll in December 1902, P. R. Lowe in April 1905 and January 1907, and W. J. Ansorge in April 1909, whilst David Bannerman had assistants there from as early as 1913, and G. H. Wilkins called at São Vicente on 29 October 1921. On 16–18 September 1912, Robert Cushman Murphy and Portuguese taxidermist José Gonçalves Correia collected on São Vicente for the American Museum of Natural History. Impressed by the number of seabirds, Correia returned between 4 May and 26 July 1922. He collected a large series of seabirds but also made detailed notes on their breeding and habits. For sheer numbers of birds, the most important collection was obtained between December 1923 and May 1924 by the Blossom South Atlantic Expedition. This expedition, sponsored by the Cleveland Natural History Museum, took more than 1,300 skins of 72 species, including several migrants new to the archipelago. They were originally housed in Cleveland, but many were exchanged with New York and subsequently a large number were transferred to the Yale Peabody Museum, in New Haven.

More recently, on 24 July 1951, W. R. P. Bourne commenced a two-month stay in the islands on São Vicente. He also visited Santiago and Brava, and his most important discovery was the breeding of Purple Herons inland on Santiago. These were described later by de Naurois and named *Ardea purpurea bournei* in Bourne's honour. During the 1960s it was Frenchman René de Naurois who dominated the ornithology of the archipelago. He visited the islands five times between 1962 and 1969 and studied virtually all of the breeding species, resulting in a flow of publications. David and Mary Bannerman resided in the islands from January to March 1966, as preparation for the fourth volume of *Birds of the Atlantic Islands*.

Thereafter, in 1969, 1970 and 1972 Jamie Vieira dos Santos conducted the final collecting expeditions, during which he obtained 380 skins of 51 species. In 1988 Kees Hazevoet began his visits to the islands, which resulted in the publication in 1995 of a detailed checklist to the birds of the Cape Verdes, since updated on a near-annual basis with papers reporting new, rare and threatened species in the archipelago. Since then, many birdwatchers have visited the islands and many have, in various ways, contributed to our knowledge of the Cape Verde avifauna.

Birdwatching areas

This section is intended to form a general guide to some of the more important birdwatching areas in the region. Although not as comprehensive as the many trip reports now available, details are given here of good areas in which to search for the endemics, plus other sites for general birding and for looking for migrants and accidental visitors.

CANARY ISLANDS

This archipelago consists of seven main islands, plus three smaller ones and a few islets and rock stacks. The islands boast six endemic species and three shared with other islands, as well as an impressive 29 endemic subspecies. Most birdwatchers will base themselves on Tenerife, as this is the only island to have five of the six endemic species and all three Macaronesian specialities. Further details can be found in *A Birdwatcher's Guide to the Canary Islands* (Clarke & Collins 1996).

TENERIFE Of the endemics, Canary Islands Chiffchaff is the most widespread, occurring everywhere from the hotel gardens of Playa de las Americas and Los Cristianos to the subalpine zone of Cañadas del Teide. Tenerife Goldcrest is more specialised in its habitat requirements, being generally confined to forested areas where it is a common resident, although more often heard than seen. The three near endemics, Canary, Berthelot's Pipit and Plain Swift are all widespread and should be found without difficulty, but many Plain Swifts depart the islands in October–January. They can be much harder to locate at this time, but are usually still present along the north coast or at higher altitudes within Las Cañadas. The three star species, Blue Chaffinch, Bolle's Pigeon and Laurel Pigeon, are confined to specific habitats and require detailed information on their locations; they are unlikely to be found by casual visitors.

Blue Chaffinch, also present but much rarer on Gran Canaria, is confined to mature Canary Pine *Pinus canariensis* and is best looked for at the barbeque area, 11 km from Vilaflor, on the road to Las Cañadas del Teide National Park. This area, known as Las Lajas, supports good numbers of the species and, although sometimes slightly elusive, this is the best area on Tenerife to find this unique bird. Try to avoid visiting at the weekend, as this is a popular recreational area with local people and it can be very crowded and noisy. Another bird usually found with ease at this site is the endemic *canariensis* subspecies of Great Spotted Woodpecker. Blue Chaffinch also occurs at a number of other sites on Tenerife, including La Caldera at Aguamansa, Zona Recreativo de Chio and Zona Recreativa de Ramon Caminero, but at the first two of these is often more difficult to locate.

Bolle's Pigeon is the commoner and more widespread of the two endemic pigeons on Tenerife. Most visiting birders should have no problem finding this species in the laurel forest at Monte del Agua (or Erjos), which is accessed via the rough track opposite the casa forestal at Erjos, on the road between Santiago del Teide and Icod de los Vinos. Another area where the species is usually seen relatively easily is the lookout at Mirador de Pico Inglés, on the Anaga Peninsula, north-east of La Laguna. Other areas include Chanajiga and Pijaral, but in these areas the pigeon is harder to find.

Laurel Pigeon occurs at Monte del Agua or Pico del Inglés, and although less easily found than the previous species, perseverence at either place should produce sightings. Another area is the road from Los Realejos to Icod el Alto; scan the forested valley and hillside before reaching Icod el Alto. Perhaps the most reliable site on Tenerife is the small lookout on the north coast, west of Puerto de la Cruz, known as Mirador Las Grimonas. Looking inland from this lookout should produce excellent views of this endemic.

Tenerife also has some headlands from which seawatching can be productive. Punta de Teno in the north-west is one such; Little Shearwater can be seen from here, even in winter. As with all seawatching sites, early morning and the last couple of hours before dark are best. Punta de la Rasca, in the south, is another excellent site for the avid seawatcher; even the rarer petrels have been seen from there. A third headland, Punta de Abona on the east coast, has also produced Little Shearwater and Bulwer's Petrel, but this is the least known of the three sites.

Migrants and vagrants can appear just about anywhere, but southern Tenerife has some particularly good areas to search. Neighbouring Golf del Sur and Amarilla Golf & Country Club can be magnets under certain weather conditions, as they are the largest expanses of greenery in this part of the island. A wide variety of species has been reported from both areas and they are particularly productive in the autumn, when Buff-breasted Sandpiper, Pectoral Sandpiper and American Golden Plover are almost annual. The pool behind the beach south of El Médano is another site that attracts shorebirds, while the surrounding scrub harbours a very small population of Trumpeter Finch. The small reservoir at Roquito del Fraile, reached by walking from the road to Punta de la Rasca, is the best site on Tenerife for wildfowl in winter. Species such as Lesser Scaup and Ring-necked Duck have been reported, as well as many shorebirds and even a few migrant passerines. This site is worth checking at any season, as it is the largest expanse of fresh water in the south of the island.

LA GOMERA The closest island to Tenerife, La Gomera is easily reached from Los Cristianos by ferry, and this is the best way to see the various seabirds that breed in the archipelago, although specialities such as Bulwer's Petrel, Little Shearwater and, in particular, Madeiran Storm-petrel are all quite scarce and by no means easy to see. The best time of year for Bulwer's Petrel is May–August, whilst Little Shearwater is present year-round, and Madeiran Storm-petrel (the rarest of the three) is a winter-breeder most often recorded in August/September and April/May. La Gomera offers an easier alternative than Tenerife for birders eager to see Laurel Pigeon. Either take your hired car from Tenerife on the ferry, reserve a car which can be waiting for you in San Sebastian, where the ferry docks, or use of one of the local taxis, but remember to agree a price including waiting time in advance. The relative cost of these options is very similar and depends mainly on the ferry charges at the time of your visit. The best area for the pigeons is on the road to Hermigua. After passing several short tunnels you reach a long tunnel, beyond which used to be a bar on the right called La Carbonera. The bar is now closed but both species of pigeon can be seen from the chained-off car park in front, by viewing the forested hillsides on the opposite side of the road. Views here are likely to be fairly distant, so *c.* 1 km further on take the left turn signposted to Monte El Cedro; anywhere along this road should, with patience, provide good flight views of the pigeons, and sometimes perched birds too.

Other areas worth exploring if time permits are the reservoirs behind San Sebastian at Chejelipes, the laurel forests around Chorros de Epina and, particularly, the reservoir near Vallehermosa, the Presa de la Encantadora. The latter is perhaps the best area on the island to look for winter waterfowl, with Ring-necked Duck one of the more unusual species recorded here.

FUERTEVENTURA The speciality here is the Canary Islands Stonechat, which is found nowhere else in the world. The species occurs throughout the island but is not common and its preferred habitat is vegetated barrancos (dry river gullies). It can be found at Barranco de la Torre (south of the airport, near Las Salinas), Embalse de Los Molinos and in Valle de Tarajalejo, to name but three areas. The other main attraction for visiting birders is the healthy population of the endemic subspecies of Houbara Bustard. This widespread species is nowhere common and can be very difficult to locate. Probably the best area to look is the plain inland of Cañadas del Rio, by Costa Calma, in the south of the island. If this fails, try the plains south of El Cotillo, north-west of Tindaya, north-west of La Oliva or around Triquivijate. Persistence at any of these sites should be rewarded, but for a large bird Houbara can be remarkably hard to find. Other species of interest include Cream-coloured Courser and Black-bellied Sandgrouse, both of which are found in similar areas to the Houbara, and Egyptian Vulture. The vulture is now rare on the island and has declined dramatically in recent years. Though still widespread, it is most frequently encountered in the hills around Betancuria in the centre of the island.

Any location on the island with fresh water is worth a look regardless of season, but such areas come into their own in spring and autumn when migrants include herons and crakes, as well as passerines in the surrounding vegetation. During winter they attract small numbers of wildfowl and have an impressive list of vagrants to their credit. The best sites are Embalse de Los Molinos and Presa de Las Peñitas, although the smaller ponds at Rosa de Catalina Garcia and Rosa del Taro (on the road from Casas de Angeles to Triquivijate) have also hosted some excellent birds in recent years. Not quite as good are the pools in Barranco de la Torre, but these still have potential for rarities and are worthy of exploration. Other areas to check are any coastal vegetation, particularly in spring and autumn, e.g. the plantation at Costa Calma, the gardens of the various hotels and the tamarisks in Barranco de la Torre or around Las Lajitas. Finally, it is always an idea to check coastal areas for waders and gulls; the three best appear to be south of Caleta de Fustes and at Salinas del Carmen, just south of the airport, and the most extensive, the beach at Playa de Sotavento, between the Hotel Sol Gorriones and Risco del Paso. At certain times the retreating tide leaves coastal lagoons which can hold large numbers of waders and gulls, and huge numbers of migrant terns have been seen here. This area is probably a regular wintering site for Slender-billed Gull as well as a good place for godwits, plovers and sandpipers most of the year.

LANZAROTE Much like Fuerteventura, Lanzarote is mostly scrubby desert: species diversity on the two islands is very similar, albeit with a few differences in status. On Lanzarote, Canary Islands Stonechat is only a rare visitor, Black-bellied Sandgrouse is rare and its breeding is unconfirmed, Egyptian Vulture is very rare with fewer than five pairs, and Cream-coloured Courser is far more local and uncommon. On the other hand, recent surveys have shown the population of Houbara to be much greater than previously thought. The best areas for the latter are the plains near Playa Blanca, Sóo, Tahiche and Playa Quemada.

One of the best areas for birding is the area of saltpans and the tidal lagoon at Janubio in the south-west. This site is famous as the only area in the Canaries where Black-winged Stilts breed regularly, and is a good area for shorebirds and occasional waterfowl. Some of the most unusual species on the Canaries list have been recorded here, including Smew and Bufflehead. Another area worth checking at any time is the waterfront at Arrecife, which has attracted a variety of shorebirds, gulls and terns, whilst on Avenida Generalísimo Franco is a small colony of Cattle Egrets with an occasional pair of Little Egrets and a feral pair of Sacred Ibises. In the north of the island are the spectacular cliffs of Riscos de Famara, probably the most reliable site in the eastern islands for Barbary Falcon; try the Mirador del Río or Mirador de Guinate. Eleonora's Falcon, which breeds on the islets north of Lanzarote, is sometimes seen in this area in May–September, but a sighting is by no means certain. For Barbary Partridge, an early morning search of the golf course at Tahiche is one of the most reliable ways to see this species.

GRAN CANARIA This island boasts its own endemic subspecies of Blue Chaffinch (*polatzeki*), and Great Spotted Woodpecker (*thanneri*). The latter is a common resident in stands of Canary Pine, but the Blue Chaffinch is now on the verge of extinction and special permission is needed to enter the area where it still occurs. To do this, contact Francisco del Campo by telephone ([00 34] 637-724319) or e-mail (francisco.campo@eresmas.net); he can arrange the necessary permits. Other species include resident breeding populations of both Eurasian Tree and House Sparrows, which do not breed on any of the other islands, Red-legged Partridge (which also breeds only on this island) and the introduced Common Waxbill, which is commonest on Gran Canaria. Egyptian Vulture formerly bred, whilst the current status of Cream-coloured Courser is uncertain, though it used to breed in numbers, and Houbara was recorded in the 19th century and possibly bred in the south of the island.

Some areas worth visiting include the small reserve at La Charca de Maspalomas which has a small coastal lagoon and is a good area for migrant landbirds and waterbirds; Punta de la Sardina for seawatching; the various freshwater reservoirs for waterfowl (particularly in winter); and the coast near Arinaga in the south-east of the island, where Lesser Short-toed Lark and Trumpeter Finch occur.

LA PALMA La isla bonita (the beautiful island, as it is known in Spanish) certainly lives up to its name, and the lush laurel forests harbour the largest population of Laurel Pigeons in the archipelago. It is also the only island in the group to boast Red-billed Chough as a resident breeder. The endemic subspecies of African Blue Tit (*palmensis*) and Common Chaffinch (*palmae*) are fairly common and widespread. Probably one of the best known birding areas is the forested area around Los Tilos, a site where both Laurel and Bolle's Pigeons can usually be found with ease, and the African Blue Tit and Common Chaffinch occur around the bar. Other species of interest include the Canarian endemics Tenerife Goldcrest and Canary Islands Chiffchaff, plus the near-endemic Plain Swift, Berthelot's Pipit and Canary. The island also attracts some unusual species, particularly Nearctic visitors in autumn; it hosted the only Western Palearctic record of Louisiana Waterthrush in November 1991.

Other sites worth a visit include Laguna de Barlovento, the reservoirs at Punta Gorda and Los Llanos de Aridane, the tidal pools near the airport and, for seawatching, the southern tip of the island at Punta Fuencaliente. La Palma is relatively unknown to the visiting birder and further visits will undoubtedly yield much more information on this lovely island.

EL HIERRO The smallest of the main islands, with fewest facilities for the tourist. Both Laurel and Bolle's Pigeons have been discovered breeding in the last 20 years, although numbers are not comparable to those on other islands. The main areas to look are the laurel forests at El Golfo and, particularly, Mount Sabinosa. El Hierro also hosts the Canarian endemics Canary Islands Chiffchaff and Tenerife Goldcrest, the near-endemic Plain Swift, Berthelot's Pipit and Canary, plus endemic subspecies of African Blue Tit (*ombriosus*) and Common Chaffinch (*ombriosa*). The island is the least visited by visiting birdwatchers and little is known about the best areas for birding. The southern tip of the island, at Restinga, is presumably good for seawatching and has the potential for rarities, as it demonstrated in February 1992 when a Glaucous-winged Gull was photographed there, and again in September 2001, when it hosted an immature White-winged Tern. There is very little fresh water on the island and so any irrigation reservoirs are worth looking at.

MADEIRA

The Madeiran archipelago comprises two main islands, Madeira and Porto Santo, as well as the smaller islands and islets of the Desertas and the Salvages, the latter being rather difficult to visit without ample time and money.

MADEIRA Most visitors will be based on the main island of Madeira, which has the international airport and the best infrastructure; it is also home to the archipelago's three endemic species and has three near-endemics shared with other islands in the region. The endemic Trocaz Pigeon can be found virtually anywhere in and around the remaining areas of native laurel forest, but one area to concentrate on is at Ribeira Frio. From this attractive village, take the track following the levada (water canal) to Balcões, where there is a lookout across the laurel forest. Given sufficient time, the rewards can be very good and the views of the birds often excellent. The walk from the village is also productive for the endemic Madeira Firecrest, and for the endemic subspecies of Common Chaffinch (*maderensis*). The third Madeiran endemic is Zino's Petrel, a critically endangered species and one of the rarest seabirds in the world. The breeding haunts of this bird are the precipitous valleys around Pico Arieiro, which should only be visited with a local guide. Amilcar Vasconcelos is one such guide, and regularly takes people to the area where Zino's Petrels breed. At night it is often possible to hear the birds returning to their burrows, but seeing them in the moonlight is another matter. Amilcar can be contacted by e-mail (geral@madeiradventure.com), telephone ([00 34] 964-541793) or fax ([00 34] 291-253038). There is also an entrance fee to the ecological park. The three near-endemic species on Madeira are Plain Swift, Berthelot's Pipit and Canary. The swift is a partial migrant, common and widespread in summer but scarce and more local in winter. The pipit is rather local but not uncommon in appropriate habitat such as the grassland and rocky scree near Pico de Arieiro. Finally, the Canary occurs throughout Madeira but is less common on Porto Santo.

Other areas for general birding include the beach and harbour in Funchal, which have attracted many vagrants, particularly gulls and terns. Ponto da Cruz, west of Funchal, makes a good spot for seawatching, as it forms the most southerly tip of Madeira. The man-made reservoir at Santo da Serra, north of the airport, occasionally hosts rare waders and ducks. At Machico, the second-largest town on Madeira, areas to search include the beach and river outflow, both of which have hosted migrants. Ponta da São Lorenço can be productive under those rare conditions suitable for migrants to arrive, but at other times Berthelot's Pipit and Rock Sparrow might be the only rewards of a visit. At the opposite end of the main island is Ponta do Pargo, which though too high for seawatching has Spectacled Warbler near the lighthouse and migrants occur under the right conditions. To the north-west is Porto Moniz, which unlike Ponta do Parga is excellent for seawatching. Inland western Madeira is dominated by the plains of Paúl da Serra, an area worthy of a visit at any season but particularly in winter when species such as Northern Lapwing and Common Starling can be found. It is also a known haunt of Berthelot's Pipit, and various migrants have been recorded. Finally, the small lagoon at Lugar de Baixo, near Ponta del Sol, attracts waterfowl, waders and herons, and is worth checking at any time of the year, although it is probably best in winter.

PORTO SANTO Approximately 35 km north-west of Madeira is the small island of Porto Santo. The most noteworthy birding site is the pool at Tanque, which is the largest area of standing water on the island and consequently attracts a variety of residents and migrants. Some of the rarest birds in the archipelago have been recorded here including Semipalmated Plover and American Cliff Swallow in 2001, and a Reed Bunting in 2002; this site also produced the first breeding record of Common Moorhen for the archipelago.

For seabird fanatics, probably the best option is the Porto Santo ferry that runs from Funchal. All of the breeding seabirds and many migrants have been recorded, but obviously some of the rarer species, such as Zino's Petrel, are not regular. Another option for seabirding is to take one of the many charter boats that operate out of Funchal; it is possible to join some of the fishing trips as an observer at a reduced rate. Finally, there are organised trips to the Desertas but these are irregular and enquiries should be made as soon as you arrive on Madeira. If you intend to take such a trip, seasickness medication is recommended, as the seas around the Desertas can be very rough. For those keen to reach the Salvages, a boat must be chartered and permission obtained from the relevant authorities, as these islands form part of a marine reserve.

THE AZORES

This group of nine islands possesses just one endemic, the Priôlo or Azores Bullfinch. The bullfinch is confined to more remote parts of São Miguel, and the only area to look for this species is around Pico da Vera. Probably the best place is the track between Povoação and Nordeste, in the vicinity of a small, yellowish-coloured, wooden hut. The only other Macaronesian endemic on the Azores is the Canary, which is common and widespread. Other birds of great interest to visiting birdwatchers are the resident and migrant seabirds and the endemic subspecies, while there is always the chance of finding an American vagrant. Seabirds are best looked for from the inter-island ferries, probably the best of which is that making the crossing between Terceira and Faial (calling at São Jorge and Pico), which takes around 8 hours in total. Land-based seawatching can also be productive, but obviously the ferry trips are better for the rarer and more pelagic species. Rarities of American origin occur on a regular basis and can appear almost anywhere in the archipelago, but certain areas are worthy of concentrated effort. Non-passerines can turn up at any time of the year but are most frequent from August to November, while passerines are best looked for in October–November.

SANTA MARIA This island is rarely visited by birdwatchers but it has an endemic subspecies of Goldcrest (*sanctaemariae*) and the only breeding Sooty Terns in the Western Palearctic, on the Ilhéu da Vila. These can sometimes be seen from the main island.

SÃO MIGUEL The largest island in the archipelago, it also has the largest population and a good infrastructure. The main birdwatching areas are the volcanic crater lakes at Furnas, Sete Cidades and Lagoa do Fogo, but smaller lakes are also worth visiting, particularly in autumn or winter. The harbour at Ponta Delgada, the island's capital, has attracted unusual shorebirds, ducks and gulls, and the coast around Mosteiros, in the north-east, is another good area for migrants/vagrants, as well as being a summer haunt of Roseate Terns, and one of the best seawatching points on the island.

TERCEIRA The second-most populated island in the archipelago and site of the famous quarry at Cabo da Praia, which has attracted more American shorebirds than any other area on the eastern side of the Atlantic. Often, the different species are discovered in small groups, but occasionally flocks of birds such as White-rumped Sandpipers have been recorded. The area has a list of 17 species of American waders plus other rarities, including Nearctic ducks and gulls. The nearby harbour and bay of Praia da Vitória are also very good for vagrant gulls, with large flocks of Ring-billed Gulls reported. Recently the area hosted the first Forster's Tern for the islands. Other sites include the freshwater Lagoa do Ginjal, the small unnamed pond in Praia da Vitória near the Residential Branco, and the harbour at Angra do Heroísmo.

FAIAL The main areas worth checking include the harbour at Horta, the beach and bay at Porto Pim, and the lake at La Caldeira. Don't forget to seawatch from the lower slopes of Monte da Guia, Horta. Recent sightings on the island include Semipalmated Plover, Belted Kingfisher, Snowy Egret, Rough-legged Buzzard and Red-breasted Merganser, proving that it has potential for vagrants.

PICO The most famous site is the area of tidal pools and rank coastal vegetation in front of the seawall at Lajes do Pico. This is yet another area that attracts Nearctic vagrants and is noted not only for waterfowl and shorebirds, but also herons. In the west of the island is the main town and port of Madalena, and although this area looks rather unattractive for birds, it has had its share of interesting species in recent years, including Little Blue Heron and Double-crested Cormorant, as well as attracting unusual gulls, such as Glaucous and Iceland, in winter. The three freshwater lakes of Capitão, Caiado and Paul are certainly worth a visit in autumn/winter, but they are often fogbound. The *inermis* subspecies of Goldcrest can usually be located in vegetation near the airport.

SÃO JORGE Again this island has a site famous for attracting vagrants, a brackish lagoon with some reedy areas at Fajã dos Cubres. In the centre of the island are many small freshwater ponds that could attract interesting birds in autumn/winter, but at other seasons are probably devoid of birds. Ilhéu do Topo, off the east end of São Jorge, is a reserve for seabirds such as Roseate Tern and it has hosted one of the few Western Palearctic records of Willet.

GRACIOSA Birders only rarely visit Graciosa, which has no permanent fresh water and is of less interest than others of the central group. Ilhéu da Praia is an important breeding site for seabirds, in particular Madeiran Storm-petrel and Roseate Tern, and it has hosted the only Bermuda Petrel in the Western Palearctic, as well as Sooty, Bridled and Royal Terns, and one of the few Peregrine Falcons to be recorded in the Azores. The harbour at Santa Cruz has held such species as Bufflehead and Wood Duck in the past, indicating the island's vagrancy potential.

FLORES The westernmost of the islands, Flores unsurprisingly possesses a long list of Nearctic vagrants, including a few landbirds as well as the usual waterbirds. Due to its location, birds can turn up just about anywhere on the island but one favoured area is the coast at Fajã Grande, the westernmost point in the Western Palearctic. The rather flat headland is a mix of rocky coastline and agricultural fields, with a couple of small ponds at the base of a heavily vegetated cliff. Another good area lies in the centre of the island, where the seven main lakes attract the visiting birdwatcher. These lakes (Negra, Comprida, Seca, Branca, Raso, Funda and Lomba) have all hosted Nearctic species and have the potential for many more in the future. American Black Duck has bred and there is some hybridisation with Mallard, as a group of 20 hybrids are often present. It is also well worth checking the various habitats around the main town of Santa Cruz, including the airport for roosting gulls and occasional terns, the harbour for ducks, cormorants and egrets, and the trees in the town for passerines. Even the rooftops are worth checking, as a Killdeer was found in such a location in autumn 1998. Apart from Santa Cruz, it is also worth searching Ponta Delgada, Fajãzinha and Lajes, as all have recorded Nearctic vagrants in the recent past.

CORVO The smallest and most isolated of the main islands, the main site is the two small lakes at Caldeirão, on Monte Grosso, but on this island check anything that flies as birds turn up almost anywhere. In 2005 this tiny island turned up no fewer than 16 species new for Macaronesia, most of them Nearctic passerines (see page 331).

CAPE VERDE ISLANDS

This former Portuguese colony lies some 600 km north-west of Senegal. It comprises ten main islands and many smaller islets. There are four endemic species and numerous endemic subspecies, some of which are now on the verge of extinction, plus a fine range of breeding seabirds.

SAL Most visitors arrive on Sal, since the main international airport for the archipelago is located here. Although not always conspicuous, it is possible to see your first Cape Verde endemic within the airport terminal, Iago Sparrow. If time permits, a visit to nearby Pedra da Lume saltworks is an essential part of any birding itinerary. This area is undoubtedly the premier destination on Sal and, in winter and during migration, there can be many shorebirds at this location. The area also attracts other migrants, the endemic *alexandri* race of Common

Kestrel, and Bar-tailed Lark. In the south lies the main tourist development of Santa Maria, where an area of unused saltworks and waste ground inland of the town has Cream-coloured Courser and many Bar-tailed Larks. Hoopoe Lark has recently been confirmed as breeding on Sal but the population is very small, so any sightings should be reported.

BOAVISTA One of the most famous sites for birding in the Cape Verdes is Rabil lagoon, which over the years has attracted many unusual species, including several new to the Cape Verde list. It is also one of the best places in the Western Palearctic to look for Intermediate Egret. The surrounding semidesert has many larks including Black-crowned Sparrow-lark and Hoopoe Lark. In the south is the deserted settlement of Curral Velho, where occasionally there is a seasonal lagoon, but the main reason for visiting this area is to view the offshore islet of Ilhéu de Curral Velho, where the last breeding Magnificent Frigatebirds in the Western Palearctic maintain a toehold. In the north is the settlement of Baia das Gatas, from where you can negotiate a rather expensive 'lift' for the few hundred metres offshore to the Ilhéu dos Pássaros, where White-faced Storm-petrel breeds. Remember, it is necessary to stay on the island overnight so bring food, water and a sleeping bag.

SANTIAGO This, the largest island in the archipelago, is home to the largest remaining population of the endemic Cape Verde Warbler, plus good numbers of the endemic Cape Verde Swift and Iago Sparrow, the only colonies of the critically endangered endemic subspecies of Purple Heron, *bournei,* and the endangered Cape Verde subspecies of Common Buzzard (*bannermani*), Peregrine (*madens*) and Barn Owl (*detorta*). There are also important populations of Grey-headed Kingfisher, Common Waxbill and Helmeted Guineafowl, all of which are widespread on the island. The warbler is easy to find around Boa Entrada, near Assomada, which is also one of only two locations for the Purple Heron, and a known haunt of the Barn Owl. Another location for the warbler is São Jorge dos Órgãos, where it seems particularly common around the botanic gardens. Cape Verde Buzzard can also be found in this area, and a pair usually breeds on the cliffs overlooking the village. Probably the best area for the Purple Heron is at Banana de Ribeira Montanha, *c.* 5 km from Pedra Badeja, just off the road to João Teves. Breeding commences in September but sometimes the birds have a second brood and young are still present in the nesting tree in March/April, but this depends on the amount of rainfall during the previous wet season. Banana is also a site for the Barn Owl; look for them flying along the valley at dusk. Between Assomada and Tarrafal the road passes through a range of hills where both Cape Verde Buzzard and Cape Verde Peregrine have been recorded, while the latter has also been seen along the cliffs near Praia and at São Jorge dos Órgãos; although widespread it is now highly endangered and rarely reported. A couple of areas worth checking for migrants are the harbour at Praia, where Western Reef Egret is often found, and the coastal lagoons south of Pedra Badeja. The amount of water in these is dependent on recent rains and they can be completely dry. In Tarrafal, the pond by the Baia Verde Hotel is another area where migrants concentrate due to the presence of fresh water, and the water purification tanks to the south have produced some interesting birds in recent years.

SÃO NICOLAU The main reason to visit this island is to arrange transport to the uninhabited islet of Raso where the endemic Raso Lark occurs. This bird has one of the most restricted ranges of any species on the planet. Normally, one can find boats willing to take you across from the harbour at Tarrafal, but be warned this could prove expensive. If visiting Raso, it would be worthwhile to spend a night on neighbouring Branco where many of the archipelago's seabirds breed. White-faced and Madeiran Storm-petrels and Cape Verde Shearwaters are usually easily found, but the Cape Verde race of Little Shearwater (*boydi*) and Bulwer's Petrel are a lot more difficult to locate. The crossing between Tarrafal and Raso can be good for seabirds with Cape Verde Macaronesian Shearwater and Fea's Petrel the main targets, but the smaller storm-petrels are possible and there are plenty of Cape Verde Shearwaters. Back on São Nicolau, Cape Verde Peregrine has been reported from Tarrafal in the past and the headland at Ponta do Barril can be productive for seawatching with Fea's Petrel regularly seen from here. In 1998 the endemic Cape Verde Warbler was rediscovered on the island after around 28 years without a record.

SÃO VICENTE The star attraction is the sewage ponds just south-west of the island's capital, Mindelo. This site has been a major source of rare birds, particularly Nearctic ducks and shorebirds, in recent years and several species new to the Cape Verde list have been found there. There was also a record here of Cape Verde Kite a few years ago.

SANTO ANTÃO This, the westernmost of the group, was formerly the last stronghold of the Cape Verde Kite. Unfortunately, none have been seen in the last couple of years. The island is also a nesting place for Fea's Petrel and these birds can often be seen by seawatching from Vila de Ribeira Grande, preferably in the evening, although early mornings can also produce sightings.

MAIO, FOGO, BRAVA AND SANTA LUZIA These islands are very infrequently visited by birders and hence few data are available on where to find birds on any of them. Santa Luzia is also uninhabited, so a visit must be well planned in advance.

How to use this book

This book is divided into two main sections – plates and species accounts. Each section contains cross-references to the other for ease of use. The illustrations on the plates are generally in systematic order, and illustrate the plumages that are most likely to be encountered by the visiting birder on the islands. As well as showing virtually every species that has occurred on the four archipelagos, great attention has been placed on ensuring that every one of the endemic taxa has been illustrated – in some cases for the first time in a field guide. Concise caption text opposite the plates contains important identification information, generally in the order **adult** (abbreviated to **ad**) **male**, **adult female**, **juvenile** (abbreviated to **juv**), with subspecies slotted in by sex/age. The length of the bird in cm is included for comparative purposes. The range of the bird within the region is given, with the Canary Islands abbreviated to **CI**, Madeira abbreviated to **Md**, the Azores abbreviated to **Az** and the Cape Verde Islands abbreviated to **CV**.

SPECIES ACCOUNTS

The second main section of the book provides detailed accounts of the identification and distribution of each species that has been recorded on the islands, with brief information on the biology where relevant. English and scientific names follow the '*British Birds*' *List of Western Palearctic Birds*, which incorporates revisions up to 2005 as published by the BOURC Taxonomic Subcommittee in *Ibis* (147: 821-826). This includes several important changes with regard to the Atlantic Islands avifauna, such as the treatment of Madeira Firecrest as a full species distinct from Firecrest, and the division of Little Shearwater into several species, one of which is now called Macaronesian Shearwater. Where alternative names are regularly used, these are included in parentheses within the main title, and in a lighter font, for example (NORTHERN) **PINTAIL**.

The heading also includes Spanish and Portuguese names. Nomenclature is debated and revised as keenly in these languages as in English; Spanish names are based on the Sociedad Española de Ornitología's 2005 *List of the Birds of Spain*, although some of these have been replaced by terms that are more familiar to Spanish birders. Portuguese names are based on *Nomes Portugueses das Aves do Paleárctico Ocidental* (Costa et al. 2000), though many of the recommendations in this work are yet to be generally accepted; in such cases vernacular names are used.

The species texts begin by stating either that the species is monotypic or, for polytypic species, by listing the subspecies that occur in the region (where known). This is followed by the following sections:

Measurements Each species text has one or two sets of measurements; **L** gives the length of the adult bird, while **WS** is the wingspan. Both metric and imperial measurements are given, although imperial measurements have been rounded to the nearest half-inch.

Identification This section highlights important identification features, with particular emphasis on the plumages most likely to be seen in the region. As with the plate caption text, this deals with sex- and age-related plumage variations as well as seasonal differences. Separation from possible confusion species is also discussed.

Voice Calls and, where appropriate, songs that may be encountered in the region are discussed.

Range The regular breeding and wintering world ranges of each species are provided.

The following three sections appear for species that breed in the region, including residents and migrant breeders:
    Distribution This section describes the breeding distribution of the species within the Atlantic Islands.
    Breeding A discussion of the breeding biology of the species, including nest sites, clutch size, and usual laying and fledging dates.
    Habitat The preferred habitats of the species on the islands.

Taxonomic notes These are included for a number of species. The notes discuss recent revisions of taxonomic status, and possible future changes. For example, it seems likely that the Canary Islands Robin, currently treated as a subspecies, will soon be split from European Robin; similarly, some authorities consider a number of Cape Verde endemic taxa as full species, such as the Cape Verde Purple Heron, but this is not widely accepted. We have kept this and most others as subspecies, but some of these may well be promoted to full species status in the future.

Status After a brief summary of the species' status in the region, a more detailed statement is given for each of the four archipelagos. Breeding population figures are given where relevant. Where a vagrant species has occurred on only a few occasions, all the known records are listed individually, with locality and dates (where available). The total number of records is given for rarities of more frequent occurrence.

SOURCES OF RECORDS

The vast majority of species recorded in the region are either migrants or vagrants. As regards the latter we have tried to include as many reports as possible for the sake of completeness, and in certain instances this has involved species that are not officially recognised as having occurred by the rarities committee assessing records for that particular archipelago (mainly the Canary Islands – see below). These records include specimens collected many years ago which have remained unconfirmed, sight records not submitted by the observers who found them, or records that are awaiting acceptance. All these records are included in the book to give a fuller picture of the species that have occurred within the region, but no records are listed where there is doubt about the identification of the species. With ever-increasing observer coverage it is clear that many more vagrants will be found in Macaronesia in the coming years, especially on the Azores, where no fewer than 21 new species for the archipelago were found in 2005 (see Appendix A on page 331).

CANARY ISLANDS A considerable number of records of rarities claimed for the Canary Islands are not on the official list for the islands. Where a species is already on the list for the archipelago, no attempt has been made to identify the officially accepted records, and therefore some of the records listed in this book may not be included in the official totals for a particular species. However, species that have not yet been formally admitted to the list are marked with an asterisk (*) in the main text. The sources of these as-yet unaccepted Canary Islands records are summarised in the table below.

Table 1. *List of species not on the official Canary Islands list. See page 32 for the key and notes.*

Species	Code	Species	Code
Red-throated Diver	AC, AA, OR	Purple Sandpiper	BA, AC, AA, PO
Little Grebe	BA, AC, AA	Great Snipe	BA, AC, AA
Great Crested Grebe	AC, AA	Long-billed Dowitcher	AA, PO
Fulmar	AC, AA	Slender-billed Curlew	AC, AA
Fea's Petrel[1]	AA	Solitary Sandpiper[4]	OR
Cape Verde Shearwater[2]	AA, PO	Red-necked Phalarope	AC, AA
Balearic Shearwater	AA, OR	South Polar Skua	PO
Brown Booby	AA	Pallas's Gull	AA, PO
Magnificent Frigatebird	AC, AA	Herring Gull	BA, AA, PO
Shag	AA	Iceland Gull[5]	AA, OR
Black Stork	AC, AA	Royal Tern	AA, PO
White-faced Whistling Duck	AC, AA	White-cheeked Tern	AC, AA
Mute Swan	AA, PO	White-winged Tern[6]	AC, AA, OR
Pink-footed Goose	BA, AC, AA	Forster's Tern	AA
Greater White-fronted Goose	AA	Little Auk	AC, AA
Falcated Duck	AC, AA	Razorbill	BA, AC, AA
Long-tailed Duck	AA	Pin-tailed Sandgrouse	AC, AA
Smew	AA	Namaqua Dove[7]	AA, P
Bufflehead	AA	Red-necked Nightjar	AC, AA
Honey-buzzard	AC, AA, PO	Hawk Owl	AA
Black-shouldered Kite	AA,	Tawny Owl	BA, AC,
White-tailed Eagle	BA, AC, AA	Eagle Owl	AA
Golden Eagle	AC, AA	Marsh Owl	AC, AA
Lanner Falcon	AC,	White-rumped Swift	AA, PO
Red-knobbed Coot	BA, AC	Chimney Swift	AA
Water Rail	AC	Common Kingfisher	BA, AC, AA
Lesser Moorhen	AC	Blue-cheeked Bee-eater	BA, AC, AA
Little Bustard	BA, AC	Dupont's Lark	AA
Egyptian Plover	BA, AC	Short-toed Lark	AC, AA
Killdeer[3]	OR	Crested Lark	AC, AA
Lesser Sand Plover	AA	Thekla Lark	AA
Pacific Golden Plover	AA	Plain Martin	AA
Western Sandpiper	AC,	Rock Pipit	AC, AA
Least Sandpiper	AA, PO	Water Pipit	AC, AA
Sharp-tailed Sandpiper	AA	Olive-backed Pipit	AA, PO

Dipper	AA	Penduline Tit	AA
Alpine Accentor	AA	Isabelline Shrike	AA
Rufous Bush Robin	AC, AA	Red-backed Shrike	BA, AC, AA
Thrush Nightingale	AA	Nutcracker	AA
Isabelline Wheatear	AC, AA	Rook	AA
Mourning Wheatear	AA	Carrion Crow[9]	OR
Rock Thrush	AA	Brown-necked Raven	AA
Mistle Thrush	AA, PO	Spotless Starling	AC, AA
Savi's Warbler[8]	OR	Snowfinch	BA, AC, AA
Aquatic Warbler	BA, AC, AA	Blackpoll Warbler[10]	OR
Ménétries's Warbler	AC, AA	Cirl Bunting	AC, AA
Lesser Whitethroat	AA	Rock Bunting	AC, AA
Hume's Warbler	AA	Cretzschmar's Bunting	AC, AA
Icterine Warbler	AA	Little Bunting	AC, AA, PO
Eurasian Nuthatch	AA	Reed Bunting[11]	OR, PO

Key

The sources of these records are as follows:

BA = *Birds of The Atlantic Islands, Volume 1* (Bannerman 1963)

AC = *Las Aves de Canarias, 3rd Edition* (Pérez Padrón 1983)

AA = *Aves del Archipiélago Canario* (Martin & Lorenzo 2001)

PO = Personal observation

OR = Other records submitted to the author, usually supported by photographs

Notes

1. Included as *Pterodroma* sp.
2. Included as subspecies of Cory's Shearwater.
3. Killdeer was seen and photographed by Guido Rastig near Caleta de Fustes.
4. Solitary Sandpiper was reported by Dean Harlow.
5. There have been two records of Iceland Gull and there are photos of the first at least, seen by Maciej Zimowski at Puerto de Santiago on Tenerife. The second was seen by Barry Lancaster at various location in the south of Tenerife, and was also seen by Tim Norris. Photographs of the first bird have been submitted to the Spanish Rarities Committee.
6. White-winged Black Tern was photographed by Leo Boon on El Hierro.
7. Namaqua Dove was seen and photographed by R. J. Lowe. The photo has been submitted to the Spanish Rarities Committee.
8. Savi's Warbler was reported by Dave Foster.
9. Carrion Crow was reported by Geoff Hunt.
10. Blackpoll Warbler was reported by Tim Hutchinson.
11. Reed Bunting now has at least four records; the first was seen by Andrew Lawson, the second by Dan Lupton, one was photographed by Barry Lancaster and one was a personal sighting.

MADEIRA Rarity records for Madeira mainly follow the published list for the archipelago, but a few recent sightings awaiting acceptance by the Portuguese rarities committee have been included. These are marked with an asterisk in the main text.

THE AZORES There is no official list yet for the Azores but rarity records are assessed by the Portuguese rarities committee. Many of the recent 'firsts' for the Azores are awaiting acceptance by the committee, but these have been included in this book where there was good reason to do so; they are marked with an asterisk in the main text. A pre-1903 record of Edible-nest swiftlet Collocalia fuciphaga has been excluded from the book; its identification and provenance are unproven, and the specimen untraceable

CAPE VERDE ISLANDS The published list for Cape Verde (Hazevoet 1995) has been followed, together with regular updates published by the author, with the exception of Great Blue Heron, Green-winged Teal and Woodchat Shrike, which are awaiting assessment.

SUBMISSION OF UNUSUAL RECORDS

As with all rarities found within the Western Palearctic, we hope that any observer fortunate enough to encounter an unusual species on the Atlantic Islands will submit their records to the following organisations, or to the author who will forward them to the appropriate committee. The author will also be pleased to receive any information concerning the birds of any of these archipelagos. Please write to Tony Clarke, C/ Republica Dominicana, Nº 61B, Barrio de Fatima, 38500 Güímar, Tenerife, Spain, or by email to bluechaffinch@btopenworld.com. Records of rarities from the islands should be submitted to the following addresses:

Canary Islands
J. I. Dies
SEO/Birdlife Valencia
Av. Los Pinares 106
46012 El Saler
Valencia
SPAIN

or by e-mail to rarezas@seo.org. Rarities submission forms can be downloaded from the website www.rarebirdspain.net

Azores and Madeira
SPEA
Comité Português de Raridades
Rua da Vitória, 53, 3-Esq
1100-618 Lisboa
PORTUGAL

or by email to raridades@spea.pt. Rarities submission forms can be downloaded from the website www.spea.pt

Cape Verde Islands
There is no official committee assessing records for Cape Verde, but records should be submitted to:

Dr C. J. Hazevoet
Museu e Laboratorio Zoologico e Antropologico
Museu Nacional de Historia Natural
Rua da Escola Politecnica 58
1250-102 Lisboa
PORTUGAL

or by email to cjhazevoet@fc.ul.pt

PLATE 1   DIVERS AND GREBES

1 GREAT NORTHERN DIVER *Gavia immer*　　**L** 61–91 cm　　Text p. 173

The largest diver of the region with a large head and a heavy bill. Vagrant to the Az and CI.
1a Ad non-breeding: Resembles Black-throated Diver but larger with hindneck and crown darker than upperbody. Dark half-collar on the lower neck emphasised by the white indentation above it. Bill greyish-white with dark tip and dark culmen.
1b Ad breeding head: Head and neck mainly black with a white striped patch on neck sides. Bill dark blackish.

2 BLACK-THROATED DIVER *Gavia arctica*　　**L** 58–73 cm　　Text p. 173

Larger than Red-throated Diver and swims with bill held horizontal. Vagrant to the CI.
2a Ad non-breeding: Often shows a distinct white rear flank patch. At least half the neck is dark when seen side-on. Appears full-chested when swimming.
2b Ad breeding head: Head and back of neck grey. Black and white stripes on neck sides. Throat black.

3 RED-THROATED DIVER *Gavia stellata*　　**L** 53–69 cm　　Text p. 173

The smallest diver. Often swims with bill held above the horizontal. Vagrant to the Az and CI.
3a Ad non-breeding: Grey-brown above with fine white spotting. Grey-brown crown does not extend below the eye. Neck mainly white. Appears flat-chested when swimming.
3b Ad breeding head: Face and neck sides grey. Rufous throat patch. Crown and rear neck finely streaked black-and-white.

4 GREAT CRESTED GREBE *Podiceps cristatus*　　**L** 46–51 cm　　Text p. 174

The largest grebe of the region with a long pale bill in winter. Vagrant to the Az and CI.
4a Ad non-breeding: Back, hindneck and crown brown. Face and foreneck white. Flanks pale brownish. Dark loral stripe and white above the eye are diagnostic.
4b Ad breeding head: Unmistakable with the chestnut-and-black head plumes.

5 PIED-BILLED GREBE *Podilymbus podiceps*　　**L** 30–38 cm　　Text p. 173

A small stocky grebe with a thick heavy bill. Vagrant to the Az and CI.
5a Ad non-breeding: Like a large Little Grebe with a thick yellowish bill, sometimes with a hint of a black band.
5b Ad breeding head: Mostly grey-brown with a black forehead and throat. Bill whitish with a thick black band.

6 SLAVONIAN GREBE *Podiceps auritus*　　**L** 31–38 cm　　Text p. 174

Medium sized with contrasting black-and-white non-breeding plumage. Bill short and straight. Vagrant to the Az and Md.
6a Ad non-breeding: Distinct black cap contrasts with white cheeks. Flat crown peaks at rear. Back rather flat and rear end not so prominently fluffed-up as in Black-necked Grebe.
6b Ad breeding head: Black head, chestnut neck and erect golden head plumes.

7 BLACK-NECKED GREBE *Podiceps nigricollis*　　**L** 28–34 cm　　Text p. 172

Medium sized with black-and-white non-breeding plumage. Bill fine-pointed and uptilted. Irregular in the Canaries in winter. Vagrant to the Az and Md.
7a Ad non-breeding: Peak of head in the middle of the crown. Cheeks dark. Back more arched and rear end more fluffed-up than in Slavonian Grebe.
7b Ad breeding head: Head and neck black, with drooping yellow cheek feathers.

8 LITTLE GREBE *Tachybaptus ruficollis*　　**L** 25–29 cm　　Text p. 174

The smallest grebe of the region. Rear end frequently fluffed-up. Vagrant to the CI and Md.
8a Ad non-breeding: Brown above and buff below, with a fine straight bill.
8b Ad breeding head: Black cap, chestnut cheeks and prominent yellowish spot on the gape.

35

PLATE 2 PETRELS

1 (NORTHERN) FULMAR *Fulmarus glacialis*   **WS** 102–112 cm   Text p. 175
A fairly large tubenose with broad, straight, stiffly held, round-tipped wings, a thick neck and a large head. Vagrant to the Az and CI.
1 **Ad upperside:** Mantle, rump and upperwings steel grey, with a pale basal patches on primaries. Head and neck white. Thick yellow bill. Tail pale.

2 TRINDADE PETREL *Pterodroma arminjoniana*   **WS** 88–102 cm   Text p. 175
A medium-sized tubenose with long wings and predominantly dark plumage. Polymorphic. Vagrant to the Az.
2a **Ad light morph underside:** White below with pale undersides to flight-feathers. Grey patches on breast-sides.
2b **Ad dark morph underside:** Dark brown with white on primary bases extending narrowly across the secondaries.

3 FEA'S PETREL *Pterodroma feae*   **WS** 92–95 cm   Text p. 176
A medium-sized seabird with dynamic flight combining towering arcs and strong glides. Breeds on the Desertas, in Md and in the CV. Vagrant to the Az and CI.
3a **Ad underside:** Body white except for grey patches on breast-sides. Underwings mostly dark except for a white patch on the axillaries.
3b **Ad upperside:** Grey with darker primaries, a dark M pattern across the wings and a pale grey tail.
3c **Ad head:** Bill more robust than Zino's Petrel.

4 ZINO'S PETREL *Pterodroma madeira*   **WS** 83–86 cm   Text p. 176
Very similar to Fea's Petrel but smaller with a finer bill. Breeds only on Md.
4a **Ad underside:** As Fea's Petrel but dark bar across the underwing broadens at the carpal joint.
4b **Ad upperside:** As Fea's Petrel but the hand is shorter, the wing-tip blunter and the open M mark is less pronounced.
4c **Ad head:** Bill less robust than in Fea's Petrel.

5 SOFT-PLUMAGED PETREL *Pterodroma mollis*   **WS** 86–92 cm   Text p. 176
The South Atlantic equivalent of the previous two species and very similar to both. Doubtful vagrant to Md.
5a **Ad underside:** Distinguished from its North Atlantic counterparts by the presence of a complete breast band. Even if incomplete this breast-band is bolder and broader on this species.
5b **Ad upperside:** Very similar to both Fea's and Zino's Petrels, but the upper tail is distinctly darker.

6 BERMUDA PETREL *Pterodroma cahow*   **WS** 89 cm   Text p. 175
A relatively pale gadfly petrel. Vagrant to the Az.
6a **Ad upperside:** Greyish-brown with blackish-brown cap and nape. Indistinct pale band at base of tail.
6b **Ad underside:** Mainly white with dusky breast-side patches. Underwing white with black margins and a short diagonal bar on the underwing-coverts.

7 BULWER'S PETREL *Bulweria bulwerii*   **WS** 68–73 cm   Text p. 177
Sooty-brown with long wings and a long tapering tail. Breeds on all four archipelagos.
7 **Ad upperside:** Uniform dark sooty-brown with a pale carpal bar, long pointed tail and long angled wings.

37

PLATE 3   SHEARWATERS

1 CORY'S SHEARWATER *Calonectris diomedea*   **WS** 100–125 cm   Text p. 177

The largest shearwater of the region. Flies on bowed wings using long glides close to the sea. Pale yellow bill obvious. Breeds on the Az, Md and the CI. Occurs on passage in the CV.
1a Ad upperside: Grey brown upperparts with darker wings and tail. Variably white uppertail-coverts.
1b Ad underside: Mostly white. Dark border to the underwing, most extensive on the primaries.

2 CAPE VERDE SHEARWATER *Calonectris edwardsii*   **WS** 90–110 cm   Text p. 178

Smaller than Cory's Shearwater with a more slender, dark-tipped, grey bill. Breeds in the CV. Vagrant to the CI.
2a Ad upperside: Cap and mantle darker than in Cory's Shearwater.
2b Ad underside: Almost identical to the underside of Cory's Shearwater but the dark margin to the underwing is darker and more extensive.

3 GREAT SHEARWATER *Puffinus gravis*   **WS** 100–118 cm   Text p. 178

A large, dark-capped shearwater. Flies with stiffer and more rapid wing beats than either of the *Calonectris* species. Probably regular on passage in all four archipelagos.
3a Ad upperside: Distinct dark cap and pale collar, obvious white uppertail coverts. Bill dark.
3b Ad underside: White with a dirty brown patch on the belly, dusky undertail-coverts and variable dark on the axillaries and the median and lesser coverts.

4 SOOTY SHEARWATER *Puffinus griseus*   **WS** 94–109 cm   Text p. 178

The only large dark shearwater of the region, dark brown with silvery underwings. Regular on passage in the Az and Md. Vagrant to the CI and CV.
4   Ad upperside: Uniformly dark brown.

5 MANX SHEARWATER *Puffinus puffinus*   **WS** 76–85 cm   Text p. 179

A medium-sized, black-and-white shearwater. Breeds in small numbers in the Az and CI and more commonly in Md. Vagrant to the CV.
5a Ad upperside: Uniformly sooty-black. Cap extends below eye.
5b Ad underside: Uniformly white, including undertail-coverts.

6 BALEARIC SHEARWATER *Puffinus mauretanicus*   **WS** 76–89 cm   Text p. 179

Larger, shorter-tailed and slower-flying than Manx Shearwater. Vagrant to the Az, Md and CI.
6a Ad upperside: Brown, generally merging into paler underparts without sharp demarcation.
6b Ad underside: Variable. Typically dusky on vent, undertail-coverts, axillaries and neck, with dirty-white belly and underwings.

39

PLATE 4   MACARONESIAN SHEARWATER AND STORM-PETRELS

1 MACARONESIAN SHEARWATER *Puffinus baroli*   **WS** 58–67 cm   Text p.179

A small black-and-white shearwater with rapid flight action. Race *baroli* breeds in the Az, Mad and the CI; race *boydi* breeds only in the CV.
1a Ad nominate upperside: Black with an indistinct silvery-grey panel on the secondaries. The white on the side of the face does not extend below the eye.
1b Ad nominate underside: Underparts and underwings completely white except for a thin dark border to the underwing.
1c Ad *boydi* upperside: Browner than *baroli* with the dark cap extending below the eye.
1d Ad *boydi* underside: Similar to *baroli* but the undertail-coverts are mostly dark brown.

2 WHITE-FACED STORM-PETREL *Pelagodroma marina*   **WS** 41–43 cm   Text p. 180

A large grey-brown, black and white storm-petrel with very long legs and broad rounded wings. In flight often appears to bounce off the surface of the sea. Breeds in Md (Ilhéus Selvagens), the CI and CV. Vagrant to the Az.
2   Ad *hypoleuca* upperside: Wings dark with a pale bar on the greater coverts, grey-brown crown and mantle and a paler greyish rump that can appear white at times. Face, underparts and underwing-coverts white except for a dark mask around the eyes and a grey patch on the side of the breast.

3 WHITE-BELLIED STORM-PETREL *Fregatta grallaria*   **WS** 44 cm   Text p. 181

The only black/brown and white storm-petrel of the region with extensive white on the underparts. Vagrant to seas north of the CV.
3   Ad *leucogaster* underside: Head and upper breast black. Rest of underparts white and this extends across the central underwing. The upperside is similar to other storm-petrels of the region: mostly blackish-brown with a white rump, but has a paler greyer mantle and upperwing-coverts.

4 EUROPEAN STORM-PETREL *Hydrobates pelagicus*   **WS**: 36–39 cm   Text p. 181

A small black-and-white storm-petrel with a rapid fluttering flight action. Breeds CI. Vagrant to Md.
4   Ad upperside: All blackish except for white rump that extends far onto the sides. Tail square ended. If visible the white stripe on the centre of the underwing is diagnostic.

5 WILSON'S STORM-PETREL *Oceanites oceanicus*   **WS** 38–42 cm   Text p. 180

Long-legged, with toes projecting beyond the tail in flight. Few records but probably regular on passage off all four archipelagos.
5   Ad upperside: Similar to European Storm-petrel but slightly larger, has a more obvious pale carpal bar and the white rump almost encircles the rear end.

6 LEACH'S STORM-PETREL *Oceanodroma leucorhoa*   **WS** 45–48 cm   Text p. 181

A large and long-winged storm-petrel with a forked tail. Occurs on passage in the CI. Vagrant to Md and the Az.
6   Ad upperside: Long angled wings with a distinct pale carpal bar, extensive white rump patch with an indistinct central divide, and a deeply forked tail. This last feature is only visible from certain angles.

7 SWINHOE'S STORM-PETREL *Oceanodroma monorhis*   **WS** 44–46 cm   Text p. 181

Within the region this is the only all sooty-brown storm-petrel. Vagrant to Md (may have bred).
7   Ad upperside: Completely sooty-brown except for a paler carpal bar. Tail forked.

8 MADEIRAN STORM-PETREL *Oceanodroma castro*   **WS** 42–45 cm   Text p. 182

A large long-winged storm-petrel with a notched tail. Nests on all four archipelagos.
8   Ad upperside: Similar to Leach's Storm-petrel but the covert bar is less prominent, the tail shows little or no fork, the rump patch is wider than its length and the white is often extensive on the sides.

41

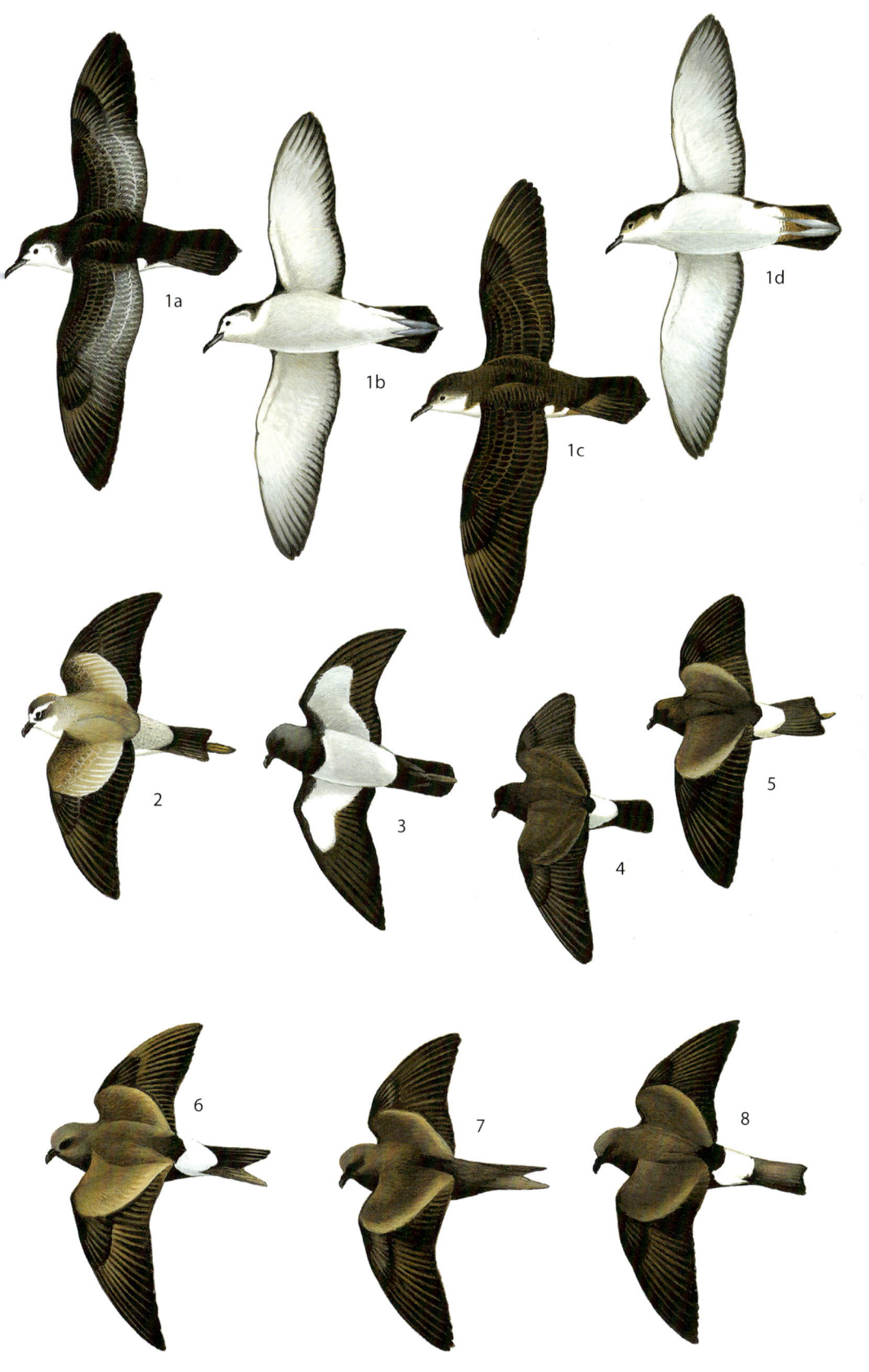

PLATE 5 TROPICBIRDS, BOOBIES AND FRIGATEBIRDS

1 WHITE-TAILED TROPICBIRD *Phaeton lepturus* **WS** 90–95 cm Text p. 182

A small orange-billed tropicbird. White except for a black eye patch, black outer primaries and a black band across the inner secondaries and median coverts. Vagrant to the CV.
1a **Ad:** Distinguished from Red-billed Tropicbird by black innerwing-band and orange bill.
1b **Juv:** More coarsely barred above than juvenile Red-billed, and lacks nuchal collar.

2 RED-BILLED TROPICBIRD *Phaeton aethereus* **WS** 100–110 cm Text p. 182

Black eye patch and black outer primaries contrast with white wings and tail. Black-barred upperparts. Nests on the CV. Vagrant Az (has bred) and Md. Rare but recorded with increasing regularity off the CI.
2a **Ad:** Red bill. Lacks black band across the inner wing of White-tailed Tropicbird.
2b **Juv:** Lacks tail streamers. Bill yellowish, tipped black. Black eye patch extends as a band across nape.

3 BROWN BOOBY *Sula leucogaster* **WS** 132–150 cm Text p. 183

Has contrasting plumage with clearly demarcated dark head, neck, tail and upperparts contrasting with pale underparts. Breeds on the CV. Vagrant to the Az and CI.
3a **Ad:** Chocolate-brown head, neck and upperparts contrast sharply with white lower breast, belly and underparts.
3b **Juv:** Duller brown with white areas mottled brownish.

4 RED-FOOTED BOOBY *Sula sula* **WS** 124–142 cm Text p. 183

Red feet diagnostic. Has several morphs and intermediates occur. Juvenile uniform brown or grey with yellowish-grey feet. Vagrant to the CV.
4a **Ad pale morph:** White. Resembles small Northern Gannet but has black secondaries, as well as black primaries. Black carpal patch below.
4b **Ad brown morph:** Brown or greyish brown.

5 NORTHERN GANNET *Morus bassanus* **WS** 165–180 cm Text p. 183

Large, with long angled wings, wedge-shaped tail, prominent head and bill and a cigar-shaped body. Regular in winter and on passage in the CI and Mad, scarce in winter in the Az, vagrant to the CV.
5a **Ad flight:** Chiefly white, with contrasting black wingtips
5b **Ad:** On water surface.
5c **Juv flight:** Dark grey-brown, lightly spotted white above, with white crescent on the upper tail. Paler below.
5d **Juv:** On water surface

6 MAGNIFICENT FRIGATEBIRD *Fregata magnificens* **WS** 217–244 cm Text p. 185

Unmistakable shape and buoyant gliding flight with elastic wingbeats distinctive. A tiny relict population breeds in the CV. Vagrant to the Az and CI.
6a **Ad male:** All black, with red gular pouch.
6b **Ad female:** Black with white breast-band.
6c **Juv:** Head and breast white, with an incomplete brownish breast-band.

43

PLATE 6 CORMORANTS AND PELICANS

1 (GREAT) CORMORANT *Phalacrocorax carbo*   **WS** 130–160 cm   Text p. 181

A large dark cormorant with a white throat, the white extending to the neck and upper breast in African subspecies. Occasional in winter in the Az. Vagrant to Md and the CI in winter; *lucidus* may have bred formerly in the CV.
1a Ad nominate non-breeding: Blackish, glossed bluish or greenish on the body and bronze on the wings.
1b Ad nominate flight: Appears thick-necked. Uniformly black, with contrasting white throat and yellow-orange facial skin and gular patch.
1c Ad *maroccanus/lucidus*: Foreneck and upper breast white.
1d Juv nominate: Brownish above and dirty white below.

2 DOUBLE-CRESTED CORMORANT *Phalacrocorax auritus*   **WS** 122–137 cm   Text p. 182

Smaller and more lightly built than Cormorant with slenderer bill and smaller, less angular head. Vagrant to the Az, but recorded increasingly frequently in recent years
2a Ad: Facial and gular skin brighter orange. Lacks white throat of Cormorant.
2b Ad flight: Neck appears thinner and with a more pronounced kink than in Cormorant.
2c Juv: Brownish above. Neck and breast whitish but belly dark, unlike immature Cormorant.

3 (EUROPEAN) SHAG *Phalacrocorax aristotelis*   **WS** 90–105 cm   Text p. 182

A small, round-headed, slender-necked cormorant. Vagrant to Md and the CI.
3a Ad breeding: All dark, glossed greenish, with a small crest on the forehead. Yellow gape.
3b Ad flight: Slender neck and rounded head distinguish it from larger cormorants.
3c Juv: Darker than nominate Cormorant, with a whitish throat and upper breast.

4 (GREAT) WHITE PELICAN *Pelecanus onocrotalus*   **L** 140–175 cm **WS** 265–290 cm   Text p. 182

A large black-and-white pelican. Vagrant to the CV.
4a Ad flight: Blackish primaries and secondaries contrast with white wing-coverts and body.
4b Juv flight: Head and neck dark. Breast and belly whitish. Underwing mainly dark with a whitish stripe along the middle.

5 PINK-BACKED PELICAN *Pelecanus rufescens*   **L** 125–132 cm **WS** 265–290 cm   Text p. 183

A relatively small pelican, with poorly contrasting plumage in comparison to White Pelican. Vagrant to the CI.
5a Ad: Greyish-white. Bill pale yellowish-white.
5b Ad flight: Blackish primaries, grey secondaries and grey tail that shows little contrast with rest of plumage.
5c Juv: Brown mottled. Bill duller.
5d Juv flight: As adult but tail dark brown, contrasting with pale rump.

3c

2b

2a
2c

3a
3b

1b

1d
1c

1a

4a

4b

5d

5a

5b

5c

45

PLATE 7   HERONS I

1 EURASIAN BITTERN *Botaurus stellaris*   **L** 70–80 cm   Text p. 185
A secretive, cryptically-patterned heron. Flight feathers barred black. Flies on deeply bowed, rounded wings. Vagrant to the Az, Md and CI.
1a   **Ad:** Has black crown and malar stripe.
1b   **Juv:** Less strongly marked. Crown and malar stripe brown.

2 AMERICAN BITTERN *Botaurus lentiginosus*   **L** 60–85 cm   Text p. 186
Smaller than Eurasian Bittern with unbarred, plain blackish flight feathers. Vagrant to Az and CI.
2a   **Ad:** Crown, nape and sides of head rufous. Neck boldly streaked rufous.
2b   **Juv:** Rufous crown less extensive. Lacks dark malar stripe.

3 LEAST BITTERN *Ixobrychus exilis*   **L** 28–36 cm   Text p. 186
Smaller than Little Bittern. Chestnut greater coverts are diagnostic, as are chestnut trailing edge to inner primaries and outer secondaries, and chestnut tips to greater primary-coverts. Vagrant to the Az.
3a   **Ad male:** Cap, back and scapulars black.
3b   **Ad female:** Cap, back and scapulars dark brown. Neck and underparts more streaked.
3c   **Juv:** Mantle and scapulars rufous, tipped paler.

4 LITTLE BITTERN *Ixobrychus minutus*   **L** 33–38 cm   Text p. 186
A very small, dark heron showing conspicuous pale cream wing patches. Nominate is a scarce migrant to Md and the CI, a rare migrant in the Az and a vagrant to the CV; *payesii* is a vagrant to the CI.
4a   **Ad male nominate:** Crown and upperparts black. Wing patches pale cream. Head and neck-sides pale.
4b   **Ad female nominate:** Crown brown. Upperparts brown streaked black. Underparts more streaked.
4c   **Juv nominate:** Resembles dull, heavily-streaked female. Wing patch streaked and less contrasting.
4d   **Ad male** *payesii***:** As nominate male but sides of head and neck rich chestnut.
4e   **Ad female** *payesii***:** As nominate female but less streaked below.

5 DWARF BITTERN *Ardeirallus sturmii*   **L** 30 cm   Text p. 186
Diminutive. Very dark above. Vagrant to the CI.
5a   **Ad male:** Dark slate grey above. Pale below very heavily streaked black.
5b   **Juv:** Duller and browner above, more rufous below. Upperparts show pale feather tips.

6 (BLACK-CROWNED) NIGHT HERON *Nycticorax nycticorax*   **L** 58–65 cm   Text p. 187
A stocky heron with crepuscular habits. Regular migrant in the CI. Scarce migrant in Md. Vagrant to the Az and CV.
6a   **Ad:** Black crown and mantle contrast with grey wings and whitish-grey underparts.
6b   **Juv:** Dark brown with buff streaks. Conspicuously spotted white above.

7 GREEN-BACKED HERON *Butorides virescens*   **L** 40–48 cm   Text p. 187
A small, dark, long-billed heron. Vagrant to the Az.
7a   **Ad:** Dark cap. Sides of head and neck rich chestnut. Wings tinged greenish. Bluish-green lanceolate feathers on mantle and scapulars.
7b   **Juv:** Browner. Neck less rufous and heavily streaked. White tips to wing coverts.

8 SQUACCO HERON *Ardeola ralloides*   **L** 44–47 cm   Text p. 187
Buffish or brown but revealing contrasting white wings, rump and tail in flight. Irregular migrant in the Ci. Vagrant to the Az, Mda and CV.
8a   **Ad breeding:** Warm buff. Black and white nuchal plumes are lacking in non-breeding plumage, which is darker brown above.
8b   **Juv:** Greyish brown, with heavily streaked head and neck.

9 CATTLE EGRET *Bubulcus ibis*   **L** 48–53 cm   Text p. 188
A small, stocky, short-necked egret. Bill yellow. Often associates with livestock. Breeds locally in the CI. Common in winter in the CV. Irregular in Md. Vagrant to the Az.
9a   **Ad:** White, with buff crown, breast and mantle. Legs reddish.
9b   **Ad non-breeding:** All white. Legs blackish.

47

PLATE 8   HERONS II

1 LITTLE BLUE HERON *Hydranassa caerulea*   **L** 58–63 cm   Text p. 188
Adult unmistakable. Juvenile and immature could be confused with other egrets but distinguished by bill and leg coloration. Vagrant to the Az.
1a Ad: Uniform slaty-blue with purple-tinged head and neck.
1b Juv: Two-toned bill, dusky primary tips and yellow-green legs.

2 TRICOLORED HERON *Hydranassa tricolor*   **L** 63–68 cm   Text p. 188
Very long-necked and long-billed. Very active when feeding. Vagrant to the Az.
2a Ad non-breeding: Upperparts and neck blue-grey, contrasting with white underparts. Foreneck edged rufous.
2b Juv: Neck, sides of head and tips of wing-coverts rufous. Foreneck white.

3 SNOWY EGRET *Egretta thula*   **L** 55–65 cm   Text p. 189
Resembles a Little Egret but stockier with yellow facial skin. Vagrant to the Az.
3 Ad non-breeding: Lores yellow. Black legs have variable yellow stripe at rear.

4 BLACK HERON *Hydranassa ardesiaca*   **L** 44–47 cm   Text p. 189
A slender all-black heron with diagnostic feeding behaviour. Vagrant to the CV.
4a Ad breeding: All black, with elongated plumes on crown, nape, breast and mantle. Feet bright orange.
4b Ad canopy-feeding: Distinctive feeding posture, wings shading head.

5 WESTERN REEF EGRET *Egretta gularis*   **L** 55–65 cm   Text p. 188
Resembles Little Egret but bill usually yellowish or brown, deeper at base and slightly curved. Legs dark but not black. Dimorphic. Vagrant to all four archipelagos.
5a Ad dark morph: Entirely dark slate-grey except for white throat patch.
5b Ad white morph: Entirely white.

6 LITTLE EGRET *Egretta garzetta*   **L** 55–65 cm   Text p. 189
A small elegant egret with a slender black bill, black legs and yellow feet. Breeds commonly on the CV and locally on CI. Common in winter and on passage in the CI and regular on Md and the Az.
6a Ad breeding: Lores orange to greyish-green. Elongated plumes on nape, mantle, scapulars and breast.
6b Juv: As non-breeding adult but bill, legs and feet duller.

7 INTERMEDIATE EGRET *Egretta intermedia*   **L** 65–72 cm   Text p. 189
Larger than Little Egret with black feet and a yellow bill. Less graceful than Great White Egret and gape line ends below eye. Vagrant to the CV.
7 Ad non-breeding: Bill and lores yellow. Black feet. Black legs with yellowish tibiae.

8 GREAT WHITE EGRET *Egretta alba*   **L** 85–102 cm   Text p. 190
A large, graceful egret with a sinuous, slender, often-kinked neck. Gape line extends behind eye. Legs and feet dark. Vagrant to the Az, CI and CV.
8 Ad non-breeding: Bill yellow.

9 PURPLE HERON *Ardea purpurea*   **L** 78–90 cm   Text p. 191
Smaller and more slender than Grey Heron, dark-capped with a rufous, black-streaked neck. Nominate is regular but scarce on passage in the CI, irregular in Md and a vagrant to the Az; *bournei* is endemic to the CV.
9a Ad nominate: Cap black. Mantle and wing-coverts dark grey, tinged purple.
9b Juv nominate: Brown above with broad pale brown fringes to mantle, scapulars and upperwing coverts.
9c Ad *bournei*: Much paler than nominate with less black on neck, chest and belly. Underparts white and pale chestnut.

49

PLATE 9   HERONS, STORKS AND IBISES

1 GREAT BLUE HERON *Ardea herodias*   **L** 110–125 cm   Text p. 190
Larger than Grey Heron and with a heavier bill. Distinguished from it in all plumages by chestnut thighs and carpal patches. Vagrant to the Az, CI and CV.
1a Ad: Darker than Grey Heron. Carpal patches black and chestnut.
1b Juv: Browner above and buffier below than juvenile Grey Heron.

2 GREY HERON *Ardea cinerea*   **L** 90–98 cm   Text p. 190
A large greyish heron. Fairly common visitor to all the archipelagos. Has bred Az and CV.
2a Ad: White crown and black nuchal plumes. Black and white carpal patches. Yellow bill.
2b Juv: Foreneck and crown grey.

3 WHITE STORK *Ciconia ciconia*   **L** 100–115 cm   Text p. 191
White, with contrasting black flight feathers. Scarce passage migrant in the CI. Vagrant to Md.
3   Ad: Bill and legs red.

4 BLACK STORK *Ciconia nigra*   **L** 95–100 cm   Text p. 191
Dark, with a contrasting white breast and belly. Vagrant to the Az, Md and the CI.
4a Ad: Black above and white below. Bill and legs red.
4b Juv: Dusky above. Bill and legs greenish.

5 GLOSSY IBIS *Plegadis falcinellus*   **L** 55–65 cm   Text p. 192
All-dark. Vagrant to all four archipleagos.
5a Ad: Dark brown, glossed purple and green on the wings and tail.
5b Juv: Duller. Paler below, with pale streaks on head and neck.

6 BALD IBIS *Geronticus eremita*   **L** 70–80 cm   Text p. 192
A large, stocky and short-legged ibis. Vagrant to the Az and CV.
6a Ad: Bill, legs and bare skin on head red. Plumage black, glossed purple and green on the wings. Shaggy nuchal crest.
6b Juv: Darker, with feathered head. Legs and bill greyish.

7 SACRED IBIS *Threskiornis aethiopicus*   **L** 65–75 cm   Text p. 192
Dark head, neck and rear contrast with white plumage. Feral birds breed in the CI.
7a Ad: Head and neck black and unfeathered. Loose lower scapulars and tertials comprise black rear end.
7b Juv: Head and neck feathered black and white. Lacks tail plumes.

51

PLATE 10 SPOONBILLS AND FLAMINGOS

1 EURASIAN SPOONBILL *Platalea leucorodia*　　　　**L** 80–90 cm　　　　Text p. 192
White with large spatulate bill. Scarce on passage and in winter in the CI. Scarce in winter in the CV. Vagrant to Md and the Az.
1a Ad breeding: Bill black with yellow tip. Legs black. Bushy crest. Buff breast patch.
1b Juv: Legs and bill pinkish. Black tips to outer primaries.

2 ROSEATE SPOONBILL *Ajaia ajaja*　　　　**L** 69–87 cm　　　　Text p. 193
Unmistakable. The only pink spoonbill. Unconfirmed vagrant to the Az.
2a Ad: Unfeathered greenish head. Bright pink wings and body.
2b Juv: Duller. Head feathered white.

3 LESSER FLAMINGO *Phoenicopterus minor*　　　　**L** 80–105 cm.　　　　Text p. 193
A small flamingo with a uniformly coloured bill. Possible vagrant to the CI.
3a Ad: Pink, but may appear white if faded. Bill, facial skin and eye dark red.
3b Juv: Dull greyish-brown. Bill dark grey.

4 GREATER FLAMINGO *Phoenicopterus roseus*　　　　**L** 125–155 cm　　　　Text p. 193
Large flamingo with large two-tone bill. Vagrant to the CI. Former breeder in the CV.
4a Ad: Pinkish-white with contrasting red wing-coverts and black flight feathers. Bill pink with dark tip.
4b Juv: Greyish. Bill grey with dark tip.

53

PLATE 11 WHISTLING DUCKS, SWANS AND GEESE

1 FULVOUS WHISTLING DUCK *Dendrocygna bicolor* **L** 45–53 cm Text p. 193
A long-legged, pale-vented duck with an erect posture. Vagrant to the Az.
1 **Ad:** Dark brown, fringed rufous above. Rufous buff below with contrasting white and black flank streaks.

2 WHITE-FACED WHISTLING DUCK *Dendrocygna viduata* **L** 45–53 cm Text p. 194
A long-legged, dark-vented duck with an erect posture. Vagrant to the CI.
2a **Ad:** Dark, with contrasting white face.
2b **Juv:** Duller. Face greyish-buff.

3 MUTE SWAN *Cygnus olor* **L** 125–155 cm Text p. 194
Unmistakable. Vagrant to the Az and CI.
3a **Ad:** White. Orange bill with black basal knob.
3b **Juv:** Grey-brown with greyish bill.

4 CANADA GOOSE *Branta canadensis* **L** 55–110 cm Text p. 195
A large goose with a black head and neck and contrasting white chin and head sides. Vagrant to the Az.
4a **Ad:** Neatly barred above and on flanks. Head and neck glossy black with pure white throat marking.
4b **Ad flight:** Upperwings and mantle uniform grey-brown. Upper breast pale.
4c **Juv:** Duller and barring more uneven. Throat marking tinged brown.

5 BARNACLE GOOSE *Branta leucopsis* **L** 58–71 cm Text p. 196
A neat black-necked goose with a mainly white head. Vagrant to the Az and CI.
5a **Ad:** Neatly barred above and on flanks. Head and neck glossy black.
5b **Ad flight:** Upperwings and mantle light grey. Upper breast black.
5c **Juv:** Duller. Flank barring diffuse.

6 BRENT GOOSE *Branta bernicla* **L** 55–66 cm Text p. 196
A small dark goose with a pale belly and rear. Vagrant to the Az and CI.
6a **Ad** *hrota*: Plain dark grey above. Small white neck crescents on sides of black neck.
6b **Juv:** Duller. Wing coverts tipped white. White neck crescents absent in youngest birds.
6c **Juv flight:** Upperwings and mantle dark grey. Head, neck and upper breast black.

55

PLATE 12 GEESE AND SHELDUCKS

1 BEAN GOOSE *Anser fabalis* **L** 66–84 cm Text p. 194

Large goose with a dark head and neck. Legs and feet orange. Long orange-and-black bill. Vagrant to the Az and Md.
1a Ad: Dark above with pale feather edgings. Flanks lightly barred.
1b Ad flight: Tail narrowly edged white. Dark above.
1c Juv: Flanks unbarred. Legs and feet duller.

2 PINK-FOOTED GOOSE *Anser brachyrhynchus* **L** 66–75 cm Text p. 194

A compact, short-necked, dark goose with pink legs. Short pink and black bill. Vagrant to the Az, Md and the CI.
2a Ad: Mantle and upperwing-coverts grey, contrasting with dark brown neck. Flanks dark with pale barring.
2b Ad flight: Tail broadly edged white. Pale grey above.
2c Juv: Flanks unbarred. Legs and feet duller.

3 (GREATER) WHITE-FRONTED GOOSE *Anser albifrons* **L** 65–78 cm Text p. 195

A compact, short-necked, dark goose with orange legs and feet, and a pink (*albifrons*) or yellow (*flavirostris*) bill. Vagrant to the Az, Md and the CI.
3a Ad: White patch at base of bill. Large dark patches on belly.
3b Ad flight: Paler above than Bean Goose, with less obvious pale wingbars.
3c Juv: Lacks white frontal patch and dark belly patches. Legs greyish-yellow.

4 GREYLAG GOOSE *Anser anser* **L** 75–90 cm Text p. 195

A large, bulky, thick-necked goose. Strikingly pale on upper forewings. Head and neck pale. Legs and feet pink. Bill orange. Vagrant to the Az, Md and the CI.
4a Ad: Barred above and on flanks. Indistinct dark spots on belly.
4b Ad flight: Wings broad. Pale forewings contrast with dark hindwings and back.
4c Juv: Pale feather edges give scaly appearance above. Lacks flank bars and belly spots.

5 SNOW GOOSE *Anser caerulescens* **L** 65–84 cm Text p. 195

A sturdy goose. Dimorphic. Vagrant to the Az.
5a Ad white morph: White with black primaries. Legs, feet and bill pink.
5b Ad white morph flight: Unmistakable. Entirely white except for contrasting black primaries.
5c Juv white morph: Upperparts pale greyish-brown. Legs, feet and bill dark grey.

6 RUDDY SHELDUCK *Tadorna ferruginea* **L** 61–67 cm Text p. 196

A large orange-brown duck with contrasting black-and-white wings. Breeds in the CI. Vagrant to the Az and Md.
6a Ad male breeding: Black neck collar (lost in winter).
6b Ad male flight: White wing-coverts contrast with blackish flight feathers.
6c Ad female: Duller. Whitish face. Lacks neck collar.
6d Ad female flight: As for male, with white wing-coverts contrasting with blackish flight feathers.

7 COMMON SHELDUCK *Tadorna tadorna* **L** 58–67 cm Text p. 196

A large black-and-white duck with a green-glossed black head and neck, and a chestnut-brown breast-band. Vagrant to Md and the CI.
7a Ad male breeding: Bright red bill with swollen red basal knob.
7b Ad male breeding flight: White coverts contrast with black flight feathers, scapulars and median belly line.
7c Ad female: Lacks bill knob. Breast-band narrower.
7d Ad female flight: As with male, white coverts contrast with black flight feathers, scapulars and median belly line.

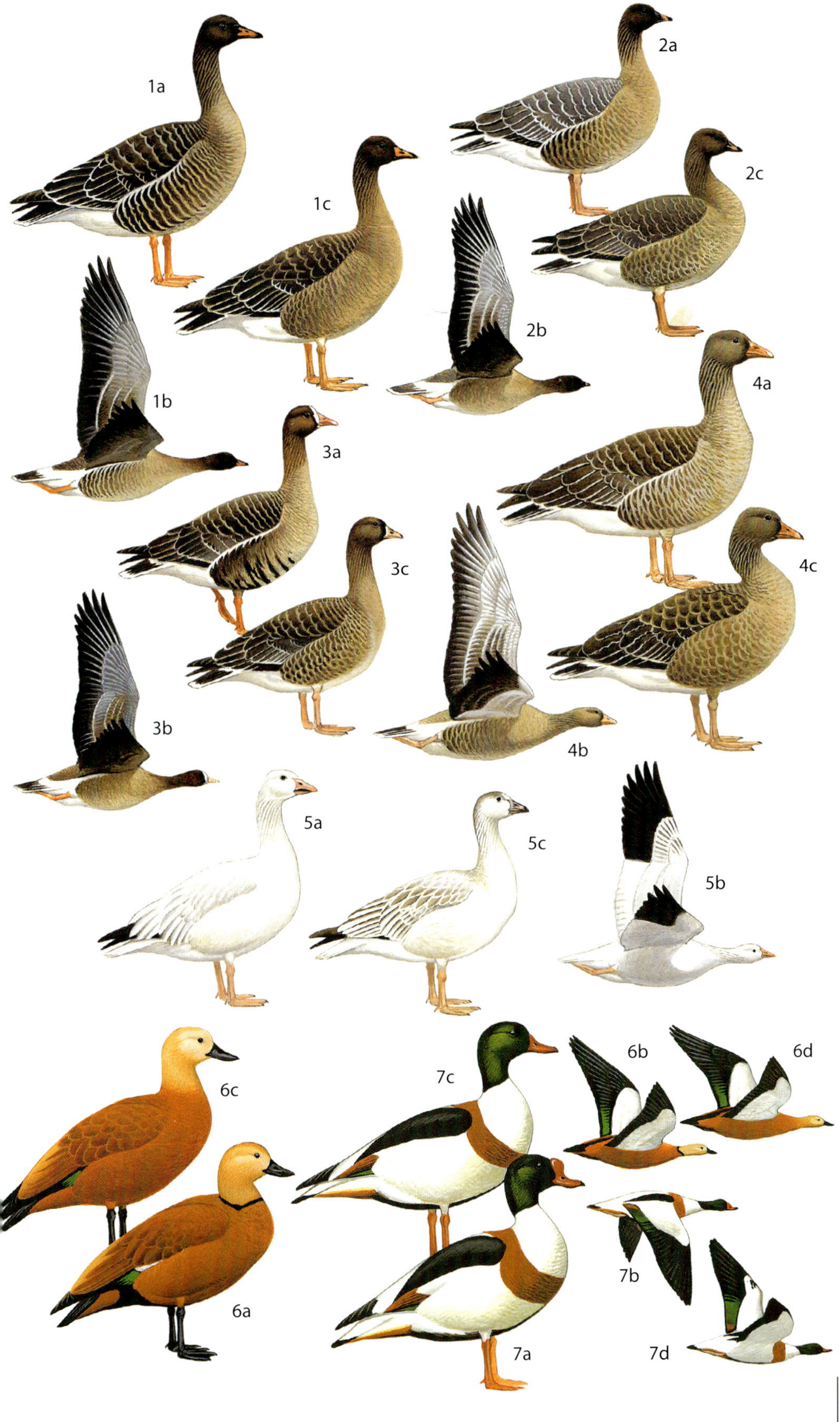
1a
2a
2c
1c
2b
1b
4a
3a
3c
4c
3b
4b
5a
5c
5b
6b
6d
7c
6c
7b
6a
7a
7d

PLATE 13   DABBLING DUCKS I

1 WOOD DUCK *Aix sponsa*   **L** 43–51 cm   Text p. 197
A compact, large-headed, long-tailed duck with a short bill. Vagrant to the Az and CI.
1a Ad male: Crested. Blackish above, glossed green and blue. White marks on head, neck and flanks.
1b Ad male flight: Appears mainly dark, with white tips to secondaries above. Underwing-coverts mottled white.
1c Ad female: Predominantly greyish-brown. Mottled on flanks. Broad white patch around eye.
1d Ad female flight: As male. Eye patch often visible.

2 EURASIAN WIGEON *Anas penelope*   **L** 45–51 cm   Text p. 197
A short-necked, round-headed, small-billed duck with a pointed tail. Scarce winter visitor to the CI. Vagrant to the Az, Md and CV.
2a Ad male: Head chestnut with yellow forehead and central crown. Pinkish breast.
2b Ad male flight: Large white patch across forewing. Black-bordered green speculum. White belly.
2c Ad female: Dark. Rufous brown or greyish.
2d Ad female flight: Look for shape and white belly. Greyish median underwing-coverts and axillaries.

3 AMERICAN WIGEON *Anas americana*   **L** 45–56 cm   Text p. 197
Structure as Eurasian Wigeon but slightly larger-headed. Vagrant to the Az, CI and CV.
3a Ad male: Unmistakable. Head mottled grey and white with creamy-white forehead and central crown. Broad, dark, green-glossed stripe extends back from around eye to nape. Body pinkish-brown with black-and-white rear.
3b Ad male flight: Wing pattern similar to male Eurasian Wigeon but usually distinguishable by head markings and pinkish-brown body.
3c Ad female: Similar to female Eurasian Wigeon but rufous-brown breast and flanks contrast more strongly with greyish head and neck.
3d Ad female flight: Similar to female Eurasian but has white median underwing-coverts and axillaries.

4 GADWALL *Anas strepera*   **L** 46–56 cm   Text p. 198
A neatly plumaged duck with a small white speculum. Scarce winter visitor to the CI. Vagrant to the Az.
4a Ad male: Predominantly grey with contrasting black rear. Head brownish.
4b Ad male flight: Inner wing marked black and chestnut.
4c Ad female: Brownish with darker feather centres. Orange sides to bill.
4d Ad female flight: White belly.

5 MALLARD *Anas platyrhynchos*   **L** 50–65 cm   Text p. 199
A large dabbling duck with a large, white-bordered blue speculum and bright orange legs. Scarce on the CI and occasional in Md. Regular in winter in the Az (has bred).
5a Ad male: Unmistakable. Green head and upper neck separated by white collar from purplish-brown lower neck and breast.
5b Ad male flight: Speculum and head and neck markings conspicuous.
5c Ad female: Brown-streaked with a dark cap and eyestripe and paler supercilium.
5d Ad female flight: Speculum conspicuous. Buff belly.

6 (NORTHERN) PINTAIL *Anas acuta*   **L** 51–66 cm   Text p. 200
Long-necked, with a pointed tail. Irregular in winter in the CI. Vagrant to the Az, Md and CV.
6a Ad male: Unmistakable. Head and rear neck choclate-brown. White breast extends as white stripe along sides of head. Body greyish with contrasting black rear and very long central tail-feathers.
6b Ad male flight: Small head on long, narrow neck and long, pointed tail. Secondaries edged white.
6c Ad female: Mainly pale grey, mottled brown. Head and neck plain pale brown. Grey bill.
6d Ad female flight: Similar shape to male but lacks long central tail-feathers. Secondaries edged white.

7 (AMERICAN) BLACK DUCK *Anas rubripes*   **L** 53–61 cm   Text p. 199
Mallard-like but very dark, with contrasting paler head and white underwing-coverts. Speculum purplish. Vagrant to the Az (has bred) and CI.
7a Ad male: Bill yellow.
7b Ad male flight: Speculum lacks white of Mallard. White underwings contrast with dark body.
7c Ad female: As male but plumage shows less contrast. Bill olive or greenish-yellow.
7d Ad female flight: As male.

59

PLATE 14 DABBLING DUCKS II

1 (NORTHERN) SHOVELER *Anas clypeata*   **L** 44–52 cm   Text p. 201
Long, broad bill is diagnostic. Vagrant to the Az and Md and a scarce winter visitor to the CI.
1a **Ad male:** Head green, breast white, flanks and belly chestnut.
1b **Ad male flight:** Blue forewing.
1c **Ad female:** Pale brown, streaked and mottled darker.
1d **Ad female flight:** Grey forewing. Belly dark and green speculum lacks white trailing edge.

2 EURASIAN TEAL *Anas crecca*   **L** 34–38 cm   Text p. 198
Common winter visitor to the Az and CI, scarce on Md and a vagrant to the CV.
2a **Ad male:** Head chestnut, with buff-edged green eye patch. Body greyish, with buff-and-black rear.
2b **Ad male flight:** Head and neck look dark, contrasting with grey body.
2c **Ad female:** Mottled and streaked brownish, with white-edged green speculum. White patches on tail sides. Orange base to grey bill. Legs grey.
2d **Ad female flight:** White wingbar and green speculum obvious.

3 GREEN-WINGED TEAL *Anas carolinensis*   **L** 36 cm   Text p. 199
Vagrant to the Az, CI and CV.
3a **Ad male:** As male Eurasian Teal but shows diagnostic vertical white line on breast sides and lacks horizontal white body stripe. Buff edges to green eye patch indistinct.
3b **Ad male flight:** Distinguishable by above features in favourable conditions.

4 FALCATED DUCK *Anas falcata*   **L** 48–54 cm   Text p. 198
Vagrant to the CI.
4a **Ad male:** Chiefly grey with contrasting green- and purple-glossed, maned head, buff-and-black rear and drooping, elongated black and grey tertials.
4b **Ad male flight:** Upper forewing pale grey. Speculum green-glossed with white leading edge.
4c **Ad female:** Rich brown, with dark chevron markings below.
4d **Ad female flight:** Similar to male.

5 GARGANEY *Anas querquedula*   **L** 37–41 cm   Text p. 200
Larger and longer-billed than Eurasian Teal. Scarce on passage in the CI. Vagrant to the Az, Md and CV.
5a **Ad male:** Broad white crescent extends from above eye to nape. Head, neck and breast dark pinkish-brown, separated from grey flanks by a vertical white line. Elongated black-and-white scapulars.
5b **Ad male flight:** Forewing blue-grey above, with green speculum bordered white.
5c **Ad female:** Face pattern more prominent than in Eurasian Teal, with pale loral spot. Tail lacks white edges. Bill all grey. Legs grey.
5d **Ad female flight:** Above shows narrow white line on the greater coverts and a broad white trailing edge to the secondaries. Speculum brownish. Belly white.

6 BLUE-WINGED TEAL *Anas discors*   **L** 37–41 cm   Text p. 199
Slightly larger than Eurasian Teal. Both sexes have blue forewings. Vagrant to the Az, CI and CV.
6a **Ad male:** Head dark blue-grey, with a conspicuous white facial crescent in front of the eye.
6b **Ad male flight:** Bright blue forewings.
6c **Ad female:** Plain grey-brown, with pale loral spot. Tail lacks white edges of female Eurasian.
6d **Ad female flight:** Pale blue forewings. Lacks white trailing edge to secondaries of female Eurasian.

7 MARBLED DUCK *Marmaronetta angustirostris*   **L** 39–42 cm   Text p. 201
A small, pale, sandy-brown duck. Sexes similar. Vagrant to the Az, CI (has bred) and CV.
7a **Ad:** Flanks and upperparts pale-spotted. Breast and undertail-coverts with darker brown barring.
7b **Ad flight:** Looks very pale. Palest on secondaries above. Primary-coverts darker brown.

8 RED-CRESTED POCHARD *Netta rufina*   **L** 53–57 cm   Text p. 202
A bulky duck with a rounded head. Vagrant to the CI.
8a **Ad male:** Head orange and bill bright red. Breast and rear black contrasting with white flanks.
8b **Ad male flight:** Broad white wingbar.
8c **Ad female:** Brown above with paler breast and flanks. Crown and nape darker brown with very pale head- and neck-sides. Bill greyish with a pinkish spot near the tip
8d **Ad female flight:** Broad white wingbar.

1c
1b
1d
1a
2b
2c
2a
2d
3b
3a
4d
4c
4b
4a
5c
5b
5a
5d
6c
6a
6b
6d
7a
7b
8c
8b
8a
8d

PLATE 15 DIVING DUCKS

1 COMMON POCHARD *Aythya ferina*   **L** 42–49 cm   Text p. 202
A compact diving duck with a large head and bill and sloping forehead. Uncommon winter visitor to the CI. Vagrant to the Az, Md and CV.
1a Ad male: Head and neck chestnut. Breast and rear black, contrasting with pale grey wings and body.
1b Ad male flight: Pale greyish wings contrast with dark head, neck and rear.
1c Ad female: Profile distinctive. Flanks and back greyish, mottled brown. Head, breast and neck browner. Pale loral patch, eye-ring and stripe behind eye.
1d Ad female flight: Wings grey with pale greyish wingbar, contrasting with browner head and neck.

2 FERRUGINOUS DUCK *Aythya nyroca*   **L** 38–42 cm   Text p. 203
A very dark diving duck with contrasting blackish rear and pure white undertail-coverts. Striking white wingbars. Vagrant to the Az, CI and CV.
2a Ad male: Rich chestnut on head, neck and breast, blackish above. Striking white irides.
2b Ad male flight: Very dark with contrasting white wingbars and white belly.
2c Ad female: Duller. Irides dark.
2d Ad female flight: Similar to male but duller.

3 TUFTED DUCK *Aythya fuligula*   **L** 40–47 cm   Text p. 203
A small, dark diving duck with contrasting paler sides. Head large and crested on the rear crown. Regular winter visitor to the CI. Vagrant to the Az, Md and CV.
3a Ad male: White flanks contrast with black upperparts. Crest long and drooping.
3b Ad male flight: Black with white upperwing bar and white underwings and belly. Black back.
3c Ad female: Largely dark brown, with paler brown flanks. Crest a short tuft.
3d Ad female flight: As male but browner.

4 GREATER SCAUP *Aythya marila*   **L** 42–51 cm   Text p. 203
Larger than Tufted Duck, with an uncrested, large, rounded head. Black on bill tip restricted to nail. White upperwing bar extends across primaries. Vagrant to the Az, Md and CI.
4a Ad male: Head green-glossed black. Neck, breast and rear black. Flanks white. Back finely vermiculated grey.
4b Ad male flight: Pale back and greyish forewings.
4c Ad female: Brown with greyish flanks and darker back. Broad white band around bill base.
4d Ad female flight: Paler than female Tufted Duck.

5 LESSER SCAUP *Aythya affinis*   **L** 38–46 cm   Text p. 204
Resembles a small Greater Scaup but smaller-billed and thinner-necked. Head peaks at rear of crown. White upperwing bar confined to secondaries. Vagrant to the Az, CI and CV.
5a Ad male: Head purplish-glossed black. Back more coarsely vermiculated than in Great Scaup.
5b Ad male flight: White upperwing bar becomes grey across primaries.
5c Ad female: Shows less white around bill base than female Greater Scaup.
5d Ad female flight: As male but browner.

6 RING-NECKED DUCK *Aythya collaris*   **L** 37–46 cm   Text p. 202
A small diving duck with a distinctive large head and peaked rear crown. Grey bill has broad white subterminal band. Vagrant to the Az, CI, Md and CV.
6a Ad male: Resembles Tufted Duck but lacks a crest. Flanks grey with a white crescent in front. Narrow white band at bill base.
6b Ad male flight: Distinguished from male Tufted Duck by grey wingbar.
6c Ad female: Suggests small female Common Pochard but has pale lores, white subterminal bill band and upperparts darker than flanks. White eye-ring.
6d Ad female flight: Darker than female Common Pochard, with grey wingbar contrasting with browner forewing above.

63

PLATE 16 SEA DUCKS

1 COMMON EIDER *Somateria mollissima*   **L** 50-71 cm   Text p. 204

A large, short-necked sea duck with a large wedge-shaped bill. Vagrant to the Az.
1a Ad male: Unmistakable. Predominantly white above and black below, with a black crown and pale green patches on the nape and neck-sides.
1b Ad male flight: Black-and-white appearance distinctive.
1c Ad female: Rufous-brown with dark barring. Bill with pale tip and feathering along the culmen shorter than on the bill-sides. Profile distinctive.
1d Ad female flight: Heavy brown body with contrasting pale underwings. Dark speculum with white border.

2 KING EIDER *Somateria spectabilis*   **L** 47-63 cm   Text p. 204

Smaller and shorter-billed than Common Eider. Vagrant to the Az.
2a Ad male: Unmistakable. Largely black, with pink breast, pale blue head, large orange frontal shield and red bill.
2b Ad male flight: Black wings, belly and rear contrast with white forewing and rear flank patches and pale foreparts.
2c Ad female: Warmer brown and flank markings angled, not barred as in Common Eider. Bill dark and feathering along the culmen longer than on the bill sides.
2d Ad female flight: Warmer rufous brown than female Common Eider.

3 LONG-TAILED DUCK *Clangula hyemalis*   **L** 36-62 cm including tail   Text p. 205

A small sea duck with a pointed tail and all-dark wings contrasting with a white body. Flight fast, with shallow wingbeats. Vagrant to the Az, Md and CI.
3a Ad male winter: Unmistakable. Largely white with black breast and rear neck patch and elongated, black central tail-feathers.
3b Ad male winter flight: Dark brown wings and back contrast with largely white foreparts and underparts, and white upper mantle and scapular patches.
3c Ad female winter: Brownish above and white below, with brown lower neck and breast. Head and upper neck white, with dark brown crown and an obvious dark brown patch on the upper neck-sides.
3d Ad female winter flight: All-dark wings and upperparts contrast with white underparts. Neck spot usually visible.

4 COMMON SCOTER *Melanitta nigra*   **L** 44-54 cm   Text p. 205

A dark sea duck with paler primaries. Vagrant to the Az, Md and CI.
4a Ad male: All-black plumage. Bill has yellow patch on culmen.
4b Ad male flight: Black with paler primaries.
4c Ad female: Dark brown with paler cheeks and foreneck.
4d Ad female flight: Dark brown with paler primaries.

5 SURF SCOTER *Melanitta perspicillata*   **L** 45-56 cm   Text p. 205

A dark, large-billed sea duck with uniformly dark wings. Vagrant to the Az and Md.
5a Ad male: Plumage black with contrasting white forehead and nape patches. Swollen orange bill with white base.
5b Ad male flight: Black with white head markings.
5c Ad female: Dark brown, usually with pale patches in front of and behind the eye and sometimes a pale nape patch. Large bill.
5d Ad female flight: Dark brown with paler head markings.

65

PLATE 17 SAWBILLS AND ALLIES

1 BUFFLEHEAD *Bucephala albeola* **L** 32–39 cm Text p. 206
A tiny, small-billed, large headed duck. Vagrant to the Az and CI.
1a Ad male: Largely black above and white below. Head mainly black: glossed dark green, purple and bronze, with a large white patch on the sides and rear.
1b Ad male flight: Large white innerwing patches and white rear head marking conspicuous.
1c Ad female: Head smaller than male, brown, with a white patch behind and below the eye. Overall greyish-brown with whitish belly.
1d Ad female flight: Upperwings dark brown with a white patch on the inner secondaries. White patch behind and below eye.

2 COMMON GOLDENEYE *Bucephala clangula* **L** 42–50 cm Text p. 206
A small, stocky diving duck with a large rounded head. Adults have striking yellow irides. Vagrant to the Az and Md.
2a Ad male: Black above and white below. Head green-glossed black with a circular white loral patch.
2b Ad male flight: Large head prominent. Black primaries, wing bases, back, head and tail contrast with white inner wings, scapular patches, neck and underparts.
2c Ad female: Grey with a chocolate-brown head and white collar.
2d Ad female flight: Large head prominent. White belly and white patch on inner wing.

3 HOODED MERGANSER *Mergus cucullatus* **L** 42–50 cm Text p. 206
A small, long-tailed merganser with a large, erectile, bushy crest on the rear crown. Vagrant to the Az and CI.
3a Ad male: Unmistakable combination of large white patches on the side of the black head, white breast with two black bands, orangey flanks and black upperparts with white shaft-streaks to the tertials.
3b Ad male: Crest can be raised or lowered.
3c Ad male flight: Very dark above, with white bar on greater coverts.
3d Ad female: Dark grey-brown with brown crest. Thin yellow bill.
3e Ad female flight: As male but browner.

4 SMEW *Mergellus albellus* **L** 38–44 cm Text p. 206
A small, compact merganser. Vagrant to the CI.
4a Ad male: Mainly white with a short, loose crest on crown and nape. Black mantle, black loral mask and black line either side of nape.
4b Ad male flight: Strikingly black and white, with white oval patches on forewings.
4c Ad female: Body predominantly greyish. Head dark chestnut with contrasting white chin, throat and lower cheeks.
4d Ad female flight: White oval wing patches smaller than in male.

5 RED-BREASTED MERGANSER *Mergus serrator* **L** 52–58 cm Text p. 207
An elongated duck, with a long, thin, red bill and a shaggy crest. Vagrant to the Az and CI.
5a Ad male: Black head glossed green, separated by white collar from rufous-brown, black-streaked breast. Black area with large white spots between breast-sides and grey flanks.
5b Ad male flight: Conspicuous white patch on inner wing above.
5c Ad female: Mostly greyish with a rusty-brown head and upper neck.
5d Ad female flight: As male but white patch smaller, confined to secondaries and greater coverts.

67

PLATE 18   HONEY-BUZZARDS AND KITES

1 (WESTERN) HONEY-BUZZARD *Pernis apivorus*   **WS** 125–145 cm   Text p. 207

Buzzard-like but structure distinctive: wings long, tail long and narrow, and head and neck slim and projecting. Plumage highly variable but all have dark carpal patches below and clearly barred flight feathers and tail. Vagrant to the CI and Md.
1a Ad medium morph: 'Cuckoo-like' head, grey in males and brown in females. Cere grey.
1b Ad medium morph underside: Broad terminal tail bar and two narrower bars together at tail base. Broad dark trailing edge to wing.
1c Ad medium morph upperside: Distinguishing features as underside.
1d Juv dark morph: Cere yellow. Tail bars evenly spaced.
1e Juv pale morph underside: Secondaries darker than primaries. Narrow dark trailing edge to wing. Tail bars evenly spaced.
1f Juv dark morph underside: Distinguishing features as juvenile pale morph.
1g Juv dark morph upperside: Distinguishing features as juvenile pale morph.

2 BLACK-SHOULDERED KITE *Elanus caeruleus*   **WS** 75–85 cm   Text p. 207

A kestrel-sized kite with broad, pointed wings and a squarish tail. Vagrant to the CI.
2a Ad hovering: Grey above and white below. Black forewing above. Red eye in black eye-patch.
2b Ad underside: White, with dusky primaries.
2c Juv: Mantle, scapulars, greater coverts and flight feathers tipped white. Breast streaked brownish. Eyes pale yellowish.

3 (AMERICAN) SWALLOW-TAILED KITE *Elanoides forficatus*   **WS** 119–136 cm   Text p. 207

Unmistakable. Black and white with a long forked tail. Vagrant to the CI.
3a Ad: Head and body white. Rest of plumage dark greyish-black, glossed purple.
3b Ad upperside: White head contrasts with uniformly dark upperside.
3c Ad underside: Head, body and underwing-coverts white.

4 BLACK KITE *Milvus migrans*   **WS** 135–170 cm   Text p. 208

Dark brown, with angled wings and long, shallow-forked tail. Formely bred CV but possibly extinct. Rare passage migrant and winter visitor in CI. Vagrant to Md.
4a Ad: Unstreaked below with contrasting pale head. Unspotted above.
4b Ad underside: Uniform dark brown, with paler primary bases and head.
4c Ad upperside: Pale brown panel across upperwing-coverts.
4d Juv: Head and body streaked. Pale tips to upperwing-coverts.
4e Juv underside: Paler than adult, with streaked body and whiter primary bases.
4f Juv upperside: Pale upperwing-coverts, with white tips to greater coverts.

5 RED KITE *Milvus milvus*   **WS** 155–180 cm   Text p. 209

Larger and longer-winged than Black Kite, with a long, deeply-forked, rufous tail. Vagrant to the CI: bred formerly.
5a Ad: White head and rufous-brown body both streaked darker. Upperparts rufous-brown with dark feather centres.
5b Ad underside: White primary bases conspicuous. Orange-rufous tail tipped black at fork tips. Deep rufous undertail-coverts.
5c Ad upperside: Pale panel across upperwing-coverts, more obvious than in Black Kite.
5d Juv: Paler than adult. White tips to greater coverts.
5e Juv underside: Body paler brown than adult. Pale undertail-coverts.
5f Juv upperside: White tips to greater coverts.

6 CAPE VERDE KITE *Milvus fasciicauda*   **WS** 134–152 cm   Text p. 208

Slightly smaller than Red Kite, with shorter and more rounded wings and shallower tail fork. Endemic to CV, where close to extinction.
6a Ad: Less strongly marked than Red Kite. Paler below with narrower shaft-streaks.
6b Ad underside: Primary bases greyish-white.
6c Ad upperside: Tail shows 8-10 dark bars.

69

PLATE 19   **VULTURES, SNAKE EAGLES AND HARRIERS**

1 EGYPTIAN VULTURE *Neophron percnopterus*   **WS** 155–170 cm   Text p. 209

Large, with a wedge-shaped tail and long, narrow bill. Resident on the CV but declining. Also resident and declining in the eastern CI, extinct in the western islands. Vagrant to Md and the Az.
1a **Ad:** Dirty white with yellow facial skin. Flight feathers black.
1b **Ad underside:** White, including wedge-shaped tail, with contrasting black primaries and secondaries.
1c **Ad upperside:** Similar to underside but primaries and secondaries have white centres.
1d **Juv:** Dark brown, particularly on neck and breast. Facial skin grey.
1e **Juv underside:** Dark brown with contrasting pale tips to tail feathers.
1f **Juv upperside:** Dark brown, with pale tips to upperwing-coverts and tail feathers.

2 SHORT-TOED (SNAKE) EAGLE *Circaetus gallicus*   **WS** 170–190 cm   Text p. 210

A large and large-headed pale eagle. Hovers when hunting. Vagrant to the CI.
2a **Ad:** Pale brown above. White below, barred brown. Bib dark chocolate-brown (white in immatures).
2b **Ad underside:** White, with brown bars on primaries, secondaries and tail.
2c **Juv:** Paler than adult. Bib rufous-brown.
2d **Ad upperside:** Greyish brown mantle and upperwing-coverts contrast with darker flight feathers. Three tail bars.
2e **Juv underside:** Similar to adult but markings less distinct.
2f **Juv upperside:** Upperparts show greater contrast with flight feathers than in adult.

3 HEN HARRIER *Circus cyaneus*   **WS** 105–125 cm   Text p. 210

Larger and broader-winged than Montagu's and Pallid Harriers. Wingtip shows five 'fingers'. Rare and irregular on passage and in winter in the CI. Vagrant to Md (*cyaneus*) and the Az (*hudsonicus*).
3a **Ad male:** Pale grey above, including neck and upper breast, with white underparts. Black wing tips.
3b **Ad male flight:** Black confined to wingtips. Dark trailing edge to wing. Underside white.
3c **Ad female:** Brown above, buffish-white streaked brown below. Uppertail-coverts white.
3d **Ad female underside:** Wings and tail barred. Body buffish-white streaked brown.
3e **Juv underside:** Wings and tail barred. Secondaries darker. Body and underwing buff streaked brown.

4 (WESTERN) MARSH HARRIER *Circus aeruginosus*   **WS** 120–135 cm   Text p. 210

A large broad-winged harrier. Wing tip shows five 'fingers'. Scarce but regular on passage and in winter in the CI. Rare winter visitor to the CV. Vagrant to Md and the Az.
4a **Ad male:** Dark brown above with grey secondaries and black wingtips. Head and neck pale.
4b **Ad male upperside:** Brown mantle and upperwing coverts separated from black wingtips by blue-grey secondaries and primary bases.
4c **Ad male underside:** Whitish below, contrasting black wingtips and brown or rufous belly.
4d **Ad female:** Very dark brown with creamy-white crown, chin and forewing.
4e **Ad female upperside:** Dark brown, paler on primaries. Tail brown or rufous-brown.
4f **Ad female underside:** Dark brown body and underwing-coverts; paler grey-brown flight feathers.
4g **Juv upperside:** Darker than female. Crown and chin yellowish-white.

5 PALLID HARRIER *Circus macrourus*   **WS** 100–125 cm   Text p. 211

Slightly broader-winged than Montagu's Harrier. Wing tip shows four 'fingers'. Vagrant to the CI.
5a **Ad male:** Very pale grey above and white below. Black restricted to central primaries.
5b **Ad male flight:** Black primary wedge contrasts with very pale wings and body.
5c **Ad female:** Brown above and whitish, streaked brown below. Broad dark brown collar.
5d **Ad female underside:** Darker than female Montagu's, especially secondaries and secondary-coverts.
5e **Juv underside:** Body and underwing-coverts rufous. Strong head pattern including pale collar above dark brown nape and neck-sides.

6 MONTAGU'S HARRIER *Circus pygargus*   **WS** 105–125 cm   Text p. 211

Very slim-winged, with buoyant flight. Wingtip shows four 'fingers'. Melanistic birds occur. Scarce but regular on passage in the CI. Vagrant to Md and the CV.
6a **Ad male:** Dark grey above and white below, flanks streaked rufous. Black wingtips
6b **Ad male flight:** Grey with black wing tips. Two black bars across secondaries below and one above.
6c **Ad female:** Brown above and whitish streaked brown below. Smaller brown collar than female Pallid.
6d **Ad female underside:** Paler than female Pallid, especially on secondaries.
6e **Juv underside:** Body and underwing-coverts rufous or brown. Lacks head markings of juv Pallid.

71

PLATE 20 HAWKS AND BUZZARDS

1 (NORTHERN) GOSHAWK *Accipiter gentilis*   **WS** 95–125 cm   Text p. 211
Powerful and broad-winged with a long, rounded tail. Vagrant to the CI.
1a Ad male upperside: Much smaller than female. Often bluer-grey above.
1b Ad female: Grey-brown above, white with dark barring below. Prominent white supercilium.
1c Ad female underside: Wings longer and more pointed than in Sparrowhawk.
1d Juv male: Brown above. White below, with broad brown streaks comprised of drop-shaped spots.
1e Juv underside: Streaked underparts distinguish from (much smaller) juvenile Sparrowhawk, which is barred below.
1f Juv upperside: Brown with broadly barred flight feathers and tail.

2 EURASIAN SPARROWHAWK *Accipiter nisus granti*   **WS** 60–80 cm   Text p 212
A small, dashing hawk with short, rounded wings. Long tail has square corners. Breeds on Md. Residents and migrants occur in the CI.
2a Ad male *nisus*: Very small. Blue grey above. White, barred rufous below.
2b Ad male *granti*: Darker above and more heavily barred below than *nisus*.
2c Ad male *granti* upperside: Blue-grey with barred tail.
2d Ad male *granti* underside: Wings and tail barred. Body and underwing-coverts barred rufous.
2e Ad female: Grey-brown above and white, barred grey, below. White supercilium.
2f Ad female upperside: Grey-brown with long, barred tail.
2g Ad female underside: Dense bars on body, wings and tail. Body and underwing-coverts barred grey.

3 COMMON BUZZARD *Buteo buteo*   **WS** 115–135 cm   Text p 212
A medium-sized raptor, with broad, rounded wings are held in a shallow V when soaring. Plumage variable but typical individuals predominate in the region. Common resident in the Az (*rothschildi*), Md (nominate) and CI (*insularum*). Rare and declining resident in the CV (*bannermani*).
3a Ad nominate: Dark brown above; brown with dark bars below. Barred tail and paler breast-band.
3b Ad *bannermani*: Slightly paler and less rufous than the other island races. Upperparts grey-brown with paler feather fringes.
3c Ad *insularum*: Browner above and more streaked below.
3d Ad *rothschildi*: Darker than the other island races.
3e Ad nominate underside: Dark brown body and underwing coverts contrast with whitish flight feathers. Often has pale breast-band. Barred tail with broader terminal band. Wings have black trailing edge.
3f Ad nominate upperside: Brown with barred tail showing broad terminal band. Wings have black trailing edge.
3g Juv nominate upperside: Tail lacks dark terminal band. Wings lack black trailing edge.
3h Juv nominate underside: Tail lacks dark terminal band. Wings lack black trailing edge. Underparts streaked rather than barred as in adults.

4 LONG-LEGGED BUZZARD *Buteo rufinus*   **WS** 115–125 cm   Text p. 213
Subspecies in our region is typically pale-headed in all plumages. Vagrant to Md and the CI.
4a Ad *cirtensis* upperside: Tail unbarred, orange-rufous, paler at base. Pale primary bases.
4b Ad *cirtensis* underside: Rufous underwing-coverts and body, dark rufous on flanks and belly. Unbarred orange-rufous tail. Dark trailing edge to wing. Dark carpal patches on some.
4c Juv *cirtensis*: Browner than adult, lacking rufous tones. Tail barred.
4d Juv *cirtensis* upperside: Underwing-coverts and body brownish, darkest on flanks. Tail greyish and finely-barred.

5 ROUGH-LEGGED BUZZARD *Buteo lagopus*   **WS** 130–150 cm   Text p. 213
Variable but typically pale with a white tail base and dark tail barring. Vagrant to the Az.
5a Ad male underside: Mainly white with dark trailing edge to wing, dark carpal patches and narrow black terminal tail bands. Belly pale.
5b Juv pale morph: Grey-brown above. Head, breast and thighs whitish, streaked darker. Dark belly. Feathered legs.
5c Juv pale morph underside: Largely pale with contrasting blackish carpal patches and dark belly. Diffuse trailing edge and tail bands.
5d Juv pale morph upperside: Pale primary bases. Tail white with single, broad, dark, terminal band.
5e Juv dark morph underside: Body and wing-coverts blackish.

73

PLATE 21 EAGLES AND OSPREY

1 WHITE-TAILED (SEA) EAGLE *Haliaetus albicilla* **WS** 190–250 cm Text p. 209
Very large, with broad wings, short wedge-shaped tail and massive bill. Vagrant to the CI.
1a Ad: Brown, with paler head and neck. White tail. Bill yellow.
1b Ad underside: Brown with white tail.
1c Ad upperside: As underside.
1d Juv: Darker brown, especially on head and neck. Bill grey.

2 GOLDEN EAGLE *Aquila chrysaetos* **WS** 190–240 cm Text p. 213
A large brown eagle, with long, square-ended wings and a long tail. Vagrant to the CI.
2a Ad: Dark brown, with paler upperwing-coverts. Nape golden-brown. Tail grey, barred darker.
2b Ad upperside: Pale nape and upperwing-coverts. Body brown. Flight feathers and tail grey, barred darker.
2c Ad underside: Brown body and underwing coverts contrast with greyer flight feathers. Tail grey with black terminal band.
2d Juv: More uniform brown. White tail base.
2e Juv upperside: Brown with contrasting black-tipped, white tail. Small white patches on outer wing.
2f Juv underside: White tail with black terminal band and white bases to primaries and outer secondaries.

3 BOOTED EAGLE *Aquila pennata* **WS** 115–135 cm Text p. 214
A small, compact eagle with long, square-ended wings and a long, squared tail. Dimorphic, with pale morph commonest. Regular but scarce on passage and in winter in the CI. Vagrant to Md.
3a Ad pale morph: Brown above and white below. Black flight feathers. Paler, more rufous-brown below.
3b Ad upperside: Dark tail and flight feathers contrast with pale upperwing-coverts and scapulars. White uppertail-coverts.
3c Ad pale morph underside: Blackish flight feathers and grey tail contrast with white body and underwing coverts. Head and neck variably streaked rufous-brown.
3d Juv dark morph: Brown above with black flight feathers. Pale brown upperwing-coverts have white tips.
3e Juv upperside: Similar to adult but wing has pale trailing edge. Upperwing-coverts have white tips.
3f Juv dark morph underside: Brown body and underwing coverts contrast with greyer flight feathers and tail. Pale base to inner primaries. Pale trailing edge to wing.

4 BONELLI'S EAGLE *Aquila fasciata* **WS** 145–175 cm Text p. 214
A powerful, medium-sized eagle with broad wings and a long, square-ended tail. Vagrant to CI.
4a Ad: Grey-brown above. Underparts white with black streaks. Grey tail with broad black terminal band.
4b Ad upperside: Grey-brown. Barred, grey tail with black terminal band. Characteristic white patch on upper mantle.
4c Ad underside: White body and forewings contrast with black band from carpals across inner underwings.
4d Juv: Brown above and paler rufous-brown below, with brownish-black flight feathers.
4e Juv upperside: Brownish. Tail evenly barred, lacking broad terminal band.
4f Juv underside: Rufous-brown body and underwing-coverts contrast with greyer flight feathers and tail. Dark tips to primary-coverts.

5 OSPREY *Pandion haliaetus* **WS** 145–170 cm Text p. 214
Distinctive with its long, angled wings and short tail. Dark upperparts contrast with white underparts. Scarce resident and regular migrant in the CI. Resident in the CV. Vagrant to the Az and Md.
5a Ad: Uniformly dark brown above and white below. White crown contrasts with dark patch behind eye. Dark breast-band, especially in females.
5b Ad upperside: Plain dark wings contrast with white head. Tail with dark terminal band.
5c Ad underside: Black tips to greater-coverts. Tail with dark terminal band.
5d Juv: Upperpart feathers tipped white, giving scaly appearance.
5e Juv upperside: Upperparts brown with white feather tips. Crown streaked dark. Tail evenly barred.
5f Juv underside: Pattern less distinct than in adult. Tail evenly barred.

75

PLATE 22 FALCONS I

1 LESSER KESTREL *Falco naumanni*   **WS** 60–70 cm   Text p. 215
Smaller and paler than Common Kestrel. Vagrant to the CI, Md and the Az.
1a Ad male: Head, tail and greater coverts blue-grey. Lacks moustachial stripe. Upperparts unspotted rufous. Pinkish-buff below with small black spots.
1b Ad male flight: Blue-grey upper greater coverts. White underwings.
1c Ad female: Paler than female Common Kestrel and lacks cheek stripe.
1d Ad female flight: Paler than Common Kestrel. Flight feathers, especially primaries, less barred.

2 COMMON KESTREL *Falco tinnunculus*   **WS** 60–75 cm   Text p. 215
A medium-sized falcon with long wings and tail. Often hovers. Nominate is a vagrant to the Az and CV and a scarce migrant in the CI and Md. Other races resident in CI, Md and the CV (see text).
2a Ad male nominate: Rufous upperparts and wing-coverts with black spots. Blue-grey head with dark moustachial stripe.
2b Ad male nominate flight: Tail unbarred blue-grey, with black subterminal band.
2c Ad male *neglectus*: Resembles female of other races. Upperparts and tail barred. Head brownish-grey.
2d Ad male *dacotiae*: Very pale.
2e Ad male *canariensis*: Darker than nominate with heavier spotting below.
2f Ad male *alexandri*: Resembles *neglectus* but larger, paler and more spotted, and tail and head greyer.
2g Ad female nominate: Chestnut-brown above, with dark barring. Buffish below with dark brown spots. Dark moustachial and cheek stripes.
2h Ad female nominate flight: Dark brown outer wing contrasts with rufous inner wing and back. Tail greyish with dark barring.
2i Ad female *alexandri*: Paler than female nominate.

3 AMERICAN KESTREL *Falco sparverius*   **WS** 50–60 cm   Text p. 216
A small kestrel with distinctive head markings. Vagrant to the Az.
3a Ad male: Wings blue-grey, wing has black spots on coverts. Mantle bright rufous, with dark barring. Tail bright rufous with black subterminal bar.
3b Ad male flight: Blue-grey wings contrast with rufous back and tail.
3c Ad female: All rufous above with black barring. Tail barred black.

4 (WESTERN) RED-FOOTED FALCON *Falco vespertinus*   **WS** 60–75 cm   Text p. 214
A small, dark, active falcon. Vagrant to the Az, Md and the CI.
4a Ad male: Dark grey with chestnut thighs, vent and undertail-coverts. Legs and cere orange-red.
4b Ad male flight: Grey body and blackish underwing-coverts contrast with flight feathers below. Primaries silvery-grey above.
4c Ad female: Blue grey above, barred darker. Orange-buff below with dark spots. Crown and nape orange, face pale with dark eye patches and moustachial stripe.
4d Ad female flight: Barred grey flight feathers, tail and upperparts contrast with orange-buff body and underwing coverts.
4e Juv: Much darker than female. Brown above and streaked dark brown below.
4f Juv flight: Very dark. Underwing has dark trailing edge. Buff body boldly streaked dark brown.

5 MERLIN *Falco columbarius*   **WS** 55–65 cm   Text p. 216
A very small, compact falcon. Vagrant to the Az, Md and CI.
5a Ad male: Blue-grey above. Rusty orange below, streaked brown. Moustachial stripe indistinct.
5b Ad male flight: Blue-grey above with dark outer wings. Underwings lightly barred. Black subterminal band to grey tail.
5c Ad female: Dark brown above. Buffish-white below, streaked brown. Tail prominently and evenly barred.
5d Ad female flight: Tail prominently and evenly barred. Brown above, with pale-barred flight feathers. Underwings evenly barred.

77

PLATE 23 FALCONS II

1 (EURASIAN) **HOBBY** *Falco subbuteo* **WS** 65–85 cm Text p. 217
A swift-like, highly aerial, dashing falcon. Rare and irregular on passage in the CI. Vagrant to Md.
1a Ad: dark slate-grey above. White below with hevily streaked breast and belly. Thighs and vent rufous.
1b Ad flight: Uniform slate-grey above, including tail, darker on wingtips. Underwings closely barred.
1c Juv: Browner above with pale buffish feather edges. Yellowish-buff below, very heavily streaked breast and belly. Thighs and vent buff.
1d Juv flight: Brownish-grey, with pale tips to coverts above. Uniform barred underwing and streaked yellowish-buff body.

2 ELEONORA'S FALCON *Falco eleonorae* **WS** 85–100 cm Text p. 217
A slim, dark, long-tailed and long-winged falcon. Highly aerial. Nests in colonies. Local summer resident in CI. Vagrant to Md and the CV.
2a Ad dark morph: Entirely blackish-brown.
2b Ad light morph: Dark greyish-brown above. Rufous-buff below, streaked darker. White cheek and dark moustachial stripe.
2c Ad dark morph flight: Entirely blackish-brown above. Blackish underwing-coverts contrast with paler flight feather bases.
2d Ad light morph flight: Blackish underwing-coverts contrast with paler bases of flight feathers and rufous-buff body.
2e Juv: Browner above than pale-morph adult. Underparts are rich buff with fine dark streaks.
2f Juv flight: Resembles juvenile Hobby but undersides of flight feathers have paler bases and underwing-coverts and wingtips are darker.

3 LANNER FALCON *Falco biarmicus* **WS** 90–115 cm Text p. 217
A large, relatively long-winged and long-tailed falcon with a pale head. Vagrant to the CI.
3a Ad *erlangeri*: Pale blue-grey above with pale edges to upperwing-coverts and mantle feathers. Crown and nape pale buff. Narrow moustachial stripe. Almost white below, spotted dark on breast and barred on flanks and thighs.
3b Ad flight: Fairly uniform blue-grey above, with barred tail. Very pale below, with dark barring across underwing-coverts.
3c Juv: Light brown above. More heavily streaked than adult on the breast and belly. Pale head.
3d Juv flight: Compared with juvenile Peregrine and Barbary Falcons has darker underwing-coverts, paler bases to the primaries, is warmer brown above and has an evenly barred tail.

4 BARBARY FALCON *Falco pelegrinoides* **WS** 80–100 cm Text p. 218
Recalls a pale Peregrine, with a rufous nape and narrower moustachial stripes. Resident in the CI. Vagrant to Md.
4a Ad: Blue-grey above with a rusty nape. Underparts white with fine dark barring and a rufous wash on the breast and belly.
4b Ad flight: Very pale above and below with darker wingtips. Tail has dark subterminal band.
4c Juv: Brown above. Buffish-white below with fine streaking on lower breast and belly. Throat and upper breast unstreaked.
4d Juv flight: Brown above. Pale finely-streaked breast and belly. Underwings paler than juvenile Peregrine. Tail has dark subterminal band.

5 PEREGRINE FALCON *Falco peregrinus* **WS** 85–120 cm Text p. 218
Large, dark above and pale below, with dark crown and prominent moustachial stripe. Nominate a rare and irregular winter visitor to the CI and a vagrant to Md and the Az. Resident in the CV (*madens*).
5a Ad nominate: Dark blue-grey above. White below with fine black barring from the lower breast to undertail-coverts and fine black spotting on the upper breast. White throat, chin and cheeks pure white, contrasting with the almost black cap, nape, upper cheeks and broad moustachial stripe.
5b Ad *madens*: Browner above, with rufous wash to the head. Underparts have pinkish-buff wash.
5c Ad nom flight: Lower back, rump and uppertail-coverts paler than upperparts. Tail uniformly barred.
5d Juv: Dark brown above with pale feather edging. Creamy-buff below, heavily streaked dark brown from chest to vent.
5e Juv flight: Brown above with pale feather edgings. Heavy streaked below on body and underwing coverts. Tail uniformly barred.

79

PLATE 24 GAMEBIRDS AND COOTS

1 RED-LEGGED PARTRIDGE *Alectoris rufa*   **L** 32–34 cm   Text p. 218

Medium-sized gamebird with a white throat. Breeds on Gran Canaria in the CI and in Md. Formerly present in the Az.
1 **Ad:** Black border to white throat becomes heavy black streaking on the breast.

2 BARBARY PARTRIDGE *Alectoris barbara*   **L** 32–34 cm   Text p. 219

Medium sized gamebird with a grey throat. Breeds in the CI.
2 **Ad:** Reddish-brown neck band covered with white spots and a reddish-brown central crown.

3 COMMON QUAIL *Coturnix coturnix*   **L** 16–18 cm   Text p. 219

Small, compact almost tail-less gamebird with streaked brown upperparts. Breeds on all four archipelagos.
3a **Ad male:** Black centre to the throat, striped head pattern and white barring on the flanks.
3b **Ad female:** Pale throat, less pronounced head pattern and more diffuse barring on the flanks.

4 HELMETED GUINEAFOWL *Numida meleagris*   **L** 60–65 cm   Text p. 220

A large fat gamebird with a small bare head and a prominent bony casque on the top. Breeds in the CV. Has bred in the CI.
4 **Ad:** All dark grey plumage completely covered in white spots.

5 COMMON COOT *Fulica atra*   **L** 36–38 cm   Text p. 223

Stocky, all-black waterbird with conspicuous white frontal shield. Breeds in the CI and has bred in the Az. Winter visitor to Md.
5 **Ad:** Pinkish tinge to the bill and the feathering between the gape and the frontal shield reaches a sharp point.

6 AMERICAN COOT *Fulica americana*   **L** 31–37 cm   Text p. 224

A coot with a dark band across the bill. Vagrant to the Az.
6 **Ad:** Subterminal blackish band on the bill, dark reddish spot on top of the frontal shield and white sides to undertail-coverts.

7 RED-KNOBBED COOT *Fulica cristata*   **L** 38–42 cm   Text p. 224

A coot with knobs above the frontal shield. Vagrant to the CI.
7 **Ad:** Two small red nodules on forehead. Blue-grey tinge to bill. Feathering between gape and frontal shield gently rounded.

81

PLATE 25 RAILS AND CRAKES

1 WATER RAIL *Rallus aquaticus* L 23–28 cm Text p. 220
A long-billed rail with barred flanks and pale undertail-coverts. Vagrant to the Az, CI and Md.
1 Ad: Face, foreneck, breast, fore-flanks and central belly slate-grey. Flanks heavily barred black and white. Bill red with darker culmen.

2 SPOTTED CRAKE *Porzana porzana* L 22–24 cm Text p. 220
A plump, short-necked crake with a short, stubby bill. Crown streaked. Rare but regular on passage in the CI. Vagrant to the Az, Md and the CV.
2a Ad: Olive-brown above with blackish feather centres, streaked and spotted white. Chin, throat, head- and upper-breast-sides blue-grey, spotted white. Broad brown stripe on ear-coverts and neck-sides. Flanks brown, heavily barred black and white. Undertail-coverts bright buff. Bill yellowish with orange-red base.
2b Imm: Duller and browner. Lacks the grey face. Chin and throat dirty white. Bill brownish.

3 SORA *Porzana carolina* L 20–23 cm Text p. 220
A plump, short-necked crake with a short, stubby bill. Crown unstreaked, with a dark central stripe. Vagrant to the Az.
3a Ad: Similar to Spotted Crake but bill yellow, lacking red base. Face and central throat black. Unspotted grey head, neck and breast. Lacks V-shaped marks on tertials and inner greater coverts of Spotted Crake.
3b Juv: Duller. Lacks grey coloration and the black 'mask'. Plain buffish-brown below. Bill greenish. Undertail coverts more extensively buff than in adult.

4 LITTLE CRAKE *Porzana parva* L 18–20 cm Text p. 221
A small crake with a long primary projection and a red bill-base. Scarce and irregular on passage in the CI. Vagrant to the Az and Md.
4a Ad male: Blue-grey below, unbarred on flanks. Olive-brown, streaked black above.
4b Ad female: Buffish below, unbarred on flanks.
4c Juv: Pale. Chest less barred than in juvenile Baillon's Crake.

5 BAILLON'S CRAKE *Porzana pusilla* L 17–19 cm Text p. 221
A small crake with a short primary projection. Lacks red at bill-base. Rare and irregular on passage in the CI. Vagrant to the Az and Md.
5a Ad: Resembles Little Crake but upperparts richer brown, with more extensive white streaking extending onto wing-coverts and tertials. Flanks heavily barred black and white.
5b Juv: Darker and more uniformly barred below than juvenile Little Crake.

6 CORN CRAKE *Crex crex* L 27–30 cm Text p. 222
A large crake of grasslands. Rufous wings diagnostic. Vagrant to the Az, Md and CI.
6 Ad: Upperparts buff-brown with dark feather centres. Underparts are a mixture of buff, grey and white with prominent rufous-and-white barring on flanks

7 (AFRICAN) BLACK CRAKE *Limnocorax flavirostra* L 20–23 cm Text p. 221
A short tailed crake with a broad-based conical bill. Vagrant to Md.
7 Ad: All-black plumage, yellow bill and red legs.

83

PLATE 26 MOORHENS AND GALLINULES

1 COMMON MOORHEN *Gallinula chloropus*   **L** 32–35 cm   Text p. 222

A buoyant, dark waterbird, with white-streaked flanks and white sides to the undertail-coverts. Regular in winter in the Az (has bred). Occasional visitor to Md. Scarce breeder in the CI, where commoner in winter. Vagrant to the CV (bred formerly).
1a Ad: Slate grey below and brownish above. Red frontal shield. Red bill with yellow tip.
1b Juv: Duller. Bill grey.

2 LESSER MOORHEN *Gallinula angulata*   **L** 22–24 cm   Text p. 222

A small moorhen with indistinct white flank streaks. Vagrant to the CI.
2a Ad: Resembles adult Common Moorhen but flank streaks indistinct. Bill yellow with red culmen.
2b Juv: Olive-brown above with buffy-brown head-sides and underparts. Flanks greyer. Central underparts white from throat to belly

3 ALLEN'S GALLINULE *Porphyrula alleni*   **L** 22–24cm   Text p. 223

A very small gallinule, smaller than Common Moorhen. Vagrant to the Az, Md and the CI.
3a Ad: Head, neck and underparts bright purplish-blue. Upperparts bright olive-green. Bill red.
3b Juv: Duller, with 'scaly' pattern to brown upperparts.

4 AMERICAN PURPLE GALLINULE *Porphyrula martinica*   **L** 30–36 cm   Text p. 223

Larger and more colourful than Common Moorhen. Vagrant to the Az, Md and the CI.
4a Ad: Head, neck and underparts bright purplish-blue. Upperparts bright olive-green. Bill red with yellow tip..
4b Juv: Upperparts plain olive-brown, with a green tone to the wings. Underparts buffish.

85

PLATE 27 CRANES AND BUSTARDS

1 COMMON CRANE *Grus grus* L 110–120 cm Text p. 224
A large, grey crane. Vagrant to the Az, Md and the CI.
1a Ad: Grey, with black-and-white markings on head and upper neck. Bare red patch on hindcrown.
1b Ad flight: Dark flight feathers contrast with generally grey plumage. Neck outstretched in flight.
1c Juv: Brownish-grey, with brownish head and upper neck.

2 SANDHILL CRANE *Grus canadensis* L 88–95 cm Text p. 224
Smaller than Common Crane. Vagrant to the Az.
2a Ad: Grey, with bare red skin on lores and forecrown.
2b Ad flight: Primaries black. Shows less contrast than Common Crane.
2c Juv: Mainly brown above, greyish below. Browner than juvenile Common Crane.

3 LITTLE BUSTARD *Tetrax tetrax* L 40–45 cm Text p. 225
A small bustard which shows extensive white on the wing in flight. Vagrant to Md and the CI.
3a Ad male breeding: Head grey. Neck black with two striking white collars.
3b Ad male breeding flight: Largely white wings contrasting with browner upperwing coverts and black wingtips.
3c Ad female: Generally brown, with darker vermiculations.

4 HOUBARA BUSTARD *Chlamydotis undulata* L 55–65 cm Text p. 225
Large, with a long, slender neck. White on upperwings only on outer primary bases. Resident in parts of eastern CI.
4a Ad *fuertaventurae*: Deep cinnamon above, heavily vermiculated blackish-brown. White below. Black-and-white neck frill. White central crown and crest.
4b Ad flight: Cinnamon upperparts contrast with mainly black flight feathers and white primary patches.

87

PLATE 28 OYSTERCATCHERS, STILTS, COURSERS AND ALLIES

1 EURASIAN OYSTERCATCHER *Haematopus ostralegus* L 40–46 cm Text p. 225
A plump, black-and-white wader with a red bill and pink legs. Rare and irregular on passage and in winter in the CI. Vagrant to the Az, Md and CV.
1   Ad breeding: Black above and white below.

2 CANARY ISLANDS (BLACK) OYSTERCATCHER *Haematopus meadewaldoi* L 43 cm Text p. 226
An all-black oystercatcher. Formerly resident in the CI. Now extinct.
2   Ad: Entirely black, with orange bill and red legs.

3 BLACK-WINGED STILT *Himantopus himantopus* L 35–40 cm Text p. 226
An elegant black and white wader with extremely long pink legs and a long, thin, black bill. Resident in the CV. Uncommon but regular on passage in the CI (has bred). Vagrant to the Az and Md.
3a Ad male: White with black wings. Black on head and neck more developed in males.
3b First winter: Brownish black above with pale feather fringes.

4 (PIED) AVOCET *Recurvirostra avosetta* L 42–45 cm Text p. 226
A black-and-white wader with an upturned, thin, black bill. Rare and irregular on passage and in winter in the CI. Vagrant to the Az, Md and CV.
4   Ad: Mainly white, with black wing patches and black crown and hindneck. Legs blue-grey.

5 STONE-CURLEW *Burhinus oedicnemus* L 40–44 cm Text p. 227
A large, cryptically-patterned species with large yellow eyes. Nominate is a vagrant to the Az and Md. Resident in the eastern CI (*insularum*) and in the central and western CI (*distinctus*).
5a Ad nominate: Pale brown above, streaked darker. Pale below, with dark streaks on buffy breast. Conspicuous white wingbar, bordered black above and below.
5b Ad *insularum*: Ground colour sandy-pink rather than brown.
5c Ad *distinctus*: Dark streaking heavier than in nominate and ground colour paler.

6 EGYPTIAN PLOVER *Pluvianus aegyptius* L 19–21 cm Text p. 227
A compact wader with unmistakable plumage. Vagrant to the CI.
6   Ad: Combination of blue-grey, black, white and buff plumage unique.

7 CREAM-COLOURED COURSER *Cursorius cursor* L 21–24 cm Text p. 227
An elegant, sandy-coloured, plover-like bird. Resident on the CI (nominate race) and CV (*exsul*). Vagrant to Md.
7a Ad: Black eyestripes and white supercilia meet as a V on nape. Rear crown grey.
7b Juv: Head markings indistinct. Spotted above and streaked on neck and upper breast.

8 COLLARED PRATINCOLE *Glareola pratincola* L 23–26 cm Text p. 228
A short-legged, long-winged, tern-like species with a short bill. Uncommon on passage in the CI. Vagrant to the CV.
8a Ad: Chin and throat creamy with a prominent black border. Bill base red. Upperparts and breast grey-brown. Underparts white. Tail deeply forked.
8b Juv: Scaly patterned above, with a spotted breast-band. Bill all black.

89

PLATE 29 SMALLER PLOVERS

1 LITTLE RINGED PLOVER *Charadrius dubius*   **L** 14–17 cm   Text p. 228

A small ringed plover with pale legs and an indistinct wingbar. Favours freshwater habitats. Resident in the CI and also occurs on passage. Rare breeder in Md. Vagrant to the Az and CV.
1a Ad breeding: Shows narrow white line above black forehead band. Bill dark. Orbital ring yellow. Legs pinkish-grey.
1b Ad non-breeding: Black markings of breeding plumage become brown. Orbital ring less distinct.
1c Juv: Upperparts show pale feather edgings. Supercilium buffish. Legs yellowish.

2 (COMMON) RINGED PLOVER *Charadrius hiaticula*   **L** 18–20 cm   Text p. 228

A plump ringed plover with bright orange or yellow legs and a prominent white wingbar. Lacks orbital ring. Regular and fairly common on passage and in winter in all four archipelagos, commonest in the CI.
2a Ad breeding: Bill orange with black tip. Broad black breast-band.
2b Ad non-breeding: Black markings become greyish-brown. Supercilium and forehead buffy-brown.
2c Juv: Upperparts show pale feather edgings. Supercilium white. Legs dull orange.

3 SEMIPALMATED PLOVER *Charadrius semipalmatus*   **L** 17–19 cm   Text p. 229

Smaller and shorter-billed than Ringed Plover. Thin orbital ring. Vagrant to the Az, Md and CV.
3a Ad non-breeding: Breast-band narrower than in Ringed Plover.
3b Juv: Similar to Common Ringed Plover but usually dark loral patches do not reach gape.

4 KILLDEER *Charadrius vociferus*   **L** 23–26 cm   Text p. 229

A large, long-tailed ringed plover. Double breast-band is diagnostic. Vagrant to the Az, Md and CI.
4a Ad breeding: Brown above and white below. White collar and forehead. White supercilium behind eye.
4b Ad flight: Bright rufous lower back, rump and uppertail-coverts. Tail wedge-shaped. Broad white wingbar.
4c Juv: Similar but shows pale feather edging above.

5 KENTISH PLOVER *Charadrius alexandrinus*   **L** 15–17 cm   Text p. 229

A small, pale ringed plover with dark legs and black bill. Dark patches on sides of neck do not form a complete breast-band. Resident on all four archipelagos. Commonest in the CI where also occurs on passage.
5a Ad male breeding: Crown and nape rufous. Black line across forehead and black line through eye.
5b Ad female: Rufous and black markings of male are greyish-brown.
5c Juv: Upperparts show pale feather edgings. Small breast patches.

6 LESSER SAND PLOVER *Charadrius mongolus*   **L** 19–21 cm   Text p. 230

Slightly larger than Common Ringed Plover, with long dark legs and a short, black bill. Vagrant to the CI.
6a Ad *pamirensis* breeding: Forehead and face mask black. Deep rufous breast-band.
6b Juv: Duller, as adult winter, with pale feather edgings above.

7 (EURASIAN) DOTTEREL *Charadrius morinellus*   **L** 20–22 cm   Text p. 230

Pale supercilium, pale breast-band and plain upperparts lacking wingbars are distinctive. Rare passage migrant and winter visitor in the CI. Vagrant to the Az and Md.
7a Ad breeding: Long white supercilia meet at nape. White breast-band bordered black. Lower breast and flanks chestnut, with black belly.
7b Ad non-breeding: Much duller. Broad buff supercilia meet at nape.
7c Juv: Dark feather centres above give scaly appearance.

91

PLATE 30 LARGER PLOVERS AND LAPWINGS

1 AMERICAN GOLDEN PLOVER *Pluvialis dominica* L 24–28 cm Text p. 230
Slightly smaller but longer-legged than Eurasian Golden Plover. Underwings greyish-brown. Long primary projection. Vagrant to the Az, CI and CV.
1a Ad breeding: White border to black underparts does not extend to flanks. Undertail-coverts black.
1b Ad non-breeding: Much greyer than other golden plovers. Whitish supercilium prominent.
1c Juv: Much greyer than other golden plovers. Whitish supercilium prominent.

2 PACIFIC GOLDEN PLOVER *Pluvialis fulva* L 23–26 cm Text p. 231
A relatively small, long-legged and long-billed plover, with greyish-brown underwings. Vagrant to the CI.
2a Ad breeding: White border to black underparts extends along flanks. Undertail coverts black. Upperparts more coarsely marked than in Eurasian Golden Plover.
2b Juv: As adult winter but has more prominent gold spotting on upperparts, more obvious supercilium and brighter yellowish-buff breast. Juvenile American Golden Plover is much greyer.

3 EURASIAN GOLDEN PLOVER *Pluvialis apricaria* L 26–29 cm Text p. 231
A relatively short-legged golden plover with diagnostic white underwings. Scarce and irregular on passage and in winter in the CI. Vagrant to the Az and Md.
3a Ad breeding: White border to black underparts extends along flanks. Undertail-coverts white.
3b Ad non-breeding: Less brightly-patterned above. Whitish underparts; breast mottled golden-brown.
3c Juv: Breast and flanks barred grey-brown.

4 GREY PLOVER *Pluvialis squatarola* L 27–30 cm Text p. 231
A large, plump plover with diagnostic black axillaries. Regular on all four archipelagos, especially in winter.
4a Ad breeding: Silvery-grey spangled above. Black below with broad white sides to the breast.
4b Ad non-breeding: Greyish above with white feather edgings. White below. Heavy black bill.
4c Juv: Resembles adult winter, but upperparts darker with pale yellowish spangling and more obvious streaking on breast and flanks.

5 SPUR-WINGED LAPWING *Vanellus spinosus* L 25–28 cm Text p. 232
A long-legged plover with white cheek patches and black breast, flanks and tail. Vagrant to the CV.
5 Ad breeding: Upperparts uniform pale brown. Crown and hindneck black.

6 SOCIABLE LAPWING *Vanellus gregarius* L 27–30 cm Text p. 232
A long-legged lapwing with a prominent pale supercilium. Upperwing pattern distinctive: brown upperparts contrast with white secondaries and black primaries. Vagrant to the CI.
6a Ad breeding: Crown black and forehead white. White supercilia meet in a shallow V on nape. Belly black and chestnut.
6b Adult non-breeding: Markings less distinct and lacks dark belly.

7 WHITE-TAILED LAPWING *Vanellus leucurus* L 26–29 cm Text p. 232
Tail entirely white. Legs yellow and very long. Vagrant to the CI.
7a Ad breeding: Greyish-brown above, darker on the chest. White below. Striking black-and-white upperwing.
7b Juv: Breast whitish, mottled grey. Upperparts coarsely dark-spotted.

8 NORTHERN LAPWING *Vanellus vanellus* L 28–31 cm Text p. 232
A lapwing with a prominent crest, dark breast-band and broad, rounded wings. Regular winter visitor to the Az, Md and CI. Vagrant to the CV.
8a Ad breeding: Glossy bronze-green above. Broad black band across upper breast. Otherwise white below with orange undertail.
8b Ad non-breeding: Head-sides buff. Chin and throat white. Buff fringes to scapulars and coverts.
8c Juv: Very short crest. Upperpart feathers have conspicuous buff fringes.

9 UPLAND SANDPIPER *Bartramia longicauda* L 28–32 cm Text p. 242
Small head on long, thin neck and very long tail, with short, straight bill. Vagrant to the Az, Md and CI.
9a Ad: Tertials olive-brown with dark bars. Yellow legs.
9b Juv: Pale feather edgings above give more scaly appearance. Tertials dark brown with buff notches

93

PLATE 31 SANDPIPERS AND STINTS

1 (RED) **KNOT** *Calidris canutus*   **L** 23–25 cm   Text p. 233
A plump, short-legged sandpiper with a short, thick bill. Scarce but regular on passage and in winter in the CI, regular but rarer in Md. Rare but possibly regular in the Az. Vagrant to the CV.
1a Ad summer: Face and underparts bright rufous.
1b Ad winter: Appears grey and nondescript. Legs greyish-green.
1c Juv: Black subterminal crescents and white fringes to upperparts give scaly appearance.

2 SANDERLING *Calidris alba*   **L** 20–21 cm   Text p. 233
A small, pale sandpiper often seen running along the surf edge on sandy beaches. Regular visitor to all four archipelagos, especially in winter.
2a Ad summer: Head, upper mantle and breast chestnut. Dark streaking on head and upper mantle, and dark spotting on breast.
2b Ad winter: Very pale grey above and white below, with black legs and bill.
2c Juv: Upperparts strongly marked with black spots. Crown and back streaked black.

3 SEMIPALMATED SANDPIPER *Calidris pusilla*   **L** 13–15 cm   Text p. 233
Very small. Bill shorter, straighter and blunter-tipped than in next two species. Almost annual in the Az. Vagrant to Md, the CI and CV.
3a Ad summer: Upperparts dark brown with yellowish-buff fringes: lack rufous. Scapulars are black-centred. Wing-coverts and tertials plain grey-brown or brown. White below with a greyish breast-band of dark shaft-streaks.
3b Ad winter: Grey above and white below. Streaks on breast-sides indistinct.
3c Juv: Dull, lacking prominent V on mantle and shows no chestnut in scapulars or on tertial fringes as in Little Stint.

4 WESTERN SANDPIPER *Calidris mauri*   **L** 14–17 cm   Text p. 234
Very small. Bill long, fine and slightly decurved at tip. Vagrant to the Az, Md and CI.
4a Ad summer: Ear-coverts, crown-sides and both upper and lower scapulars rufous. White below, heavily streaked on breast with arrowhead marks on sides extending to flanks.
4b Ad winter: Grey above and white below. Streaks on breast distinct and may extend along flanks.
4c Juv: Rufous on upper scapulars forms a V bordering the mantle, contrasting with grey lower scapulars and wing-coverts.

5 LITTLE STINT *Calidris minuta*   **L** 12–14 cm   Text p. 234
Very small. Bill short, straight and fine-tipped. Regular on passage and in winter in the CI and CV. Vagrant to the Az and Md.
5a Ad summer: Face, cheeks, neck and upper breast rufous with brown streaking. Scapulars and most wing-coverts rufous-fringed. Obvious pale V on sides of mantle. Rest of underparts white.
5b Ad winter: Grey, mottled darker, above and on breast-sides. White below.
5c Juv: Upperparts with obvious rufous fringes to dark feather centres and prominent pale V on sides of mantle. Tertials fringed rufous.

6 TEMMINCK'S STINT *Calidris temminckii*   **L** 13–15 cm   Text p. 234
Very small with a longish tail. Bill longer and finer than in Little Stint and slightly decurved. Yellowish legs distinguish it from previous three species, which have black legs. Rare and irregular on pasage and in winter in the CI. Vagrant to the Az, Md and CV.
6a Ad summer: Plain grey-brown above with some black-centred mantle feathers and scapulars fringed pale rufous. White below, washed grey on upper breast and with gorget of brown streaks.
6b Ad winter: Uniform grey-brown above and on breast-sides. White below.
6c Juv: As adult winter but browner. Wing-coverts and scapulars buff-fringed with dark subterminal lines.

95

PLATE 32 SANDPIPERS

1 LEAST SANDPIPER *Calidris minutilla* L 11–12 cm Text p. 235
Very small. Legs pale but distinguished from Temminck's Stint by stronger markings and grey, not white, tail edges. Rare but probably annual in the Az. Vagrant to the CI and CV.
1a Ad summer: Dark brown above, fringed rufous and grey. White below except for brown on breast.
1b Ad winter: Grey-brown above with darker feather centres. Streaking on breast less distinct.
1c Juv: Narrow mantle V. Split supercilium. Dark mantle and scapulars fringed chestnut, tertials and wing-coverts browner with chestnut fringes. Underparts white, washed buff and streaked brown on breast.

2 WHITE-RUMPED SANDPIPER *Calidris fuscicollis* L 15–18 cm Text p. 235
Slightly smaller than Dunlin. Very long wings project well beyond tail. Prominent white supercilium and white uppertail-coverts. Regular in autumn in the Az. Vagrant to Md and the CI.
2a Ad summer: Underparts white, streaked dark on breast and flanks. Mantle, scapulars and tertials black-centred fringed paler with chestnut, grey and buff. Crown, ear-coverts and scapular fringes washed pale chestnut. Coverts grey-brown, edged whitish.
2b Ad winter: Grey above and white below with dark shaft-streaks on upperparts, greyish-suffused breast and fine streaks on breast and upper flanks.
2c Juv: Sides of mantle have fairly prominent white lines. Upper scapulars dark-centred with chestnut fringes, lower scapulars much greyer with whitish fringes. Cap chestnut.

3 BAIRD'S SANDPIPER *Calidris bairdii* L 14–17 cm Text p. 235
Slightly smaller than Dunlin. Very long wings project well beyond tail. Supercilium less distinct than in White-rumped Sandpiper. Possibly regular in autumn in the Az. Vagrant to the CI.
3a Ad summer: Greyish buff above with dark feather centres, lacking rufous tones. Buffish breast with brown streaks. Flanks unstreaked.
3b Ad winter: Buffish-brown above and breast-band with pale feather edges. Otherwise white below.
3c Juv: Scaly appearance above, with buffish feather edges. Head and breast buffish with dark streaks.

4 PECTORAL SANDPIPER *Calidris melanotos* L 19–23 cm Text p. 236
Slightly larger than Dunlin with a heavily streaked breast sharply demarcated from the white belly. Bicoloured bill short and slightly decurved. Legs yellowish. Vagrant to the Az, Md and CI.
4a Ad summer: Dark brownish above with contrasting fringes varying from chestnut to buffish.
4b Ad winter: Similar but duller and less rufous.
4c Juv: Brighter, with prominent white Vs on mantle and scapulars. Supercilium more distinct.

5 SHARP-TAILED SANDPIPER *Calidris acuminata* L 17–21 cm Text p. 236
Slightly smaller and slightly longer-legged than Pectoral Sandpiper, with a flatter, rufous crown, shorter neck and shorter, uniformly dark, bill. Supercilium prominent. Vagrant to the Az and CI.
5a Ad summer: Upperparts dark-centred with chestnut and buffish fringes. White below with head, neck and upper breast suffused buff. Dark streaks on neck and upper breast become chevrons on flanks.
5b Ad winter: Much duller and less well marked on underparts, markings reduced to some fine streaking on foreneck and greyish breast.
5c Juv: Dark brown above with bright chestnut, white or buff fringes. White below with a strong orange-buff wash on breast. Gorget of fine streaks across upper breast.

6 CURLEW SANDPIPER *Calidris ferruginea* L 18–23 cm Text p. 236
Larger and longer-legged than Dunlin, with a long, decurved bill. Regular on passage and more rarely in winter in the CI. Common on passage and in winter in the CV. Scarce on passage in the Az and Md.
6a Ad summer: Head, neck and most underparts striking brick red. Uppertail-coverts white.
6b Ad winter: Grey above and white below. Breast-sides greyish.
6c Juv: As winter adult but browner and uniformly scaly above. Breast suffused buffish.

7 DUNLIN *Calidris alpina* L 16–22 cm Text p. 237
A small wader. Rump and central uppertail-coverts dark. Regular on passage in the CI, scarcer in winter. Scarce visitor to the Az and Md. Rare and irregular on passage and in winter in the CV.
7a Ad *alpina* summer: Black belly diagnostic. Upperparts fringed rust-red.
7b Ad winter: Grey above and on breast-sides, white below.
7c Juv: Spotting on flanks and breast-sides diagnostic.

97

PLATE 33 SANDPIPERS, SNIPE AND WOODCOCK

1 PURPLE SANDPIPER *Calidris maritima*   **L** 20–22 cm   Text p. 237

A dark, dumpy wader of rocky shores. Legs yellow or orange. Bill yellow-based in winter. Vagrant to the Az, Md and CI.
1a Ad summer: Dark brownish-black above with extensive white and chestnut fringes. White supercilium. White below with heavy dark streaking on breast and flanks.
1b Ad winter: Head, breast and upperparts dark slate-grey with greyish fringes to wing-coverts. White below breast with dark streaked flanks.
1c Juv: As breeding adult but less rufous above and has neat, scaly appearance to wings and scapulars.

2 BUFF-BREASTED SANDPIPER *Tryngites subruficollis*   **L** 18–20 cm   Text p. 237

Resembles a small Reeve (female Ruff) but lacks wingbar and white sides to rump. Vagrant to the Az and CI.
2a Ad: Face and most underparts buff. Breast-sides dark-spotted. Scaly brown-and-buff above. Short black bill and obvious dark eye. Legs bright yellow.
2b Juv: Similar to adult but buff on belly less extensive and paler. Coverts appear scalier.

3 RUFF *Philomachus pugnax*   **L** m: 26–32 cm; f: 20–25 cm   Text p. 238

A bulky, long-necked, small-headed wader. Legs usually yellow or orange. White oval patches border uppertail-coverts.
3a Ad male summer: Head and neck plumes: variably black, rufous, orange or white, diagnostic. Regular but uncommon on passage in the CI and CV and occasional in Md. In winter also rare in the CI and regular in the CV. Accidental in the Az.
3b Ad male winter: Pale grey-brown above, with dark blackish-brown feather centres and buff fringes on wings and mantle. Breast variably buff. Belly and undertail white. Pale feathering around bill base.
3c Ad female: Similar to male winter but smaller.
3d Juv: More scaly above. Buff coloration on head, neck and breast stronger. Legs greenish or brownish.

4 JACK SNIPE *Lymnocryptes minimus*   **L** 17–19 cm   Text p. 238

A very small, secretive snipe. Crown lacks central stripe. Tail wedge-shaped, lacking white sides. Rare and irregular on passage and in winter in the CI. Vagrant to the Az, Md and CV.
4 Ad: Mantle with two broad yellowish-buff stripes on each side. Prominent split supercilium. Streaked (not barred) flanks and breast.

5 COMMON SNIPE *Gallinago gallinago*   **L** 25–27 cm   Text p. 239

A typical snipe, short-legged and long-billed, with a white, unbarred belly. Resident Az population augmented by migrants in winter. Regular on passage and in winter in the CI and, less commonly, in Md. Vagrant to the CV.
5 Ad: Crown dark brown with distinct buff central stripe. Prominent supercilium. Mantle striped buff. Breast streaked brownish. Flanks barred dark blackish-brown.

6 WILSON'S SNIPE *Gallinago delicata*   **L** 25–27 cm   Text p. 239

Resembles Common Snipe but lacks its warm plumage tones, appearing much colder, with the dark areas blacker and pale areas whiter. Vagrant to the Az and CI.
6 Ad: Supercilium may show a distinct bulge in front of eye. Axillaries and underwing-coverts have dark barring equal to or broader than white barring, producing a darker looking underwing than in Common Snipe.

7 GREAT SNIPE *Gallinago media*   **L** 27–29 cm   Text p. 239

A large, bulky, relatively short-billed snipe. Tail has prominent white sides and secondaries lack white trailing edge of Common Snipe. Vagrant to Md and the CI.
7 Ad: Breast and flanks are more intensely barred than in Common Snipe, barring often extending onto belly. White tips to wing-coverts form obvious white lines on folded wing

8 (EURASIAN) WOODCOCK *Scolopax rusticola*   **L** 33–35 cm   Text p. 240

A stocky, pigeon-sized, broad-winged snipe of wooded habitats. Resident but declining in the Az, Md and the CI. Some migrants also occur in the CI.
8 Ad: Plumage is an intricate mix of black, white and brown bars, vermiculations and spots.

99

PLATE 34 DOWITCHERS, GODWITS AND CURLEWS

1 SHORT-BILLED DOWITCHER *Limnodromus griseus* L 25–29 cm Text p. 239
Snipe-sized. Very similar to Long-billed Dowitcher. Juveniles readily separable but adults most easily identified by call. Vagrant to the Az.
1a Ad *griseus* summer: Orange-brown below, dark-spotted on neck-sides and barred on flanks and undertail-coverts. Belly whitish.
1b Ad winter: Grey above and whitish below. Throat and upper breast grey and finely speckled. Flanks barred grey.
1c Juv: Dark blackish-brown above with broad rusty-buff fringes. Scapulars and tertials usually have conspicuously pale internal markings. Underparts washed buffish-orange on foreneck and breast, with distinct dark spotting on breast-sides.

2 LONG-BILLED DOWITCHER *Limnodromus scolopaceus* L 24–26 cm Text p. 240
Snipe-sized. Very similar to Short-billed Dowitcher. Juveniles readily separable but adults most easily identified by call. Vagrant to the Az and CI.
2a Ad summer: More extensive and darker orange-red below than Short-billed Dowitcher and has more densely spotted foreneck and barred upper breast.
2b Ad winter: Usually slightly darker grey above and on breast than Short-billed Dowitcher, with no fine speckling on breast.
2c Juv: Scapulars, inner wing-coverts and tertials have narrow reddish-buff fringes and any internal markings are usually lacking or at most very inconspicuous.

3 BLACK-TAILED GODWIT *Limosa limosa* L 36–44 cm Text p. 240
Large, with a very long bill, a striking white wingbar, square white rump and black terminal tail band. Scarce on passage and in winter in the Az. Regular on passage in the CI, rare in winter. Occasional in Md. Vagrant to the CV.
3a Ad male summer: Neck and breast brick red. Flanks barred. Bill-base orange.
3b Ad winter: Plain grey-brown above, white below, with white supercilium. Black-and-white tail.

4 BAR-TAILED GODWIT *Limosa lapponica* L 37–41 cm Text p. 241
Shorter-legged than Black-tailed Godwit, with an upturned bill. Tail barred. Lacks wingbar. Scarce and irregular on passage and in winter in the Az. Regular on passage and in winter in the CI. Scarce on passage in Md. Regular in winter in the CV.
4a Ad male summer: Entire underparts brick red. Flanks unbarred. Bill all dark.
4b Ad winter: Brownish-grey above streaked darker. Breast and neck washed greyish-brown with some fine dark streaking.

5 SLENDER-BILLED CURLEW *Numenius tenuirostris* L 36–41 cm Text p. 242
An extremely rare small curlew. Bill relatively short and fine-tipped. Possible vagrant to the Az and CI.
5 Ad: Lower-breast-sides and fore flanks marked with distinct blackish spots.

6 WHIMBREL *Numenius phaeopus* L 40–46 cm Text p. 241
Small. Dark crown divided by a pale median stripe. Regular on all four archipelagos, particularly on passage and in winter; *hudsonicus* vagrant to the Az and CV.
6a Ad: Dark olive-brown above, most feathers notched, spotted or fringed whitish or buffish. Whitish below, suffused buff on breast and neck, streaked dark.
6b Ad flight: Rump and lower back white.
6c Ad *hudsonicus*: Head pattern bolder than in Whimbrel and buffier generally, particularly below, with less distinct streaking.
6d Ad *hudsonicus* flight: Rump and lower back brown.

7 EURASIAN CURLEW *Numenius arquata* L 50–60 cm Text p. 242
A large wader, with a very long, decurved bill and plain-looking face. Regular but scarce on passage and in winter in the CI. Occasional in Md. Vagrant to the Az and CV.
7 Ad: Generally grey-brown, streaked or blotched darker.

101

PLATE 35 *TRINGA* SANDPIPERS

1 SPOTTED REDSHANK *Tringa erythropus*   **L** 29–32 cm   Text p. 243

Larger, more elegant and longer-billed than Common Redshank and lacks white on wings. Regular but scarce on passage and in winter in the CI. Vagrant to the Az, Md and CV.
1a Ad summer: Sooty-black with fine white spotting above.
1b Ad winter: Grey above and white below, with a white supercilium.
1c Juv: Darker than adult winter with white feather edging above. Underparts extensively barred.

2 COMMON REDSHANK *Tringa totanus*   **L** 27–29 cm   Text p. 243

Medium-sized wader with red or orange legs. White lower back and secondaries.
2a Ad summer: Has irregular coarse dark markings on both upper- and underparts.
2b Ad winter: Plain grey-brown above and on breast with some pale barring on flanks and undertail.
2c Juv: Browner above, heavily spotted and notched buff. Heavily streaked and barred dark brown below. Legs yellow-orange.

3 MARSH SANDPIPER *Tringa stagnatilis*   **L** 22–25 cm   Text p. 243

Slim and elegant, with a fine, straight bill and long yellow or greenish legs. Vagrant to the Az, CI and CV.
3a Ad summer: Brownish-grey above, with black spots and bars.
3b Ad winter: Plain grey above and white below. Distinct white supercilium.
3c Juv: Darker above than adult, finely marked with buff notches giving a speckled appearance.

4 (COMMON) GREENSHANK *Tringa nebularia*   **L** 30–34 cm   Text p. 244

Larger than Common Redshank with a stouter, upturned bill and grey-green legs. Regular on passage in the Az and Md. Regular on passage and scarce in winter in the CI. Regular on passage and in winter in the CV.
4a Ad summer: Brownish-grey above with a variable number of dark-centred scapulars. Head, neck and breast heavily streaked and barred. White below.
4b Ad winter: More uniform, paler grey above with an indistinct scaly appearance. Streaking on head and neck less extensive and bold.
4c Juv: Dark grey-brown above with fine buffy-white feather fringes. Tertial edges more heavily marked than in adult and more extensively streaked below.

5 GREATER YELLOWLEGS *Tringa melanoleuca*   **L** 29–33 cm   Text p. 244

Similar to Greenshank but lacks white wedge on back and legs are bright yellow. Vagrant to the Az, CI and CV.
5a Ad summer: Dark blackish-brown above with some whitish spots and notches. Head, neck and breast very heavily streaked, forming spots and chevrons on flanks, lower breast and even upper belly.
5b Ad winter: Browner above with fine pale spotting, notches and fringes. Less streaked below.
5c Juv: Dark brownish above with many pale whitish spots and notches. Head and neck strongly streaked brown.

6 LESSER YELLOWLEGS *Tringa flavipes*   **L** 23–25 cm   Text p. 244

Smaller, more elegant, longer-necked and longer-legged than Common Redshank and lacks white wedge on back and white on wings. Dark bill straight and fine. Legs bright yellow. Vagrant to all four archipelagos.
6a Ad summer: Dark blackish-brown above with some whitish spots and notches. Wing-coverts dull grey-brown with fine, pale fringes. Head, neck and breast heavily streaked dark brown, with relatively little streaking on flanks. Otherwise white below.
6b Ad winter: Paler with less intense and more diffuse streaking.
6c Juv: As adult winter but darker and browner above with more contrasting pale spots.

103

PLATE 36 *TRINGA* SANDPIPERS AND WILLET

1 GREEN SANDPIPER *Tringa ochropus*   **L** 21–24 cm   Text p. 245

A sturdy sandpiper, dark above and white below with blackish underwings and a white rump. Legs greenish. Regular on passage and in winter in the CI. Vagrant to the Az, Md and CV.
1a Ad summer: Very dark olive-brown above, finely spotted white. Head and neck heavily streaked dark. Prominent white supercilium in front of eye.
1b Ad winter: Less spotted above. Head and breast paler grey-brown and less heavily streaked.
1c Juv: Resembles winter adult but spotting above is buffish.

2 WOOD SANDPIPER *Tringa glareola*   **L** 19–21 cm   Text p. 245

More elegant than Green Sandpiper with longer, yellow legs, a more finely barred tail and pale underwings. Scarce but regular on passage in the CI, rare in winter. Regular on passage and in winter in the CV. Vagrant to the Az and Md.
2a Ad summer: Dark brown above coarsely spangled whitish. White supercilium and white throat contrast with dark-streaked head and neck, and dark-barred flanks.
2b Ad winter: Less boldly spangled above. Breast washed grey and less boldly streaked.
2c Juv: Resembles winter adult but spotted buff above, with brownish mottling on neck and breast.

3 TEREK SANDPIPER *Xenus cinereus*   **L** 22–25 cm   Text p. 246

Grey, with short, yellow legs and a long, upturned bill. Vagrant to the CI.
3a Ad summer: Upperparts and breast grey. Dark carpal area. Dark line on scapulars. Obscure white supercilium.
3b Ad winter: Plainer, almost unmarked grey above.
3c Juv: Darker grey-brown above. Greyer tertials and coverts, fringed pale buff and with rather faint dark subterminal bars.

4 WILLET *Catoptrophorus semipalmatus*   **L** 33–41 cm   Text p. 247

A robust, greyish wader revealing a startling black-and-white wing pattern in flight. Vagrant to the Az.
4a Ad summer: Grey-brown above barred whitish-buff and dark brown. Neck, breast and flanks heavily streaked and barred.
4b Juv: Slightly darker above with buff fringes and dark subterminal line to most feathers. Tertials and rear scapulars notched with buff.

105

PLATE 37 SANDPIPERS AND TURNSTONE

1 COMMON SANDPIPER *Actitis hypoleucos*   **L** 19–21 cm   Text p. 246

A short-necked and short-legged wader which bobs its rear end habitually. Olive-brown above and white below, with obvious dark breast-side patches. Legs usually greenish-grey. Regular on passage and in winter in Md, the CI and the CV. Vagrant to the Az.
1a Ad summer: Slightly glossed bronze above, with darker streaks and subterminal marks on some larger feathers. Breast-side patches form an almost complete breast-band. Bill black.
1b Ad winter: Plainer above and breast-side patches smaller.
1c Juv: Wing-coverts barred pale buff and dark brown. Most upperpart feathers pale-fringed. Distinct dark and buff notches on tertials.

2 SPOTTED SANDPIPER *Actitis macularius*   **L** 18–20 cm   Text p. 246

Very similar in most plumages to Common Sandpiper, but wingbar less distinct on inner wing above. Regular in autumn and often in winter in the Az. Vagrant to Md, the CI and the CV.
2a Ad summer: White underparts boldy spotted black. Bill pinkish with dark tip.
2b Ad winter: Greyer above and shorter-tailed than Common Sandpiper with smaller breast-side patches. Bill pinkish with dark tip.
2c Juv: Plainer above and shorter-tailed than juvenile Common Sandpiper. Coverts more strongly marked. Tertials lack pale- and dark-spotted fringes.

3 STILT SANDPIPER *Calidris himantopus*   **L** 18–23 cm   Text p. 237

Long-legged, long-necked and long-winged, with a long, decurved bill. Legs greenish or yellowish. Vagrant to the Az.
3a Ad summer: Unmistakable given combination of barred underparts, rather dark pale-fringed upperparts and rusty ear-coverts and crown, separated by a whitish supercilium.
3b Ad winter: Plain grey-brown above with white fringes to larger wing-coverts and tertials. White below with rather indistinct streaking on breast extending to flanks and undertail-coverts. Distinct white supercilium.
3c Juv: Mantle, scapulars and tertials dark brownish-black with cinnamon fringes, but these soon fade to off-white. Underparts whitish with indistinct streaks on breast, flanks and undertail-coverts. Breast washed buff. Supercilium whitish.

4 (RUDDY) TURNSTONE *Arenaria interpres*   **L** 21–26 cm   Text p. 247

A stocky wader with a black-and-white wing pattern, stout conical bill and short orange legs. Regular on all four archipelagos, especially on passage and in winter. The most abundant wader in the CI.
4a Ad summer: Mostly chestnut-orange above. Head and breast strongly patterned black-and-white. White rear underparts.
4b Ad winter: Much duller. mottled blackish and dull brown above. Dark grey-brown breast with pale patches on sides.
4c Juv: Similar to adult winter but wing and mantle feathers fringed buff giving scaly appearance above.

107

PLATE 38 PHALAROPES AND SKUAS

1 WILSON'S PHALAROPE *Phalaropus tricolor* **L** 22–24 cm Text p. 247
Legs, neck and bill longer than other phalaropes. Square white rump. Vagrant to the Az, Md and the CI.
1a Ad female summer: Throat and supercilium white. Crown and hindneck pale grey, with broad black eye-stripe reaching onto the neck-sides where it becomes deep chestnut. Foreneck rufous. Legs black.
1b Ad winter: Pale grey above, white below, with a white supercilium extending down grey neck-sides.
1c Juv: Darker brown upperparts than ad winter, with pale buff fringes to feathers. Legs yellowish-flesh.

2 RED-NECKED PHALAROPE *Phalaropus lobatus* **L** 18–19 cm Text p. 248
A delicate phalarope with a fine, pointed bill. Vagrant to the Az, CI and CV.
2a Ad female summer: Crown, hind neck, mantle and cheeks dark grey. Chin and throat white. Orange-red on neck. Dark above with buff lines on mantle sides and scapulars.
2b Ad winter: White below and grey above with an obvious black mask. The whitish edges to larger feathers form lines along mantle sides and scapulars,
2c Juv: Blackish-brown above. Buff fringes form stripes on mantle sides and scapulars. Initially, neck and breast washed greyish-buff.

3 GREY (RED) **PHALAROPE** *Phalaropus fulicarius* **L** 20–22 cm Text p. 248
Sturdy phalarope with a relatively thick bill. Recorded off all four archipelagos.
3a Ad summer: Face white. Chestnut-red below. Dark blackish-brown above, fringed rufous and buff. Bill mainly yellow.
3b Ad winter: Crown and nape dark greyish-black. Black mask. Grey above. Unstreaked on mantle.
3c Juv: Similar to juvenile Red-necked but narrow stripes above. Breast and neck washed pinkish-buff.

4 POMARINE SKUA *Stercorarius pomarinus* **L** 46–51 cm Text p. 248
A deep-chested, relatively broad-winged skua. Regular on passage in all four archipelagos.
4a Ad pale morph: Central tail feathers with spoon-shaped extensions. Dark cap and yellowish neck-sides. Most have a dark upper breast band. Lower breast and belly white, vent dark. Wings dark brown with pale primary-base flashes above and below.
4b Juv intermediate morph: head and body greyish-brown. Underwings heavily barred and with a double white flash, formed by pale bases to the primaries and primary-coverts.

5 LONG-TAILED SKUA *Stercorarius longicaudus* **L** 48–53 cm Text p. 250
A small, slim skua with slender wings and buoyant flight. Recorded rarely on passage from all four archipelagos but probably commoner than records suggest.
5a Ad: Very long tail projection. Cap black. Prominent contrast between grey upperparts and black flight feathers and tail. Pale below, greyer on belly. Lacks white wing-flashes on either surface.
5b Juv dark morph: Sooty-brown, heavily-barred on underwing and tail-coverts.

6 ARCTIC SKUA *Stercorarius parasiticus* **L** 41–46 cm Text p. 249
A compact skua with long, pointed wings and swift, falcon-like flight. Scarce on passage in all four archipelagos and also reported in winter from the CI and CV.
6a Ad pale morph: Short tail projection. Cap dark. Neck tinged yellowish. Whitish below with a variable dark breast-band. Wings dark brown with pale primary-base flashes above and below.
6b Juv intermediate morph: Warm brown with a distinct rusty tinge below, particularly prominent in barring on underwing and undertail-coverts. Rusty neckband. Pale primary-base flashes above and below.

7 GREAT SKUA *Stercorarius skua* **L** 53–58 cm Text p. 251
A large skua, recalling a large, dark gull. Pale primary-base flashes above and below. Scarce but regular on passage and in winter in the Az, Md and the CI. Recorded from the CV and probably also regular there.
7a Ad: Rufous-brown with variable dark cap. Neck- and breast-sides streaked yellowish. Blotchy below. Wing-coverts paler than blackish flight feathers. White wing flashes very prominent.
7b Juv: Similar to adult but darker and more uniform, with narrower white primary flashes

8 SOUTH POLAR SKUA *Stercorarius maccormicki* **L** 51–54 cm Text p. 251
Similar to Great Skua but smaller and greyer, with narrower wings. Vagrant to the Az and CI.
8a Ad dark morph: Similar to Great Skua but note smaller head and grey-brown underside.
8b Ad pale morph: Pale greyish to pale brown on head and underparts, and sometimes on mantle. Tail, rump and wings uniformly dark with pale primary-base flashes, creating a two-toned appearance.

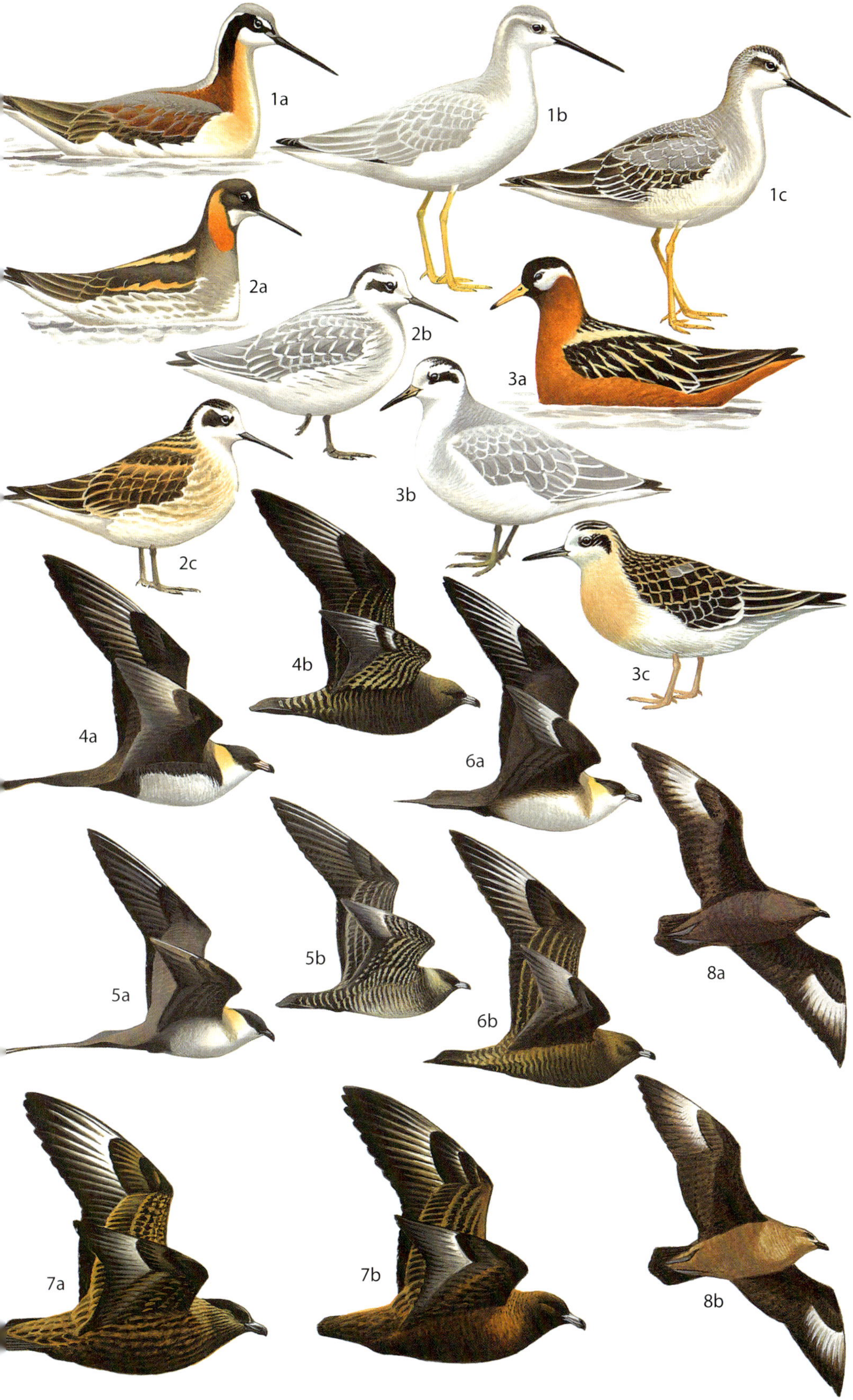
1a
1b
1c
2a
2b
3a
2c
3b
4b
3c
4a
6a
8a
5b
5a
6b
7a
7b
8b

PLATE 39 GULLS I

1 MEDITERRANEAN GULL *Larus melanocephalus*   **L** 36–38 cm   Text p. 251
A small gull, with relatively broad wings and a short, stout bill. Rare and irregular in the CI in winter. At least occasional in winter in Md. Vagrant to the Az.
1a Ad breeding: Hood black, with white eye-crescents. Red bill with black subterminal band.
1b Ad winter: Very pale grey above and otherwise white, with black mask behind eye.
1c Ad winter flight: Looks almost entirely white, with contrasting black eye mask.
1d 1st-winter: Head white with black mask behind eye. Bill blackish with pale base. Mantle plain grey.
1e 1st-winter flight: Blackish outer primaries and band across secondaries contrast with pale grey wing centres above. Underwing coverts white. White tail with narrow blackish subterminal band.

2 LAUGHING GULL *Larus atricilla*   **L** 36–41 cm cm   Text p. 252
A small gull with a distinctive long bill. Dark grey above and white below. Vagrant to the Az and CI.
2a Ad breeding: Mantle and upperwings dark grey with blackish primaries. Hood black, with white eye-crescents. Dark red bill with black subterminal band. Legs dark red.
2b Ad winter: Hood replaced by greyish patch on ear-coverts that may extend across nape and crown.
2c Ad winter flight: Upperwings show very little white.
2d 1st-winter: Mantle and scapulars dark grey. Hindneck, breast and flanks greyish, contrasting with white forehead, throat and belly. Dark patch on ear-coverts may extend over crown. White eye crescents.
2e 1st-winter flight: Outer wing and secondaries blackish, the latter with white tips. Tail greyish with a broad blackish subterminal band.

3 FRANKLIN'S GULL *Larus pipixcan*   **L** 32–36 cm   Text p. 253
Smaller, shorter-legged and shorter-billed than Laughing Gull and with larger white eye-crescents. Dark grey above and white below. Vagrant to the Az, Md and CI.
3a Ad breeding: Mantle and upperwings dark grey with large white tips to blackish primaries. Hood black, with white eye-crescents. Red bill with black subterminal band. Legs bright red.
3b Ad winter: Black half hood extending from behind eyes over rear crown. Bill black. Legs dark.
3c Ad winter flight: White band separates black primary tips from grey bases. Grey centre to white tail.
3d 1st-winter: Half-hood as adult winter, grey mantle and scapulars grey. Primaries blackish.
3e 1st-winter flight: Dark bar on secondaries. Tail grey; black band does not reach outermost rectrices.

4 SABINE'S GULL *Larus sabini*   **L** 27–32 cm   Text p. 253
A long-winged gull with a striking tricoloured upperwing pattern. Vagrant to all four archipelagos.
4a Ad breeding: Hood dark grey with black border. Bill black with yellow tip.
4b Ad winter: Hood reduced to dark eye-crescents and variable area on nape and hindneck.
4c Ad winter flight: White inner primaries and secondaries form white triangles on upperwings, contrasting with dark grey forewings and black outer primaries.
4d 1st-winter: Head, hindneck and breast sides greyish brown. Back grey. Inner wing-coverts brown.
4e 1st-winter flight: White inner primaries and secondaries form white triangles on upperwings, contrasting with brown forewings and black outer primaries. Tail white with black terminal band.

5 LITTLE GULL *Larus minutus*   **L** 25–27 cm   Text p. 253
Very small and short-legged. Rare and irregular in winter and on passage in CI. Vagrant to Az and Md.
5a Ad breeding: Black hood extends onto upper neck. Bill black.
5b Ad winter: Dusky patch on the crown. Dark ear-spot and eye-crescents. No black in wings.
5c Ad winter flight: Wings grey above with white trailing edge and contrastingly blackish below.
5d 1st-winter: Dark blackish-brown crown and ear spots. Hindneck and mantle grey. Primaries black.
5e 1st-winter flight: Striking W pattern on upperwings. Dark bar on secondaries. Underwing white.

6 BLACK-HEADED GULL *Larus ridibundus*   **L** 34–37 cm   Text p. 254
A small gull with a white wedge on the outer primaries above and dusky underwing tips. Bill mainly red. Regular in winter in Md and the CI and, less commonly, in the Az and CV.
6a Ad breeding: Hood chocolate brown.
6b Ad winter: Pale grey above with dark ear spot. Bill red with black tip.
6c Ad winter flight: Wings grey above with white outer primaries and black primary tips.
6d 1st-winter: Dark ear spot, often indistinctly extending across rear crown. Dusky eye-crescents.
6e 1st-winter flight: Dark secondaries and median wing-coverts. Tail white with black terminal band.

1a
1b
1c
1d
1e
2a
2b
2c
2d
2e
3a
3b
3c
3d
3e
4a
4b
4c
4d
4e
5a
5b
5c
5d
5e
6a
6b
6c
6d
6e

PLATE 40 GULLS II

1 PALLAS'S GULL *Larus icthyaetus* L 57–61 cm Text p. 251
A large gull with a sloping forehead and a long, heavy bill. Vagrant to Md and the CI.
1a Ad breeding: Black hood with distinct white eye crescents. Bill yellow with black subterminal band and reddish tip.
1b Ad breeding flight: Wings mainly white, with a black crescent across the outer primaries.
1c Ad winter: Hood reduced to dark patches behind the eyes, sometimes extending across rear crown. Legs yellow.
1d 1st-winter: Head and body white with dark marks around eyes. Dusky patch behind eye extends diffusely over rear crown. White eye-crescents. Dark patch on lower hindneck contrasts with grey mantle. Dark-tipped pale bill. Legs greyish.
1e 1st-winter flight: Outer primaries blackish-brown. Pale grey-brown panel in mid-wing. Tail white with black terminal band.

2 AUDOUIN'S GULL *Larus audouinii* L 48–52 cm Text p. 255
Smaller, slimmer and more elegant than other large gulls, with proportionately long wings and short tail. Sloping forehead. Vagrant to Md and the CI.
2a Ad: Red bill with black subterminal band and yellow tip. Legs grey-green. Black primaries with white tips. Dark eye.
2b Ad flight: Pale grey above. Black primaries with small white tips.
2c 1st-winter: Pale grey mantle, spotted darker. Wings brownish, black on primaries. Head and hindneck mottled greyish brown. Bill grey with dark tip. Legs dark grey.
2d 1st-winter flight: Darker wings contrast above with grey back. Tail mainly black with U-shaped white rump patch.

3 SLENDER-BILLED GULL *Larus genei* L 42–44 cm Text p. 254
Similar to Black-headed Gull but neck very long, head long with sloping forehead and bill long. Probably regular in the eastern CI (has bred). Vagrant to the CV.
3a Ad breeding flight: Lacks hood. Head white. Underparts often tinged pink. Bill dark red.
3b Ad winter: Whiter below. Head white or with very faint dark ear spot. Bill paler red.
3c 1st-winter: Bill shape distinctive. Ear spot indistinct. Bill pale red with darker tip. Legs pale orange.
3d 1st-winter flight: Profile distinctive. Pattern like 1st-winter Black-headed Gull but less clearly marked.

4 RING-BILLED GULL *Larus delawarensis* L 43–47 cm Text p. 255
Similar to Common Gull but larger and with a heavier bill. Regular in winter in the Az. Scarce winter visitor to Md. Vagrant to the CI and CV.
4a Ad winter: Bill yellow with broad black band. Irides pale. Head and neck mottled brownish.
4b Ad winter flight: Paler grey above than Common Gull. Small mirrors on outer two primaries.
4c 1st-winter: Mantle and scapulars pale grey. Head and underparts white with brownish spotting on nape/hindneck, and crescents, spots and bars on breast-sides and flanks. Bill pink with a broad dark tip.
4d 1st-winter flight: Tail band less neat than in first winter Common Gull. Uppertail-coverts strongly barred brownish. Upperwing is dark on outer primaries with a dark bar on secondaries and predominantly dark coverts. Pale grey midwing panel above. Irides dark.
4e 2nd-winter flight: Pale grey above with brownish-black primary-coverts and outer primaries. Tail white, often with residual black tail band.

5 COMMON GULL *Larus canus* L 40–42 cm Text p. 256
A compact gull with a rounded head and a fairly short bill. Irides always dark. Scarce and irregular in winter in the CI. Vagrant to the Az and Md.
5a Ad winter: Dark grey above with heavy mottling on head and hindneck. Bill greenish-yellow with thin dark band.
5b Ad winter flight: Large white mirror on outer two primaries.
5c 1st-winter: Similar to first winter Ring-billed Gull but mantle darker grey and markings on head and underparts less distinct. Bill greyish with dark tip.
5d 1st-winter flight: Clear-cut black tail band. Dark grey upper midwing panel less contrasting than in Ring-billed Gull.
5e 2nd-winter flight: Grey above with brownish-black primary-coverts and outer primaries. Large white mirror on outer two primaries. Tail white.

113

PLATE 41 GULLS III

1 LESSER BLACK-BACKED GULL *Larus fuscus*   **L** 52–67 cm   Text p. 256

A large gull, generally dark above and pale below. Regular on passage and in winter in the Az, Md and, especially, in the CI. Vagrant to the CV.
1a Ad winter nominate: Uniformly black above. Head and neck white or only sparsely mottled. Legs yellow.
1b Ad winter nominate flight: Wings uniformly black above except for one small white mirror on outermost primary.
1c Ad winter *graellsii*: Dark grey above. Head and neck heavily mottled greyish-brown. Legs yellow.
1d Ad winter *graellsii* flight: Wings dark grey with blackish, white-tipped primaries and two small white mirrors.
1e 1st-winter: Head and underparts whitish, streaked dark grey-brown. Dark ear coverts. Brownish-grey above with darker feather centres. Bill black.
1f 1st-winter flight: Primaries and primary-coverts blackish brown. Dark bar on secondaries. Dark underwing-coverts. Rump whitish. Tail has solid blackish subterminal band.

2 YELLOW-LEGGED GULL *Larus michahellis*   **L** 55–67 cm   Text p. 258

A large gull, dark above and pale below. Resident (*atlantis*) in the Az, Md and CI, with nominate birds also occurring in the CI in winter. Probably a scarce winter visitor to the CV.
2a Ad winter *atlantis*: Dark grey above and white below. Variable fine streaking on head, particularly on crown and ear-coverts. Legs yellow.
2b Ad winter *atlantis* flight: Wings dark grey with extensive black tips. White spots on outer primaries and one or two white mirrors.
2c 1st-winter: Head and neck whitish, with dark ear-coverts. Neck and underparts marked brownish. Dark brown above with darker feather centres. Bill black. Legs pinkish.
2d 1st-winter flight: Flight feathers dark brown, with paler panel on inner primaries. Underwing coverts dark brown. Rump and tail whitish, with distinct blackish tail band.

3 AMERICAN HERRING GULL *Larus smithsonianus*   **L** 53–65 cm   Text p. 257

Similar to Herring Gull but immature birds much darker. Vagrant to the Az and Md.
3a 1st-winter: Head pale greyish brown contrasting with dark marks on hindneck/upper mantle and fairly uniform brown underparts. Mantle and scapulars with some grey feathers and dark subterminal markings. Bill greyish with black tip.
3b 1st-winter variant: Head whitish. Bill pink with a black tip.
3c 1st-winter flight: Wings dark brown with paler inner primaries. Underwing coverts darker and more uniform than in first winter Herring Gull. Rump heavily barred dark brown to greyish-brown. Undertail-coverts distinctly but narrowly barred.

4 HERRING GULL *Larus argentatus*   **L** 56–64 cm   Text p. 257

A large gull with pink legs. Vagrant to Md and the CI.
4a Ad winter: Upperparts pale grey. Head and neck heavily streaked and spotted brown.
4b Ad winter flight: Wings pale grey above with black tips, white trailing edges and large white mirrors on outer primaries.
4c 1st-winter: Rather plain brown, paler on head. Flight feathers darker.
4d 1st-winter flight: Upperwing has conspicuous pale panel on inner primaries. Underwing-coverts paler than in similar species. Rump dark-spotted and darker than in Yellow-legged Gull, with less contrasting tail band.

5 GREAT BLACK-BACKED GULL *Larus marinus*   **L** 64–78 cm   Text p. 260

A very large gull with a heavy, deep bill. Legs pale pink. Scarce but regular in the Az, chiefly in winter. Scarce and irregular in winter in the CI. Vagrant to Md.
5a Ad: Blackish above with with large white spots on black primaries.
5b Ad flight: Black upperwings with broad white trailing edges and large white tips to outer primaries.
5c 1st-winter: Head and neck whitish. Underparts whitish streaked dark brown. Upperparts chequered, with dark feather centres. Bill black, developing pinkish base.
5d 1st-winter flight: Upperwings barred and mottled, with darker flight feathers and an indistinct pale window on the inner primaries. Underwing-coverts contrast with paler undersides to flight feathers. Rump whitish. Tail with several narrow bars.

115

PLATE 42 GULLS IV

1 ICELAND GULL *Larus glaucoides* **L** 52–60 cm Text p. 259
A compact, pale gull, slightly smaller than Herring Gull. Legs pink. Regular in small numbers in winter in the Az. Vagrant to Md and the CI.
1a Ad winter: Upperparts pale grey, otherwise all white except for brown streaking on head and neck.
1b Ad winter flight: Upperwings and mantle pale grey, with white trailing edge to wings.
1c 1st-winter: Milky coffee-coloured, with pale primaries and barred upperwing-coverts, mantle, rump and tail. Bill predominantly dark.
1d 1st-winter flight: Pale unbarred flight feathers contrast with finely barred coverts, mantle and tail.
1e 2nd-winter: Largely white with less barring than first winter. Head and underparts often have some coarse streaking. Distal half of bill black, remainder pinkish or grey.

2 GLAUCOUS GULL *Larus hyperboreus* **L** 62–68 cm Text p. 259
A very large, heavy-bodied pale gull with a large bill. Vagrant to the Az, Md and CI.
2a Ad winter: Pale grey above with white tips to primaries and tertials. Head and neck heavily mottled brown.
2b Ad winter flight: Pale grey wings with broad white trailing edges. Head markings give hooded appearance.
2c 1st-winter: Milky coffee-coloured, streaked pale brown above. Primaries paler. Bill pink with distal third black.
2d 1st-winter flight: Milky coffee coloured with paler flight feathers, often showing small dark tips.
2e 2nd-winter: Paler than first-winter blotched darker. Bill pink with distal third black except for pale tip.

3 GLAUCOUS-WINGED GULL *Larus glaucescens* **L** 60–66 cm Text p. 260
A large-headed, broad-winged and heavy-bodied pale gull. Bill heavy. Vagrant to the CI.
3a Ad winter: Grey above and white below, with grey, white-tipped primaries. No black in wings. Head and neck mottled brown.
3b Ad winter flight: Upperwings show white mirror on outermost primary and broad white trailing edge.
3c 1st-winter: Greyish-brown, finely barred darker. Head and hindneck whitish. Some pale grey feathers on mantle. Bill black.
3d 1st-winter flight: Uniform greyish-brown with faint darker markings and barred rump. White undersides to flight feathers contrast with underwing-coverts.

4 (BLACK-LEGGED) KITTIWAKE *Rissa tridactyla* **L** 38–40 cm Text p. 261
A small gull with long wings and a slightly notched tail. Legs short. Regular in the Az in winter. Rare winter visitor to Md and uncommon in winter in the CV. Irregular winter visitor to the CI.
4a Ad winter: Head white with a dark ear spot. Rear crown and hindneck pale grey. Upperparts grey, with black primary tips. Underparts, rump and tail white. Bill yellowish. Irides dark. Legs black.
4b Ad winter flight: Grey of upperwing coverts darker than on flight feathers, giving two-toned appearance. Wingtips almost solid black.
4c 1st-winter: Head white with a dark ear spot. Rear crown and hindneck pale grey, with black half-collar often reduced or replaced by dark grey. Mantle grey with black bar across upperwing-coverts and black primaries. Bill black. Legs dark.
4d 1st-winter flight: Upperwing tricoloured, with grey coverts, black carpal bar and outer wing forming W pattern, and whitish trailing edge to inner wing. Mantle grey. Tail white with black terminal bar.

5 BONAPARTE'S GULL *Larus philadephia* **L** 28–30 cm Text p. 254
Smaller, shorter-legged and neater than Black-headed Gull, with a black bill and pale underwings. Vagrant to the Az and CI.
5a Ad breeding: Hood black.
5b Ad winter: Darker grey above than Black-headed Gull. Legs pink. Black ear spot.
5c Ad winter flight: Upperwing pattern as Black-headed Gull but undersides of primaries white.
5d 1st-winter: Greyer above than juvenile Black-headed Gull, with blacker wing markings. Black ear spot.
5e 1st-winter flight: Upperwing markings blacker than in juvenile Black-headed Gull, with more black on primary-coverts.

117

PLATE 43 TERNS I

1 GULL-BILLED TERN *Gelochelidon nilotica* **L** 35–38 cm Text p. 261
Similar to Sandwich Tern in size. Long-legged, with a short, thick black bill. Vagrant to the Az, Md and the CV. Rare and irregular on passage in the CI.
1a Ad breeding flight: Grey above and white below, with an uncrested black cap.
1b Ad winter: Black cap reduced to a patch behind the eye. Grey upperwings with dark trailing edge to primaries.
1c 1st-winter flight: As adult winter with faint dark secondary bar and dusky primary-coverts above.

2 CASPIAN TERN *Hydroprogne caspia* **L** 47–54 cm Text p. 261
A very large, broad-winged, short-tailed tern with a very thick red bill. Extensive dark underside to primaries. Vagrant to Md, the CI and CV.
2a Ad breeding flight: Extensive black cap. Bill red with dark tip.
2b Ad winter: Cap remains extensive but forehead streaked white.
2c Juv flight: Faint scaly markings on upperparts. Lacks dark carpal bar. Blackish cap streaked white. Bill less red, with dusky tip.

3 ROYAL TERN *Sterna maxima* **L** 45–50 cm Text p. 262
A very large tern with a large orange bill. Undersides of primaries pale. Vagrant to the Az, CI and CV.
3a Ad breeding flight: Extensive black cap, black forehead soon becoming white. Crest more prominent than in Caspian Tern.
3b Ad winter: Forehead and crown white, with black reduced to a shaggy nape patch.
3c Juv flight: Dark carpal bar. Pale mid-wing panel distinctive. Head pattern as adult winter.

4 LESSER CRESTED TERN *Sterna bengalensis* **L** 35–37 cm Text p. 262
Similar to Sandwich Tern in size and shape but bill entirely orange. Vagrant to Md and the CI.
4a Ad breeding flight: Full black cap. Long, slender bright orange bill. Rump, tail centre and uppertail-coverts pale grey.
4b Ad winter: Forehead and forecrown white.
4c Juv flight: Upperwing pattern less contrasting than in Royal Tern. Bill yellowish-orange.

5 SANDWICH TERN *Sterna sandvicensis* **L** 36–41 cm Text p. 262
A large tern with a slender bill. Fairly common on passage and in winter in the CI. Also recorded in winter in the Az, Md and CV.
5a Ad breeding flight: Cap black, with a shaggy nuchal crest. Bill black with yellow tip. Pale grey above with dusky outer primaries and white trailing edge to upperwing. Rump, tail and underparts white.
5b Ad winter: Forehead and forecrown white.
5c 1st-winter flight: As adult winter with dusky juvenile feathers retained on wing-coverts, secondaries and tail.

6 ROSEATE TERN *Sterna dougallii* **L** 33–38 cm Text p. 263
A small, slender tern with a long bill and very long tail. Breeds in the Az and, rarely, Md. Vagrant to the CI (has bred) and CV.
6a Ad breeding flight: Bill black with red base. Very pale grey above. White below, often suffused pinkish on breast. Lacks dark trailing edge to primaries below.
6b Ad winter: Forehead white. Bill black. Looks very white overall.
6c 1st-winter flight: Similar to first winter Common Tern. Note structural differences.

7 LITTLE TERN *Sternula albifrons* **L** 22–24 cm Text p. 264
A very small, active and fast-flying tern. Rare and irregular on passage in the CI. Vagrant to the Az and Md. Perhaps regular in small numbers on passage and in winter in the CV.
7a Ad breeding flight: Crown, nape and eye-stripe black. Forehead and shallow V above eye, rest of head and underparts white. Pale grey above with black outer primaries and whiter rump and tail. Bill yellow with black tip. Legs yellow-orange.
7b Ad winter: Lores white and white forehead more extensive. Bill dark.
7c 1st-winter flight: Crown browner. Upperpart feathers of upperparts tipped brown. Dark carpal bar on upperwing.

119

PLATE 44 TERNS II

1 COMMON TERN *Sterna hirundo* **L** 31–35 cm Text p. 263
Larger, longer-billed and longer-legged than Arctic Tern, with a more elongated head. Breeds in the Az, Md and the CI. Common on passage in the CI. Vagrant to the CV but perhaps under-recorded.
1a Ad breeding flight: Only inner primaries translucent. Greyish belly paler than upperparts. Outer primaries show diffuse dark trailing edge below. Distinct dark wedge on upperside of outer primaries. Bill red with dark tip.
1b Ad winter: Forehead and underparts white. Dark carpal bar. Bill all black.
1c 1st-winter flight: Dark carpal bar obvious. Narrow grey bar on the secondaries, with a distinct pale mid-wing panel. Primaries dusky above.

2 ARCTIC TERN *Sterna paradisaea* **L** 33–35 cm Text p. 263
A small, round-headed, short-billed tern with very short legs. Vagrant to all four archipelagos but perhaps under-recorded.
2a Ad breeding flight: Resembles Common Tern but underparts greyer and white cheeks well demarcated from breast. All primaries translucent. Outer primaries show neat dark trailing edge below. Bill coral red without black tip.
2b Ad winter: Forehead and underparts white. Bill black.
2c 1st-winter flight: Similar to 1st winter Common Tern but lacks dark secondary bar and has fainter carpal bar.

3 WHITE-CHEEKED TERN *Sterna repressa* **L** 32–34 cm Text p. 264
Recalls Common Tern but much greyer. Vagrant to the CI.
3a Ad breeding flight: Darker grey and more uniform than Common Tern. Obvious white facial streak separates black cap from grey underparts.
3b Ad winter flight: Forehead, breast and belly white. Rump, vent and tail grey.

4 SOOTY TERN *Onychoprion fuscata* **L** 43–45 cm Text p. 265
A large tern with very dark upperparts. Vagrant to the Az, Md and the CI but has recently bred in the Az and may have bred Md.
4a Ad flight: Uniform blackish above and white below. Broad white forehead patch does not reach eye.
4b Juv flight: Entirely dark sooty-brown, except for pale lower belly, underwings and undertail-coverts. Whitish tips to upperwing-coverts and on mantle form long narrow bars.

5 BRIDLED TERN *Onychoprion anaethetus* **L** 35–38 cm Text p. 265
Smaller than Sooty Tern with more contrasting plumage. Vagrant to the Az (has bred).
5a Ad flight: Black cap contrasts with greyish neck. Upperparts greyish-brown, contrasting with darker flight feathers. Narrow white forehead patch extends behind eye.
5b Juv: Crown much paler. Upperpart feathers with buff and dark fringes giving scaly appearance, although some are much darker brown and markings less obvious.

6 WHISKERED TERN *Chlidonias hybrida* **L** 23–25 cm Text p. 265
Smaller than Common Tern, with a short, slightly forked grey tail. Scarce and irregular on passage in the CI. Vagrant to the Az and Md.
6a Ad breeding flight: Body and upperwings mainly dark grey. Black cap, white cheeks and red bill. Underwings and undertail white.
6b Ad winter flight: Very pale grey above and white below, with black-streaked rear crown and black patch behind eye.
6c Juv flight: Rump, tail and wings pale grey. No carpal bar. Mantle and scapulars gingery-brown, barred black.

7 FORSTER'S TERN *Sterna forsteri* **L** 33–36 cm Text p. 264
A white tern with a black face mask. Vagrant to the Az, Md and CI.
7a 1st-winter: White with black eye patch and ear-coverts. Tertials have dark centres. Legs red.
7b 1st-winter flight: Pale grey above, with slightly darker primaries. No carpal bar. Rump white. Tail centre grey above.

121

PLATE 45 TERNS III AND AUKS

1 BLACK TERN *Chlidonias niger*   **L** 22–24 cm   Text p. 266
A small tern with mainly grey upperparts and a slightly forked tail. Scarce and irregular on passage and in winter in the CI. Vagrant to the Az and Md.
1a Ad breeding flight: Head and underparts black, except for white vent. Mantle and wings dark grey. Rump and tail paler grey-brown. Upperwings grey with darker outer primaries. Underwing-coverts pale.
1b Ad winter flight: Cap solid black. Rump grey. Dark breast-side patches distinctive.
1c Juv flight: Black cap extensive. Rump grey. Dark breast-side patches. Saddle lightly barred.

2 WHITE-WINGED (BLACK) TERN *Chlidonias leucopterus*   **L** 20–23 cm   Text p. 266
Slightly smaller and shorter-billed than Black Tern, with a slightly-notched tail. Vagrant to Md and the CI.
2a Ad breeding flight: Black body and underwing-coverts contrast with white upperwing-coverts and white rump and tail.
2b Ad winter flight: Similar to adult winter Black Tern but crown paler, rump whitish and lacks dark patches on breast-sides.
2c Juv flight: Saddle darker than in juvenile Black Tern. Rump whitish and lacks dark patches on breast-sides.

3 LITTLE AUK *Alle alle*   **L** 17–19 cm   Text p. 267
A tiny auk with a very short stubby bill. Probably regular in winter in small numbers in the Az. Vagrant to Md and the CI.
3a Ad winter: Black above. Head-sides, throat and breast white, with a partial dark breast-band. Scapulars streaked white.
3b Ad winter flight: Underwings dark. Inner secondaries white-tipped.

4 COMMON GUILLEMOT *Uria aalge*   **L** 38–43 cm   Text p. 266
A large auk, brownish above and white below. Vagrant to the CI.
4a Ad winter: Dark brownish above and white below. Head-sides mainly white with a dark line reaching back from and below eye.
4b Ad winter flight: Underwing-coverts and axillaries dusky and upper flanks often streaked brownish.

5 BRÜNNICH'S GUILLEMOT *Uria lomvia*   **L** 39–43 cm   Text p. 267
Thicker billed than Common Guillemot and blacker above. Vagrant to the Az.
5a Ad winter: Dark area on head-sides more extensive than in Common Guillemot and finishes well below eye. Pale gape stripe.
5b Ad winter flight: Flanks unstreaked. Underwing-coverts and axillaries white.

6 RAZORBILL *Alca torda*   **L** 37–39 cm   Text p. 267
Smaller than guillemots, with a thicker neck and deeper, stubby bill. Vagrant to the CI.
6a Ad winter: Black above and white below. Head sides pale, lacking dark line of Common Guillemot.
6b Ad winter flight: Underwing-coverts white contrasting more with black flight feathers than in guillemots.

7 (ATLANTIC) PUFFIN *Fratercula arctica*   **L** 26–30 cm   Text p. 267
A compact auk with a large conical bill. In winter regular in the Az and scarce and irregular in the CI. Vagrant to Md.
7a Ad winter: Mainly black above and white below. Cheeks dusky grey. Dark band across upper breast.
7b Ad winter flight: Underwings grey.

123

PLATE 46 SANDGROUSE AND DOVES

1 BLACK-BELLIED SANDGROUSE *Pterocles orientalis*   L 33–35 cm   Text p. 268
A large, broad-winged sandgrouse with a conspicuous black belly patch. Resident in the eastern CI.
1a Ad male: Head and breast grey. Throat orange with black centre. Upperparts spotted golden-brown.
1b Ad male flight: White underwing-coverts contrast with black belly and black undersides to flight feathers.
1c Ad female: Breast finely-spotted yellow-brown. Upperparts vermiculated.
1d Ad female flight: As male.

2 PIN-TAILED SANDGROUSE *Pterocles alchata*   L 31–39 cm   Text p. 268
Smaller and slimmer than Black-bellied Sandgrouse, with a long tail projection. Vagrant to the CI.
2a Ad male breeding: Mantle, nape, crown and neck band olive-green, spotted yellow on mantle. Breast reddish-brown, bordered black. Wing-coverts black-barred green. Black bib and eye stripe. Lower back, rump and tail heavily vermiculated.
2b Ad male breeding flight: Belly and underwing-coverts white, contrasting with black undersides to flight feathers.
2c Ad female: Heavily vermiculated above. Face and breast-band pale brown. Double black band on neck. White bars on wing-coverts.
2d Ad female flight: As male.

3 AFRICAN COLLARED DOVE *Streptopelia roseogrisea 'risoria'*   L 29–30 cm   Text p. 270
Smaller and paler than similar Eurasian Collared Dove. Introduced to the CI.
3 Ad: Plumage very similar to Eurasian Collared Dove but tail darker. Best distinguished by voice.

4 EURASIAN COLLARED DOVE *Streptopelia decaocto*   L 31–33 cm   Text p. 270
A sandy ring-necked dove. Recent colonist of the CI. Vagrant to Md.
4a Ad: Pale greyish-brown above with blackish-brown primaries. Head and underparts pinkish-grey with greyish belly and undertail. Black semi-collar with white borders on hindneck.
4b Ad flight: Sandy body contrasts with dark flight feathers and grey transverse band across wings. Tail with white corners above. Undertail darker with contrasting white tip.

5 (EUROPEAN) TURTLE DOVE *Streptopelia turtur*   L 26–28 cm   Text p. 271
A small, slim, fast-flying dove with a dark-centred, white-rimmed tail. Summer resident in Md and the CI. Vagrant to the Az and CV.
5a Ad: Scapulars and upperwing-coverts with dark with bright rufous fringes. Head grey. Pale below, tinged pinkish. Black-and-white striped neck patch.
5b Ad flight: Looks dark with contrasting white belly and white-rimmed dark tail..

6 LAUGHING DOVE *Streptopelia senegalensis*   L 25–27 cm   Text p. 271
Longer-tailed and shorter-winged than Turtle Dove, with more extensive white on tail corners. Resident in the CI (Fuerteventura).
6 Ad: Head, neck and breast distinctly vinous-pink. Belly and vent white. Unspotted reddish-brown above. Inner upperwings blue-grey, wing tips dark brown. Tail centre grey-brown above, the outer feathers with blackish bases and prominent white tips.

7 NAMAQUA DOVE *Oena capensis*   L 26–28 cm   Text p. 271
Tiny, with a very long graduated tail and rufous wing patches. Vagrant to the CI and CV.
7a Ad male: Forehead, throat and breast black. Bill red and yellow.
7b Ad male flight: Inner primaries have large rusty-brown patches.
7c Ad female: Lacks male's black head and breast markings. Bill duller.

125

PLATE 47   PIGEONS

1 TROCAZ PIGEON (LONG-TOED PIGEON) *Columba trocaz*   **L** 38–40 cm   Text p. 269
Similar to Bolle's Pigeon but paler. Endemic to Md.
1a Ad: Largely grey, with vinaceous breast. Silvery-grey neck patch.
1b Ad flight: Flight feathers and tail almost black, the tail with a broad, grey subterminal band.

2 BOLLE'S PIGEON *Columba bollii*   **L** 35–37 cm   Text p. 269
Large, with two-toned upperwings and a black terminal tail band. Endemic to Tenerife and the western CI.
2a Ad: Dark blue-grey with darker flight feathers. Tail dark, with a broad, pale grey subterminal band. Bill red with orange tip.
2b Ad flight: Grey upperwing coverts contrast with darker flight feathers and primary-coverts. Tail banded black-grey-black.

3 (WHITE-TAILED) LAUREL PIGEON *Columba junionae*   **L** 37–38 cm   Text p. 270
Large and dark, with a conspicuous white terminal tail band. Endemic to Tenerife and the western CI.
3a Ad: Upperwings sooty-brown. Underparts dark vinaceous. Bill off-white with reddish base.
3b Ad flight: Pale tail, with white terminal band except on central rectrices, contrasts with dark wings and body.

4 ROCK DOVE/FERAL PIGEON *Columba livia*   **L** 31–34 cm   Text p. 268
The most widespread pigeon in the region, common on all four archipelagos. Many show characters of the semi-domestic form (Feral Pigeon).
4a Ad: Wild type has head and neck dark grey, with metallic green neck patch. Mantle, upperwings and underparts paler grey. Dark band on tail tip. Two black bands across folded wings.
4b Ad flight: Black trailing edge to wing above, with second bar across secondaries. Tail grey with black terminal band. Rump white.

5 (COMMON) WOOD PIGEON *Columba palumbus*   **L** 40–42 cm   Text p. 268
Large, with distinct white upperwing crescents and white neck patches. Common in the central and eastern Az (*azorica*). Extinct in Md (*maderensis*). Vagrant to the CI (nominate).
5a Ad nominate: Grey above and vinous-pink below with white-and-green neck patches. Black band on tail tip.
5b Ad nominate flight: Transverse white crescents across upperwings, white neck patches and dark tail band conspicuous.
5c Ad *azorica*: Darker than nominate particularly on head and rump. Deeper vinous tones on breast.

127

PLATE 48 PARROTS, CUCKOOS AND NIGHTJARS

1 ROSE-RINGED PARAKEET *Psittacula krameri*   **L** 38–42 cm   Text p. 272

A medium sized, mainly green parrot with a long, graduated tail. Introduced and breeding in the CI. Also present in the Az, Md and the CV and may be becoming established there.
1   Ad male: Bill red. All green except for black bib and black and pinkish neck ring.

2 MONK PARAKEET *Myiopsitta monachus*   **L** 33 cm   Text p. 272

A medium-sized green parrot with a greyish face and breast. Introduced and breeding in the CI.
2   Ad: Green above. Yellow band on belly. Bill horn-coloured.

3 GREAT SPOTTED CUCKOO *Clamator glandarius*   **L** 38–40 cm   Text p. 273

A large, long tailed, crested cuckoo. Scarce on passage and in winter in the CI. Vagrant to Md and the CV.
3a   Ad: Crest silvery-grey. Grey-brown above, boldly spotted white. White edges to tail.
3b   Imm: Black hood with indistinct crest. Darker above than adult. Bright rusty-brown primaries obvious in flight.

4 COMMON CUCKOO *Cuculus canorus*   **L** 32–34 cm   Text p. 273

A falcon-like cuckoo with a long, wedge-shaped tail and barred underparts. Scarce on passage in the CI. Vagrant to the Az, Md and the CV.
4a   Ad male: Upperparts, head and breast grey. White below, barred darker.
4b   Imm brown morph: Head and breast with dark barring. Upperparts with fine pale bars.

5 BLACK-BILLED CUCKOO *Coccyzus erythrophthalmus*   **L** 27–31 cm   Text p. 271

A small slim cuckoo with a dark bill. Vagrant to the Az.
5   Ad: Uniform dull brownish above. Tail underside grey with narrow white tips.

6 YELLOW-BILLED CUCKOO *Coccyzus americanus*   **L** 28–32 cm   Text p. 271

A small slim cuckoo with a yellow-based bill. Vagrant to the Az.
6   Ad: Tail underside blackish with broad white tips. Rufous patch on primaries, most obvious in flight.

7 EUROPEAN NIGHTJAR *Caprimulgus europaeus*   **L** 26–28 cm   Text p. 277

A medium-sized, dark nightjar. Vagrant to the Az, Md and the CI.
7a   Ad: Cryptically patterned black, grey, brown and whitish. Nape grey.
7b   Ad male flight: White spots on primaries and tail corners. Lesser coverts blackish. Pale bar across upperwing-coverts.

8 RED-NECKED NIGHTJAR *Caprimulgus ruficollis*   **L** 30–32 cm   Text p. 277

A large nightjar with a rather large head. Vagrant to Md and the CI.
8a   Ad: Throat, upper breast and nape band rusty. Large white throat patches.
8b   Ad flight: Larger and longer tailed than European Nightjar. All wing-coverts with broad pale tips.

9 COMMON NIGHTHAWK *Chordeiles minor*   **L** 23–25 cm   Text p. 277

Wings more pointed than the two previous species. Tail forked. Vagrant to the Az and CI.
9a   Ad: Unbarred black primaries. Large white throat patch.
9b   Ad flight: White spots on primaries. Prominent pale mid-wing panel.

129

PLATE 49 OWLS

1 BARN OWL *Tyto alba* **L** 33–35 cm Text p. 274
A long-winged, long-legged owl with a pale, heart-shaped facial disc and dark eyes. Resident in the western and central CI (nominate), eastern CI (*gracilirostris*), Md (*schmitzi*) and the CV (*detorta*).
1a Ad nominate: Underparts and face almost pure white, except for a few dark spots on chest and flanks. Golden-buff above, finely streaked, mottled and vermiculated grey and brown.
1b Ad *gracilirostris*: Smaller and smaller-billed than nominate, more orange-buff and more heavily spotted below. Also, central facial disk more rufous and upperparts darker.
1c Ad *detorta*: Darker and browner above than *gracilirostris*, with a much larger and longer bill.

2 (EURASIAN) SCOPS OWL *Otus scops* **L** 19–20 cm Text p. 274
A very small owl with short ear tufts. Scarce and irregular on passage in the CI. Occasional in Md. Vagrant to the CV.
2 Ad grey morph: Greyish, with a pale shoulder line and slightly paler face and underparts. Intricately patterned with black streaks, white spots and dark vermiculations.

3 SNOWY OWL *Bubo scandiacus* **L** 53–66 cm Text p. 275
A very large, predominantly white owl. Extreme vagrant to the Az.
3 Ad female: White, heavily barred and spotted brown.

4 (NORTHERN) HAWK OWL *Surnia ulula* **L** 36–39 cm Text p. 275
Medium-sized with a long, graduated tail and rather large, flat-topped head. Vagrant to the CI.
4 Ad: Dark grey-brown above, spotted white. Whitish below barred brown. Facial disk mostly white with thick blackish borders on foreneck.

5 TAWNY OWL *Strix aluco* **L** 37–39 cm Text p. 275
A compact, medium-sized owl with a large, rounded head. Vagrant to the CI.
5 Ad: Rufous-brown to greyish-brown above, paler below, streaked, mottled and vermiculated darker. Two pale false eyebrows. Pale line on shoulder. Pale tips to secondary-coverts create an indistinct wingbar.

6 LONG-EARED OWL *Asio otus* **L** 35–37 cm Text p. 276
A slim, medium-sized owl with long ear-tufts. Resident in the Az (nominate) and CI (*canariensis*).
6a Ad nominate: Facial disc pale ochraceous-tawny with dark blackish rim. Rich buffy-brown above, with dark streaking, spotting and mottling, and row of whitish spots on scapulars. Paler below, streaked darker. Eyes yellowish-orange.
6b Ad nominate flight: Upperwing has large yellowish-buff patch above at base of primaries and 4–5 narrow black bands on primary tips and secondaries. Tail with 6–8 narrow bars.
6c Ad *canariensis*: Distinctly smaller and darker than nominate, and has darker reddish-orange eyes.

7 SHORT-EARED OWL *Asio flammeus* **L** 37–39 cm Text p. 276
Similar to Long-eared Owl but paler and with very small ear-tufts. Scarce and irregular in winter in the CI but considered frequent in Md. Vagrant to the Az and CV.
7a Ad: Yellow-brown to pale ochraceous-buff above, heavily streaked and spotted dusky. Pale yellow-tawny to ochre-whitish below, distinctly streaked brown.
7b Ad flight: White trailing edge to inner primaries and secondaries. Wingtip almost all black. Tail with 4–5 narrow bars.

8 MARSH OWL *Asio capensis* **L** 35–37 cm Text p. 276
Similar to Short-eared Owl but much darker. Vagrant to the CI.
8a Ad: Breast, upperparts and upperwing-coverts uniform dark brown. Eyes dark. Lower breast and belly pale, finely barred darker.
8b Ad flight: Very dark. Pale orange-brown patches on upsides of primaries.

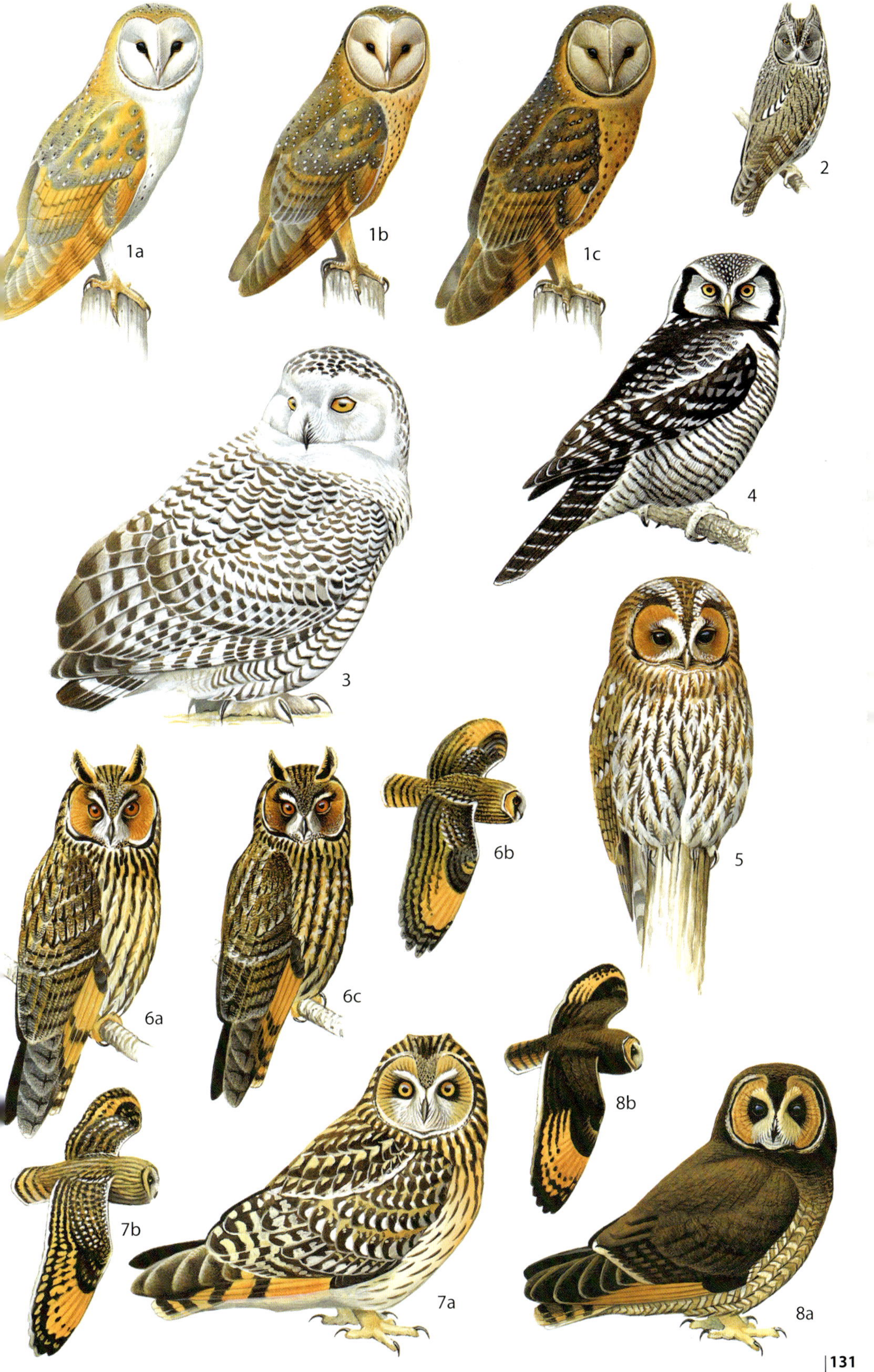
1a
1b
1c
2
3
4
5
6a
6b
6c
7a
7b
8a
8b

PLATE 50 SWIFTS

1 CHIMNEY SWIFT *Chaetura pelagica* **WS** 31–32 cm Text p. 277
A small swift with a short cigar-shaped body. Vagrant to the Az and CI.
1   **Ad below:** All dark with a slightly paler throat. Very short, square-ended tail with fine tail spines.

2 CAPE VERDE SWIFT *Apus alexandri* **WS** 34–35 cm Text p. 278
A small all-dark swift with a distinctive buzzing call. Endemic to the CV.
2   **Ad below:** Dark grey-brown. Paler than Plain Swift, with a more conspicuous pale throat patch.

3 PLAIN SWIFT *Apus unicolor* **WS** 38–39 cm Text p. 278
A medium-sized all dark swift. Slightly smaller, slimmer and narrower winged than Common Swift. Endemic breeding species in Md and the CI. Vagrant to the CV.
3   **Ad below:** All dark. When throat patch is visible it is small, greyish and indistinct.

4 COMMON SWIFT *Apus apus* **WS** 42–48 cm Text p. 278
A medium-sized all dark swift with a forked tail and long, sickle-shaped wings. Common on passage and scarce breeder in the CI. Occasional on passage in Md and regular but uncommon on passage and in winter in the CV. Vagrant to the Az.
4a   **Ad below:** All dark sooty-brown with a small, conspicuous white throat patch.
4b   **Ad above:** All dark blackish. Wingtip pointed. Head smaller than in Pallid Swift.

5 PALLID SWIFT *Apus pallidus* **WS** 42–46 cm Text p. 276
Resembles Common Swift but grey-brown with a paler panel on the secondaries. Common on passage and scarce breeder in the CI. Scarce breeder in Md. Vagrant to the CV.
5a   **Ad below:** Wings blunt tipped. Head broad with a large white throat patch.
5b   **Ad above:** Dark brown with slightly paler secondaries.

6 ALPINE SWIFT *Apus melba* **WS** 54–60 cm Text p. 279
A large, long-winged brown swift with a white belly. Regular but ucommon on passage in the CI and occasional on passage in Md. Vagrant to the CV.
6   **Ad below:** White belly and white chin separated by brown breast-band. Underwing-coverts darkest brown.

7 LITTLE SWIFT *Apus affinis* **WS** 34–35 cm Text p. 279
A small swift with a short, square-ended tail. Outer wing rather broad. Throat white. Vagrant to Md and the CI.
7   **Ad above:** Broad white rump patch extends onto flanks.

8 WHITE-RUMPED SWIFT *Apus caffer* **WS** 34–36 cm Text p. 279
Smaller than Common Swift with a slim rear end and rather narrow pointed wings. Tail deeply forked. Vagrant to Md and the CI.
8   **Ad above:** Tail often held closed so appears long and pointed. Narrow white rump patch. Pale-tipped secondaries.

133

PLATE 51   KINGFISHERS, BEE-EATERS, ROLLERS, HOOPOE AND WOODPECKERS

1 GREY-HEADED KINGFISHER *Halcyon leucocephala*   **L** 21–22 cm   Text p. 279
A medium-sized kingfisher with a large red bill. Resident on Santiago, Brava and Fogo in the CV.
1   Ad: Head and breast pale grey to white. Belly and vent chestnut-orange. Back black. Predominantly blue wings and tail. White patches at base of primaries visible in flight.

2 COMMON KINGFISHER *Alcedo atthis*   **L** 16–17 cm   Text p. 280
A small kingfisher with a black bill. Vagrant to Md and the CI.
2   Ad: Crown, nape and wings metallic greenish-blue. Mantle, rump and tail brilliant shining blue. Orange-chestnut below and on ear patch. White chin and rear neck-sides.

3 BELTED KINGFISHER *Ceryle alcyon*   **L** 28–35 cm   Text p. 280
A large kingfisher, blue-grey above and largely white below, with a white collar. Vagrant to the Az.
3a Ad male: Single grey breast band.
3b Ad female: Two breast bands: upper band grey and lower chestnut and extending down flanks.

4 BLUE-CHEEKED BEE-EATER *Merops persicus*   **L** 27–31 cm   Text p. 280
Largely green, with much longer central tail feathers than European Bee-eater. Vagrant to the CI and CV.
4   Ad: Throat red-brown with little yellow on chin. Narrow black face mask bordered bluish-white above and below.

5 EUROPEAN BEE-EATER *Merops apiaster*   **L** 27–29 cm   Text p. 280
Very colourful, with elongated central tail feathers. Regular on passage in the CI. Scarce on passage in Md. Vagrant to the Az and CV.
5   Ad: Crown, nape and inner wing chestnut, yellowish-white shoulder patches. Chin and throat yellow, bordered black below. Rest of underparts greenish-blue.

6 EUROPEAN ROLLER *Coracias garrulus*   **L** 30–32 cm   Text p. 281
Large and predominantly blue. Rare and irregular on passage in the CI and occasional in Md. Vagrant to the Az and CV.
6   Ad: Head, underparts and wing panel turquoise. Flight feathers and tail darker blue-black. Mantle chestnut-brown.

7 BROAD-BILLED ROLLER *Eurystomus glaucurus*   **L** 29–30 cm   Text p. 281
A dark rufous-chestnut bird with a short, broad bill. Vagrant to the CV.
7   Ad: Mantle, crown, nape, face and upperwing-coverts dark rufous-chestnut. Flight feathers dark blue. Deep lilac below. Tail, rump and undertail-coverts blue. Bill yellow.

8 HOOPOE *Upupa epops*   **L** 26–28 cm   Text p. 281
An unmistakable pinkish, black and white bird with an erectile crest, long bill and floppy flight on broad wings. Common resident and passage migrant in the CI. Locally resident in Md (Porto Santo). Vagrant to the Az and CV.
8   Ad: Head, neck, upper mantle and underparts pinkish-brown. Rest of upperparts, wings and tail patterned black and white.

9 (EURASIAN) WRYNECK *Jynx torquilla*   **L** 16–17 cm   Text p. 281
Small, dark and long-tailed, with mottled and barred greyish-brown plumage. Scarce and irregular on passage in the CI. Vagrant to Md.
9   Ad: Generally greyish, mottled brown and buff, with a conspicuous dark patch on mantle. Throat and upper breast barred, with arrowhead marks on lower breast and flanks.

10 GREAT SPOTTED WOODPECKER *Dendrocopos major*   **L** 22–23 cm   Text p. 282
A pied woodpecker with large white shoulder patches. Resident in the CI in Tenerife (*canariensis*) and Gran Canaria (*thanneri*).
10a Ad male *canariensis*: Prominently pied above and off-white below, with black breast-side patches. Belly dark brown or grey-brown. Vent orangey-red.
10b Ad male *thanneri*: Paler below than *canariensi*.

135

PLATE 52 LARKS

1 BLACK-CROWNED SPARROW-LARK *Eremopterix nigriceps* L 10–11 cm Text p. 282
Small and finch-like, with a short, conical bill. Resident in the CV.
1a Ad male: Black below and sandy above. Pied head pattern.
1b Ad female: Sandy plumage with some indistinct streaking on crown, nape and breast. Dark centres to median-coverts.

2 BAR-TAILED (DESERT) LARK *Ammomanes cincturus* L 14–15 cm Text p. 282
Small and sandy with a striking tail pattern. Resident in the CV. Vagrant to the CI.
2a Ad: Completely unstreaked lark, Dark primary tips contrast with pale rufous tertials.
2b Ad tail: Pale red-brown, with a clear-cut, blackish terminal bar.

3 (GREATER) HOOPOE LARK *Alaemon alaudipes* L 18–20 cm Text p. 283
A large lark with long legs, long tail and a long, slightly decurved bill. Resident in the CV. Vagrant to the CI.
3a Ad rufous morph: Sandy above with grey nape. Prominent dark spots on breast. Pale supercilium, dark eye stripe and dark moustachial stripe.
3b Ad flight: Broad wings strikingly marked black-and-white.

4 DUPONT'S LARK *Chersophilus duponti* L 17–18 cm Text p. 281
A very secretive grey-brown lark with a long, slightly decurved bill. Vagrant to the CI.
4a Ad: Upperparts and breast heavily streaked, otherwise white below. Faint crown-stripe.
4b Ad flight: Plain grey-brown above. Tail relatively short.

5 CALANDRA LARK *Melanocorypha calandra* L 18–19 cm Text p. 281
A large, stocky lark with a heavy yellowish-brown bill and dark breast-side patches. Vagrant to Md and the CI.
5 Ad: Grey-brown above with dark streaks. Mainly white below. Wings have white trailing edges and blackish undersides. White outer tail feathers

6 (GREATER) SHORT-TOED LARK *Calandrella brachydactyla* L 13–14 cm Text p. 284
A small, short-billed lark with dark breast-side patches. Regular on passage in the CI. Vagrant to Md and the CV.
6 Ad: Sandy. Unstreaked below. Little or no primary projection.

7 LESSER SHORT-TOED LARK *Calandrella rufescens* L 13–14 cm Text p. 284
A small greyish lark with a streaked breast and a short bill. Resident in the CI.
7 Ad: Greyish-brown above with darker streaking. White below with small dark streaks on breast. Long primary projection.

8 CRESTED LARK *Galerida cristata* L 17–19 cm Text p. 284
Greyish, with a long, straight bill and an erectile crest. Vagrant to the Az and CI.
8 Ad: Greyish-brown above with dark-streaked mantle. Buffish-white below with dark breast and flank streaks.

9 THEKLA LARK *Galerida theklae* L 15–17 cm Text p. 285
Smaller and greyer than Crested Lark, with a shorter bill and less spiky, fuller crest. Vagrant to the CI.
9 Ad: More finely streaked below than Crested Lark. Rusty uppertail-coverts contrast with grey rump.

10 (EURASIAN) SKYLARK *Alauda arvensis* L 17–18 cm Text p. 285
Medium-sized with a longish tail and a short crest. Regular in winter in CI and Md. Vagrant to the Az.
10 Ad: Brownish above, heavily streaked darker. Buff on breast and white on belly, breast and flanks streaked. White outer tail feathers. Whitish trailing edge to secondaries.

11 RASO LARK *Alauda razae* L 12–13 cm Text p. 285
Smaller, shorter-winged, shorter-tailed and longer-billed than Skylark. Endemic to Raso in the CV.
11a Ad male: Bill large, stout and strongly curved.
11b Ad female head: Bill smaller than in male.

137

PLATE 53 SWALLOWS AND MARTINS

1 PURPLE MARTIN *Progne subis*   L 19–20 cm   Text p. 285
A very large, dark hirundine. Vagrant to the Az.
1a Ad male: Uniform glossy violet-blue.
1b Ad female: Duller than male, with a grey forehead and grey collar. Smudgy grey-brown below.
1c Juv: Dull with pale underparts. Whitish belly contrasts with greyish throat.
1d Ad male flight above: Uniformly dark with shallow-forked tail.

2 PLAIN MARTIN *Riparia paludicola*   L 10.5–11.5 cm   Text p. 286
The smallest hirundine of the region, smaller and paler than Sand Martin. Vagrant to the CI and CV.
2a Ad flight above: Grey-brown with paler rump.
2b Ad flight below: White below but chin dusky. No breast-band. Underwings plain brownish.

3 SAND MARTIN *Riparia riparia*   L 12–13 cm   Text p. 286
A small hirundine with a forked tail. Regular on passage in the CI. Vagrant to the Az, Md and CV.
3a Ad flight above: Uniform grey-brown.
3b Ad flight below: White throat and belly separated by grey-brown breast-band. Underwings dark grey-brown with contrastingly darker underwing-coverts.

4 (EURASIAN) CRAG MARTIN *Ptyonoprogne rupestris*   L 14–15 cm   Text p. 286
A large, grey-brown martin with a shallow-forked tail.
4a Ad flight above: Grey with prominent white tail spots.
4b Ad flight below: Greyish below, paler on breast and belly. Dusky, faintly streaked throat. Darker vent and very dark underwing coverts.

5 BARN SWALLOW *Hirundo rustica*   L 17–19 cm   Text p. 286
A slim swallow with long, slender tail streamers. Common on passage and occasional in winter in the CI (has bred) and also regular on passage and in winter in the CV. Vagrant to the Az. Occasional in Md.
5a Ad flight above: Blue-black with darker flight feathers. White spots on rectrices.
5b Ad flight below: White with darker flight feathers, dark blue breast-band and deep red forehead and throat.

6 RED-RUMPED SWALLOW *Cecropis daurica*   L 16–17 cm   Text p. 287
Blunter-winged than Barn Swallow with thicker tail streamers. Scarce on passage in the CI. Vagrant to the Az, Md and CV.
6a Ad flight above: Blue-black with browner wings and tail. Rump pale rust-coloured. Nape dark rusty-brown.
6b Ad flight below: Pale buff with narrow dark streaking. Square-cut black undertail coverts.

7 (COMMON) HOUSE MARTIN *Delichon urbicum*   L 13.5–15 cm   Text p. 287
Smaller and more compact than Barn Swallow with a short, forked tail. Common on passage in the CI. Occasional in Md. Scarce on passage and in winter in the CV. Vagrant to the Az.
7a Ad flight above: Glossy blue-black, with a contrasting white rump.
7b Ad flight below: White, with darker flight feathers and tail.

139

PLATE 54 PIPITS

1 RICHARD'S PIPIT *Anthus richardi*   L 18 cm   Text p. 288
A large pipit with a very long hindclaw and strong bill. Lores pale. Vagrant to the CI.
1a Ad: Brown above, heavily streaked dark. Breast and flanks buff, also streaked dark. Wing-coverts broadly fringed buff.
1b Juv: Darker above than adult. Median and greater coverts, and tertials, thinly-fringed whitish.

2 TAWNY PIPIT *Anthus campestris*   L 17 cm   Text p. 288
Similar to Richard's Pipit but paler and with a short hindclaw. Lores dark. Scarce but regular on passage in the CI. Vagrant to Md and the CV.
2a Ad: Greyish-sandy above with some indistinct streaking. Buffish-white below, with a few streaks on breast-sides. Median coverts contrastingly dark with broad buff-white tips when fresh.
2b Juv: Upperparts and upper breast streaked.

3 BERTHELOT'S PIPIT *Anthus berthelotii*   L 14 cm   Text p. 288
Similar in size to Meadow Pipit but with a dirty-white supercilium. Common resident in Md and the CI.
3 Ad: Dull greyish and almost unstreaked above, lacking olive tones. Dirty white below with narrow, clear, dark streaking on breast. Two pale wingbars.

4 OLIVE–BACKED PIPIT *Anthus hodgsoni*   L 14.5 cm   Text p. 288
Recalls Tree Pipit but more olive above and mantle almost unstreaked. Vagrant to the CI.
4 Ad: Crown less heavily streaked in centre than Tree Pipit but bordered by a heavy dark stripe. Supercilium broad, buffy in front of eye and creamy-white above and behind. Prominent pale spot above a distinct dark spot on rear ear-coverts.

5 TREE PIPIT *Anthus trivialis*   L 15 cm   Text p. 289
Slightly larger and less strongly-marked than Meadow Pipit. Regular on passage and occasional in winter in the CI. Vagrant to Md and the CV.
5 Ad: Olive-brown above with streaked mantle. Rump unstreaked. White below, strongly washed buff on breast and flanks, which are finely streaked dark.

6 MEADOW PIPIT *Anthus pratensis*   L 14.5 cm   Text p. 289
A small, olive-toned pipit with an unstreaked rump. Scarce but regular in winter in the CI. Vagrant to the Az and Md.
6 Ad: Olive-green to greyish above, boldly streaked darker. Off-white to buff below. Streaking on flanks more prominent than in Tree and Berthelot's Pipits, and supercilium less distinct.

7 RED-THROATED PIPIT *Anthus cervinus*   L 15 cm   Text p. 289
Similar to Meadow Pipit in size but very heavily streaked black above and below, including uppertail-coverts and rump. Mainly irregular on passage in the CI but winters locally. Vagrant to the Az and Md.
7a Ad summer: Throat, and sometimes breast, red.
7b 1st winter: Streaks prominent. White stripes on mantle and two white wingbars. Pronounced malar stripe.

8 ROCK PIPIT *Anthus petrosus*   L 16.5–17cm   Text p. 290
A large, dark pipit with a long, dark bill and dark legs. Vagrant to the CI.
8 Ad: Dark dusky-olive above, appearing almost unstreaked. Dirty buffish-white below with long, broad smudgy streaks extending to rear flanks. Outer tail feathers greyish. Supercilium short and indistinct.

9 WATER PIPIT *Anthus spinoletta*   L 16.5–17 cm   Text p. 290
Paler than Rock Pipit, with white outer tail feathers. Legs dark. Vagrant to the CI.
9a Ad summer: Mantle almost unstreaked brownish-grey. Head and nape ash-grey. White supercilium. Underparts whitish, washed pink on throat and breast. Breast-sides and upper flanks streaked indistinctly.
9b Ad winter: Brownish above with two white wingbars. White supercilium. Whitish below heavily streaked dark.

141

PLATE 55 WAGTAILS, DIPPER, CATBIRD AND ACCENTORS

1 YELLOW WAGTAIL *Motacilla flava*   **L** 17 cm   Text p. 290

A relatively short-tailed wagtail. Several subspecies are regular on passage in the CI. Vagrant to Md and the Az.
1a Ad male *flava* breeding: Greenish above and yellow below. Head blue-grey with white supercilium, white submoustachial stripe and yellow throat.
1b Ad female *flava*: Much duller than male, with white throat.
1c Ad male *iberiae* breeding head: Crown and nape dark ashy-grey. Lores and ear-coverts blackish. White throat and supercilium.
1d Ad male *thunbergi* breeding head: Crown and nape grey. Ear-coverts blackish. Throat yellow.
1e Ad male *flavissima* breeding head: Mainly yellow. Crown, nape and ear-coverts greenish.
1f Ad male *feldegg* breeding: Head, ear-coverts and nape black.
1g Ad female *feldegg* head: Head, ear-coverts and nape dark grey.

2 CITRINE WAGTAIL *Motacilla citreola*   **L** 17 cm   Text p. 291

Longer-tailed than Yellow Wagtail. All plumages have two broad white wing bars. Vagrant to the CI.
2a Ad male breeding nominate: Head and underparts yellow. Black lower nape and neck-sides. Mantle grey.
2b Ad female head: Crown, nape and ear-coverts dirty greyish. Ear-coverts bordered yellow.
2c Juv head: Grey crown and nape. Pale lores and pale border to ear-coverts.

3 WHITE WAGTAIL *Motacilla alba*   **L** 18 cm   Text p. 291

Pied, with a black bib. Regular but scarce in winter in the CI and Md. Vagrant to the Az and CV.
3 Ad non-breeding nominate: Grey above and white below. Cap black. Forehead, ear-coverts, cheeks and throat white.

4 GREY WAGTAIL *Motacilla cinerea*   **L** 18–19 cm   Text p. 291

Long-tailed with a yellow vent. Resident in the CI (*canariensis*), Az (*patriciae*) and Md (*schmitzi*). The subspecies on Az and Md are very similar to *canariensis*; *schmitzi* is darker above with less white on the tail feathers, while *patriciae* has a slightly longer bill. Some nominate birds may also winter in the CI.
4a Ad male breeding nominate: Grey above and yellow below. White supercilium and moustachial stripe. Throat black.
4b Ad female nominate: Throat paler or white. Whiter below than male.
4c Ad male breeding *canariensis*: Underparts deep yellow and facial markings more contrasting.
4d Ad female *canariensis*: Resembles nominate female.

5 (WHITE-THROATED) DIPPER *Cinclus cinclus*   **L** 18 cm   Text p. 292

Dumpy and short-tailed. Usually associated with running water. Vagrant to the CI.
5 Ad: Adult mainly black-brown with contrasting white throat and breast.

6 GREY CATBIRD *Dumetella carolinensis*   **L** 21 cm   Text p. 292

Long tailed and largely slate-grey. Vagrant to the CI.
6 Ad: Slate-grey except for black cap and tail, and chestnut undertail-coverts

7 DUNNOCK *Prunella modularis*   **L** 14.5 cm   Text p. 292

Recalls a sparrow but with a fine bill. Vagrant to the CI.
7 Ad: Head and breast slate grey, browner on crown and ear-coverts. Otherwise mostly rufous-brown. Mantle heavily streaked and flanks finely streaked.

8 ALPINE ACCENTOR *Prunella collaris*   **L** 18 cm   Text p. 292

Larger and more colourful than Dunnock. Vagrant to the CI.
8 Ad: Head, breast and belly greyish. White throat barred black. Flanks streaked brownish-orange. Brownish-grey above with heavy streaking on mantle. Two white wingbars.

143

PLATE 56 WREN, ROBINS, NIGHTINGALES AND REDSTARTS

1 (WINTER) **WREN** *Troglodytes troglodytes*   **L** 9–10 cm   Text p. 292

Very small, with a long, fine bill and a tiny, often-cocked tail. Vagrant to Md and the CI.
1   Ad: Dull rufous-brown above, with pale off-white supercilium. Paler buff-brown below. Black vermiculations on back, rump, wings, tail, flanks and vent.

2 RUFOUS BUSH ROBIN *Cercotrichas galactotes*   **L** 15 cm   Text p. 293

Rufous above and pale below with a long rufous tail which is often cocked. Vagrant to Md and the CI.
2   Ad: Sandy-brown above. Off-white below with a buff wash on breast. Prominent facial pattern. Orange-brown rump and tail, the latter broadly tipped with white spots and subterminal black bar.

3 (EUROPEAN) **ROBIN** *Erithacus rubecula*   **L** 14 cm   Text p. 293

Orange face, throat and breast distinctive. Nominate is resident in the Az, Md and western CI and also occurs in winter in the eastern CI; *superbus* is resident in the central CI (Tenerife and Gran Canaria).
3a Ad nominate: Upperparts olive-brown, separated from orange bib by a narrow blue-grey band. Wings and tail also olive-brown.
3b Ad *superbus*: Darker brown above than nominate. Bib redder and rest of underparts whiter. Obvious white eye-ring. Blue-grey bib margin wider.

4 THRUSH NIGHTINGALE *Luscinia luscinia*   **L** 16.5 cm   Text p. 293

Duskier than Common Nightingale. Vagrant to the CI.
4   Ad: Greyer brown above than Common Nightingale, with a darker brown tail. Underparts diffusely mottled grey-brown on breast and flanks. Dusky malar stripe.

5 COMMON NIGHTINGALE *Luscinia megarhynchos*   **L** 16.5 cm   Text p. 294

Plain-looking with a longish rufous tail. Rare but regular on passage in the CI. Vagrant to Md and the CV.
5   Ad: Warmer brown above with a contrasting orange-brown rump and tail. Whitish below, washed pale buff on breast.

6 BLUETHROAT *Luscinia svecica*   **L** 14 cm   Text p. 294

Recalls Robin but has white supercilium and red tail patches. Vagrant to the CI.
6a Ad male nominate: Blue bib bordered below with black, white and rufous. Bib centre red.
6b Ad male *cyanecula*: As nominate but bib centre white.
6c Ad female nominate: Bib often plain white, with black streaking on breast. Black malar stripe.
6d Ad female *cyanecula*: As nominate.
6e Ad behind: Tail blackish with rufous base.

7 BLACK REDSTART *Phoenicurus ochruros*   **L** 14.5 cm   Text p. 294

A dark greyish redstart. Regular on passage and in winter in the CI. Vagrant to the Az, Md and CV.
7a Ad male: Dark greyish above and below, except greyish-white central belly and orange-buff undertail. Face blacker. Prominent white wing panel. Rusty-red tail.
7b Ad female: More uniform smoky-grey than male and lacks darker face and wing panel.

8 COMMON REDSTART *Phoenicurus phoenicurus*   **L** 14 cm   Text p. 295

Much paler than Black Redstart, especially below. Occurs regularly on passage in the CI and occasionally in Md. Vagrant to the Az and CV.
8a Ad male: Face black with white forehead. Crown and back ash-grey. Breast orange. Rump and tail rufous.
8b Ad male flight: Note bright orange of breast and tail.
8c Ad female: Brown above with a narrow white eye-ring. Buffy-white below with an orange-washed breast.

145

PLATE 57 CHATS AND WHEATEARS

1 WHINCHAT *Saxicola rubetra*   **L** 12.5 cm   Text p. 295

Paler than other chats, with a broad, pale supercilium. Regular on passage in the CI. Vagrant to the Az, Md and CV.
1a Ad male: Head-sides blackish. Supercilium and malar stripe white. Orange-buff on throat and breast with white belly and vent. Dark brown above with darker feather centres. Tail dark with white sides.
1b Ad fem: Duller with browner cheeks and less contrasting head pattern. Paler and less orange below.

2 CANARY ISLANDS STONECHAT *Saxicola dacotiae*   **L** 12.5 cm   Text p. 295

Recalls Stonechat but has a pale supercilium. Endemic to Fuerteventura.
2a Ad male: Dark brown above with blacker cheeks and streaked mantle. Narrow white supercilium, white throat and half-collar. Orange-buff patch on breast, pale buff flanks and white belly. Tail all-dark.
2b Ad female: Rather nondescript with only a faint buffish wash on breast. Paler above than male with less distinct supercilium and no white half-collar.

3 COMMON STONECHAT *Saxicola torquatus*   **L** 12.5 cm   Text p. 296

Dark-headed, with orange underparts. Scarce and irregular in winter in the CI. Vagrant to Md.
3a Ad male: Head blackish with broad white half-collar. Extensive orange below. Brownish-black above with small whitish rump and all-dark tail.
3b Ad female: Duller with a brownish head and throat, and smaller half-collar.

4 ISABELLINE WHEATEAR *Oenanthe isabellina*   **L** 16.5 cm   Text p. 296

A large, long-legged, rather featureless pale wheatear. Vagrant to Md and the CI.
4a Ad: Sandy-brown above, white supercilium and dark lores. Whitish below washed sandy on breast and flanks. Darker alula contrasts with paler wing.
4b Ad tail: Broad black terminal band and white base.

5 NORTHERN WHEATEAR *Oenanthe oenanthe*   **L** 14.5–15.5 cm   Text p. 296

Predominantly greyish above. Occurs in passage in the CI especially but also in the Az and Md. Vagrant in winter to the Az. Possibly regular in winter in the CV.
5a Ad male nominate: Bluish-grey above with a white rump. White supercilium. Mask and wings black.
5b Ad male *leucorhoa*: Larger than nominate with larger bill, longer legs and darker, richer, more uniform buff underparts.
5c Ad female nominate: Duller and browner than male.
5d Ad female *leucorhoa*: Resembles a larger, buffish nominate female.
5e Ad tail: White except for black central rectrices and tip, forming a T pattern.

6 BLACK-EARED WHEATEAR *Oenanthe hispanica*   **L** 14.5 cm   Text p. 297

Slimmer than Northern Wheatear and brown or buffish above. Scarce and irregular on passage in the CI.
6a Ad male breeding pale-throated morph: Resembles male Northern but crown, mantle and breast deep buff. Throat white.
6b Ad male breeding dark-throated morph, head: Throat black.
6c Ad female dark-throated morph: Resembles male but duller. Dark throat and face feathers pale-tipped.
6d Ad tail: Whiter than in other wheatears, white almost reaching tip along the sides.

7 DESERT WHEATEAR *Oenanthe deserti*   **L** 14–15 cm   Text p. 297

Tail almost entirely black. Vagrant to Md and the CI.
7a Ad male: Recalls black-throated male Black-eared Wheatear but black on throat joins that on wings.
7b Ad female: Paler than female Northern Wheatear.
7c Ad tail: Almost entirely black except for bases of outer rectrices. Rump white.

8 MOURNING WHEATEAR *Oenanthe lugens*   **L** 14.5 cm   Text p. 297

A dark-winged wheatear. Vagrant to the CI.
8a Ad male: White forehead, crown and nape contrast with black throat, head-sides, mantle and wings. White below with buffish undertail-coverts.
8b Ad female: Crown, upperparts and lesser wing-coverts sandy-grey, contrasting with blackish face and throat. White below.
8c Ad tail: Similar to Northern Wheatear.

147

PLATE 58   THRUSHES

1 (RUFOUS-TAILED) ROCK THRUSH *Monticola saxatilis*   **L** 18.5 cm   Text p. 298
A small, red-tailed thrush. Vagrant to Md and the CI.
1a Ad male: Head, mantle, scapulars and rump blue-grey. Mid-back white. Bright orange-chestnut below. Orange-rufous uppertail-coverts and tail, with exception of brown central rectrices.
1b Ad female: Brownish, intensely mottled, scaled and barred black, brown and buff. Orange-rufous uppertail-coverts and tail

2 BLUE ROCK THRUSH *Monticola solitarius*   **L** 22 cm   Text p. 298
Darker than Rock Thrush and has dark tail. Vagrant to the CI.
2a Ad male: Uniform dark slaty-blue with darker, blacker wings and tail.
2b Ad female: Grey-brown above. Paler below, distinctly scaled buff or buff-white.

3 WOOD THRUSH *Hylocichla mustelina*   **L** 19 cm   Text p. 298
A small thrush resembling a rufous Song Thrush. Vagrant to the Az and Md.
3 Ad: Crown and nape rusty. Brown above. White below with large black spotting on throat-sides, breast, fore-belly and flanks. Narrow but obvious white eye-ring.

4 GREY-CHEEKED THRUSH *Catharus minimus*   **L** 18.5 cm   Text p. 298
A small, dark thrush. Vagrant to the Az.
4 Ad: Dull grey-brown above. Greyish-white below. Small, dark spots on throat and breast. Face distinctly greyish with a narrow and indistinct pale eye-ring.

5 RING OUZEL *Turdus torquatus*   **L** 23–24 cm   Text p. 298
A large black thrush with a pale breast crescent. Vagrant to Md and the CI (may winter Fuerteventura).
5a Ad male: Black with a mostly yellow bill. Breast crescent white. Pale fringes to flight feathers
5b Ad female: Browner with more distinct scaling on underparts. Breast crescent smaller and less distinct. Bill duller.

6 (COMMON) BLACKBIRD *Turdus merula*   **L** 24–25 cm   Text p. 299
A uniformly dark thrush. Common resident in the Az (*azorensis*) and in Md and the CI (*cabrerae*).
6a Ad male *cabrerae*: Black with yellow bill and eye-ring. Smaller and blacker than nominate.
6b Ad male *azorensis*: As *cabrerae*.
6c Ad female *cabrerae*: Dark blackish-brown, darker than *azorensis* and with limited pale area on throat.
6d Ad female *azorensis*: As *cabrerae* but less dark and with larger pale area on throat.

7 DARK-THROATED THRUSH *Turdus ruficollis*   **L** 24 cm   Text p. 299
A strikingly plumaged dark-throated thrush. Vagrant to Md.
7 Ad male *atrogularis*: Brownish-grey above. Face, throat and breast black. Otherwise white below with slight streaking on flanks.

8 FIELDFARE *Turdus pilaris*   **L** 25.5 cm   Text p. 299
A large, strikingly-patterned thrush. Vagrant to the Az. Very scarce in winter in Md and the CI.
8 Ad: Head and rump grey. Mantle and wing-coverts dark chestnut. Tail black. Warm buff on throat, breast and flanks. Black spots on throat and breast and black chevrons on flanks.

9 REDWING *Turdus iliacus*   **L** 21 cm   Text p. 300
Small and dark with red flanks and underwing coverts. Vagrant to the Az, very scarce in winter in Md and the CI.
9 Ad: Brown above with prominent white supercilium. Mainly white below. Dark underpart streaks.

10 MISTLE THRUSH *Turdus viscivorus*   **L** 27 cm   Text p. 300
A large spotted thrush. May breed Md. Vagrant to the Az and CI.
10 Ad: Grey brown above and white below, with irregular black spotting. Tail has white corners.

11 SONG THRUSH *Turdus philomelos*   **L** 22 cm   Text p. 300
A small spotted thrush. Regular in winter in the CI. Vagrant to the Az, Md and CV.
11 Ad: Grey-brown above. Breast and flanks buffish with black-brown 'arrowhead' spotting.

149

PLATE 59 WARBLERS I

1 ZITTING CISTICOLA *Cisticola juncidis* L 10 cm Text p. 301
Tiny, with a short, rounded tail. Vagrant to the CI and CV.
1 Ad: Pale brown above with heavy dark streaking on crown, nape and mantle. Buffish-white below. Tail brown with prominent black-and-white tips to all but the central rectrices.

2 (COMMON) GRASSHOPPER WARBLER *Locustella naevia* L 12.5 cm Text p. 301
A very secretive, streaked and spotted warbler. Vagrant to Md, the CI and CV.
2 Ad: Olive-brown above with darker spotting and streaking from crown to back. Rump olive-brown with faint streaks. Tail warmer brown, faintly barred. Pale below with slight streaking on breast-sides and undertail-coverts.

3 RIVER WARBLER *Locustella fluviatilis* L 13 cm Text p. 301
Brown, with prominent pale tips to the long undertail-coverts. Vagrant to Md.
3 Ad: Unmarked dark brown above with a short yellowish-buff supercilium. Off-white below, faintly mottled greyish on throat and upper breast. Vent and undertail-coverts brownish-buff, with broad dull whitish tips often forming distinct crescents.

4 SAVI'S WARBLER *Locustella luscinioides* L 14 cm Text p. 301
Recalls Reed Warbler but has very long undertail-coverts. Legs brownish-pink. Vagrant to the CI and CV.
4 Ad: Uniform dull rufous-brown. Tail faintly barred above. Short and faint buff supercilium. Throat-sides, breast and flanks to undertail-coverts pale brownish. Chin, throat and belly buffy-white. Pale tips to undertail-coverts.

5 MOUSTACHED WARBLER *Acrocephalus melanopogon* L 12.5 cm Text p. 302
Recalls Sedge Warbler but more rufous and shorter-winged. Vagrant to the CV.
5 Ad: Crown densely streaked blackish. Square-ended white supercilium. Dark cheeks contrast with white throat.

6 AQUATIC WARBLER *Acrocephalus paludicola* L 13 cm Text p.302
Recalls Sedge Warbler but has a pale median crown-stripe and pale lores. Vagrant to the CI.
6 Ad: Prominently striped head, with long pale supercilium and central crown-stripe, and dark, almost black, lateral crown-stripes. Prominent pale 'tramlines' bordering mantle. Rump and flanks finely streaked.

7 SEDGE WARBLER *Acrocephalus schoenobaenus* L 13 cm Text p. 302
A faintly-streaked warbler with a prominent supercilium and dark lores. Vagrant to Md and the CI.
7 Ad: Crown darker than streaked mantle. Unstreaked deep buff rump. Unstreaked below.

8 CAPE VERDE (SWAMP) WARBLER *Acrocephalus brevipennis* L 13.5 cm Text p. 302
Rather nondescript. Endemic to the CV.
8 Ad: Larger than European Reed Warbler, with a noticeably longer bill, longer tail and very short wings. Grey-brown above and greyish-white below. Short, thin and indistinct greyish supercilium.

9 MARSH WARBLER *Acrocephalus palustris* L 13 cm Text p. 303
Very similar to European Reed Warbler but colder toned. Vagrant to Md.
9 Ad: More olive above than European Reed Warbler, and whiter below, washed buff or pale yellowish and lacking any rufous tones on flanks. Has whitish tips to primaries and pale borders to tertials.

10 (EUROPEAN) REED WARBLER *Acrocephalus scirpaceus* L 12.5 cm Text p. 303
Plain-coloured, with a longish bill and flat forehead. Scarce and irregular on passage in the CI. Vagrant to Md.
10 Ad: Earth-brown above with a more rufous rump and uppertail-coverts. Buff below.

11 GREAT REED WARBLER *Acrocephalus arundinaceus* L 19 cm Text p. 303
Very large warbler with a prominent bill. Vagrant to Md and the CI.
11 Ad: Brown above with a rufous-tinged rump. Whitish below with some indistinct streaking on upper breast.

151

PLATE 60   WARBLERS II

1 TRISTRAM'S WARBLER *Sylvia deserticola*   L 12.5 cm   Text p. 305
Recalls Subalpine Warbler but wings rufous. Vagrant to the CI.
1a Ad male: Bluish-grey above. Throat, breast and flanks vinous-brown. Paler pink to off-white belly and vent. Orange-rufous wing-patch. Faint whitish submoustachial stripe and fairly broad white eye-ring.
1b Ad female: Generally duller and whitish belly is more extensive.

2 SPECTACLED WARBLER *Sylvia conspicillata*   L 12.5 cm   Text p. 305
Resembles a miniature Common Whitethroat. Resident in Md, the CI and the CV.
2a Ad male: Head slate-grey, with black lores, prominent white eye-ring, white chin and grey-centred white throat. Mantle and wings also bright rufous-brown. Pinkish below with paler central belly and vent.
2b Ad female: Less intensely coloured, with paler lores and olive-brown replacing the slate-grey above.

3 SUBALPINE WARBLER *Sylvia cantillans*   L 12 cm   Text p. 306
Greyish, with a white moustachial stripe. Scarce regular migrant in the CI. Vagrant to Md and the CV.
3a Ad male: Blue-grey above with dark grey-brown wings and tail. Prominent white moustachial stripe. Chin, throat, breast and flanks pink-chestnut, with a paler, almost white-centred belly. Outer tail feathers white.
3b Ad female: Much paler, browner above and whiter below, only faintly washed pink.

4 MÉNÉTRIES'S WARBLER *Sylvia mystacea*   L 13.5 cm   Text p. 306
Suggests a poorly-marked Sardinian Warbler. Vagrant to the CI.
4a Ad male: Hood blackish, paler than in male Sardinian Warbler. Base of lower bill pink or straw.
4b Ad female: Sandy-grey above. Differs from female Sardinian Warbler in having the head concolorous with mantle. Buffish on breast and flanks but throat almost white.

5 SARDINIAN WARBLER *Sylvia melanocephala*   L 13 cm   Text p. 306
Long-tailed, with noticeable red eye-ring. Resident in the CI. Vagrant to Md and the CV.
5a Ad male: Head glossy black contrasting with white throat. Dark grey above, tail black edged white.
5b Ad female: Head and face grey, throat white. Upperparts, flanks and undertail-coverts brownish.

6 AFRICAN DESERT WARBLER *Sylvia deserti*   L 11.5 cm   Text p. 307
Very small with a pale yellow iris and straw-coloured legs and bill. Vagrant to Md, the CI and CV.
6 Ad: Pale sandy above and whitish below. Rump and tail rufous. White outer tail feathers.

7 ORPHEAN WARBLER *Sylvia hortensis*   L 15 cm   Text p. 307
A robust warbler, suggesting a large blackcap. Vagrant to Md and the CI.
7a Ad male: Grey-black hood with contrasting pale irides. Dusky-brown above, whitish pink below.
7b Ad female: Browner above and duskier below. Irides usually dark.

8 LESSER WHITETHROAT *Sylvia curruca*   L 12.5–13.5 cm   Text p. 307
Greyer than Common Whitethroat. Vagrant to Md and the CI.
8 Ad: Grey-brown above, dull white below. White sides to dusky tail. Blackish mask on ear-coverts.

9 COMMON WHITETHROAT *Sylvia communis*   L 14 cm   Text p. 307
Rufous-winged and white-throated. Regular on passage in the CI. Vagrant to Md.
9a Ad male: Brown above with a grey head, white throat and pinkish-buff wash to the underparts. Tertials and wing coverts broadly fringed rufous. Tail grey with white edges.
9b Ad female: Head and mantle brown. More buffish below, lacking pink.

10 BLACKCAP *Sylvia atricapilla*   L 13 cm   Text p. 308
A black-capped, greyish warbler without any white in the tail. Nominate is regular on passage and in winter in the CI; *heineken* is resident in the CI and Md and *gularis* is resident in the Az and C (both extremely similar to the nominate).
10a Ad male nominate: Dusky-brown above and greyish-buff below. Cap black.
10b Ad female nominate: Similar to male but with reddish-brown cap.

11 GARDEN WARBLER *Sylvia borin*   L 14 cm   Text p. 308
A medium-sized greyish warbler. Scarce but regular on passage in the CI. Vagrant to Md.
11 Ad: Grey-brown above and buffish-white below, washed grey on breast-sides. Dark eyes and grey legs.

153

PLATE 61 WARBLERS III

1 YELLOW-BROWED WARBLER *Phylloscopus inornatus* L 10 cm Text p. 308
Smaller than chiffchaffs, with a long, broad supercilium and two wingbars. Vagrant to the Az, Md and CI.
1 **Ad:** Olive-green above and whitish below. Long yellowish-white supercilium extends almost to nape. Long, dark eyestripe. Two off-white wingbars, off-white fringes to tertials and pale tips to primaries.

2 HUME'S (YELLOW-BROWED) WARBLER *Phylloscopus humei* L 10 cm Text p. 309
Similar to Yellow-browed Warbler but duller. Vagrant to the CI.
2 **Ad:** Greyer above than Yellow-browed Warbler, supercilium and underparts buffish-white, upper wingbar less distinct and bill more uniformly dark.

3 WESTERN BONELLI'S WARBLER *Phylloscopus bonelli* L 11.5 cm Text p. 309
Strikingly white below with yellow feather-edging above. Scarce but regular on passage in the CI. Vagrant to Md and the CV.
3 **Ad:** Plain grey-brown with only a very indistinct pale supercilium. Lower back and rump greenish-yellow. Strikingly silky white below. Wing-coverts, wing feathers and tail edged greenish-yellow.

4 WOOD WARBLER *Phylloscopus sibilatrix* L 12–13 cm Text p. 309
Large and much longer-winged than chiffchaffs. Regular on passage in the CI. Vagrant to Md.
4 **Ad:** Green above; dark brown wings and tail fringed bright yellow-green. Supercilium yellow and eyestripe dark. Ear-coverts, chin, throat and upper breast yellow. Rest of underparts almost pure white.

5 COMMON CHIFFCHAFF *Phylloscopus collybita* L 10–11 cm Text p. 309
A small, greenish warbler, usually with dark legs. Regular on passage and in winter in the CI. Irregular and scarce on passage in Md. Vagrant to the Az and CV.
5a **Ad nominate:** Dull greenish above. Off-white below, washed yellowish on breast and flanks. Legs usually black. Wings and tail feathers dark brown, finely fringed olive-green.
5b **Ad** *abietinus***:** Very similar to nominate, but with greyish wash on crown, nape and mantle; underparts generally whiter than nominate.

6 CANARY ISLANDS CHIFFCHAFF *Phylloscopus canariensis* L 10–11 cm Text p. 310
The endemic, resident chiffchaff of CI. Nominate occurs on western and central islands, *exsul* formerly on Lanzarote, now extinct; *exsul* was paler than nominate, with yellowish underparts and very dark legs.
6 **Ad nominate:** Similar to Common Chiffchaff but with shorter wings and slightly longer tail. Strong brownish-buff tone below, particularly on breast and flanks. Darker and more brownish above than Common Chiffchaff and the pale supercilium is more distinct and slightly longer.

7 IBERIAN CHIFFCHAFF *Phylloscopus ibericus* L 10–11 cm Text p. 310
Similar to Common Chiffchaff; the downward-inflected call note is distinctive. Recorded in the CI.
7 **Ad:** Whiter on belly than Common Chiffchaff, with lemon-yellow vent and undertail-coverts.

8 WILLOW WARBLER *Phylloscopus trochilus* L 10.5–11.5 cm Text p. 310
Typically brighter than the chiffchaffs, with pale legs. Regular on passage in the CI but irregular in Md. Vagrant to the Az and CV, although perhaps under-recorded in the latter.
8 **Ad:** Brownish-green above. Yellowish white on breast and throat, belly white. Whitish supercilium.

9 MELODIOUS WARBLER *Hippolais polyglotta* L 13 cm Text p. 304
Yellowish and large-billed, lacking any distinctive markings. Scarce migrant in the CI. Vagrant to Md.
9 **Ad:** Brownish above and yellowish below, with a short supercilium and pale lores. Legs brownish. Short primary projection.

10 ICTERINE WARBLER *Hippolais icterina* L 13 cm Text p. 304
Similar to Melodious Warbler but longer-winged. Vagrant to Md and the CI.
10 **Ad:** Greenish-olive to distinctly greyish above. Yellow to almost white below. Long primary projection. Pale edges to tertials and secondaries form pale wing panel. Legs grey.

11 WESTERN OLIVACEOUS WARBLER *Hippolais opaca* L 13–13.5 cm Text p. 304
A greyish, large-billed warbler. Squarish tail has white edges. Rare on passage in CI. Vagrant to the CV.
11 **Ad:** Grey-brown above. Off-white below, often washed buff or greyish on flanks and breast.

155

PLATE 62 GOLDCRESTS, FLYCATCHERS, NUTHATCH AND PENDULINE TIT

1 GOLDCREST *Regulus regulus*   **L** 9 cm   Text p. 311

Tiny; olive-green above and off-white below. Resident in the Az, where the three races are all longer-billed and darker than the nominate.
1a Ad male *azoricus*: Orange-and-yellow crown-stripe bordered black. Face greyish with prominent pale eye-ring. Double pale wingbar and pale tips to tertials. Dark olive green above and olive-buff below.
1b Ad female *azoricus*: As male but lacks orange in crown.
1c Ad male *sanctaemariae*: Palest race. Yellowish-olive above and whitish below.
1d Ad male *inermis*: Dark brownish-olive above.

2 TENERIFE GOLDCREST (TENERIFE KINGLET) *Regulus teneriffae*   **L** 9 cm   Text p. 311

Very similar to Goldcrest but with more black on head. Endemic to Tenerife and the western CI.
2a Ad male: Orange and yellow crown-stripe bordered black. Black crown-stripes meet on forehead, unlike Goldcrest. Also relatively long-billed, less white on tertials and darker buff below.
2b Ad female: Yellow crown-stripe bordered black. Black crown stripes meet on forehead, unlike Goldcrest.
2c Juv: Lacks adult head markings.

3 MADEIRA FIRECREST *Regulus madeirensis*   **L** 9 cm   Text p. 311

Very similar to Goldcrest but has a white supercilium and black eyestripe. Resident in Md.
3a Ad male: Whiter below than Goldcrest with bronzy patches on neck-sides. Central crown orange. Black lateral crown-stripes meet on forehead. Supercilium ends just behind eye (shorter than in nominate).
3b Ad female: As male but central crown yellow.
3c Juv: Lacks adult crown markings but has white supercilium and dark eyestripe.

4 SPOTTED FLYCATCHER *Muscicapa striata*   **L** 14.5 cm   Text p. 311

A brown flycatcher with a streaked breast. Regular on passage in the CI. Vagrant to Md and the CV.
4 Ad: Greyish-brown above and off-white below, with indistinct streaking on breast, throat, forehead and forecrown.

5 RED-BREASTED FLYCATCHER *Ficedula parva*   **L** 11.5 cm   Text p. 312

A small brownish flycatcher with white tail sides. Vagrant to the Az, Md and the CI.
5a Ad male: Head greyish. Throat and upper breast reddish-orange.
5b Ad female: Dull brown above and off-white below, washed buff on breast. Narrow buff eye-ring.

6 PIED FLYCATCHER *Ficedula hypoleuca*   **L** 13 cm   Text p. 312

A small flycatcher with a conspicuous white wing patch. Regular on passage in the CI and at least occasional in Md. Vagrant to the CV.
6a Ad male breeding: Black above and white below with small white patch on forehead. White sides to base of tail. White patch on wing on tertials and inner greater coverts.
6b Ad female: Brown above and less white on wings than male. Lacks white on forehead.

7 EURASIAN NUTHATCH *Sitta europaea*   **L** 14 cm   Text p. 313

Short-tailed, with a long, chisel-like bill. Vagrant to the CI.
7 Ad: Blue-grey above with long black eyestripe curving down neck-sides. Cheeks and throat white. Otherwise orangey-buff below with darker, more chestnut rear flanks and undertail-coverts.

8 PENDULINE TIT *Remiz pendulinus*   **L** 11 cm   Text p. 314

Small, with a pointed, conical bill. Vagrant to the CI.
8 Juv: Head plain grey-brown. Mantle paler brown with contrasting dark tertials. Off-white below.

157

PLATE 63 TITS, ORIOLES AND SHRIKES

1 AFRICAN BLUE TIT *Cyanistes ultramarinus* **L** 11.5 cm Text p. 312

More blackish on crown with darker, more slate-blue upperparts, and a longer bill than European Blue Tit. Four races resident in the CI: *degener* (eastern islands), *teneriffae* (central islands), *palmensis* (La Palma) and *ombriosus* (Hierro).
1a Ad *degener*: Paler and greyer above than other races, with a broad white wingbar and obvious white tertial fringes.
1b Ad *ombriosus*: Greenish tones above. Indistinct greyish-white wingbar.
1c Ad *palmensis*: Greenish tones above. Wingbar and tertial fringes more pronounced than in *ombriosus* and paler yellow below with white belly.
1d Ad *teneriffae*: No wingbar and very indistinct pale tertial fringes. Dark slaty-blue above and deep yellow below.
1e Juv *teneriffae*: Crown and back greenish. Face yellow.

2 GOLDEN ORIOLE *Oriolus oriolus* **L** 24 cm Text p. 314

Thrush-like with dark wings and a yellow rump. Scarce but regular on passage in the CI. Vagrant to the Az and Md.
2a Ad male: Bill pink. Head and body bright yellow. Wings and tail black.
2b Ad female: Bill pinkish. Mainly green above. Greyish-white below with pale yellow flanks and vent, streaked brown.

3 ISABELLINE SHRIKE *Lanius isabellinus* **L** 17.5 cm Text p. 314

A rufous-brown shrike with a rufous tail. Vagrant to the CI.
3 Juv: Rump and tail rufous. Face mask brownish. Faintly barred below.

4 RED-BACKED SHRIKE *Lanius collurio* **L** 17 cm Text p. 314

Dark brown or rufous-backed, with a dark tail and face mask. Vagrant to the CI.
4a Ad male: Crown, nape and rump grey. Broad black mask. Mantle and wing-coverts reddish-brown. White below with pinkish tone. Tail black with white sides at base
4b Ad female: Coarsely vermiculated off-white below. Mantle dark brown. Crown grey-brown. Small brown mask.
4c Juv: Resembles female but more heavily vermiculated above.

5 SOUTHERN GREY SHRIKE *Lanius meridionalis* **L** 24–25 cm Text p. 315

A large, long-tailed, grey-and-white shrike. Resident in the CI (*koenigi*). Vagrant to the CV (*elegans*).
5a Ad *koenigi*: Medium to dark grey above. Pale greyish breast and flanks contrast with white chin and throat.
5b Juv *koenigi*: Grey-brown above and distinctly vermiculated below.

6 WOODCHAT SHRIKE *Lanius senator* **L** 18 cm Text p. 315

A compact shrike with pale scapular patches. Regular on passage in the CI. Vagrant to Md and the CV.
6a Ad male nominate: Crown and nape red-brown. Mask, forehead and back black. Predominantly black wings with white primary patches. Large white patch on scapulars. Generally white below. Tail black, edged white. Uppertail-coverts white.
6b Ad male *badius*: As nominate but lacks white patch at base of primaries.
6c Juv nominate: Resembles juvenile Red-backed Shrike but greyer above, with a pale rump and pale scapular patches.

159

PLATE 64 CROWS AND ALLIES

1 (SPOTTED) **NUTCRACKER** *Nucifraga caryocatactes* **L** 32 cm Text p. 315
Spotted, with a long, pointed bill. Vagrant to the CI.
1 **Ad:** Dark chocolate-brown body, with black wings, white-tipped black tail and white vent. Entire body, except crown, covered in white spots.

2 RED-BILLED CHOUGH *Pyrrhocorax pyrrhocorax* **L** 39–40 cm Text p. 315
Broad-winged, with a curved red bill. Resident in the CI (La Palma).
2a **Ad:** Glossy black except for long, downcurved red bill and pinkish-red legs and feet.
2b **Ad flight:** Wings uniformly broad, blunt-tipped and deeply 'fingered'.

3 WESTERN JACKDAW *Corvus monedula* **L** 33 cm Text p. 316
A small crow with a grey nape. Vagrant to the CI.
3 **Ad:** Dark grey, with paler nape and neck-sides. Forehead and crown noticeably darker than nape. Eye greyish-white.

4 ROOK *Corvus frugilegus* **L** 46–47 cm Text p. 316
Black, with a long-pointed bill. Vagrant to the CI.
4 **Ad:** Glossy black. Bare skin around base of bill greyish-white.

5 CARRION CROW *Corvus corone* **L** 47 cm Text p. 316
Smaller than Common Raven, with rounded tail. Vagrant to the Az, Md and the CI
5a **Ad:** All black, with fairly stout bill.
5b **Ad flight:** Wings relatively broader and shorter than in Common Raven and tail rounded.

6 COMMON RAVEN *Corvus corax* **L** 64 cm cm Text p. 316
Very large and heavy-billed. Resident in the CI. Vagrant to Md.
6a **Ad:** All glossy black. Bill heavy. Feathers on throat elongated and pointed.
6b **Ad flight:** Long, wedge-shaped tail distinctive.

7 BROWN-NECKED RAVEN *Corvus ruficollis* **L** 50–52 cm Text p. 316
A smaller and slimmer raven with a bronze-glossed nape. Resident in the CV. Vagrant to the CI.
7 **Ad:** Shorter-tailed than Common Raven, with a comparatively thin bill and less obvious throat hackles.

161

PLATE 65 STARLINGS AND SPARROWS

1 COMMON STARLING *Sturnus vulgaris* L 21 cm Text p. 317
Short-tailed, with a long, pointed bill. Resident in the Az (*granti*) and locally in the CI (nominate). Winter visitor to Md and the CI. Vagrant to the CV.
1a Ad breeding nominate: Black with a pointed yellow bill and pinkish legs. Obvious green and purple sheen to the plumage, with some yellowish spotting, particularly above and on vent.
1b Ad breeding *granti*: Darker than nominate. [inferred from plate. Text gives no differences. Suggest omit]
1c Juv nominate: Dirty brown with a paler throat. Indistinct streaking on underparts. Dark bill.
1d Juvenile *granti*: Darker than nominate.

2 SPOTLESS STARLING *Sturnus unicolor* L 21 cm Text p. 318
Similar to Common Starling but more uniformly coloured. Vagrant to Md and the CI.
2a Ad breeding: All black. Pink legs brighter than in Common Starling.
2b Juv: Darker than juvenile Common Starling.

3 ROSE-COLOURED STARLING *Sturnus roseus* L 21 cm Text p. 318
Typical starling shape but distinctive in all plumages. Vagrant to Md and the CI.
3a Ad male: Pink body. Head, wings and tail black. Long crest on nape.
3b Juv: Resembles juvenile Common Starling but a paler milky-tea colour, with a whiter rump and shorter, yellowish bill.

4 COMMON MYNA *Acridotheres tristis* L 24 cm Text p. 318
Large and brownish, with a yellow bill and eye-patch. Introduced to the CI (Tenerife).
4 Ad: Predominantly brown with a glossy black head. Wings black with white primary patch. Tail black with white tip. Undertail and central belly white.

5 HOUSE SPARROW *Passer domesticus* L 14–15 cm Text p. 318
Sturdy, with a short, stout bill. Resident in the Az, CI and CV.
5a Ad male breeding: Dull greyish-white below. Black bib extending onto upper breast. Crown and forehead grey with a chestnut border. Nape also chestnut. Rump grey. Brown above with dark streaking.
5b Ad female: Dull greyish-brown below. Brown above, streaked black. Dirty buffish supercilium.

6 SPANISH SPARROW *Passer hispaniolensis* L 15 cm Text p. 319
Similar to House Sparrow but males strikingly marked. Resident in Md, the CI and CV.
6a Ad male breeding: Crown chestnut. Broken white supercilium. Chin, throat and breast black. Otherwise white below, heavily and boldly streaked black on flanks and lower breast. Mantle, back and scapulars also heavily and boldly streaked black.
6b Ad female: Very similar to female House Sparrow but faintly streaked below.

7 IAGO SPARROW *Passer iagoensis* L 13 cm Text p. 319
A small, typical sparrow. Endemic to the CV.
7a Ad male: Forehead and crown blackish, becoming grey on nape and upper mantle. Face whitish, black eyestripe, narrow white stripe above lores, and a rich chestnut line from eye around rear of ear-coverts. Whitish below except for neat black bib. Brown above, streaked black, with chestnut scapulars and broad white tips to median coverts.
7b Ad female: Similar to female House Sparrow but smaller.

8 (EURASIAN) TREE SPARROW *Passer montanus* L 14 cm Text p. 319
Rufous-crowned with a black cheek spot. Local resident in the CI.
8 Ad breeding: Cap chestnut, head-sides white, black spot on ear-coverts. Whitish collar, small neat black bib. Brown above streaked black. Dull greyish or dirty white below.

9 ROCK SPARROW *Petronia petronia* L 14 cm Text p. 319
A stocky and large-headed sparrow with a strong pale supercilium. Resident in Md and locally in the CI.
9 Ad: Crown dark blackish-brown with a pale, broad central stripe, bordered by prominent creamy supercilium and dark eyestripe. Grey-brown above streaked darker. Whitish below streaked brown. Pale subterminal spots on rectrices.

163

PLATE 66 WAXBILLS AND FINCHES

1 RED-BILLED QUELEA *Quelea quelea*   L 13 cm   Text p. 320
Sparrow-like, with a strong, conical bill. Vagrant to the CI and CV.
1a Ad male breeding: Bill bright red. Face and throat black. Crown, nape and breast tawny.
1b Ad female: Lacks black face and throat. Crown and nape browner. Pale supercilium.
1c Juv: Resembles female but bill is brownish-pink.

2 RED-CHEEKED CORDON-BLEU *Uraeginthus bengalus*   L 13 cm   Text p. 321
Small, long-tailed and blue-rumped. Recorded from the CI and CV, but probably escapes.
2a Ad male: Brown above and mainly blue below. Tail blue. Red patch on ear coverts. Bill pinkish.
2b Ad female: Lacks red patch. Less blue below.
2c Juv: Browner than female. Bill blackish.

3 COMMON WAXBILL *Estrilda astrild*   L 11 cm   Text p. 321
Very small with a long tail. Breeds locally in all four archipelagos, all or most populations of captive origin.
3a Ad male: Grey-brown above and buff below. Pinkish belly. Red stripe through eye, bright red bill.
3b Ad female: Resembles male but usually paler.
3c Juv: Bill blackish. Orange eye patch.

4 EUROPEAN SERIN *Serinus serinus*   L 11 cm   Text p. 323
Very small, streaked with a very short bill. Localy resident in the CI. Vagrant to Md.
4a Ad male: Head and breast bright yellow; yellow-green crown, nape, and ear-coverts. Belly and flanks white, streaked black on breast-sides and flanks. Mantle green with dark streaks; yellow rump.
4b Ad female: Similar to male but duller and more extensively streaked.
4c Juv: Browner and duller than female. Lacks yellow.

5 (ATLANTIC) CANARY *Serinus canaria*   L 13 cm   Text p. 323
Larger and longer-tailed than European Serin. Resident in the Az, Md and CI.
5a Ad male: Head yellow with dusky ear-coverts. Mainly yellow below, streaked blackish on flanks.
5b Ad female: Duller than male: greyer on head and breast, rump duller and underparts less yellow
5c Juv: Browner than female, lacking yellow or green tones. Rump pale brown, streaked darker.

6 (EUROPEAN) GREENFINCH *Carduelis chloris*   L 15 cm   Text p. 323
Plump finch with a stout bill. Resident and winter visitor in the CI. Breeds locally in the Az and Md.
6a Ad male: Yellowish-green below, greyish-green above. Prominent yellow base to tail sides.
6b Juv: Brownish above, mottled darker. Paler below with distinct streaking.

7 (EUROPEAN) GOLDFINCH *Carduelis carduelis*   L 12.5 cm   Text p. 324
A slender finch with a broad yellow wing-bar, white rump, black tail with white spots and a conical white bill. Resident in the Az, Md and CI. Has bred in the CV.
7a Ad: Face red. Throat and head-sides white. Crown and nape black. Mantle and scapulars brown.
7b Juv: Lacks adult head markings. Streaked above and below.

8 (EUROPEAN) SISKIN *Carduelis spinus*   L 12 cm   Text p. 324
A streaked, greenish finch. Scarce and irregular in winter in the CI. Vagrant to the Az and Md.
8a Ad male: Mantle green with black streaks. Cap and chin black. Rump and tail-base yellow. Wings black with two prominent yellow wingbars. Yellow below with dark-streaked flanks.
8b Ad female: Duller. Lacks black crown and more extensively streaked.
8c Juv: Similar to female but duller, with more extensive streaking on throat and rump.

9 (COMMON) LINNET *Carduelis cannabina*   L 13–14 cm   Text p. 324
A streaked, brownish finch. Resident in Md (*guentheri*), the eastern CI (*harterti*) and the western and central CI (*meadewaldoi*); Md race inseparable from *meadewaldoi* in the field.
9a Ad male *meadewaldoi*: Mantle, back and wing-coverts chestnut-brown. Head grey. Small crimson forehead patch, and a more extensive one on breast.
9b Ad male *harterti*: Paler above than *meadewaldoi*.
9c Ad female *meadewaldoi*: Duller, lacks crimson patches and streaked on mantle and underparts.
9d Ad female *harterti*: Paler than the other races.

165

PLATE 67 FINCHES

1 COMMON CHAFFINCH *Fringilla coelebs*   L 14.5 cm   Text p. 321

Sparrow-sized, with two white wingbars. Resident in the Az (*moreletti*), Md (*maderensis*) and the CI (*canariensis, ombriosa* and *palmae*). Nominate occasionally reaches CI.
1a Ad male *canariensis*: Deep slate-blue above with a black forehead, dull yellowish-green rump, black wings with prominent white bars on median coverts and tips of greater coverts. Peachy-buff below.
1b Ad female *canariensis*: Dull grey-brown above with an olive tone to rump and bluish-grey uppertail-coverts. Throat dull peachy-buff with throat- and breast-sides washed brownish.
1c Ad male *palmae*: Similar to *canariensis* but rump concolorous with back, slightly paler above and less blackish on head. Breast is pinkish, otherwise more extensively white below.
1d Ad male *ombriosa*: Very similar to *palmae* but has a greenish rump and less white on underparts.
1e Ad male *maderensis*: Similar to *canariensis* but lower mantle, scapulars, back and rump bright green.
1f Ad male *moreletti*: Resembles *maderensis* but has more extensive green upperparts, browner breast, less vinous belly and less white in outer tail feathers.
1g Ad male nominate: Underparts and face vinaceous-red. Crown and nape slaty blue-grey. Mantle and back chestnut-brown. Lower back and rump green.

2 BLUE CHAFFINCH *Fringilla teydea*   L 16 cm   Text p. 322

Larger and more uniformly coloured than Common Chaffinch. Resident endemic in the CI on Gran Canaria (*polatzeki*) and Tenerife (nominate).
2a Ad male nominate: Slaty-blue except for paler whitish belly and striking white vent. Wings dark with two pale bluish-grey wingbars. Bill rather metallic steel-blue.
2b Ad female nominate: Dull olive-brown above and slightly paler grey-brown below. Wingbars buffish or whitish.
2c Ad male *polatzeki*: Smaller than nominate, with greyer plumage, a blacker band over base of bill and whiter wingbars.

3 BRAMBLING *Fringilla montifringilla*   L 14 cm   Text p. 322

Recalls Common Chaffinch but orange below and with a white rump. Vagrant to Az, Md and the CI.
3a Ad male winter: Head, neck, mantle and back blackish with paler greyish-brown fringes. Nape and head-sides pale grey. Underparts, lesser coverts, and scapulars orange. Bill pale yellowish with a darker tip.
3b Ad female winter: Similar to male but with a paler head, duller orange underparts and more diffuse spotting on flanks.

4 COMMON CROSSBILL *Loxia curvirostra*   L 16.5 cm   Text p. 325

A stout finch with a large head and thick neck. Mandibles crossed. Vagrant to the CI.
4a Ad male: Dull red, brightest on rump. Wings and tail brownish.
4b Ad female: Greyish-green with diffuse streaking. Rump yellowish-green. Wings and tail brownish.

5 AZORES BULLFINCH *Pyrrhula murina*   L 16 cm   Text p. 325

Greyish-brown, with contrasting cap, wings and tail. Endemic to the Az (São Miguel).
5 Ad: Grey-brown above, paler and buffier on cheeks, rump and below. Cap, throat, wings and tail black with a purple-blue gloss. Greyish tips to greater coverts. Outer web of shortest tertial usually orange-pink.

6 HAWFINCH *Coccothraustes coccothraustes*   L 18 cm   Text p. 326

A large finch with a massive bill and broad, pale wingbars. Vagrant to Md.
6a Ad male: Throat and lores black. Head and underparts rich brown, nape grey. Tail black with broad white tip. Wings black, glossed blue on secondaries. Median and greater coverts form large white patch.
6b Ad female: Duller than male and shows pale grey panel on secondaries.

7 TRUMPETER FINCH *Bucanetes githagineus*   L 12.5–15 cm   Text p. 325

A small, pale, large-headed finch. Resident in the CI. Vagrant to the CV.
7a Ad male: Head grey. Thick-based red bill and red forehead. Body plumage unstreaked sandy grey-brown, tinged dark pink on rump, wings and below.
7b Ad female: Unstreaked sandy-grey with pinkish tips to wing-coverts and on tail-sides. Bill pinkish.

8 (WHITE-WINGED) SNOWFINCH *Montifringilla nivalis*   L 17 cm   Text p. 320

Large, with long, white, black-tipped wings. Vagrant to the CI.
8 Ad summer: Head grey with dark lores; mantle and scapulars brown. Much white in wings and tail.

167

PLATE 68   NORTH AMERICAN WARBLERS AND SPARROWS

1 BLACKPOLL WARBLER *Dendroica striata*   **L** 13.5 cm   Text p. 326
Short-tailed but with long undertail coverts and long wings. Vagrant to the CI.
1   **1st-winter:** Head and upperparts olive-green. Supercilium, throat and breast pale yellow. Faintly streaked above and below.

2 MAGNOLIA WARBLER *Dendroica magnolia*   **L** 12 cm   Text p. 326
White subterminal tail-band is diagnostic. Vagrant to the Az.
2   **1st-winter:** Crown, ear coverts and nape grey. Grey neck-band. Mainly greenish above and yellow below, faintly streaked. Two white wingbars.

3 YELLOW WARBLER *Dendroica petechia*   **L** 12.5 cm   Text p.326
Rather plain, with a stout bill and noticeable dark eye. Vagrant to the Az and Md.
3   **1st-winter:** Brownish above and buffy-white or greyish below, with pale tertial edges.

4 YELLOW-RUMPED WARBLER *Dendroica coronata*   **L** 13.5 cm   Text p. 326
Yellow rump distinctive. Vagrant to the Az and CI.
4   **1st-winter:** Brownish above, streaked darker. Two pale wingbars. Yellow rump and flank patches. Throat white.

5 AMERICAN REDSTART *Setophaga ruticilla*   **L** 13 cm   Text p. 327
A round-winged warbler with a long, yellow-based tail. Vagrant to the Az and Md.
5   **Ad female:** Head greyish. Olive above and white below. Yellow patches on wings, tail and sides of the breast.

6 LOUISIANA WATERTHRUSH *Seiurus motacilla*   **L** 15 cm   Text p. 327
Pipit like, with a short tail and striking supercilium. Vagrant to the CI.
6   **Ad:** Dark brown above and whitish below, streaked dark on throat, breast and flanks. White supercilium broadens behind eye. Flanks and undertail buff. Throat unstreaked. Legs bright pink.

7 NORTHERN WATERTHRUSH *Seiurus noveboracensis*   **L** 15 cm   Text p. 327
Very similar to previous species. Vagrant to the Az.
7   **Ad:** Smaller-billed and more heavily streaked below than Louisiana Waterthrush. White supercilium tapers at rear. Legs usually dark flesh.

8 SAVANNAH SPARROW *Passerculus sandwichensis*   **L** 14 cm   Text p. 327
Small and streaked, with yellow loral spot. Vagrant to the Az.
8   **Ad:** Pale sandy-brown to dark brown, variably streaked above and below. Yellow loral stripe and fore-supercilium. Bright pinkish legs. Whitish median crown-stripe.

9 BOBOLINK *Dolichonyx oryzivorus*   **L** 18 cm   Text p. 330
A large bunting with spike-tipped tail feathers and a conical bill. Vagrant to the Az.
9   **Ad winter:** Buffy-brown above with heavy dark streaking on head and upperparts. Whitish 'braces' on mantle-sides. Yellowish-buff below with fine streaks on flanks and breast-sides. Pale central crown-stripe and whitish supercilium.

169

PLATE 69 BUNTINGS

1 SNOW BUNTING *Plectrophenax nivalis* L 16–17 cm Text p. 328
Large white wing patches distinctive. Irregular in winter in the Az. Vagrant to Md and the CI.
1a Ad male winter: A complex mixture of orange-brown, white and black.
1b Ad female winter: Duller than male, with a less distinct white supercilium and less white in wing.

2 CIRL BUNTING *Emberiza cirlus* L 16 cm Text p. 328
Streaked, with striped head-sides. Vagrant to the CI.
2a Ad male breeding: Face yellow and black-striped. Throat black. Crown dark greyish-green, streaked darker. Two breast bands: yellow and greyish-green, with chestnut patches on breast-sides.
2b Ad female: Dull and streaky, with a greyish rump, faint yellowish wash below and rufous-tinged breast-sides.

3 ROCK BUNTING *Emberiza cia* L 16 cm Text p. 328
Long-tailed, with a dark-striped grey head. Vagrant to the CI.
3a Ad male: Head and upper breast pale grey with black border to ear-coverts and black lateral crown-stripes. Orangey lower breast, flanks and belly.
3b Ad female: Similar to male but head pattern duller and often faintly streaked on breast-sides and flanks.

4 HOUSE BUNTING *Emberiza striolata* L 13.5 cm Text p. 329
Rufous-brown with a finely streaked grey head. Vagrant to the CI.
4a Ad male *sahari*: Head, neck and upper breast greyish. Pale supercilium, dark eyestripe, white submoustachial stripe and black moustachial stripe. Throat and breast streaked blackish but otherwise unmarked above and below.
4b Ad female *sahari*: Duller overall than male, with a browner head and breast.

5 ORTOLAN BUNTING *Emberiza hortulana* L 16.5 cm Text p. 329
Medium sized, with a pale yellowish-white eye-ring and pale submoustachial stripe. Vagrant to the CI.
5a Ad male: Head greenish-grey with yellow submoustachial stripe and throat. Greenish-grey breast sharply demarcated from rufous-buff belly and flanks. Brownish above, heavily streaked darker.
5b Ad female: Similar to male but duller and streaked on crown and breast.

6 CRETZSCHMAR'S BUNTING *Emberiza caesia* L 16 cm Text p. 329
Smaller than Ortolan Bunting, with a rufous-brown rump. Vagrant to the CI.
6a Ad male: Head and breast blue-grey with orange-chestnut lores, throat and submoustachial stripe, and pale eye-ring. Rest of underparts rufous-chestnut.
6b Ad female: Duller than male, with streaking on crown, nape and breast.

7 LITTLE BUNTING *Emberiza pusilla* L 13–14 cm Text p. 329
Small, dark and streaked. Vagrant to the CI.
7a Ad: Ear-coverts chestnut. Narrow pale eye-ring. Prominent lateral crown-stripes and dark border to ear-coverts, which does not reach bill.
7b Juv: Resembles adult but crown-stripes more diffuse and head pattern duller.

8 (COMMON) REED BUNTING *Emberiza schoeniclus* L 15–16 cm Text p. 329
Streaked and dark-headed, with a prominent submoustachial stripe. Vagrant to Md and the CI.
8a Ad male breeding: Head black. Submoustachial stripe and nuchal collar white. White below with some black streaks on flanks. Wings predominantly rufous. Rump greyish and tail black with gleaming-white outer feathers.
8b Ad female breeding: Crown brown. Pale supercilium. Brown ear-coverts bordered darker. White moustachial stripe and throat, and dark malar stripe. More heavily streaked below than male.

9 CORN BUNTING *Emberiza calandra* L 18 cm Text p. 330
A large-headed, streaked, brown bunting with a large, pale bill. Resident in the CI.
9 Ad: Generally grey-brown above and whitish below, streaked darker throughout and buffish on breast and flanks. Face pattern plain apart from conspicuous blackish malar stripe.

171

RED-THROATED DIVER   *Gavia stellata*   Plate 1 (3)
Sp. Colimbo Chico   **P.** Mobelha-pequena (Mobêlha-pequena)

Monotypic. **L** 53–69 cm (21–27″); **WS** 106–116 cm (42–46″). **IDENTIFICATION** Sexes similar. Unlikely to be seen in our region in breeding plumage, when plain upperparts with grey face and neck-sides are amongst the best features. The red throat, if seen, is diagnostic, but can appear black at distance. In all plumages identified by its fine bill, upward-tilted lower mandible and habit of holding its head and bill above horizontal when swimming. Non-breeding birds largely grey-brown above and white below, with extensive white on face and neck. Grey-brown crown does not extend below eye, unlike Black-throated Diver, and upperparts usually clearly spotted white. First-winter has duskier face and neck than adult winter, and upperparts spotting less obvious than in non-breeding adult. **VOICE** Usually silent in winter and unlikely to be heard in our region. Flight call a goose-like cackling *kruk-kruk-kruk…* **RANGE** Circumpolar in arctic and subarctic, moving south in winter to *c.* 30°N. **STATUS** Very rare vagrant with only three records. **Azores** Immature, Ponta Delgada (São Miguel), 14–15 Feb 1979. **Canaries★** One, Puerto de la Cruz (Tenerife), Sep 1982; and one, Playa de Sotavento (Fuerteventura), 2 Feb 2004.

BLACK-THROATED DIVER   *Gavia arctica*   Plate 1 (2)
Sp. Colimbo Ártico   **P.** Mobelha-de-garganta-preta (Mobêlha-ártica)

G. a. arctica. **L** 58–73 cm (23–29″); **WS** 110–130 cm (43–51″). **IDENTIFICATION** Sexes similar. Like previous species, unlikely to be seen in our region in breeding plumage, when white-chequered upperparts, grey crown and hindneck, black and white stripes on neck-sides and black throat are key features. Non-breeding differs from Red-throated in its darker, more uniform upperparts, dark crown reaching below eye, deeper based and straighter bill (usually held horizontal when swimming), and often shows a white patch on rear flank, which is diagnostic but not always visible. First-winter recalls adult winter but is browner with scaly upperparts. **VOICE** Usually silent in winter, thus unlikely to be heard in our region. Generally silent in flight. **RANGE** Breeds across N Palearctic from Scotland to E Siberia. In winter migrates south to *c.*30°N. **STATUS** Very rare vagrant with just two recent records and vague mentions in the literature. **Canaries** Paterson (1997) noted it as having been captured recently in Gran Canaria, and mentioned the species for La Gomera without details. One off Playa Rosalia (Tenerife), 4–5 Dec 2002; and one close offshore off El Médano (Tenerife), 19 Jan 2004.

GREAT NORTHERN DIVER   *Gavia immer*   Plate 1 (1)
Sp. Colimbo Grande   **P.** Mobelha-grande (Mobêlha-grande)

Monotypic. **L** 61–91 cm (24–36″); **WS** 127–147 cm (50–58″). **IDENTIFICATION** Sexes similar. The largest diver in our region with a large, heavy bill, which is usually greyish but sometimes whitish, when good views required to eliminate White-billed Diver *G. adamsii* (unknown in our region). Like other divers, most likely to be seen in non-breeding plumage, which resembles Black-throated but usually lacks white flank patch, has crown and hindneck darker than rest of upperparts and has different head shape. Crown flatter and forehead more steeply angled compared to more rounded head shape of Black-throated. Also distinctly larger than black-throated, although this is probably of little use when dealing with a lone vagrant. Recorded from our region in breeding plumage when the all-black head, white-chequered upperparts and white-striped neck patch easily identify it. First-winter like adult winter but has scaly pattern to browner upperparts. **VOICE** Usually silent in winter and in flight, so unlikely to be heard in our region. **RANGE** Breeds in arctic and subarctic North America, from the Aleutians to Greenland and Iceland, moving south in winter, rarely as far as 30°N. **STATUS** Accidental, with records from three of the archipelagos. **Azores** More than 20 records, *c.*10 of these since 1998, suggesting that it is perhaps a regular winter visitor in small numbers. **Madeira** Vagrant with only two records, the first noted by Harcourt in 1855 without details; and a ♂ obtained at Machico (Madeira), 30 Nov 1945. **Canaries** Vagrant with only three records: dead juvenile, Santa Cruz (La Palma), 1970s; one, La Barranquera (Tenerife), 12 Jul 1996 (perhaps of captive origin); and one, Caleta del Sebo (La Graciosa), 30 Dec 2000.

PIED-BILLED GREBE   *Podilymbus podiceps*   Plate 1 (5)
Sp. Zampullín Picogrueso   **P.** Mergulhão caçador (Mergulhão-de-bico-grosso)

P. p. podiceps. **L** 30–38 cm (12–15″); **WS** 56–64 cm (22–25″). **IDENTIFICATION** Sexes similar. Larger and stockier than Little Grebe with which most likely to be confused. Bill is very thick and deep-based; a good ID feature in all plumages. Breeding adult has whitish bill with a broad vertical black band, a black chin and throat, and greyish-brown neck. In non-breeding plumage characteristic bill-band very faint or absent, black throat and chin become whitish and neck-sides browner. Juvenile similar to non-breeding adult but has head-sides striped. **VOICE** Silent in winter so unlikely to be heard in our region. **RANGE** Breeds over much of North and South America. Northern populations move south in winter but most winter in the breeding range. **STATUS** Accidental visitor with 12 records, but ten of these are from one archipelago and six from the same island. **Azores** Ten records, four of these in 1997–2001, and perhaps occurs annually in tiny numbers. First-winter ♀ obtained, Terceira, 24 Oct 1927; one obtained, São Miguel, 19 Oct 1954, ♂ obtained, Mosteiros (São Miguel), 19 Oct 1964; one, Jan 1969; one in breeding plumage, Lajes (Pico), 24–25 Jun 1984; immature, Lajes (Pico), 22–24 Oct 1985; one, Sete Cidades (São Miguel), 26 Nov. 1997; adult in winter plumage, Lagoa Funda (Flores), 7 Nov 1998; one, Lagoa Azul, Sete Cidades (São Miguel), 1–2 Nov 1999; one, Lagoa das Furnas (São Miguel), 19 Feb 2001; and one, Lagoa Azul, Sete Cidades (São Miguel), 2 Nov 2004–31 Jan 2005 at least. **Canaries** One record: adult, Los Silos (Tenerife), 13 & 19–20 Jan 1991.

LITTLE GREBE (DABCHICK) *Tachybaptus ruficollis*   Plate 1 (8)
Sp. Zampullín Chico (Zampullín Común)   **P.** Mergulhão-pequeno

T. r. ruficollis. **L** 25–29 cm (10–12″); **WS** 40–45 cm (16–18″). **IDENTIFICATION** Sexes similar. The smallest of the region's grebes and unlikely to be confused, except with the preceding species. Breeding plumage has chestnut cheeks and foreneck and diagnostic yellow gape patch. Non-breeder lacks black and white contrast of the 'horned' grebes, being generally brown above and buff below, but probably most easily identified by small size and tiny bill. The latter makes separation from Pied-billed Grebe rather straightforward. **VOICE** Like others in the family usually silent in winter and unlikely to be heard in our region. **RANGE** Breeds throughout Eurasia and Africa, with some winter dispersal. **STATUS** Accidental winter visitor. **Madeira** Vagrant with only one record, a ♂ obtained on Madeira on 1 Sep 1932. **Canaries★** Accidental in winter with five records: 14, Lago de Janubio (Lanzarote), 1904; up to 30, Embalse de Las Peñitas (Fuerteventura), 6 Aug–17 Sep 1957; four, Lago de Janubio (Lanzarote), 13 Jan 1967; one captured, Tenerife, 5 Oct 1973; and one, Embalse de Los Molinos (Fuerteventura), 23 Jan–14 Feb 1997.

GREAT CRESTED GREBE *Podiceps cristatus*   Plate 1 (4)
Sp. Somormujo Lavanco   **P.** Mergulhão-de-poupa (Mergulhão-de-crista)

P. c. cristatus. **L** 46–51 cm (18–20″); **WS** 85–90 cm (34–36″). **IDENTIFICATION** Sexes are similar. Breeding plumage is unmistakable with characteristic chestnut-and-black head plumes. Non-breeding bird might be confused with Red-necked Grebe *P. grisegena* (unrecorded in our region), but is larger, longer necked and whiter faced, with a prominent white stripe from the eye to bill. Also, bill pinkish (yellowish in Red-necked), and first-winter has vestiges of juvenile head-stripes. Noticeably larger than either of the 'horned' grebes, which should not pose identification pitfalls. Long, pale bill of non-breeding plumage unique amongst the region's grebes. **VOICE** Fairly quiet in winter but occasional clicks, croaks or growls may be given. **RANGE** Widespread across Eurasia, and in Africa and Australia. Winter dispersal to coasts and some northern populations move south as far as Morocco. **STATUS** Vagrant with only two records. **Azores** Two females obtained, Ponta Delgada (São Miguel), 29 May 1909. **Canaries★** Winter-plumaged adult, Arinaga (Gran Canaria), 12 Nov 1984.

SLAVONIAN GREBE (HORNED GREBE) *Podiceps auritus*   Plate 1 (6)
Sp. Zampullín Cuellirrojo   **P.** Mergulhão-de-penachos (Mergulhão-de-pescoço-castanho)

Monotypic. **L** 31–38 cm (12–15″); **WS** 59–65 cm (23–26″). **IDENTIFICATION** Sexes similar. Breeding adult has erect golden head plumes, chestnut neck, breast and flanks, a black head and dark brown upperparts. Non-breeding plumage has blackish upperparts and white underparts. Unlikely to be mistaken for Black-necked Grebe in breeding plumage, but confusion in non-breeding plumage more likely. Slavonian has a diagnostic pale bill tip and flatter crown, coming to a peak on the rear head, and back flatter and tapered. A useful plumage feature is that the black on the face of Slavonian reaches level with the eye, not below it as in Black-necked. First-winter browner than adult winter with a less well-defined border between the dark cap and white face, and sometimes has brownish foreneck. **VOICE** Usually silent away from the breeding areas so unlikely to be heard in our region. **RANGE** Breeds throughout much of the north of the Northern Hemisphere, moving south in winter. In North America winters further south than in the W Palearctic. **STATUS** Vagrant with only five records. **Azores** One obtained, Rosto de Cão, pre-1888; one obtained, Ponta Delgada (São Miguel); one obtained, near Pico de Bagacina (Terceira), 11 Nov 1965; and one obtained, Santa Cruz (Flores), 14 Mar 1966. **Madeira** One record: ♀ obtained, Funchal (Madeira), 25 Sep 1890.

BLACK-NECKED GREBE (EARED GREBE) *Podiceps nigricollis*   Plate 1 (7)
Sp. Zampullín Cuellinegro   **P.** Cagarraz (Mergulhão-de-pescoço-preto)

P. n. nigricollis. **L** 28–34 cm (11–14″); **WS** 56–60 cm (22–24″). **IDENTIFICATION** Sexes similar. Like all other grebes in the region, unmistakable in breeding plumage with black neck and long yellowish-white cheek feathers.

Non-breeding most likely confused with Slavonian Grebe, but Black-necked has a much steeper forehead giving the crown an obvious peak above the eye, the black crown clearly extends below the eye, the bill is finer and tilted up, and the bird's rear end is often fluffed up, not tapered like Slavonian. First-winter very similar to non-breeding adult with slightly browner upperparts and marginally whiter underparts and flanks. **VOICE** Usually silent in winter so unlikely to be heard in our region. **RANGE** Breeds in Eurasia east to Mongolia and China, W North America to NW Mexico, and S and E Africa. Birds from the north of the range move south in winter. **STATUS** An accidental or scarce winter visitor. **Azores** Casual winter visitor with only ten records, the most recent a ♀, Ribeira de Agua d'Alto (São Miguel), 5–7 Sep 1965, and one, Horta (Faial), 8–9 Dec 2000. **Madeira** Considered exceptional, with less than five records in the last 50 years. **Canaries** Rare and irregular winter visitor; never recorded from La Gomera.

(NORTHERN) FULMAR *Fulmarus glacialis* Plate 2 (1)
Sp. Fulmar (Fulmar Boreal) **P.** Pombalete (Fulmar-glacial)

F. g. glacialis. **L** 45–51 cm (18–20″); **WS** 102–112 cm (40–44″). **IDENTIFICATION** Sexes similar. At distance might be mistaken for a shearwater or a small Yellow-legged Gull, but smaller size, broader, straighter, stiffly held wings, shorter thicker bill and all-white head help distinguish Fulmar. Rather shearwater-like flight also distinguishes Fulmar from any species of gull found in the region. Various colour morphs: the pale morph is medium grey on the mantle, rump and upperwing with a white head, neck, underparts and tail, and the dark morph is entirely smoky-grey with a paler forehead. Intermediates between occur in clinal gradation. **VOICE** Fairly silent away from colonies and unlikely to be heard in our region. **RANGE** Breeds in the N Pacific and N Atlantic, and in adjoining arctic seas, dispersing south in winter. **STATUS** Extreme vagrant with just four records. **Azores★** Two records: one, Mosteiros (São Miguel), 28 Sep 1998; one, Mosteiros (São Miguel), 16 Oct 2002. **Canaries★** Two records: one dead, Graciosa, Mar 1983; and one, off Bajamar (Tenerife), 19 Mar 1998.

BERMUDA PETREL (CAHOW) *Pterodroma cahow* Plate 2 (6)
Sp. Petrel Cahow **P.** Freira das Bermudas

Monotypic. **L** 38 cm (15″); **WS** 89 cm (35″). **IDENTIFICATION** Sexes and all plumages similar. Upperparts greyish-brown with a blackish-brown cap and nape, and an indistinct pale band at the base of the tail. Underparts predominantly white with dusky breast-side patches. Underwing white with black margins and a short diagonal bar on the underwing-coverts. **VOICE** Usually silent at sea. At colonies gives a sweet *ooo-ee* or a fluty *tew-ee*, sometimes interspersed with a harsh croaking *quawer, borrr, awoo* or equivalent, and a dulcet *oo-ee wee wee ooo-ee*, occasionally with variations. **RANGE** Breeds only in Bermuda, on rocky islets off Castle Harbour, and recently reintroduced to Nonsuch Island (the species' former stronghold). Wintering area is unknown. **STATUS** Extreme vagrant with two very recent records involving the same individual. **Azores** One in a burrow on Vila Islet (Santa Maria), 17–21 Nov 2002, and the same individual in the same burrow, 19–21 Nov 2003.

TRINDADE PETREL *Pterodroma arminjoniana* Plate 2 (2)
Sp. Petrel de la Trindade **P.** Freira-da-Trindade

Monotypic. **L** 35–39 cm (14–15″); **WS** 88–102 cm (35–40″). **IDENTIFICATION** Sexes and all plumages similar. A large, long-winged gadfly petrel with dark underwings and a variety of colour morphs, from pale to dark. Dark morph predominantly dark brown with a white underwing patch at the base of the primaries, extending narrowly across the secondaries. Pale morph has white underparts and intermediates have white underparts with a dark breast-band. In our region unlikely to be mistaken for any other species except perhaps Sooty Shearwater, which differs in having extensive pale underwing patches, or dark-morph skuas, which usually have a very different flight action. **VOICE** Silent at sea. **RANGE** Breeds only on Trindade and Martin Vas (Brazil), in the tropical S Atlantic. Pelagic range poorly known but recently recorded several times off E North America. **TAXONOMIC NOTE** Sometimes considered conspecific with the Herald Petrel *P. heraldica*, which breeds in the Pacific. It has recently been suggested that the pale and dark morphs of Trindade Petrel be given specific status based on behavioural and molecular evidence. **STATUS** Extreme vagrant with only one record. **Azores** Dark morph, *c.* 16 km south of Pico, 18 Jul 1997 (the only W Palearctic record).

The Soft-plumaged Petrel complex: Fea's, Zino's and Soft-plumaged Petrel

Currently, both the taxonomy and identification of these taxa is in a state of flux. Concerning taxonomy, Bourne (1983) proposed to recognise all three as species, which has gained fairly widespread acceptance. However, more recently, Bretagnolle (1995) considered that the N Atlantic forms, Fea's and Zino's Petrels, should be treated as conspecifics distinct from the Soft-plumaged Petrel of the Southern Ocean, based on their vocalisations. I have retained Bourne's treatment because the separate breeding periods of Fea's and Zino's Petrels indicate their reproductive isolation. Identification is even more complicated and rests predominantly on the extent of the breast-band, the size and shape of the bill, and wing shape. Some suggested features for separating Fea's and Zino's Petrels in the field were proposed by Gantlett (1995), Howell (1996) and, more recently, by Tove (2001). It now seems just about possible to identify these two very closely related species at sea, but such a challenge is by no means easy and only birds seen very well should even tentatively be assigned to species.

FEA'S PETREL (CAPE VERDE PETREL) *Pterodroma feae* Plate 2 (3)
Sp. Petrel Gon-gon **P.** Gon-Gon

P. f. deserta (Desertas and possibly the Azores) and *P. f. feae* (Cape Verdes). **L** 36–37 cm (14–15″); **WS** 92–95 cm (36–38″). **IDENTIFICATION** Sexes and all plumages similar. A medium-sized petrel, about the same size as Manx Shearwater but with slightly longer wings. Generally grey-brown above and white below with dark grey underwings. Flight fast and erratic, frequently towering above the water. Significantly heavier (*c.* 50%) than Zino's Petrel, which should produce a much heavier bodied appearance. Also slightly larger, longer winged and with a slightly longer and considerably deeper bill. The length of the hand is noticeably longer and the wingtip more pointed in Fea's than in Zino's. Plumage differences between them are slight and relatively unproven. It has been suggested that Fea's has a more conspicuous M pattern on the upperwing and a broad black bar across the median underwing-coverts that fades or narrows at the carpal joint, and a uniform pale grey upper tail. **VOICE** At colonies gives a long, drawn-out, mournful wailing somewhat reminiscent of Tawny Owl, which ends in a hiccup that is only audible at close range. Also a continuous cackling *gon-gon*, which has earned the species its local name. Silent at sea. **RANGE** Breeds only in Madeira and the Cape Verdes. Non-breeding range incompletely known; recorded recently off the E USA and from seas around the British Isles. **DISTRIBUTION** Confined as a breeder to Bugio, in the Desertas (off Madeira), and to Santiago, Fogo, Santo Antão and São Nicolau in the Cape Verdes. **BREEDING** On Bugio, from late Jul and mid–late Aug, with young fledging mid-Dec to mid-Jan. On the Cape Verdes Jan–Apr. Lays one egg; nest usually sited under loose rocks on a ledge at high elevation, or in a burrow if sufficient soil is available. **HABITAT** Pelagic but comes ashore to breed in the high mountains of the Cape Verdes and the arid slopes of Bugio. **STATUS** Varies between archipelagos. **Azores** Two caught recently, one Jun 1990, the other Sep 1993 and retrapped Aug 1994. A very small breeding population perhaps exists. **Madeira** Population on Bugio estimated at 150–200 pairs. **Canaries★** In Jun 1998 one heard on three occasions in the Barranco del Agua (La Palma), indicating possible breeding. Otherwise, accidental in Canarian waters with *c.* 15 records, most between Tenerife and La Gomera but also between Tenerife and Gran Canaria, Gran Canaria and Fuerteventura, and also seen from land off Tenerife and Fuerteventura. **Cape Verdes** Total population estimated at 500–1,000 pairs.

ZINO'S PETREL (MADEIRA PETREL) *Pterodroma madeira* Plate 2 (4)
Sp. Petrel Freira **P.** Freira

Monotypic. **L** 32–34 cm (12–13″); **WS** 83–86 cm (33–34″). **IDENTIFICATION** Sexes and all plumages similar. Very similar to the previous species and very difficult to identify with certainty at sea. Most useful is the thinner bill but close views and/or good photographs required to evaluate this. The shorter hand and blunter wingtip, producing a distinctive wing shape, have recently been shown to be reliable identification features. The M mark on the upperwing is less pronounced than in Fea's, and Zino's has a pale mid-wing panel rather than an obvious M. Possible plumage differences include the black underwing bar becoming broader at the carpal and the darker tips to the central rectrices. **VOICE** Similar to Fea's but perhaps slightly sharper and less mournful; also a call likened to a whimpering small dog. Largely silent at sea. **RANGE** Breeds only on Madeira; non-breeding range unknown. **DISTRIBUTION** Madeira. **BREEDING** Similar to Fea's, in burrows excavated in soil protected by natural vegetation. Restricted to high mountains in C Madeira, at *c.* 1,650 m. Eggs laid May–Jun and young fledge Oct. Single-brooded and lays one egg. **HABITAT** Pelagic, only coming to land to breed. **STATUS** All records from the Madeiran archipelago. **Madeira** Population until recently placed at no more than 30–40 pairs and, despite an increase in the numbers of fledged young, still extremely conservation-dependent. In 2003 a new colony of 15 pairs was discovered near the Pico do Arieiro and in 2004 another small colony was found, bringing the total population to an estimated 65–80 pairs.

[SOFT-PLUMAGED PETREL *Pterodroma mollis*] Plate 2 (5)
Sp. Petrel Suave **P.** Freira-de-penas-lisas

P. m. mollis. **L** 33–36 cm (13–14″); **WS** 86–92 cm (34–36″). **IDENTIFICATION** Sexes and all plumages similar. Almost identical to the two previous species but fairly easily distinguished by the complete or near-complete greyish breast-band, which is always much bolder and more obvious than in Fea's or Zino's Petrels. Also, the upper tail is darker in Soft-plumaged than in either of its N Atlantic counterparts. In terms of measurements, Soft-plumaged Petrel is intermediate between Fea's and Zino's, both in body and wing lengths; any structural differences are unlikely to be useful at sea. Wing shape also intermediate between Fea's and Zino's. **VOICE** Usually silent away from the colonies and unlikely to be heard in our region. **RANGE** Breeds on Tristan da Cunha and Gough (S Atlantic); on Prince Edward, Marion, Crozet and Amsterdam (S Indian Ocean); the Antipodes, off New Zealand, and in Tasmania. Post-breeding, ranges across Southern Ocean, from South America via the Atlantic and Indian Oceans to New Zealand and the SW Pacific. **STATUS** Extreme vagrant with only one record. **Madeira** Two, between Madeira and the Desertas, 9 Sep 1986. No photographs or specimens were obtained, and thus considered 'dubious' by Zino *et al*. (1995).

BULWER'S PETREL *Bulweria bulwerii* Plate 2 (7)
Sp. Petrel de Bulwer   **P.** Alma-negra (Pardela de Bulwer)

Monotypic. **L** 26–28 cm (10–11″); **WS** 68–73 cm (27–29″). **IDENTIFICATION** Sexes and all plumages similar. An all-dark, sooty-brown petrel with long narrow wings, long tail and a prominent pale bar across the upperwing-coverts. Tail wedge-shaped but this is rarely obvious as the tail is usually held closed in flight, when the wings are held forward and clearly angled backwards at the carpal. Flight erratic and twisting and rarely more than 1–2 m above the sea's surface. A potential confusion risk is Swinhoe's Storm-petrel, but this is smaller with shorter, less angled wings and a shorter, forked tail. **VOICE** A low barking hoot, repeated at varying speeds and pitch, given at colonies. Usually silent at sea. **RANGE** Occurs in tropical and subtropical waters of the Atlantic, Pacific and Indian Oceans, but only known to breed on islands in the Atlantic and Pacific. **DISTRIBUTION** Breeds on the Canaries, the Salvage Islands, Porto Santo group and Desertas of Madeira, Raso and Ilhéu de Cima in the Cape Verdes, and on Ilhéu da Vila, off Santa Maria, in the Azores. **BREEDING** On the Azores, Madeira, Salvage Islands and Canaries does not return to colonies before mid Apr. Egg laying occurs late May-mid Jun, and fledging is in mid-late Sep. However, in the Cape Verdes the season is more protracted, from Feb to Sep. Breeds in small holes and crevices, under boulders and sometimes in holes in walls, apparently not in burrows. **HABITAT** Predominantly oceanic, usually only encountered on land at night and only during breeding season. Tends to prefer small, uninhabited islands and rocky outcrops for nesting. **STATUS** Population estimates as follows. **Azores** Probably *c*. 50 pairs. **Madeira** Approximately 10,000 pairs, the majority on the Salvage Islands and Desertas. The colony on Deserta Grande is the largest in the Atlantic and possibly the largest colony in the world. **Canaries** *c*. 1,000 pairs, of which 400–430 pairs are on islets off Tenerife, including *c*. 300 pairs on Roques de Anaga, which is the most important colony in the archipelago. **Cape Verdes** Probably no more than 100 pairs.

CORY'S SHEARWATER *Calonectris diomedea* Plate 3 (1)
Sp. Pardela Cenicienta   **P.** Cagarra (Pardela-de-bico-amarelo)

C. d. borealis. **L** 46–53 cm (18–21″); **WS** 100–125 cm (39–49″). **IDENTIFICATION** Sexes and all plumages similar. A large shearwater, greyish-brown above and white below, often with a variable amount of white on the uppertail-coverts, and a dark tail. Underwing predominantly white with a dark tip and narrow dark border. Bill mostly yellowish but, at close range, has a dark tip. In flight, wings usually held slightly bowed at the carpal to wingtip, and angled backwards from the carpal joint. Flight action best described as slow and relaxed, even languid. Subspecies *borealis* separated from the nominate by underwing pattern: *borealis* has white underwing-coverts and the underside of the flight feathers dusky, thus presenting a fairly substantial dark area at the wingtip, whereas *diomedea* has whitish inner webs to the outer primaries creating an underwing panel and reducing the dark underwing tip to a dark border. **VOICE** At colonies gives a harsh, nasal, wailing scream *kaa-ough* or *keeouwrah*. Usually silent at sea except when squabbling over food. Call of *borealis* comprises three rather short syllables. That of *diomedea* differs, being much more disyllabic and the syllables are more drawn-out. **RANGE** Breeds in the Mediterranean (*diomedea*), and on the Azores, Madeira and the Canaries (*borealis*); *diomedea* mainly winters off S Africa, while *borealis* winters off N Argentina and Brazil. Immature *borealis* spend their first summer in C and N American waters, whereas immature *diomedea* return to the Mediterranean. **DISTRIBUTION** Breeds on the Azores, Madeira (including the Salvage Islands) and the Canaries. **BREEDING** Present from mid Feb but egg laying does not start until the last week of May and continues until late Jul. Clutch one and single-brooded. Fledging is mostly mid–late Oct but can continue into early Nov. **HABITAT** Oceanic, usually only coming to land at night if breeding, when their eerie calls betray their presence. Nests under boulders and vegetation and in caves. **TAXONOMIC NOTE** With the current trend to promote many procellariiform taxa to species level, it has been suggested that the two subspecies of Cory's Shearwater be considered specifically distinct; *borealis* would then be known as Cory's Shearwater and the nominate *diomedea* as Scopoli's Shearwater. As yet there are no records of *diomedea* from our region, but it should be looked for. **STATUS** A common and widespread summer visitor, breeding in three archipelagos. **Azores** Breeds on all islands and islets, and considered abundant with an estimated 50,000–90,000 pairs. **Madeira** Breeds on the main island plus the Porto Santo group and the Desertas, and considered abundant. On the Salvage Islands the population is estimated at 13,000 pairs, which has declined from the late-19th century, when an estimated 30,000 young were slaughtered in some years. However, this colony is still one of the largest in the world. **Canaries** Breeds on all islands and islets, and considered abundant with 30,000 pairs, of which one-third breed on Alegranza; *diomedea* presumably occurs on passage, but I am unaware of any documented records. **Cape Verdes** Status poorly known but probably regular on passage.

CAPE VERDE SHEARWATER *Calonectris edwardsii* Plate 3 (2)
Sp. Pardela Cenicienta de Edwards   P. Cagarra de Cabo Verde

Monotypic. **L** 40–41 cm (16″); **WS** 90–110 cm (36–43″). **IDENTIFICATION** Sexes and all plumages similar. Smaller than similar Cory's Shearwater with a darker cap, neck and mantle. Bill also finer and dark olive-grey with a blackish culmen and tip. Underwing pattern and underparts coloration as Cory's, but wing margins darker. Size difference can be quite noticeable when the two species are together but of little use given a lone bird. Flight silhouette and action as Cory's. **VOICE** Described as a deep, laughing note *cagarrah*, from which the species acquires its local name. **RANGE** Breeds only in the Cape Verdes. Non-breeding range poorly known but presumably largely disperses to the S Atlantic, where recently recorded in S Brazil. Numbers also recorded recently in Oct off Senegal, but unknown if this reflects post-breeding dispersal or foraging trips from Cape Verdean colonies. **DISTRIBUTION** Cape Verdes: breeds only on Santiago, Brava, Santo Antão, Branco, Raso and São Nicolau, with possibly a few pairs still on Sal and Boavista. **BREEDING** Arrives at colonies late Feb/early Mar. Clutch one and single-brooded. Eggs laid Jun–Jul and young fledge late Sep–Nov. **HABITAT** Oceanic, only coming to land at night when breeding. Nests in hollows, on rocky ledges on cliffs and offshore rocks and islets, and under boulders. **STATUS** Common in summer around breeding islands, but a vagrant elsewhere. **Madeira** Record of *c.* 150, Ponta da Cruz (Madeira), 25 Mar 2001, subsequently withdrawn by observer. **Canaries*** Three records: two, between Tenerife and La Palma, 12 Oct 1987; one recovered from fish quay, Santa Cruz (Tenerife), 21 Oct 1992; and one, between Tenerife and La Gomera, 13 Oct 1998. **Cape Verdes** Population currently estimated at *c.* 10,000 pairs, although this total perhaps optimistic. An estimated 7,500 chicks are killed for food by local inhabitants annually, giving grounds for concern as to the future of this taxon, which has only recently been recognised specifically.

GREAT SHEARWATER *Puffinus gravis* Plate 3 (3)
Sp. Pardela Capirotada   P. Pardela-de-barrete (Pardela-de-bico-preto)

Monotypic. **L** 43–51 cm (17–20″); **WS** 100–118 cm (39–47″). **IDENTIFICATION** Sexes and all plumages similar. A large shearwater but, despite its name, slightly smaller than Cory's; best distinguished from latter by its browner upperparts, darker clear-cut cap, white hind neck, more prominent white band on uppertail-coverts and dark bill. Distinct dark patch on belly, dark undertail-coverts, and underwing has a variable dark diagonal line across the axillaries and underwing-coverts. In flight wings held more stiffly and wingbeats more rapid than Cory's. **VOICE** Generally silent at sea but utters a harsh squeal when squabbling for food. **RANGE** Breeds only on the Tristan da Cunha group, Gough and the Falklands (S Atlantic). Post-breeding undertakes a transequatorial, anticlockwise, journey around the Atlantic, reaching *c.* 66°N. **STATUS** Previously considered accidental, probably due to a relative lack of observers, the species is certainly regular in autumn, although it is not common, except in the Azores. **Azores** An exceptional *c.* 28,000 in the Canal de São Jorge, between São Jorge and Pico/Faial, 29 Aug 1990, is by far the largest concentration recorded in our region and probably in the W Palearctic. **Madeira** Previously considered occasional, i.e. a species that appears seasonally in unusual weather conditions; with increased observer effort in recent years it seems to be a regular passage migrant in autumn. **Canaries** Recorded almost annually in small numbers in autumn, and very rarely in spring. **Cape Verdes** Records from nearby suggest it is probably a regular visitor to Cape Verdean seas in autumn to early winter.

SOOTY SHEARWATER *Puffinus griseus* Plate 3 (4)
Sp. Pardela Sombría   P. Pardela-preta

Monotypic. **L** 40–51 cm (16–20″); **WS** 94–109 cm (34–43″). **IDENTIFICATION** Sexes and all plumages similar. A large shearwater, but smaller than both Cory's and Great Shearwaters. Easily identified by its generally dark, sooty-brown plumage with a varied amount of silvery white on the underwing-coverts. At distance and in poor light can appear all dark. Flight powerful and direct, on rather narrow, pointed wings swept back at the carpal joints in strong winds. **VOICE** Silent away from colonies. **RANGE** Breeds in the S Atlantic and Pacific, and undertakes a transequatorial post-breeding migration. **STATUS** Like Great Shearwater considered accidental, but probably more frequent than records suggest. Status varies between the archipelagos. **Azores** Since 1997 at least 12 records of 44 birds, and probably best considered a scarce migrant rather than a vagrant, as previously thought. **Madeira** First recorded in 2000 but records now suggest that it is a regular passage migrant in small numbers, particularly in autumn. **Canaries** Accidental visitor recorded on at least 15 occasions, usually Aug/Sep but also Apr/May and Jul, with records from all main islands except La Palma and El Hierro. **Cape Verdes** One, 56 km east of Maio, 15 Apr 1976, and one, between Raso and São Nicolau, 12 Apr. 2001.

BALEARIC SHEARWATER   *Puffinus mauretanicus*   Plate 3 (6)
Sp. Pardela Balear   **P.** Fura-bucho

Monotypic. **L** 30–38 cm (12–15″); **WS** 76–89 cm (30–35″). **IDENTIFICATION** Sexes and all plumages similar. Smaller than Sooty Shearwater but averages slightly larger than Manx. Plumage can vary considerably, with some approaching that of Sooty whilst others are very similar to Manx. Generally paler, duskier brown than Manx and lacks sharp contrast between upperparts and underparts of latter species. Underwing shows less white than Manx and usually has dusky axillaries. Best separated from Sooty by noticeably smaller size and presence of pale feathering on underparts. Faced with a bird in which the underparts appear all dark, the shorter wing, more pot-bellied appearance and less dynamic flight action of Balearic are useful features. **VOICE** Usually silent at sea. **RANGE** Breeds W Mediterranean, dispersing into Atlantic. **TAXONOMIC NOTE** Until relatively recently treated as a subspecies of Manx Shearwater but now generally recognised as specifically distinct. **STATUS** Vagrant but probably more frequent than records suggest. **Azores★** One, off Mosteiros (São Miguel), 27 Sep 1998; two, off Fajã Grande (Flores), 13 Oct 1999; one, off Ponta Delgada (São Miguel), 30 Oct 2000; and one, off Mosteiros (São Miguel), 19 Dec 2003, are the only records. **Madeira** First recorded 1986 but increased seawatching since late 1990s has yielded more sightings and the species is apparently regular in small numbers, particularly in autumn. **Canaries★** At least eight records: one, Puerto Rico (Gran Canaria), 13 Apr 1983; two, Punta de la Rasca (Tenerife), 6 Sep 1988; two, La Caleta (Tenerife), 11 May 1989; one, between Gran Canaria and Tenerife, 20 Jul 1989; five, Corralejo (Fuerteventura), 28 Oct 1995; two, Playa de Sotavento (Fuerteventura), 12 Nov 1996, and one there, 6 Nov 1997; one, La Gomera, 14 Sep 1999; and one, between Tenerife and La Palma, 25 Sep 2001.

MANX SHEARWATER   *Puffinus puffinus*   Plate 3 (5)
Sp. Pardela Pichoneta   **P.** Fura-bucho do Atlântico (Pardela-sombria)

Monotypic. **L** 31–36 cm (12–14″); **WS** 76–85 cm (30–33½″). **IDENTIFICATION** Sexes and all plumages similar. A medium-sized shearwater with strongly contrasting blackish-brown upperparts and white underparts. Dark cap extends below the eye and upperwing uniform. In Macaronesian Shearwater (the main confusion species) of the race *baroli*, the white face extends above the eye and the upperwing has a paler silvery panel on the secondaries. The underwing of Manx Shearwater is mainly white with a broad dark tip and trailing edge; the wing is also longer and more pointed than in Macaronesian Shearwater. Flight action a series of fast, shallow, stiff-winged wingbeats followed by a long, low glide, often with the bird swinging from side to side. **VOICE** On the breeding grounds a series of raucous, cackling and crooning calls with much variation in pitch, e.g. *cack-cack-cack-carr-hoo*. Like many other Procellariidae, usually silent at sea. **RANGE** Breeds in the N Atlantic between Feb and Aug, then undertakes a transequatorial migration. **DISTRIBUTION** Breeds in Madeira, the Azores and fairly recently discovered breeding in the Canaries. **BREEDING** Nests on ledges in walls of damp, inaccessible rocky gullies, and excavates burrows in suitable locations, often between the roots of trees. Clutch one and is single-brooded. Eggs laid Mar–Apr, and most chicks fledge Jul but some still in the nest mid Aug. **HABITAT** Pelagic, coming ashore to breed in vegetated rocky gullies in montane regions and on uninhabited islands. **STATUS** Breeds on three archipelagos but only a vagrant to the Cape Verdes. **Azores** Population estimated at 115–235 pairs, mainly on Corvo and Flores. **Madeira** A common breeder on Madeira and formerly on the Desertas and the Porto Santo group. **Canaries** Population estimated at c. 200 pairs on La Palma and <20 pairs on Tenerife. Both islands may harbour larger colonies, and the species may also nest on El Hierro and La Gomera. **Cape Verdes** Four records but probably more regular: one, at 17°40′N 21°51′W, 5 Sep 1973; four, near 14°16′N 23°17′W, 6 Oct 1973; five, near Santa Luzia, 5 Oct 1976; and two, 15°51′N 23°37′W, 31 Dec 1983.

MACARONESIAN SHEARWATER (LITTLE SHEARWATER)   *Puffinus baroli*   Plate 4 (1)
Sp. Pardela Chica   **P.** Pintainho (Pardela-pequena)

P. b. baroli (Canaries, Azores and Madeira), *P. b. boydi* (Cape Verdes). **L** 20–25 cm (8–10″); **WS** 58–67 cm (23–26″). **IDENTIFICATION** Sexes and all plumages similar. Macaronesian Shearwater of the nominate race *baroli* is a small black-and-white shearwater distinguished from Manx Shearwater by its blacker upperparts, more extensive white face, smaller size, shorter and more rounded wings, narrower dark tip and trailing edge to the underwing, and more rapid flight action, which consists of a series of stiff, flapping wingbeats followed by a short, low glide. The Cape Verde race *boydi* is separable from *baroli* by the dark crown extending onto the head-sides to just below the eye, browner upperparts, broader underwing margins and predominantly dark undertail-coverts. Best separated from Balearic Shearwater by underwing pattern, which in *boydi* lacks dusky axillaries. Obviously, size and shape are also useful in attempting to separate *boydi* from Balearic Shearwater. **VOICE** On the breeding grounds *baroli* utters a 5–8-syllable laughing call, a high-pitched and frequently repeated *ha HA ha-ha-ha HOOOOO* with the emphasis on the second note and the drawn-out final note. Flight call of *boydi* described as a crowing *karki-karroo*. **RANGE** Elsewhere the species breeds on islands in the Southern Ocean. Wintering areas are poorly known but many remain near the breeding areas. **DISTRIBUTION** In the Azores, breeds on Ilhéu da Vila, off Santa Maria, and on Ilhéu da Praia and Ilhéu de Baixo, off Graciosa. Also recorded nesting on Corvo, São Miguel (not recently) and São Jorge. In Madeira breeding confined to the Desertas and to Baixo and Cima in the Porto Santo group. On the Canaries breeding only confirmed on Tenerife and on islets off Tenerife and Lanzarote, but probably breeds on islets off all main islands. In the Cape Verdes *boydi* breeds on most islands and islets, except Maio, and is extinct on Sal. Some remain around the various archipelagos all year. **BREEDING** Season prolonged and variable, not only between, but also within archipelagos. In the Canaries it is presumed that most fledge in May/early Jun. Generally breeds earlier than most other Procellariidae in the region. Clutch contains one egg. The nest is located in a crack, a small cave, under rocks or in a burrow, and the species is single-brooded. Heard at the colonies in all months. **HABITAT** Mainly pelagic but returns to land more frequently in the non-breeding season than other Procellariiformes in our region. **TAXONOMIC NOTE** Subspecies *boydi* has been treated as a separate species, Cape Verde Little Shearwater *P. boydi* (Hazevoet 1995). Prior to 2005, the two subspecies of Macaronesian Shearwater were considered to belong to Little Shearwater *P. assimilis*; the latter is now considered to be a solely southern-hemisphere taxon. **STATUS** Breeds throughout the region. **Azores** Recent estimate of *c.* 150 pairs on the islets of Vila, Baixo and Praia. **Madeira** A common breeder on all three of the Desertas. On the Salvages 15,000 pairs were estimated in a colony on Selvagem Grande in 1981, although this was revised to *c.* 2,050 pairs in 1995. **Canaries** In 1987 there were thought to be fewer than 400 pairs. **Cape Verdes** The population of *boydi* is reportedly several thousands.

WILSON'S STORM-PETREL   *Oceanites oceanicus*   Plate 4 (5)
Sp. Paíño de Wilson   **P.** Paínho-casquilho (Casquilho)

Subspecies in our region unknown. **L** 15–19 cm (6–7½"); **WS** 38–42 cm (15–16½"). **IDENTIFICATION** Sexes and all plumages similar. Larger than European Storm-petrel, from which only safely separated by the following features, which are usually only visible at close range; the lack of a pale stripe on the underwing, a more pronounced pale carpal bar, feet extending beyond the tail in flight, yellow webs between the toes (only visible at very close range when feeding), and wing shape, namely the distinctly straighter trailing edge and smoothly curved leading edge, lacking any obvious kink at the carpal joint. Also, the white rump-band extends well onto the ventral region, unlike in European Storm-petrel. The flight is more direct than European Storm-Petrel, and lacks the bat-like quality of the latter. It is much less erratic than Leach's Storm-petrel and lacks the shearing glides of Madeiran Storm-petrel. **VOICE** Usually silent away from the breeding grounds, but feeding birds sometimes give a rapid chattering or squeaking call. **RANGE** Breeds on many subantarctic islands and the Antarctic littoral, and performs a transequatorial migration through the Atlantic, Indian and, to a lesser extent, Pacific oceans. **STATUS** Probably a regular summer visitor but difficulties in identifying this species probably account for the relatively few records. **Azores** Only three records but should occur much more regularly: specimens were obtained by Godman, 30 miles from Faial, 21 May 1865; one, presented by Carlos Bicudo de Medeiros to the Carlos Machado museum, 8 Jul 1950; and one, captured on Ilhéu do Baixo (Graciosa), 11 Jul 1990. **Madeira** Listed as an exceptional visitor with fewer than five records in the last 50 years. **Canaries** Considered accidental with just over ten records, but presumably regular in seas around the islands. **Cape Verdes** Status uncertain but extralimital records suggest that it is not uncommon in Sep–Oct in surrounding seas.

WHITE-FACED STORM-PETREL   *Pelagodroma marina*   Plate 4 (2)
Sp. Paíño Pechialbo   **P.** Paínho-de-ventre-branco (Calca-mar)

P. m. hypoleuca (Canaries and Salvages), *P. m. eadesi* (Cape Verdes). **L** 20–21 cm (8–8½"); **WS** 41–43 cm (16–17"). **IDENTIFICATION** Sexes and all plumages similar. Much easier to identify than the black-and-white storm-petrels, and given a reasonable view is unmistakable. Combination of white underparts with grey breast-side patches, white forehead and supercilium curving down behind ear-coverts to join white throat, as well as grey rump and broad dark arcs on upperwing, eliminate all other N Atlantic storm-petrels. In flight, the feet project well beyond the tail, the wings are broad and, when feeding, it characteristically bounces from side to side off the sea surface. Cape Verde race *eadesi* has a whiter forehead, an indistinct whitish hindneck collar and slightly paler upperparts, but all are difficult to observe and racial identification is probably impossible at sea. **VOICE** On the breeding grounds, a thin, short, monotonous *koo* repeated slowly over long periods. Usually silent at sea. **RANGE** Breeds in the N and S Atlantic, and in Australian and New Zealand waters. Disperses into the Atlantic, Pacific and Indian oceans in the non-breeding season. **DISTRIBUTION** Breeds on the Salvage Islands, Montaña Clara, off Lanzarote, in the Canaries, and on Branco, Ilhéu de Cima (Ilhéus do Rombo), Ilhéu dos Pássaros (Boavista) and Ilhéu Laje Branca (Maio), in the Cape Verdes. **BREEDING** Nests in sandy and soft soils where burrows easily excavated. Clutch contains one egg, which in the Cape Verdes is laid Jan–Mar but on the Canaries in Mar–early Apr. Exceptionally lays June. Fledging occurs on the Cape Verdes in May–July and on the Canaries in July. **HABITAT** Mainly pelagic, only coming ashore at night and mostly when breeding. **STATUS** Varies between archipelagos. **Azores** One record: one obtained, 6 May 1912. **Madeira** Accidental/rare passage migrant, except at the Salvages, where the breeding population has been estimated at up to 36,000 pairs. **Canaries** Breeding population estimated at 24–30 pairs, otherwise a very rare passage migrant in Canaries waters. **Cape Verdes** Population recently placed at 5,000–10,000 pairs, with the largest colonies on Ilhéu de Cima (Ilhéus do Rombo) and Ilhéu Laje Branca (off Maio).

WHITE-BELLIED STORM-PETREL *Fregetta grallaria* Plate 4 (3)
Sp. Paíño Ventriblanco   **P.** Paínho de Vieillot

Subspecies in our region unknown but *F. g. leucogaster* would seem the most likely. **L** 18 cm (7″); **WS** 44 cm (17″). **IDENTIFICATION** Sexes and all plumages similar. Fairly large blackish-brown, white-rumped storm-petrel with prominent white belly and underwing-coverts, obvious carpal bar and greyish mantle. Flight direct and follows contours of the sea, the body often swinging side to side. None of the regular N Atlantic storm-petrels has a white belly except White-faced Storm-petrel, which differs in having a white breast, patterned face, paler upperparts and a very different flight action. **VOICE** Normally silent at sea. **RANGE** Breeds on subtropical islands in the Southern Ocean, dispersing to tropical waters. **STATUS** An extreme vagrant with just one record. **At sea** One, at 23°48′N 22°01′W north of the Cape Verdes, 17 Aug 1986.

EUROPEAN STORM-PETREL *Hydrobates pelagicus* Plate 4 (4)
Sp. Paíño Común   **P.** Paínho-de-cauda-quadrada (Alma-de-mestre)

Monotypic. **L** 14–17 cm (5½–7″); **WS** 36–39 cm (14–15″). **IDENTIFICATION** Sexes and all plumages similar. The smallest storm-petrel in the region, further distinguished from other black-and-white storm-petrels by lack of a prominent carpal bar on upperwing, white band on central underwing and blackish upperparts. Almost bat-like flight action is also useful for identification. See Wilson's Storm-petrel, with which it is most likely to be confused, for other details. **VOICE** Common call, given from burrow, a quiet, purr of more than 30 notes per second ending in a hiccup-like *chikka*. At sea usually silent. **RANGE** Breeds on islands in N Atlantic and the Mediterranean, moving south in winter to the S Atlantic. **DISTRIBUTION** Breeds in the Canaries, on all islets north of Lanzarote (except Roque del Oeste), Roques de Anaga, off Tenerife, and Roques de Salmor, off El Hierro. **BREEDING** Nests Jul–Sep, under rocks, in cracks and even in caves occupied by Cory's Shearwater. Eggs laid mostly in Jul, young hatch Sep and fledge Oct. Clutch contains one egg and is single-brooded. **HABITAT** Pelagic, coming ashore only when breeding. **STATUS** Varies between archipelagos. **Madeira** Listed as exceptional with fewer than five records in the last 50 years. **Canaries** Summer visitor, breeding in small numbers, but also a passage migrant. **Cape Verdes** Unrecorded in Cape Verdean seas but records from adjacent areas suggest that it probably occurs on migration.

LEACH'S STORM-PETREL *Oceanodroma leucorhoa* Plate 4 (6)
Sp. Paíño de Leach (Paíño Boreal)   **P.** Paínho-de-cauda-forcada

O. l. leucorhoa. **L** 19–22 cm (7½–9″); **WS** 45–48 cm (18–19″). **IDENTIFICATION** Sexes and all plumages similar. A large storm-petrel with an obvious pale carpal bar, which is often more prominent than the rump. Wing shape quite distinctive, being pushed forward at the body and angled sharply back at the carpal joint. Rump less pure white than European Storm-petrel and usually split by an indistinct dark central line, but this is only visible at very close range, as is tail shape, which in this species is forked. Confusion most likely with Madeiran Storm-petrel, which see for discussion of the differences. **VOICE** Usually silent at sea. **RANGE** Breeds in the N Atlantic and N Pacific, wintering to S Africa in the Atlantic and to the equator in the Pacific. **STATUS** Records from all archipelagos. **Azores** Accidental visitor with 11 records but perhaps more common than records suggest. The most recent was a bird found exhausted in Horta (Faial), 6 Dec 2000. **Canaries** Scarce passage migrant and winter visitor with most records in late autumn. **Madeira** Vagrant but probably more common than the few records suggest. **Cape Verdes** Status uncertain but records from nearby seas suggest that it is probably not an uncommon winter visitor.

SWINHOE'S STORM-PETREL *Oceanodroma monorhis* Plate 4 (7)
Sp. Paíño de Swinhoe   **P.** Paínho de Swinhoe

Monotypic. **L** 19–20 cm (7½–8″); **WS** 44–46 cm (17–18″). **IDENTIFICATION** Sexes and all plumages similar. Similar in size and structure to Leach's Storm-petrel but differs from all other N Atlantic storm-petrels in being wholly sooty-brown. Long angled wings as in Leach's, but no white on the rump and the upperwing bar is less pronounced and less extensive. Some white on the primary shafts but this feature very difficult to observe at sea. Flight action similar to Leach's. **VOICE** In burrows gives a purring *tchir-r-r-r-r-r-r-r-r-r-r* similar to Leach's and, in flight over colonies, a chattering trill. Usually silent at sea. **RANGE** Breeds in the NW Pacific off Japan, Korea and China, migrates west into N Indian Ocean. Recent records suggest this species may breed in the N Atlantic. **STATUS** Vagrant only known from the one island group. **Madeira** All records from the Salvage Islands, where presumably a vagrant, recorded on 29 Jun 1983, 30 Jun 1988 and 23 Jul 1991, with a bird captured beside a collapsed stone wall several times in 1993–1995, on each occasion with a vascular brood-patch; this suggests that breeding was taking place nearby.

MADEIRAN STORM-PETREL *Oceanodroma castro* Plate 4 (8)
Sp. Paíño de Madeira **P.** Roquinho (Paínho da Madeira)

Monotypic. **L** 19–21 cm (7½–8″); **WS** 42–45 cm (16½–18″). **IDENTIFICATION** Sexes and all plumages similar. Differs from similar Leach's Storm-petrel in slightly shorter and broader wings, more prominent white rump and less obvious upperwing bar. Also, never shows a dark central divide on the rump, and tail only slightly forked. Flight action also different, being smoother and having more pronounced shearwater-like gliding and zigzagging. **VOICE** In flight above colonies a whistling chatter, *kair chuch-a-chuck chuk chuk*, from the burrow a guttural purring *urr-rrr-rrr-rrr-rrr*, mixed with the occasional sharp *wicka* note as a bird takes flight. Usually silent at sea. **RANGE** Breeds in the tropical and subtropical Atlantic and Pacific, dispersing over these regions when not breeding. **DISTRIBUTION** Breeds off Santa Maria and Graciosa, in the Azores, off Lanzarote and Tenerife, in the Canaries, and on Ilhéus do Rombo, Branco, Raso and off Boavista, in the Cape Verdes. **BREEDING** On the Azores, two distinct breeding seasons, hot season and cool season. Hot-season birds breed in May and cool-season birds in Sep. On Vila breeds only in the cool season, but breeding on Baixo and Praia occurs in both seasons, with the two groups occupying the same colony in Aug and early Sep. Birds on the Canaries breed Oct–Feb with most eggs laid Oct–Nov. However, on the eastern islets birds can be heard in Jul, which may be hot-season birds. Clutch size is one, and there is just one brood. **HABITAT** Highly pelagic, only coming to land to breed. **TAXONOMIC NOTE** Recent studies on the Azores have shown that birds breed during two distinct seasons, with proof of biometric and genetic differences between the populations, suggesting that two species may be involved; further work is in progress. Research on Madeira and the Cape Verdes has revealed that there are hot-season and cool-season populations in these archipelagos as well. **STATUS** Breeds throughout the region. **Azores** Nests on Vila islet off Santa Maria and on Baixo and Praia off Graciosa, with an estimated population of 800 pairs, consisting of 200 hot season, on Baixo and Praia, and 600 cool season (400 on Baixo and Praia, and 200 on Vila). **Madeira** Both hot-season and cool-season birds on the Desertas, and also breeds on Baixo and Cima off Porto Santo, and on the Salvage Islands. **Canaries** Breeds on Roque del Este, Alegranza, Roque del Oeste and Montaña Clara off Lanzarote, and on Los Roques de Anaga off Tenerife, but the population probably numbers fewer than 300 pairs in total. **Cape Verdes** Total population perhaps no greater than *c.* 1,000 pairs.

WHITE-TAILED TROPICBIRD *Phaethon lepturus* Plate 5 (1)
Sp. Rabijunco Menor **P.** Rabijunco-pequeno

Subspecies in our region unknown but *P. l. ascensionis* from the C Atlantic or *P. l. catesbyi* from the Caribbean are the only real possibilities. **L** 37 cm (14½″) plus tail streamers of 33–45 cm (13–18″); **WS** 90–95 cm (35½–37½″). **IDENTIFICATION** Sexes similar. A small tropicbird, mostly white except a black line from the gape through the eye and two black patches on the upperwing, one on the primaries and the other a diagonal line across the inner secondaries and median coverts. Only likely to be confused with Red-billed Tropicbird, which is distinctly larger. Combination of two black patches on each upperwing, orange bill and long white tail streamers diagnostic. Lack of barring on wing-coverts and mantle in adult plumage also separates this species from Red-billed Tropicbird. Juvenile similar to that of Red-billed Tropicbird but lacks nuchal collar and barring on upperparts and wing-coverts coarser and more widely spaced. **VOICE** Usually silent away from colonies so unlikely to be heard in our region. **RANGE** Breeds on islands and disperses throughout tropical oceans. **STATUS** Extreme vagrant with just one recent record. **Cape Verdes** Adult, Ilhéu de Curral Velho (Boavista), 20 Feb 1999; this is the only W Palearctic record.

RED-BILLED TROPICBIRD *Phaethon aethereus* Plate 5 (2)
Sp. Rabijunco Etéreo **P.** Rabijunco

P. a. mesonauta. **L** 90–105 cm (35½–41″) including tail streamers of 45–50 cm (18–20″); **WS** 100–110 cm (39–43″). **IDENTIFICATION** Sexes similar. Combination of red bill, long tail streamers and generally white plumage with black-barred upperparts, black outer primaries and an obvious black eye-stripe, make adult almost unmistakable. Juvenile also predominantly white but lacks tail streamers, has a yellowish bill with a black tip and the dark eye-stripe meets on the nape forming a nuchal collar. **VOICE** Common call a loud, shrill whistle given at sea. At breeding colonies an extended series of shrill, grating whistles and cackles given in display flight. Also calls from ground, delivering a similar mixture of piercing whistles, yells and screams. **RANGE** Breeds in the tropical and subtropical regions of Atlantic, E Pacific and Indian oceans. **DISTRIBUTION** Breeds on the Cape Verdes on Sal, Boavista, Santiago, Brava, Raso and Ilhéus do Rombo, but a vagrant to other island groups, although may have bred on La Gomera, Canaries, in 1988, and confirmed on Baixo islet, off Graciosa in the Azores, in Sep 1993. **BREEDING** Nests in holes, cavities and crevices in sea cliffs, and under boulders. Season protracted, perhaps peaking Jan–May. Clutch one and just one brood, but does relay after egg loss. **HABITAT** Pelagic away from colonies. **STATUS** Vagrant or accidental, but breeds in the Cape Verdes. **Azores** Two recent records: two, *c.* 150 km SW of Graciosa, late Oct 1991; and an adult incubating an egg, Baixo islet (Graciosa), Sep 1993. The 1993 record was the first breeding record in Europe. Measurements of both the bird and egg are within the range of the tropical N Atlantic subspecies *mesonauta*. **Madeira** Four records: one obtained, between Madeira and

Porto Santo, 3 Aug 1893; two, 5–6 miles off Porto Santo, 19 Sep 1966; one, *c.* 50 miles SSW of the Desertas, 22 Jul 1997; and one, between Madeira and the Desertas, 3 Sep 2001. **Canaries** Nearly 20 records, all since 1987. In recent years recorded almost annually, suggesting that a pair could be breeding in the archipelago. One of the early records concerned two, entering a hole in a cliff on La Gomera, Mar 1988. More recently two displaying, Puerto del Rosario (Fuerteventura), Apr 1998. **Cape Verdes** Population *c.* 100–125 pairs, declining because local people catch and eat them.

RED-FOOTED BOOBY *Sula sula* Plate 5 (4)
Sp. Piquero Patirrojo   **P.** Atobá-de-pés-vermelhos

S. s. sula. **L** 66–77 cm (26–30″); **WS** 124–142 cm (49–56″). **IDENTIFICATION** Sexes similar. Two colour morphs, pale and dark, as well as intergrades. Pale morph recalls Northern Gannet but much smaller, mainly white with black primaries and secondaries, and a black carpal patch on underwing. Dark morph wholly brown or greyish-brown. The white-tailed brown morph is like dark morph but has lower back, rump, tail and vent white or yellowish-white. All morphs have red feet. Juvenile streaked brownish on fledging and separated from adult brown morph by darker bill and yellowish-grey legs and feet. **VOICE** Silent at sea. **RANGE** Breeds on tropical islands in Atlantic, Pacific and Indian oceans. **STATUS** Vagrant with only one record. **Cape Verdes** A white-tailed brown morph in a colony of Brown Boobies, Ilhéu de Cima (Ilhéus do Rombo), 24 Aug 1986.

BROWN BOOBY *Sula leucogaster* Plate 5 (3)
Sp. Piquero Pardo   **P.** Alcatraz-pardo

S. l. leucogaster. **L** 64–74 cm (25–29″); **WS** 132–150 cm (52–59″). **IDENTIFICATION** Sexes similar. Adult has chocolate-brown head and upperparts contrasting with white underparts and white central underwing. Bill yellowish and legs and feet yellowish-green. Juvenile similar to adult but white areas mottled brown and the brown areas somewhat duller, whilst the bill is greyish. **VOICE** Calls normally given at colonies include a high-pitched, hissing whistle by ♂ and various grunts and honks by ♀ and young. **RANGE** Breeds in tropical Atlantic, Pacific and Indian oceans, rarely strays outside tropical and subtropical waters. **DISTRIBUTION** Breeds only in the Cape Verdes. **BREEDING** Year-round on steep or flat coastal cliffs. Eggs, downy chicks, large young and juveniles recorded in all months. Clutch contains 1–3 eggs. **HABITAT** Usually shallow seas near colonies, but occasionally wanders far from shore. **STATUS** Recorded from three archipelagos but only as a vagrant away from the Cape Verdes. **Azores** One record: one obtained, Ilhéu da Vila Franca (São Miguel), 23 May 1966, which location is incorrectly cited in Bannerman & Bannerman (1966) as being off Nordeste (G. Le Grand, pers. comm.). **Canaries*** One record: immature, Punta de Teno (Tenerife), 28 Nov 1995. **Cape Verdes** Breeds on Santiago, Brava, Ilhéus do Rombo, Raso, Boavista and possibly on Ilhéu de Rabo de Junco, off Sal. Like other seabirds in this archipelago, population has decreased due to human depredation and is currently estimated at fewer than *c.* 1,000 pairs.

NORTHERN GANNET *Morus bassanus* Plate 5 (5)
Sp. Alcatraz Atlántico   **P.** Alcatraz

Monotypic. **L** 87–100 cm (34–39″); **WS** 165–180 cm (65–71″). **IDENTIFICATION** Sexes similar. Adult easily recognised as mainly white, with black wingtips and a creamy yellow hue to the rear head. Juvenile generally grey-brown above with white spotting on upperwing; from second year resemble adults but amount of white increases with age. Confusion with Cape Gannet *M. capensis* must be considered, as the latter is a potential vagrant to the region. Cape Gannet can be identified from its third year by the all-dark tail and dark secondaries, but juvenile and second-year indistinguishable from Northern. **VOICE** Typical call a harsh *urrah* given when feeding communally. **RANGE** Breeds on rocky islands and cliffs in N Atlantic, disperses in winter as far south as coastal W Africa. **STATUS** Varies between archipelagos but is generally a scarce winter visitor. **Azores** Scarce winter visitor. **Madeira** Regular winter visitor and passage migrant in small numbers; also recorded off the Salvage Islands. **Canaries** Regular winter visitor and passage migrant. **Cape Verdes** Only two records: one ringed on Bass Rock in Scotland was recovered in the islands, 18 Apr 1958; and two immatures, at 14°16′N 23°17′W (noon position), on 6 Oct 1973. There is also a record of a second-year gannet (species unknown) north of Brava on 1 Jan 1988.

(GREAT) CORMORANT *Phalacrocorax carbo* Plate 6 (1)
Sp. Cormorán Grande   **P.** Corvo-marinho-de-faces-brancas (Corvo-marinho)

Possibly *P. c. carbo* on Azores; *P. c. maroccanus*, *P. c. sinensis* (Canaries and Madeira); *P. c. lucidus* (Cape Verdes). **L** 80–101cm (31½–40″); **WS** 130–160 cm (51–63″). **IDENTIFICATION** Sexes similar. Adult generally dark with a bluish or greenish sheen on body and bronze sheen on wings, and a white throat and thigh patch. Juvenile and immature browner above with dirty white underparts that progressively darken with age; juvenile has quite extensively white underparts but immature darker with each moult, and by second autumn is like adult. Non-breeding adult lacks white thigh patch and white on the head duller. In some subspecies white on throat continues onto neck, as in *sinensis*, and even the upper breast in *maroccanus* and *lucidus*. **VOICE** Usually silent away from the breeding colonies. **RANGE** Throughout the Old World and E Canada. **TAXONOMIC NOTE** Subspecies *lucidus* is often treated as a full species, the White-breasted Cormorant *P. lucidus*. **STATUS** Accidental in winter. **Azores** Vagrant with just four records: a specimen in Carlos Machado museum was obtained on São Miguel prior to 1888; two, Lagoa das Furnas (São Miguel), 2–10 Nov 2001 (one still present 6 Dec 2001); one, Corvo 5–8 Apr 2002; and one, Mosteiros (São Miguel), 16 Dec 2003. **Madeira** Occasional in winter; also recorded on the Salvages. **Canaries** Accidental, recorded on Lanzarote, Fuerteventura, Gran Canaria, Tenerife and El Hierro. **Cape Verdes** Possibly bred in the past but no records since one collected, Sal Rei (Boavista), 17 Mar 1924.

DOUBLE-CRESTED CORMORANT *Phalacrocorax auritus* Plate 6 (2)
Sp. Cormorán Orejudo **P.** Corvo-marinho-d'orelhas

Subspecies unknown, but nominate *P. a. uritus* is most likely. **L** 74–86 cm (29–34"); **WS** 122–137 cm (48–54"). **IDENTIFICATION** At all ages averages smaller than Great Cormorant, but difficult to separate from latter. Probably best identified by shape and colour of bare facial skin. This is more extensive in Double-crested than in Great Cormorant, and can seem more orange. On the upper lores Double-crested often has a pale orange stripe, which is not usually present in Great. The underside of the bill has no gular feathering in Double-crested, whereas Great has a prominent wedge of such feathering. Double-crested has a proportionately smaller bill than Great and it is usually pale yellow rather than ivory. Also, Double-crested has only 12 tail feathers whereas Great has 14, but this is difficult to see and one must beware Great in moult. In flight, tends to show a more pronounced neck kink, but this is difficult to judge and is not diagnostic. Juvenile Double-crested lacks the pale central belly of Great. Possibly most similar are subadult Double-crested and subadult Great of the African subspecies *maroccanus* or *lucidus*. Generally, African races of Great Cormorant have a paler central belly and the ear-coverts are more contrastingly pale compared to Double-crested. **VOICE** Silent away from colonies. **RANGE** Breeds on coasts from the Aleutians and Gulf of St Lawrence south to Cuba, Baja California and Yucatán, and inland in S Canada and N USA. Winters mostly in breeding range but inland breeders winter coastally, many on SE coasts and in the Caribbean. **STATUS** Vagrant with all records in one archipelago, where observations in recent years suggest the species may occur annually. **Azores** One, Mosteiros (São Miguel), 24–26 Oct 1991; two, reported on Faial, Nov 1995; immature, Lagoa Comprida (Flores), 8 Nov 1998; one, Madalena (Pico), 25–26 Sep 1999; one, Angra do Heroísmo (Terceira), 9 Oct 2000–23 Feb 2001 at least; one, Santa Cruz (Flores), 13 Oct 2000–12 Feb 2001, with another there 7–12 Feb 2001; one, Vila do Franco Campo (São Miguel), 3–4 Nov. 2000; one, Lagoa das Furnas (São Miguel), 12 Nov 2000; and one, Madalena (Pico), 13–15 Feb 2001. In 2002 there was an influx of at least 16, including 11 together on Flores and groups of six on Faial, five on Corvo, five on Pico and three on São Miguel. The first birds were at Lagoa Rasa and Santa Cruz (Flores), 3 Oct. 2002, and the last were five, off Pico, 23 Mar 2003 and one, Horta harbour, until 12 Apr 2003. In autumn 2003 another three, single birds on Corvo, Flores and Faial.

(EUROPEAN) SHAG *Phalacrocorax aristotelis* Plate 6 (3)
Sp. Cormorán Moñudo **P.** Galheta (Corvo-marinho-de-crista)

Subspecies unknown but *P. a. riggenbachi* is a possibility. **L** 65–80 cm (25½–31½"); **WS** 90–105 cm (35½–41"). **IDENTIFICATION** Sexes similar. The smallest *Phalacrocorax* likely to be seen in the region, with a noticeably thinner bill than either of the other species. Adult all dark with a greenish gloss to the body and a purple gloss to the wings, a small crest on the forehead and a yellow gape. Compared to Great Cormorant lacks any white on the face and head or on the thighs. Winter adult similar to breeding but less glossy and lacks the tuft on forehead. Juvenile darker than Great Cormorant, except for dark individuals of the race *P. carbo sinensis*, and has a more discrete pale throat. In all plumages the smaller, more rounded head is a useful aid to identification. **VOICE** Silent away from colonies. **RANGE** Breeds on coasts of NW Europe, the Mediterranean and NW Africa. Winters nearby, except those breeding in N Scandinavia, which move south. **STATUS** Extreme vagrant with only two records, from different archipelagos. **Madeira** One, Porto Santo, 16 Apr 1994, had apparently been there some days. **Canaries*** Two, Arinaga and Playa Formes, Gran Canaria, mid Nov to early Dec 1985.

(GREAT) WHITE PELICAN *Pelecanus onocrotalus* Plate 6 (4)
Sp. Pelícano Vulgar **P.** Pelicano-branco

Monotypic. **L** 140–175 cm (55–69"), **WS** 245–295 cm (96–116"). **IDENTIFICATION** Sexes similar. The massive bill and associated pouch make this easily identified as a pelican. The adult is mostly white but in flight reveals contrasting black primaries and secondaries on both the upperwing and the underwing although this is slightly more contrasting on the latter. The juvenile is mostly dark brown above with pale tips to the feathers; it has a dark head and neck and whitish breast and belly. In flight the juvenile has a mainly dark underwing with a whitish stripe along the middle. **VOICE** Unlikely to be heard in the region as it is usually silent away from the breeding colonies. **RANGE** Breeds discontinuously from southeastern Europe across Asia and Africa. African birds are mostly resident but the European population migrates into Africa in the winter. **STATUS** An extreme vagrant to the region with just one recent record. **Cape Verdes** One at Sal Rei (Boavista) July or August 2000. A skull held privately at Sal Rei may have been from a different individual.

PINK-BACKED PELICAN *Pelecanus rufescens* Plate 6 (5)
Sp. Pelícano Roseo-gris (Pelícano Rosado)   **P.** Pelicano-cinzento

Monotypic. **L** 125–132 cm (49–52″); **WS** 265–290 cm (104–114″). **IDENTIFICATION** Sexes similar. Easily recognised as a pelican. Identified by a combination of its relatively small size (although obviously still a large bird), pale yellowish-white bill, and overall dirty ash-grey coloration, with dark flight feathers. Breeding adult bird has a greyish-white body with a pinkish tinge. Non-breeding plumage has mantle, back, scapulars and upperwing-coverts obviously browner. Juvenile plumage is similar to non-breeding adult but has browner upperparts, dark brown rather than grey tail and darker grey secondaries. **VOICE** Usually silent away from colonies. **RANGE** Breeds over much of sub-Saharan Africa and Madagascar. **STATUS** Extreme vagrant with only one record of perhaps uncertain origin. **Canaries** One, Los Cristianos/Playa de Las Americas (Tenerife), 23 Aug 1994.

MAGNIFICENT FRIGATEBIRD *Fregata magnificens* Plate 5 (6)
Sp. Rabihorcado Magnifico   **P.** Rabiforcado (Fragata)

Monotypic. **L** 89–114 cm (35–45″); **WS** 217–244 cm (85½–96″). **IDENTIFICATION** Sexually dimorphic. Unmistakable given shape and plumage. Might only be confused with other frigatebirds, but these are very unlikely to occur in the region. A very large seabird, with extremely long, narrow pointed wings, a long, slender, hooked bill, and a long, deeply forked tail. ♂ all black except orange/red gular pouch, which is grossly inflated in display. ♀ mostly blackish-brown above with a whitish collar, white breast and upper flanks, forming an inverted V, and rest of underparts black. Immature similar to ♀ but has an all-white head and an incomplete brownish breast-band. **VOICE** Usually silent away from colonies. **RANGE** Tropical E Pacific and W Atlantic, with a tiny population in NE Atlantic. **DISTRIBUTION** Confined to just one small islet, Ilhéu de Curral Velho, off Boavista in the Cape Verdes. **BREEDING** In the Cape Verdes nests on ground, unlike elsewhere where it more commonly nests in the tops of bushes or low trees. Present on breeding grounds all year but does not breed annually, and only lays one egg, resulting in a very low reproductive rate. **HABITAT** Usually encountered in warm inshore and offshore waters where it parasitises other seabirds such as tropicbirds and boobies, as well as being sufficiently agile to catch flying fish. **STATUS** Almost extinct as a breeder in the Cape Verdes, elsewhere a vagrant. **Azores** One record, a ♂ obtained, São Miguel, Nov, prior to 1903. **Canaries*** Though mentioned by Emmerson *et al.* (1994) and Moreno (1988) the only record with details is an adult ♂, Playa de Las Americas and Los Cristianos (Tenerife), 4 Nov 1998, but given the similarity between ♂ and Ascension Island Frigatebird *F. aquila* perhaps best treated as 'frigatebird sp.' **Cape Verdes** Breeding population now almost extinct with fewer than five pairs in the archipelago.

EURASIAN BITTERN (GREAT BITTERN) *Botaurus stellaris* Plate 7 (1)
Sp. Avetoro Común   **P.** Abetouro (Abetouro-comum)

B. s. tellarsi. **L** 70–80 cm (27½–31½″); **WS** 125–135 cm (49–53″). **IDENTIFICATION** Sexes similar. Fairly large, predominantly brown heron. Upperparts heavily marked with black streaks and bars, and underparts also quite heavily streaked. Crown black, as is moustachial streak, legs green and bill greenish-yellow. Juvenile similar to adult but crown and moustachial streak brown and thus less conspicuous. Usually very secretive and most active dawn and dusk. Easily confused with American Bittern but latter has a rufous-brown crown. Confusion also possible with immature Night Heron, which is much smaller with a shorter neck and conspicuous white-spotted wing-coverts. **VOICE** Characteristic booming unlikely to be heard in the region. In flight often gives a harsh *krau* somewhat similar to a large gull. **RANGE** Breeds in isolated pockets throughout much of Europe and across C and E Asia, also S Africa. Disperses in winter depending on climate. **STATUS** Accidental in three of the archipelagos. **Azores** Six records, four from Terceira and two on São Miguel, the most recent being one shot, Angra de Heroísmo (Terceira), Nov 1965. **Madeira** One obtained, Madeira, 20 Mar 1896; one, Porto Santo, 9 Mar, 1 Apr and 3 Apr 1903; and one obtained, Santa da Serra (Madeira), 3 Dec 1956. **Canaries** Accidental in winter, when recorded from Lanzarote, Fuerteventura, Gran Canaria, Tenerife and La Palma.

AMERICAN BITTERN *Botaurus lentiginosus* Plate 7 (2)
Sp. Avetoro Lentiginoso **P.** Abetouro-americano (Abetouro-lentiginoso)

Monotypic. **L** 60–85 cm (23½–33½″); **WS** 105–125 (41¼–49¼″). **IDENTIFICATION** Sexes similar. Resembles Eurasian Bittern but smaller and thinner billed. Probably best separated from Eurasian Bittern by the rusty crown and large black neck-side patches, and also has less extensive and less prominent upperpart markings, whilst the primaries and most secondaries are plain blackish, not barred as in Eurasian Bittern. Juvenile very similar to adult but rufous crown less extensive and black neck patches absent. **VOICE** The display call is unlikely to be heard in the region but when flushed gives a nasal *haink* and in flight a throaty *kok-kok-kok*. **RANGE** Breeds over much of N America, either migrating or dispersing in the non-breeding season. **STATUS** Accidental with 12 records from two archipelagos. **Azores** Ten records: recorded from Flores (one 1897, one 1964 and two 18 May 1966), Terceira (one cared for by Colonel José Agostinho without details, one Nov 1965 and two undated specimens) and São Miguel (♂ 9 Apr 1966, one 17 Oct 1969 and one 14 Nov 1979). **Canaries** Two records: one killed, Madre del Agua, near La Laguna (Tenerife), prior to 1893; and one, Erjos Ponds (Tenerife), 10 Dec 2002–Mar 2003.

LEAST BITTERN *Ixobrychus exilis* Plate 7 (3)
Sp. Avetorillo Panamericano **P.** Socoí-vermelho (Garça-pequena-vermelha)

I. e. exilis. **L** 28–36 cm (11–14″); **WS** 40–45 cm (16–18″). **IDENTIFICATION** Sexually dimorphic. The New World counterpart of Little Bittern, with which it could be confused, but even smaller. Like Little Bittern has pale wing patches but these are darker, and also has diagnostic chestnut greater coverts. Both sexes can be distinguished from Little Bittern using this character, in flight and at rest. Other differences include a chestnut trailing edge to the inner primaries and outer secondaries, and chestnut tips to greater primary-coverts, which are only visible in flight. ♀ differs from ♂ in having black areas of upperparts dark brown and neck and underparts more streaked. Juvenile resembles ♀ but has mantle and scapulars more rufous, normally with paler tips to feathers. Also, neck often more heavily streaked, particularly in juvenile ♀. **VOICE** Alarm a harsh *kok*, which is the only call likely to be heard in the region. **RANGE** Breeds from C North America to C South America, northern populations migratory. **STATUS** Vagrant with five records all from one archipelago. **Azores** ♀, Terceira, 7 Sep 1951; ♀, Ponta Delgada (São Miguel), 27 Nov 1951; ♀, Povoação (São Miguel), 11 Sep 1952; ♂, Ponta Delgada (São Miguel), 8 Oct. 1964; and one held in a private collection on Santa Maria with no details.

LITTLE BITTERN *Ixobrychus minutus* Plate 7 (4)
Sp. Avetorillo Común **P.** Garçote (Garça-pequena)

I. m minutus and *I. m. payesii*. **L** 33–38 cm (13–15″); **WS** 52–58 cm (20½–23″). **IDENTIFICATION** Sexually dimorphic. Slightly larger than previous species but this is of little use in identification. ♂ has black upperparts except pale cream wing patches and pale underparts with indistinct breast streaking. ♀ browner, with some streaking on upperparts and more heavily streaked underparts. Has pale upperwing patches like ♂ but these are less contrasting. Juvenile resembles ♀ but is much duller with extensive dark streaking above and below, and the pale wing patch is less obvious due to dark streaking on the inner wing-coverts. Subspecies *payesii* differs in adult ♂ by having chestnut head- and neck-sides. **VOICE** ♂ advertising call a low, muffled, monotonous, croaking *hogh* repeated every 2½ seconds, sometimes for long periods. Most often given at night but can be heard at any time. Flight call a short, low *kwer* or throaty *kerack*. **RANGE** Breeds over most of Europe to W Siberia and N India (*minutus*), and sub-Saharan Africa (*payesii*). **STATUS** Nominate *minutus* recorded on all archipelagos, but *payesii* from just one. **Azores** Irregular and rare passage migrant in spring and autumn, with occasional small influxes. **Madeira** Irregular and rare passage migrant usually in spring; also recorded on the Salvage Islands. **Canaries** Bred once on Tenerife, in 1997, but otherwise a scarce passage migrant, more usual in spring, and more frequently on the eastern islands where it is probably annual. Recorded on all main islands. ♂ collected, La Palma, without details (held in the Natural History Museum, Tring) is an example of *payesii*. **Cape Verdes** One record; juvenile collected, Vila Nova Sintra (Brava), 17 Oct 1969, was originally considered to be *payesii* but has been reassigned to the nominate subspecies.

DWARF BITTERN *Ardeirallus sturmii* Plate 7 (5)
Sp. Avetorillo Plomizo **P.** Garçote-preto

Monotypic. **L** 30 cm (12″); **WS** 50 cm (20″). **IDENTIFICATION** Sexes similar. A small bittern: ♂ dark slate-grey above with a dark bill and bare blue facial skin, buffish-white throat and breast with darker buff belly, with very heavy black streaking across entire underparts, and reddish-brown irides. In flight both surfaces of the wings are uniformly dark. ♀ paler than ♂ with a more rufous belly and yellow irides. Juvenile resembles adult but is duller and browner, with buff tips to mantle and covert feathers, and underparts even more rufous than ♀. **VOICE** Rather silent but gives a soft croak on taking off. **RANGE** Breeds discontinuously throughout most of sub-Saharan Africa. An intra-African migrant, mainly present in the north of its range in the wet season. **STATUS** Vagrant with seven records and all from one archipelago. **Canaries** ♂ obtained, Laguna (Tenerife), by Cabrera in 1889 or 1890 (later identified by Meade-Waldo); one shot, Tejina Ponds (Tenerife), 1970s (held in a private collection where only recently discovered); four adults, Arrecife (Lanzarote), 12 Jan 1978; one adult and two immatures, Arrecife (Lanzarote), 10 Oct 1978; two adults, Alegranza, 24 Sep 1994; one, Aldea Blanca (Gran Canaria), 21–30 Jan 2000; and an adult ♂, Erjos Ponds (Tenerife), 23 Aug 2002–13 Apr 2003.

(BLACK-CROWNED) NIGHT HERON *Nycticorax nycticorax* Plate 7 (6)
Sp. Martinete Común **P.** Goraz

N. n. nycticorax and possibly *N. n. hoactli*. **L** 58–65 cm (23–25½″); **WS** 105–112 cm (41–44″). **IDENTIFICATION** Sexes similar. Unmistakable in adult plumage. A medium-sized, rather stocky heron with a black crown and mantle, contrasting with the whitish underparts and uniform grey wings. Two white plumes on nape may be absent. Immature could be confused with Eurasian Bittern but is smaller, and has generally brown and buff plumage with large white spots on upperparts. Also occurs in various plumages intermediate between immature and adult. First-summer resembles adult but mantle and crown grey-brown, the wings have a brownish cast, and the nape plumes are absent. **VOICE** Usually silent away from colonies except for a rather nasal, croaking *quark* given in flight. **RANGE** Breeds from S and C Europe east to Japan, also in India, SE Asia, Africa and the Americas. European population winters in sub-Saharan Africa. **STATUS** Varies between archipelagos. **Azores** Accidental with 13 records, mainly in spring and most records are from São Miguel, with single records from Santa Maria, Terceira, Graciosa and Flores. The possibility that some, particularly one, Flores, 16–19 Oct 1979, could be of the N American subspecies *hoactli* should be considered. **Madeira** Irregular passage migrant, more frequent in spring. **Canaries** Regular passage migrant, more common in spring and often noted in small groups. Also a rare winter visitor. Records from all main islands. **Cape Verdes** Vagrant with five records: immature collected, Raso, 25 Oct 1970; wing of an immature, found near Pedra Badejo (Santiago), 19 Mar 1983; two immatures, Rabil lagoon (Boavista), 13–15 Mar 1997; adult, Mindelo sewage farm (São Vicente), 31 Jan–8 Feb 1999; and another adult, Sal Rei (Boavista), 24 Feb 1999.

GREEN-BACKED HERON (GREEN HERON) *Butorides virescens* Plate 7 (7)
Sp. Garcita Verdosa **P.** Socó-mirím

B. v. virescens. **L** 40–48 cm (16–19″); **WS** 62–70 cm (24½–27½″). **IDENTIFICATION** Sexes similar. A small heron but unlikely to be confused with any of the smaller bitterns. Adult has a dark cap, rich chestnut head- and neck-sides, greenish-tinged dark wings and bluish-green lanceolate feathers on mantle and scapulars. Immature generally browner with less rufous on head and neck-sides, white spots on tips of wing-coverts, and neck more heavily streaked than adults. Immature resembles adult but duller with more streaking and less rufous on neck, but loses the pale tips to coverts during moult. **VOICE** Usually silent but gives harsh *skyow* if flushed. **RANGE** Breeds N America, mainly in E USA. Winters from Florida south to N South America. **TAXONOMIC NOTE** Some authorities consider Green-backed Heron to be a subspecies of Striated Heron *B. striata*. **STATUS** Vagrant with only seven records, all from one archipelago. **Azores** Adult, Ponta Delgada (São Miguel), 24 Oct. 1978; immature, Porto das Lajes (Flores), 19–26 Oct 1979; adult, Mosteiros (São Miguel), 17 Oct 1985; 1–2 immatures, Lajes do Pico (Pico), 22–24 Oct 1985; second-year, Sete Cidades (São Miguel), 10 Sep 1998, and a first-year there, 30 Sep 1998; and a first-winter, Santa Cruz (Flores), 21 Oct 2000.

SQUACCO HERON *Ardeola ralloides* Plate 7 (8)
Sp. Garcilla Cangrejera **P.** Papa-ratos

Monotypic. **L** 44–47 cm (17–18½″); **WS** 80–92 cm (31½–36″). **IDENTIFICATION** Sexes similar. Adult is a small, generally buff heron with long black-and-white nape feathers forming a 'mane' in breeding plumage. In flight reveals predominantly white wings, tail and rump. Non-breeding lacks elongated nape feathers, and has head, neck-sides and upper-breast-sides streaked, and upperparts darker brown. Juvenile similar to non-breeding adult but has more heavily streaked head and neck, and in flight shows dark mottling on primary tips and upperwing-coverts. **VOICE** Generally rather silent but in flight at dusk or if flushed gives a harsh, croaking *kaahk*. **RANGE** Breeds in S Europe east to SE Iran, also Africa and Madagascar. Winters in sub-Saharan Africa. **STATUS** Varies between archipelagos. **Azores** Vagrant with ten records, the most recent one reported by Scheer (1957) without details. All records on São Miguel, except one captured, Angra (Terceira), 3 Sep 1951. **Madeira** Listed as exceptional, i.e. fewer than five records in last 50 years. **Canaries** Rare and irregular passage migrant but possibly annual on eastern islands. Recorded from all main islands except El Hierro. **Cape Verdes** Vagrant with just three records: ♀ collected, Sal, 1901; one, Pedra Badejo lagoons (Santiago), late Feb 1963; and one, Tarrafal (Santiago), 29 Nov. 2001.

CATTLE EGRET *Bubulcus ibis* Plate 7 (9)
Sp. Garcilla Bueyera **P.** Carraceiro (Garça-boeira)

B. i. ibis. **L** 48–53 cm (19–21″); **WS** 90–96 cm (35½–38″). **IDENTIFICATION** Sexes similar. Adult is a small, predominantly white heron with a shorter neck than any other all-white heron. Breeding plumage has a buff crown, breast and mantle, and reddish legs (dark greenish to black at other times). Non-breeder all white with a relatively short, yellow bill. Distinguished from slightly larger Little Egret by its shorter, stouter, yellow bill, lack of contrast between feet and legs, stockier build and generally less elegant appearance. Juvenile resembles adult but initially has a duller, darker bill. **VOICE** In flight a soft croaking *ruk* or gruff *rik-rek*. A hoarse buzz heard at colonies. **RANGE** Breeds in SW and SE Europe, Africa, S Asia, N and S America and Australasia. **DISTRIBUTION** A small colony recently established in palms along the seafront in Arrecife, on Lanzarote, Canaries. Maximum number of occupied nests reached 140 before many were destroyed by order of the local council. **HABITAT** Prefers agricultural areas where often encountered around domestic animals, sometimes near fresh water and, in the Azores, recorded on coast. **STATUS** Varies, but generally a regular visitor, especially in winter. **Azores** Accidental with at least 15 records and recorded on São Miguel, Graciosa, Santa Maria, Terceira and Topo (São Jorge). **Madeira** Scarce and irregular visitor. **Canaries** Apart from those breeding on Lanzarote, regular in small numbers on passage and in winter, with records from all main islands. **Cape Verdes** Common visitor, most numerous in Dec–Apr.

WESTERN REEF EGRET *Egretta gularis* Plate 8 (5)
Sp. Garceta Sombría (Garceta Dimorfa) **P.** Garça-dos-recifes (Garça-negra)

E. g. gularis. **L** 55–65 cm (22–25½″); **WS** 90–110 (35½–43″). **IDENTIFICATION** Sexes similar. A medium-sized heron with two colour morphs, one white and the other dark. Dark morph has completely dark slate-grey plumage except a small white throat patch. White morph very similar to Little and Snowy Egrets, but best identified by thicker based and longer, brownish/yellow bill, and dark but not black legs. Juvenile dark morph paler and browner than adult, with a whitish belly, and white morph often has dark mottling but others are plain white. **VOICE** Usually silent away from colonies, except for a short, guttural, croak when flushed or in dispute over food. **RANGE** Breeds coastally in W Africa from Mauritania to Gabon, and from the Red Sea east to the Indian subcontinent. **STATUS** Accidental with records from all archipelagos. **Azores** Two records: a dark-morph specimen in Ponta Delgada museum, collected on São Miguel pre-1888; and one, Ponta Delgada (São Miguel), 24 Oct 1998. **Madeira** Vagrant: mentioned by Schmitz (1903). **Canaries** Nine records: two, La Camella (Tenerife), 30 Aug 1974; one, Puerto del Carmen (Lanzarote), 22 Apr 1977, with a group of 13 on N coast in subsequent days; one, Arinaga (Gran Canaria), 17 Mar 1988; one, various localities on Tenerife, 21 Mar–27 Jun 1988; one, El Toscón (Gran Canaria), 23 Apr 1994; one, Puerto del Carmen (Lanzarote), 17 Jan 1998; one, Embalse de Valle Molina (Tenerife), 20 Mar & 8–10 Apr1998; one, Maspalomas (Gran Canaria), May–Aug 2001 at least; and one, Amarilla golf course (Tenerife), summer 2002. **Cape Verdes** Accidental with over 20 records, which, until recently, had all been dark morphs with most records on Santiago and Boavista.

LITTLE BLUE HERON *Hydranassa caerulea* Plate 8 (6)
Sp. Garceta Azul **P.** Garça-morena

Monotypic. **L** 58–63 cm (23–25″); **WS** 95–105 cm (37½–41″). **IDENTIFICATION** Sexes similar. Adult unlikely to be mistaken. Medium-sized heron with completely slaty plumage, the head and neck with a purplish tinge, a two-toned bill and dark legs. Juvenile basically white and could be confused with any of the white egrets, but identified by the deeper based, two-toned bill, dusky primary tips and greenish-yellow legs and feet. Immature white blotched slaty and resembles immature Western Reef Heron, with the best features for identification those mentioned for juveniles. **VOICE** Usually silent but utters a harsh, croaking *gerrr* if flushed. **RANGE** Breeds throughout much of the Americas. Winters coastally in breeding range but some post-breeding dispersal, particularly of juveniles. **STATUS** Vagrant with only three records, all from one archipelago. **Azores** One, Flores, 28 Nov 1964, had been ringed in New Jersey, 28 Jun 1954; first-winter, Fajã dos Cubres (São Jorge), 18 Sep–9 Oct 1997; and a first-winter, Magdalena (Pico), 5–7 Oct 1998.

TRICOLOURED HERON *Hydranassa tricolor* Plate 8 (2)
Sp. Garceta Tricolor **P.** Garça-tricolor

Subspecies unknown but *H. t. ruficollis* is most likely. **L** 63–68 cm (25–27″); **WS** 95–105 cm (37½–41″). **IDENTIFICATION** Sexes similar. A medium-sized heron with predominantly dark plumage, a white throat stripe, foreneck bordered rufous, and white belly. Breeding adult has white plumes on rear crown and long, brown scapular plumes. Juvenile has rufous head-sides, a rufous neck with a white stripe on the foreneck and rufous tips to wing-coverts. At all ages has a relatively long, rather thin neck and bill. **VOICE** Gives a harsh croak if flushed but otherwise silent away from colonies. **RANGE** Breeds (mainly coastally) in S North America, West Indies, C America and N South America. Northern populations migratory, but nominate *tricolor* sedentary. **STATUS** Extreme vagrant with only one record. **Azores** Immature, Lajes (Pico), 22–24 Oct 1985 (the only record in the W Palearctic).

BLACK HERON *Hydranassa ardesiaca* Plate 8 (4)
Sp. Garceta Azabache **P.** Garça-preta

Monotypic. **L** 52–55 cm (20½–22″); **WS** 90–95 cm (35½–37½″). **IDENTIFICATION** Sexes similar. Unmistakable, as plumage is all black, with a black bill and legs, and contrasting bright orange-yellow feet. Juvenile resembles adult but duller and lacks any plumes on head or breast. Could be confused with dark-morph Western Reef Egret but smaller and lacks white throat of latter. **VOICE** Usually silent away from colonies and unlikely to be heard in our region. **RANGE** Breeds in sub-Saharan Africa, except the Congo Basin and Kalahari Desert. **STATUS** Extreme vagrant with just one record. **Cape Verdes** One, Ilhéu de Curral Velho (Boavista), 6 Feb–7 Mar 1985.

SNOWY EGRET *Egretta thula* Plate 8 (3)
Sp. Garceta Nívea **P.** Garça-branca-americana

Monotypic. **L** 55–65 cm (22–25½″); **WS** 84–91 cm (33–36″). **IDENTIFICATION** Sexes similar. The New World equivalent of Little Egret, and similar in being all white with black bill and legs and yellow feet. Main differences are the coloration of the lores and back of the legs. The bare facial skin on the lores is bright yellow and almost all Snowy Egrets have a variable yellowish stripe on the rear tarsus. Also slight structural differences in size, neck length, bill length (and depth) and leg length (and thickness). Snowy is usually smaller, shorter and narrower legged, shorter necked and has a thinner bill but some overlap in all these. Breeding-plumaged Snowy is further distinguished by the shorter, more numerous head plumes and reddish-pink lores. Non-breeding adult lacks head plumes and those on the scapulars and upper breast. Juvenile very similar to winter adult and probably indistinguishable in the field. **VOICE** Usually silent away from colonies but can give a hoarse, rasping call if flushed. **RANGE** Breeds in N, C and S America but only northern populations are migratory. **STATUS** Vagrant with eight records from one archipelago. **Azores** Two, Santa Cruz (Flores), 10–11 Oct 1988; one, Mosteiros (São Miguel), 11 Nov 2000; one, Porto Pim and Horta harbour (Faial), 18 Nov 2001–Mar 2002 at least, and one also there, 13–27 Nov 2003, another 21 Dec 2003–7 Apr 2004, and another 20 May–6 Oct 2004 at least; one, St Matheus (Terceira), 13 Oct 2004; and one, Praia da Vitoria (Terceira), 26 Oct 2004.

LITTLE EGRET *Egretta garzetta* Plate 8 (6)
Sp. Garceta Común **P.** Garça-branca

E. g. garzetta. **L** 55–65 cm (22–25½″); **WS** 90–105 cm (35½–41″). **IDENTIFICATION** Sexes similar. A fairly small, but long-legged, all-white heron. Legs black with contrasting yellow feet; bill black and bare facial skin grey-green (orange at onset of breeding). When breeding, develops ornamental feathers on head, chest, mantle and scapulars, lacking in non-breeding plumage. Juvenile resembles non-breeding adult but legs duller and feet less contrasting. **VOICE** At colonies gives a loud disyllabic greeting call, several harsh calls and a rather liquid, bubbling *gulla-gulla-gulla-….* When flushed utters a hoarse, croaking *aaahk*. **RANGE** Through S and C Europe to N Africa and W Asia, also Indian subcontinent, SE Asia, Australasia, Japan and sub-Saharan Africa. **DISTRIBUTION** In Cape Verdes breeding confirmed on Santa Luzia, Raso, São Nicolau, Sal and Boavista, and suspected on São Vicente, Branco, and possibly other islands. On Canaries a couple of pairs recently found breeding in Cattle Egret colony at Arrecife, Lanzarote. **BREEDING** On Cape Verdes season Mar–Jul. Clutch normally 2–3 eggs. **HABITAT** Usually nests on cliffs on or near shore, rarely inland in Cape Verde, but on Lanzarote nests in trees. Most often found on or near coast but in Canaries also around inland freshwater dams some distance from coasts. **STATUS** Regular and common visitor, and breeder. **Azores** Regular in small numbers in spring, autumn and winter. **Madeira** Occurs year-round in small numbers; also recorded on the Salvage Islands. **Canaries** Common winter visitor and passage migrant, although some present year-round. Nesting recorded recently on Lanzarote where population *c*. 20 pairs. **Cape Verdes** Resident breeder with an estimated *c*. 500 pairs in mid 1960s (Naurois 1982). Migrants perhaps also present during the boreal winter.

INTERMEDIATE EGRET *Egretta intermedia* Plate 8 (7)
Sp. Garceta Intermedia **P.** Garça-branca-intermédia

E. i. rachyrhyncha. **L** 65–72 cm (25½–28″); **WS** 105–115 cm (41–45″). **IDENTIFICATION** Sexes similar. An all-white heron, which, as its common name implies, is smaller than Great White Egret and larger than Little Egret.

Easily distinguished from Little Egret by its dark feet and yellow bill, and from Cattle Egret by its much larger size, but separation from Great White Egret is more difficult. Apart from being smaller, Intermediate Egret is best separated from Great White Egret by bare skin around gape not extending beyond eye (close views required to see this). Juvenile lacks elongated scapulars and elongated feathers on lower throat/upper breast of adult. **VOICE** A harsh, croak when flushed but otherwise usually silent away from colonies. **RANGE** Breeds sub-Saharan Africa, India and SE Asia north to Japan and through Indonesia to Australia. **STATUS** Vagrant with eight records all from one archipelago. **Cape Verdes** One obtained, Pedra Badejo lagoons (Santiago), 11 Jan 1965; four, Santa Maria (Sal), 20 Mar 1983 with one, Pedra de Lume saltpans, next day; one, Pedra Badejo lagoons (Santiago), 9 May 1989; two, Rabil lagoon (Boavista), 20–22 Mar 1996; two, Rabil lagoon (Boavista), 10–14 Mar 1997 (perhaps same individuals as in 1996); five, mouth of Ribeira de la Torre (Santo Antão), 13 Apr 1999, then between Ribeira da Garça and Chã da Igreja next day; 1–2, Rabil lagoon (Boavista), 22 Apr 2001; and one individual on sewage ponds near Mindelo (São Vicente), 25 Feb 2002–26 Dec 2003 at least.

GREAT WHITE EGRET *Egretta alba* Plate 8 (8)
Sp. Garceta Grande **P.** Garça-branca-grande

E. a. alba and *E. a. egretta*. **L** 85–102 cm (33½–40″); **WS** 140–170 cm (55–67″). **IDENTIFICATION** Sexes similar. The largest of the white egrets, similar in size to Grey Heron, with size, the large bill and proportionately longer, thinner, more kinked neck being useful features of identification. Most likely to be confused with Intermediate Egret but Great White Egret is larger and has bare skin around the gape extending beyond the eye. Size should eliminate any confusion with Little Egret or Western Reef Egret but Great White can be further separated from these by its longer legs and dark feet. In breeding plumage bill black, the scapulars have elongated plumes and tibia yellow or even reddish. Non-breeder lacks plumes on scapulars, has yellowish tibia and yellow bill. Juvenile very similar to non-breeding adult but bill duller and greyish, and legs brownish. N American race *egretta* differs from *alba* in having tibia black (rather than pinkish-yellow) in breeding plumage and yellow in non-breeding plumage. **VOICE** Usually silent away from colonies but sometimes gives a high-pitched, rattling *krr-rrr-rrr-rrrah* when flushed. **RANGE** Breeds over much of Africa, Asia, Australasia, the Americas and parts of Europe. **STATUS** Vagrant with records from three archipelagos. **Azores** Poorly documented records by F. du Cane Godman of both nominate subspecies and *egretta* on Terceira pre-1870; two (one), Paul da Praia (Terceira), pre-1947; one, Ponta dos Caetanos (São Miguel), 26 Oct 1998; and one, Lagoa Branca (Flores), 23–31 Oct 2001. **Canaries** At least ten records: a flock, Los Rodeos (Tenerife), spring/summer 1889, one of which was shot; one, Puerto de la Cruz (Tenerife), 6 Sep 1964; one, near La Santa (Lanzarote), 6 Aug 1965; one, Roquito del Fraile (Tenerife), 4 Jun–11 Aug 1986; one, near Las Salinas (Fuerteventura), 31 Jan 1992; one, Golf del Sur (Tenerife), 19 May 1997; one, near El Matorral (Fuerteventura), 24 Oct 2000; one, Maspalomas (Gran Canaria), 9–12 Apr 2002; one, El Pinque (Tenerife), 25–26 May 2002; and one, Tejina Ponds (Tenerife), 25 Dec 2004, was probably the same as the individual seen at Los Silos (Tenerife), 27–31 Jan 2005. **Cape Verdes** One in non-breeding plumage, Rabil lagoon (Boavista), 9 Mar 1999.

GREY HERON *Ardea cinerea* Plate 9 (2)
Sp. Garza Real **P.** Garça-real

A. c. cinerea. **L** 90–98 cm (35½–38½″); **WS** 160–175 cm (63–69″). **IDENTIFICATION** Sexes similar. Large heron with generally greyish plumage and a powerful yellowish bill. Adult has grey upperparts and wing-coverts, greyish-white neck with double line of dark streaks on whiter forebeck, head white with black line from eye to nape and long black nape plume. Underparts whitish with black on flanks. In flight has distinctly two-toned upperwing with flight feathers being much darker grey than coverts, and two white patches on carpal joints that appear like headlights seen front-on. Juvenile/first-winter has greyer underparts, buffish tone to foreneck and grey crown and forehead. **VOICE** Commonest call likely to be heard is a harsh, croaking *kranhk* usually given in flight. **RANGE** Breeds throughout most of Europe, Asia and sub-Saharan Africa. **STATUS** Regular visitor. **Azores** Fairly common visitor in all months but most frequent Sep–May and reported to have bred occasionally. **Madeira** Frequent visitor that has also been recorded on the Salvages. **Canaries** Common passage migrant and winter visitor recorded on all main islands in all months. Breeding suspected but not proven. **Cape Verdes** In Jun 2000 a pair bred on Santo Antão and was present again at Tarrafal in 2001, but elsewhere it is a regular visitor in small numbers. Recorded on all main islands, except Santa Luzia, and in all months, but most frequently in Sep–Apr.

GREAT BLUE HERON *Ardea herodias* Plate 9 (1)
Sp. Garza Azulada **P.** Garça-real-americana

Subspecies unknown but *A. h. herodias* is the most likely. **L** 110–125 cm (43–49″); **WS** 175–195 cm (69–77″). **IDENTIFICATION** Sexes similar. Larger and darker than Grey Heron and identified in all plumages by rufous thighs (rarely some juveniles and immatures have very pale or even white thighs) and carpal patches. Adult resembles adult

Grey Heron but has rufous thighs and carpal patches, and a pinkish tone to the head- and neck-sides. Juvenile also similar to juvenile Grey Heron but generally browner above and buffier below, and usually has characteristic rufous thighs and carpal patches, whilst crown is darker than Grey Heron and the bill more massive. **VOICE** A deep, harsh *kraak* when disturbed but otherwise mostly silent. **RANGE** Breeds N America, south to C America, the West Indies and Galápagos. Northern populations move south in winter and southern populations disperse. **STATUS** Vagrant with only five records. **Azores** One, captured on board ship near the Azores, 29 Oct 1968; *c.* 10, São Miguel, Pico and Faial, Apr 1984, the last seen on 24 Jun; one, Fajã dos Cubres (São Jorge), 14–16 Nov 1998. **Canaries** Subadult, Embalse de La Cruz Santa (Tenerife), 5–30 Dec 1998. **Cape Verdes★** One, Rabil lagoon (Boavista), 4–5 Mar 2002.

PURPLE HERON *Ardea purpurea* Plate 8 (9)
Sp. Garza Imperial **P.** Garça-ruiva (Garça-vermelha)

A. p. purpurea (all island groups) and *A. p. bournei* (Cape Verdes). **L** 78–90 cm (30½–35½″); **WS** 120–150 cm (47–59″). **IDENTIFICATION** Sexes similar. Slightly smaller and slimmer than Grey Heron, with a black crown and rufous neck with black vertical stripes on sides. In flight, feet obviously longer in proportion to those of Grey, belly darker and upperwings lack distinct contrast between coverts and flight feathers of Grey. Adult Purple has head- and neck-sides rufous with a narrow black border to the neck. Mantle dark grey and wing-coverts also dark grey with a distinct purplish-brown cast. Foreneck white with vertical lines of dark streaks at sides and belly dark rufous-chestnut. Juvenile less distinctive but browner than any Grey Heron, with broad pale brown fringes to mantle, scapulars and upperwing-coverts. Subspecies *bournei* paler overall than nominate with less black on neck, chest and belly. Also the centre of the underparts are white and pale chestnut. **VOICE** In flight, a gruff *krrek* similar to Grey Heron but higher pitched and thinner. At colonies a wide variety of harsh calls, including a guttural *craak*, a high-pitched squawk and bill clattering. **RANGE** Breeds throughout most of W Palearctic east to Kazakhstan and Iran, in S and E Asia, Indonesia, Philippines, Madagascar and E sub-Saharan Africa. European population migratory, wintering in Africa south of the Sahara and north of the equator. **DISTRIBUTION** As a breeder, *bournei* is restricted to two small colonies, at Boa Entrada and Banana, on Santiago, in the Cape Verdes. **BREEDING** Season on Santiago Aug–Mar. Eggs laid Aug–Sep, occasionally a second brood in spring. Clutch 2–4. **HABITAT** *bournei* breeds in tall trees like mango, kapok and mahogany, and usually feeds on dry hillsides. **TAXONOMIC NOTE** Hazevoet (1995) treated Cape Verde birds as a phylogenetic species, Cape Verde Purple Heron *A. bournei*. It may well deserve specific status given its morphological differences, lack of migratory habits, and geographical isolation. **STATUS** Recorded from all archipelagos but **STATUS** varies. **Azores** Vagrant with nine records. The majority, six, are from São Miguel, but also recorded on Terceira, Santa Maria and Pico. The most recent records were at Sete Cidades (São Miguel), in autumn 1999 and 2000. **Madeira** Rare and irregular visitor, the most recent record an adult, Estreito da Camara de Lobos (Madeira), 9 Jul 2000; also recorded on the Salvages. **Canaries** Regular but scarce passage migrant, principally in spring, with records from all of main islands except El Hierro. **Cape Verdes** Drastically reduced, the population of *bournei* is now Critically Endangered with only *c.* 20 birds remaining left. Subspecies *purpurea* recorded as a vagrant on six occasions. One obtained, Porto Grande (São Vicente), 4 Apr 1897; ♀ obtained, Mindelo (São Vicente), 20 Jan 1924; one, near Praia (Santiago), 28 Dec 1986; one, near Curral Velho (Boavista), 19 Sep 1988; 2–3, Ribeira de Água (Boavista), 10 and 16 Jul 1997; and one, Rabil lagoon (Boavista), 4–20 Mar 2002.

BLACK STORK *Ciconia nigra* Plate 9 (4)
Sp. Cigüeña Negra **P.** Cegonha-preta

Monotypic. **L** 95–100 cm (37½–39″); **WS** 165–180 cm (65–71″). **IDENTIFICATION** Sexes similar. Adult unmistakable given black plumage (glossed green and purple), except white lower breast, belly, undertail and axillaries; bill and legs vary from dull red to bright scarlet. Juvenile has black replaced with brown and bill and legs greenish. **VOICE** Usually silent away from nest. **RANGE** Breeds in Spain where resident and from E Europe across the Palearctic, wintering to the south, with an isolated population in S Africa. **STATUS** Vagrant with records from three archipelagos. **Azores** Immature, Ponta Delgada (São Miguel), 7 Sep 1986. **Madeira** One reported by Rev. R. T. Lowe, who was present in 1832–54, but no other details. **Canaries★** Eight records: one, various locations in N Tenerife, 16–24 Apr 1982; one, Punta de La Rasca (Tenerife), 6 Oct 1987; one, near Presa de Chira (Gran Canaria), 1–9 Mar 1994; one, Playa de Inglés (Gran Canaria), 21 Apr 1994; one, Punta de Teno (Tenerife), 19 Sep 1994; one, Playa Santiago (La Gomera), 14–18 Apr 1999; one, Barranco de la Torre (Fuerteventura), 27 Feb–6 Mar 2000; and a juvenile, near Playa de La Laja (Gran Canaria), 1 May 2000. Also included by Emmerson *et al.* (1994) for La Palma without details.

WHITE STORK *Ciconia ciconia* Plate 9 (3)
Sp. Cigüeña Común **P.** Cegonha-branca

C. c. ciconia. **L** 100–115 cm (39–45″); **WS** 175–195 cm (69–77″). **IDENTIFICATION** Sexes similar. Unmistakable, given diagnostic combination of size, white and black plumage, red legs and bill. Juvenile similar to adult but the black parts tinged brown and bill and legs duller brownish-red. Bill has a darker tip, but this only visible when close. **VOICE** Usually silent away from nest. **RANGE** Breeds in Spain and NW Africa, and in C, E and SE Europe east to the Caspian, with almost the entire population wintering in sub-Saharan Africa. **STATUS** Varies on the two archipelagos where it has been recorded. **Madeira** Vagrant but there have been no recent records. **Canaries** Scarce and irregular passage migrant, and rare winter visitor, sometimes in small flocks; on passage usually more common in spring. Records from all main islands except La Gomera.

GLOSSY IBIS *Plegadis falcinellus* Plate 9 (5)
Sp. Morito **P.** Ibis-preto (Maçarico-preto)

P. f. falcinellus. **L** 55–65 cm (22–25½"); **WS** 80–95 cm (31½–37½"). **IDENTIFICATION** Sexes similar. Dark brownish ibis with wings and tail glossed purple and green, legs dark brown to greenish-brown and bill usually grey-brown. Juvenile similar to adult, but plumage duller and underparts paler with white streaks on head and neck. **VOICE** Usually silent away from colonies but sometimes utters a crow-like call in flight. **RANGE** Very widespread, breeding in E North America, West Indies, SE Europe, E and S Africa, Madagascar, C and S Asia and Australasia. **STATUS** Vagrant recorded on all archipelagos. **Azores** Seven records: ♀ obtained, São Miguel, 29 Dec 1913; one shot, São Miguel, 22 Oct 1959; several (one obtained), Lages airfield (Terceira), early Oct 1959; one obtained, at the airport (Santa Maria), Apr or May 1963; ♀ collected, Ribeirinha (São Miguel), 7 May 1965; first year ♀ obtained, Sete Cidades (São Miguel), Apr 1968; and one, Lagoa Verde, Sete Cidades (São Miguel), 9 Sep 2004. Also a record of an ibis sp., probably this species, although White-faced Ibis *P. chihi* could not be excluded, over Ponta Delgada (São Miguel), 22 Sep 1997. **Madeira** Three records: one, Machico (Madeira), 24 Sep 1892; one, San Roque (Madeira), 19 Sep 1896; and ♂ obtained, Madeira, 24 Sep 1936. **Canaries** Martín & Lorenzo (2001) consider it a scarce and irregular passage migrant and winter visitor. However, only on Lanzarote is this species recorded regularly, and it is certainly a vagrant to other islands on which it has been recorded. Records from La Graciosa, Lanzarote, Fuerteventura, Gran Canaria and Tenerife. **Cape Verdes** Three records: ♂ obtained, São Jorge dos Orgãos (Santiago), 8 Apr 1924; five, near Terras Salgadas (Maio), 22 Oct 2000; and one, south of Rabil (Boavista), 18–22 Mar 2001.

BALD IBIS (WALDRAPP) *Geronticus eremita* Plate 9 (6)
Sp. Ibis Eremita **P.** Ibis-pelado (Ibis-calva)

Monotypic. **L** 70–80 cm (27½–31½"); **WS** 125–135 cm (49–53"). **IDENTIFICATION** Sexes similar. Unmistakable: adult is a large, dark ibis with red legs and bill, bald red head and elongated feathers on neck. Plumage mainly black glossed green and purple on wings. Immature similar to adult but lacks gloss on feathers. Juvenile easily distinguished by its dark-feathered head, greyish legs and bill, and lack of elongated neck feathers. **VOICE** Usually silent away from colonies. **RANGE** Now restricted to two coastal sites in Morocco, a feral population in Turkey and a few pairs in Syria. **STATUS** Vagrant with only four records from two archipelagos. **Azores** One obtained, Furnas (São Miguel), Feb prior to 1905 (no other details). **Cape Verdes** Immature ♀ shot, Boavista, 14 Jan 1898; one, near Curral Velho (Boavista), 2 Feb 1965; and one also there, 21 Apr 1968.

SACRED IBIS *Threskiornis aethiopicus* Plate 9 (7)
Sp. Ibis Sagrado **P.** Ibis-sagrado

T. a. aethiopicus. **L** 65–75 cm (25½–29½"); **WS** 112–124 cm (44–49"). **IDENTIFICATION** Sexes similar. Unmistakable: adult has a bare, black head and neck, black legs and a black rear end formed by loose lower scapulars and tertials, rest of plumage white. In flight has a dark tip to all flight feathers and blood red stripe along leading edge of underwing in breeding plumage. Immature has head and neck feathered black and white, and lacks plumes of adult. **VOICE** At nest gives variety of squeaks, moans and squeals and in flight a harsh croak; usually silent away from colonies. **RANGE** Breeds in sub-Saharan Africa, SE Iraq, Aldabra and Madagascar. **STATUS** Vagrant with only one record, but breeds ferally on two islands. **Canaries** Immature found dead, Puerto Lajas (Fuerteventura), 15 Mar 1991, accepted as possibly being of wild origin. However, the species is now breeding ferally on Fuerteventura (3–4 pairs) and Lanzarote (one pair), and birds have also been recorded on Gran Canaria and Tenerife.

EURASIAN SPOONBILL *Platalea leucorodia* Plate 10 (1)
Sp. Espátula **P.** Colhereiro

P. l. leucorodia. **L** 80–90 cm (31½–35½"); **WS** 115–130 cm (45–52"). **IDENTIFICATION** Sexes similar. Asleep could be mistaken for an egret, but if bill visible unlikely to be confused with any other species. Long bill with spatula-shaped tip is most obvious feature of this white, heron-like bird. Adult in breeding plumage all white except yellowish patch on breast, with black legs and black bill with a yellow tip. Head has a bushy crest on rear crown. Non-breeding adult lacks crest and breast patch. Juvenile has blackish tips to outer primaries and pale pinkish legs and bill that darken with age. **VOICE** Usually silent away from colonies. **RANGE** Breeds across Palearctic (uncommon in Europe) to *c*. 55°N; also in Indian subcontinent, Sri Lanka, Mauritania, N Somalia and islands in S Red Sea. Birds from Holland recovered in Azores and seen on Canaries and Cape Verdes, providing some evidence of the origin of those seen on the Atlantic Islands. **STATUS** Recorded from all archipelagos but **STATUS** varies. **Azores** Vagrant with nine records, most from São Miguel and the most recent an immature, Ponta Delgada, 6 Oct 1989. **Madeira** Vagrant recorded fewer than five times in the last 50 years, most recently one, Funchal (Madeira), 3 Nov 2001 and a juvenile, El Tanque (Porto Santo), 3–7 Dec 2003. **Canaries** Scarce and irregular passage migrant and winter visitor, sometimes in small groups and occasionally remaining long periods. **Cape Verdes** Rare winter visitor, possibly annual. More than 30 records, the vast majority since 1980.

[ROSEATE SPOONBILL] *Ajaia ajaja*] Plate 10 (2)
Sp. Espátula Rosada   **P.** Ajajá

Monotypic. **L** 69–87 cm (27–34½″); **WS** 127 cm (50″). **IDENTIFICATION** Sexes similar. The only species of spoonbill with pink plumage and therefore unmistakable. Adult has bright pink body and wings, with even more intensely coloured patches on lesser coverts and centre of upper breast, appearing almost red. Neck white and head unfeathered and greenish with a black nape. Immature similar to adult but head totally feathered, the intensity of the pink coloration is much reduced and brighter patches absent. **VOICE** Usually silent away from colonies. **RANGE** Breeds in S North America, West Indies, C America and over most of lowland S America. **STATUS** Extreme vagrant with a single, unconfirmed record. **Azores** Immature apparently photographed Sep 1993, but the location and even the island are unknown, and unfortunately the original photograph cannot located by the photographer or the person who identified the bird. As this would be the only record for the W Palearctic, this species is included only in square brackets.

GREATER FLAMINGO *Phoenicopterus roseus*   Plate 10 (4)
Sp. Flamenco   **P.** Flamingo (Flamingo-comum)

Monotypic. **L** 125–155 cm (49–61″); **WS** 140–165 (55–65″). **IDENTIFICATION** Sexes similar. Adult white tinged pink, with reddish wing-coverts, black flight feathers, a bright pink bill with a dark tip and long pink legs. Juvenile predominantly grey-brown with a grey bill and black tip. The possibility of escaped Chilean Flamingo *P. chilensis* must always be considered, as must the chance of a vagrant or escaped Lesser Flamingo. Chilean can be separated from Greater Flamingo by its contrasting bright red 'knees' and more extensive black tip to the bill. Lesser Flamingo, which has been recorded on the Canaries, is smaller than Greater and has a darker bill. **VOICE** In flight a honking *ka-ha* similar to Greylag Goose. When feeding groups give a cackling chatter. **RANGE** Breeds in isolated pockets from S Spain and France east to Kazakhstan, in Africa and from Middle East to Indian subcontinent; also Galápagos, Caribbean and S America. Disperses outside breeding season. **STATUS** Accidental visitor with records from two archipelagos. **Canaries** Recorded from La Graciosa, Lanzarote, Fuerteventura, Gran Canaria and Tenerife. A ringed bird on Tenerife originated from the colony at Fuentedepiedra (Málaga). **Cape Verdes** Not recorded since 16 Mar 1924, when one collected on Boavista, but certainly bred on the eastern islands until the late 1800s.

LESSER FLAMINGO *Phoenicopterus minor*   Plate 10 (3)
Sp. Flamenco Enano   **P.** Flamingo-pequeno

Monotypic. **L** 80–105 cm (31½–41″); **WS** 95–120 cm (37½–47″). **IDENTIFICATION** Sexes similar. A small flamingo with pink plumage when adult, but some more faded, appearing whitish at distance. Bill, bare facial skin and eye dark red, appearing blackish at distance. Easily separated from Greater Flamingo by darker bill lacking two-toned appearance. Immature very pale, almost white and juvenile dull greyish-brown. Both immature and juvenile distinguished from Greater Flamingo of a similar age by structure and darker bill. **VOICE** Flight call is similar to Greater Flamingo but higher pitched and more like that of Greater White-fronted Goose. **RANGE** Breeds in the Rift Valley and north of the Kalahari, with a small population in Mauritania. Disperses in the Rift Valley, S Africa and from the Banc d'Arguin to Cameroon. **STATUS** Accidental visitor to one archipelago. **Canaries** Records from Lanzarote, Fuerteventura and Tenerife, some obviously relating to birds of captive origin, but others perhaps wild.

FULVOUS WHISTLING DUCK *Dendrocygna bicolor*   Plate 11 (1)
Sp. Suirirí Bicolor   **P.** Marreca-caneleira

Monotypic. **L** 45–53 cm (18–21″); **WS** 85–93 cm (33½–36½″). **IDENTIFICATION** Sexes similar. Adult has predominantly rich rufous-buff underparts with a creamy vent and paler neck-sides with fine dark striations. Upperparts dark brown with rufous fringes to feathers. The elongated flank feathers are off-white with black outer webs creating a streaked appearance. In flight, adult has obvious white uppertail-coverts. Juvenile similar but noticeably duller and greyer, with reduced flank markings and greyish uppertail-coverts. **VOICE** Normal flight call a shorebird-like whistle *k-weeoo*, usually repeated. **RANGE** Breeds in S North America, S America, sub-Saharan Africa and S Asia. Mainly resident but some seasonal movements. **STATUS** Vagrant with just the one, recent record. **Azores** Immature, Ponta Delgada harbour (São Miguel), 29 Aug 2000.

WHITE-FACED WHISTLING DUCK   *Dendrocygna viduata*   Plate 11 (2)
Sp. Suirirí Cariblanco   **P.** Irerê

Monotypic. **L** 45–53 cm (18–21″); **WS** 86–94 cm (34–37″). **IDENTIFICATION** Sexes similar. Adult unmistakable with generally dark plumage and predominantly white head. Breast, lower neck and back chestnut, sides of body finely barred black and white, belly and tail blackish. Juvenile like a dull adult, the white 'face' being replaced with greyish-buff. Adult unlikely to be confused, but juvenile similar to juvenile Fulvous Whistling Duck, from which it is best identified by the barring on the body-sides. **VOICE** Usual call a three-note whistle *tsree-tsree-tsreeo*. **RANGE** Breeds across sub-Saharan Africa and the tropical Americas from Costa Rica to N Argentina. **STATUS** Vagrant with just one record. **Canaries★** One captured and 18 others, Los Rodeos (Tenerife), 1967–69 (*Ardeola* 20: 330–331).

MUTE SWAN   *Cygnus olor*   Plate 11 (3)
Sp. Cisne Vulgar   **P.** Cisne-mudo (Cisne-vulgar)

Monotypic.   **L** 125–155 cm (49–61″); **WS** 200–235 cm (79–92½″). **IDENTIFICATION** Sexes similar. Adult unmistakable in the region; the all-white plumage and orange bill with a black knob at the base are diagnostic. ♀ usually smaller than ♂ with a duller bill and smaller knob. In flight the wings produce a characteristic whistling sound, which is a diagnostic feature. Juvenile almost entirely grey-brown with a greyish bill. **VOICE** Usually utters various grunts and hisses, not normally vocal in flight. **RANGE** Breeds over most of C Europe, thence east in pockets across Asia to N China. Introduced in N America, S Africa, Australia and New Zealand. All European populations are mainly of feral origin and are mostly resident. **STATUS** Vagrant with just four records. **Azores** Multiple arrival in early 1963, with two Terceira (shot), eight Faial, one Graciosa and others on Pico; also reported on Graciosa, May 1972. **Canaries★** One, Cabo Blanco (Tenerife), 4 Feb–Aug 1995, had been present for *c.* 1 month according to local inhabitants, and one at Arinaga (Gran Canaria), 2–3 Apr 2000.

PINK-FOOTED GOOSE   *Anser brachyrhynchus*   Plate 12 (2)
Sp. Ánsar Piquicorto   **P.** Ganso-de-bico-curto

Monotypic. **L** 60–75 cm (23½–29½″); **WS** 135–165 cm (53–65″). **IDENTIFICATION** Sexes similar. Adult is a dark-headed 'grey goose' with distinctly greyish upperparts, deep pink legs and feet, and a pink band on the rather short, triangular bill. Very similar to *serrirostris* Bean Goose (Tundra Bean Goose) in structure and plumage but colour of feet diagnostic, although sometimes difficult to determine. Also Pink-footed has greyer upperparts, paler than the head and neck, compared to the browner upperparts of Tundra Bean Goose, which are the same colour as the head and neck. In flight Pink-footed has a grey upper forewing, which contrasts strongly with the dark primaries; in Tundra Bean Goose the forewing is much darker and the contrast considerably less. Juvenile similar to adult but lacks barring on flanks and has duller feet. **VOICE** Distinctive nasal *wink-wink* higher in pitch than corresponding call of Bean Goose, and a continual cackling in flight. **RANGE** Breeds E Greenland, Iceland and Svalbard, winters in the British Isles, Denmark, Germany, Holland and Belgium. **STATUS** Vagrant with only four records. **Azores** One obtained, São Miguel (undated); and one recovered, Terceira, 2 Oct 1953. **Madeira** One shot, Madeira, 22 Feb 1934. **Canaries★** One recovered, Lanzarote, 29 Sep 1953, had been ringed as a chick in Iceland on 26 Jul 1953.

BEAN GOOSE   *Anser fabalis*   Plate 12 (1)
Sp. Ánsar Campestre   **P.** Ganso-campestre

A. f. rossicus.   **L** 66–84 cm (26–33″); **WS** 140–175 cm (56–70″). **IDENTIFICATION** Sexes similar. Similar in all plumages to Pink-footed Goose but usually larger and darker with a less rounded head and longer bill. Adult has legs, feet and band on bill normally orange or orange-yellow but can be pink or flesh-coloured, when great care needed to separate from Pink-footed, and structural differences are useful identification aids. Juvenile similar to adult but head and neck not as dark and upperparts and flanks scalier. **VOICE** Calls similar to those of Pink-footed Goose; include a nasal *hank-hank*, deeper than the corresponding call of Pink-footed. **RANGE** Breeds in pockets across the Palearctic, wintering mainly in temperate lowlands of Eurasia. **TAXONOMIC NOTE** It has recently been proposed that the *rossicus* and *fabalis* groups be considered separate species, Tundra Bean Goose and Taiga Bean Goose respectively. **STATUS** Vagrant with just three records. **Azores** One obtained, Terceira, Oct 1968; and one, Corvo, 8 Oct 2002, assigned to the form *rossicus*. **Madeira** One apparently shot, San Roque (Madeira), Oct 1895.

(GREATER) WHITE-FRONTED GOOSE *Anser albifrons* Plate 12 (3)
Sp. Ánsar Careto Grande **P.** Ganso-de-testa-branca (Ganso-grande-de-testa-branca)

A. a. albifrons. **L** 65–78 cm (25½–31″); **WS** 135–165 cm (53–65″). **IDENTIFICATION** Sexes similar. Within our region, adult White-fronted is probably the easiest 'grey goose' to identify, given the obvious white frontal patch at the bill base, large black belly patches, orange legs and either a pink (*albifrons*) or orange (*flavirostris*) bill. Juvenile lacks white frontal patch and black belly patches, and has much duller bill and greyish-yellow legs. **VOICE** Typical call a disyllabic *kyu-yu* in flight. Also various cackling calls. **RANGE** Breeds in taiga and tundra across Russia and in Alaska, N Canada and Greenland. Winters in Europe, Middle East, China, Japan, S USA and Mexico. **STATUS** Vagrant with three records. **Azores** One (*albifrons*) shot, Caldeira do Guilherme (Terceira), 16 Oct 1965. **Madeira** One obtained, 9 Jan 1934. **Canaries★** Juvenile, Roquito del Fraile (Tenerife), 13 Dec 1987.

GREYLAG GOOSE *Anser anser* Plate 12 (4)
Sp. Ánsar Común **P.** Ganso-bravo (Ganso-comum-ocidental)

A. a. anser. **L** 75–90 cm (29½–35½″); **WS** 150–180 cm (59–71″). **IDENTIFICATION** Sexes similar. Adult is the largest and bulkiest 'grey goose'. Plumage rather uniform grey-brown and structurally this species is heavy headed and large-billed. Legs pink, a feature shared only by Pink-footed Goose, and bill orange with a white nail. In flight easily identified by pale grey underwing-coverts and extremely pale grey upper forewing. Juvenile similar but lacks indistinct spots on belly of adult and has scalier upperparts like other juvenile 'grey geese'. **VOICE** Typical call a loud series of honks in flight *aahng-ahng-ung*, deeper and more clanging than other 'grey geese'. **RANGE** Breeds widely across Eurasia. Winters south to N Africa, Middle East, N India and S China. **STATUS** Vagrant with few records. **Azores** One, unspecified location or date; one, Rosto de Cão (São Miguel), also undated; a group, Trempes (Terceira), and one, near the airport (Santa Maria), 5 Dec 1970. **Madeira** One shot, Porto da Cruz (Madeira), 12 May 1896; one, Cima, 26 Mar 1901; and one, Ribeira de São Jorge (Madeira), 25 Dec 1998. **Canaries** Recorded from Tenerife, Fuerteventura and La Palma, with most recent records one, Roquito del Fraile (Tenerife), 9 Dec 1991–14 Mar 1992; one, Embalse de Los Molinos (Fuerteventura), 23 Oct 1993–31 Jan 1994; and three, Roquito del Fraile (Tenerife), winter 2001/2002.

SNOW GOOSE *Anser caerulescens* Plate 12 (5)
Sp. Ánsar Nival **P.** Ganso-das-neves

A. c. caerulescens. **L** 65–84 cm (25½–33″); **WS** 135–170 cm (53–67″). **IDENTIFICATION** Sexes similar. Two colour morphs, one white and one dark. Adult white morph all-white except contrasting black primaries and reddish-pink legs, feet and bill. Adult dark morph has bare parts same colour as white morph but most of plumage dark grey except white head and upper neck. Juvenile white morph has bare parts dark grey and pale greyish-brown upperparts. Dark-morph juvenile almost entirely sooty-brown with dusky bare parts. **VOICE** In flight a cackling, nasal, high-pitched *la-luk*. **RANGE** Breeds in arctic tundra of N America, NW Greenland and Wrangel I., off NE Siberia. Winters mostly in S USA and N Mexico. **STATUS** Vagrant with only four records. **Azores** One shot, São Miguel (no details); one obtained, Terceira, pre-1966; one shot, Terceira, 27 Oct 1967; and three, Terceira, 1969.

CANADA GOOSE *Branta canadensis* Plate 11 (4)
Sp. Barnacla Canadiense **P.** Ganso do Canadá

Subspecies unknown. **L** 55–110 cm (22–43″); **WS** 120–190 cm (47–75″). **IDENTIFICATION** Sexes similar. Very variable in size, but plumage similar amongst different subspecies. Long black neck, black head with white throat extending onto head-sides, grey-brown upperwings and mantle, white rump, black tail, underparts grey-brown but paler than upperparts. Adult has neat pale bars on upperparts and flanks. Juvenile has pale bars less pronounced and more irregular, head and neck duller brown-black and white face patch tinged pale brown. **VOICE** Larger races give a deep disyllabic honking *ah-hank*, repeated with varying pitch. Smaller races give a higher pitched yelping or cackling. **RANGE** Breeds across N North America from the Aleutians to the Atlantic coast. Winters across S North America south to Mexico. Introduced in various European countries. **TAXONOMIC NOTE** Recent taxonomic revisions suggest that Canada Goose should be split into two polytypic species, Lesser Canada Goose *B. hutchinsii* and Greater Canada Goose *B. canadensis*. **STATUS** Vagrant with only two records. **Azores★** Two, Lagoa das Furnas (São Miguel), Nov 1998; and one, Angra do Heroísmo (Terceira), 9 Nov 2000.

BARNACLE GOOSE   *Branta leucopsis*   Plate 11 (5)
Sp. Barnacla Cariblanca   **P.** Ganso-marisco (Ganso-de-faces-brancas)

Monotypic.   **L** 58–71 cm (23–28″); **WS** 130–145 cm (51–57″). **IDENTIFICATION** Sexes similar. Adult is unmistakable, with combination of black neck and breast, mostly grey body with black bars on wing-coverts and creamy-white face. Juvenile duller with brownish-tinged upperparts and buffish-washed flanks. **VOICE** A shrill, monosyllabic, barking *gnuk* of variable speed and pitch, sometimes sounding like a yelping dog, given both in flight and on ground. Also a lower pitched more muffled *hogoog* when feeding. **RANGE** Breeds E Greenland, Svalbard, Novaya Zemlya and adjacent Vaigach I., wintering in Scotland, Ireland, NW England and the Netherlands. **STATUS** Extreme vagrant with only two records, from different archipelagos. **Azores** One shot, Furnas (São Miguel), pre-1903 (specimen in Ponta Delgada museum). **Canaries** One, Arinaga (Gran Canaria), Nov 1997 (origin unknown).

BRENT GOOSE   *Branta bernicla*   Plate 11 (6)
Sp. Barnacla Carinegra   **P.** Ganso-de-faces-pretas (Ganso-de-faces-negras)

B. b. hrota.   **L** 55–66 cm (22–26″); **WS** 115–125 cm (45–49″). **IDENTIFICATION** Sexes similar. Generally a small and very dark goose. Adult has black head, neck and breast with a small white patch on the upper-neck-sides, dark brown upperparts, whitish belly and flanks, and vent and uppertail-coverts white. Juvenile similar but head, neck and breast duller and brownish, and white neck patch (absent in very young birds) only attained as first-winter. **VOICE** A low gargling *raunk* or *rhut* in flight or when feeding. **RANGE** Breeds widely across arctic tundra. Winters mostly in USA, British Isles, Denmark to France, the Yellow Sea region and S Japan. **STATUS** Vagrant with only three records. **Azores** Immature ♀ obtained, Furnas (São Miguel), 16 Nov 1927; and adult, Porto Judeo (Terceira), 28 Oct 2004. **Canaries** First-winter, Roquito del Fraile (Tenerife), 24 Jan–14 Mar 1992.

RUDDY SHELDUCK   *Tadorna ferruginea*   Plate 12 (6)
Sp. Tarro Canelo   **P.** Pato-casarca (Pato-ferrugíneo)

Monotypic.   **L** 61–67 cm (24–26″); **WS** 120–145 cm (47–57″). **IDENTIFICATION** Sexes similar. Mostly fairly bright orange-brown with a paler more cinnamon-buff head and neck. Flight feathers and tail blackish and both surfaces of the wing-coverts white. Breeding ♂ has a black collar on neck, which is lost in eclipse plumage. ♀ duller and has contrasting whitish facial patch, and juvenile even duller with a grey-brown tone to entire plumage particularly noticeable on white forewing in flight, but much as adult by first autumn. **VOICE** Common call in flight a loud, nasal, honking *ang* or *ah-ung* and a trumpeting *pok-pok-pok-pok* before taking flight. **RANGE** Breeds in NW Africa and from E Europe across most of C Asia. In NW Africa mostly resident but dispersive, as in Turkey, but rest of population migrates south to winter. **DISTRIBUTION** Breeds only on Fuerteventura in the Canaries. **BREEDING** Little studied in our region but starts early, as young seen in mid March. Clutch 6–9, and the nest found in 1996 was placed between two shrubs *c.* 150 m from water. **HABITAT** On Fuerteventura most often encountered by one of the larger man-made reservoirs or in surrounding desert areas. **STATUS** Formerly a vagrant but now breeds regularly on Fuerteventura. **Azores** ♀ obtained, São Miguel, Sep 1892. **Madeira** Two obtained, Porto Santo, 29 Jul 1941. **Canaries** First record: ♀, Rosa de Catalina García (Fuerteventura), 15 Apr 1994. In May 1994 a pair with seven young were seen at the same site but none was found in Aug. The same pair was at the same site on 21 Dec 1994 and at least one was still present in Mar 1995, but in Jun none was found. A pair, again at Rosa de Catalina García, Oct 1995, had five young by 2 Apr 1996. Since then, pairs have bred regularly at two sites on Fuerteventura, where now almost resident. Maximum count in 1999 was up to six pairs but a marked increase in 2004 when 47 were at Embalse de Los Molinos. Recently recorded on Gran Canaria but no evidence of breeding.

COMMON SHELDUCK   *Tadorna tadorna*   Plate 12 (7)
Sp. Tarro Blanco   **P.** Tadorna (Pato-branco)

Monotypic.   **L** 58–67 cm (23–26½″); **WS** 110–135 cm (43–53″). **IDENTIFICATION** Sexes similar. Unmistakable combination of black-and-white plumage, rusty breast-band, red bill and large size. ♂ has black head and neck glossed green and bright waxy-red bill with large knob at base when breeding. ♀ similar but duller, with white mottling at base of bill and a duller bill, which lacks prominent knob. Juvenile very different and lacks bright colouring of breeding adults; it also lacks breast-band and is generally grey-brown above and white below. Eclipse plumage of adults obscures breast-band and some ♀♀ can approach Juvenile pattern closely but can always be identified by upperwing pattern. Juveniles have the coverts washed greyish and a white trailing edge to the secondaries, whereas adults have white upperwing-coverts with dark tips to the secondaries. **VOICE** Relatively silent away from breeding areas. **RANGE** Breeds coastally in NW Europe and from SE Europe across C Asia to N China. Winters south to N Africa, Pakistan and S China. **STATUS** Vagrant with only seven records. **Madeira** One shot, Porto Santo, 25 Nov 1932. **Canaries** Six records: one, Guaragacho (Tenerife), 25 Dec 1983–12 Jan 1984; two,

Playa Arinaga (Gran Canaria) 15–26 Aug 1986; two, off Punta de la Rasca (Tenerife), 21 Aug 1988; one, Los Palmitos (Gran Canaria), 12 Apr 1994; two, Chamoriscán (Gran Canaria), 27 Jan 1999; and one, Salinas del Carmen (Fuerteventura), 26 Nov–12 Dec 2004, with a second present 11 Dec.

WOOD DUCK   *Aix sponsa*   Plate 13 (1)
Sp. Pato Joyuyo   **P.** Pato-carolino

Monotypic. **L** 43–51 cm (17–20″) **WS** 70–76 cm (27½–30″). **IDENTIFICATION** Sexually dimorphic. ♂ in breeding plumage totally unmistakable: head and neck blackish glossed purple and green with a long, downcurved crest on the nape. Throat white with a complicated pattern of white stripes on head and upper neck, breast and lower neck almost purple with white freckles and vertical white and black stripes at front of buffy flanks. Upperparts mostly blackish glossed green and blue. In flight shows a narrow white trailing edge to the secondaries on an otherwise dark upperwing, and underwing also appears dark but coverts mottled white. Irides red with a red eye-ring and bill is predominantly orange-red. ♀ is only slightly crested and is predominantly greyish-brown but mottled whitish on breast and flanks. Throat white and there is a large white patch around eye. In flight both wing surfaces are similar to those of ♂, and the speculum is bluish-purple. Irides brown and bill grey. ♀ resembles ♀ Mandarin Duck *A. galericulata* (unrecorded in the region), but extensive white patch around eye best identifies it. Eclipse ♂ resembles ♀ but retains red eye and bill, has a smaller white patch around eye and usually shows some signs of distinctive head pattern of breeding plumage. Juvenile also similar to ♀ but duller with a less obvious eye-patch. **VOICE** Fairly silent but ♂ has a thin, high-pitched whistle, *jeweep* or *sweeooo*, and ♀ a penetrating squeal *oo-eek* when flushed. **RANGE** Breeds over much of the USA, extreme S Canada and in Cuba. Northern populations move south in winter as far as C Mexico. **STATUS** Vagrant with only ten records. **Azores** ♂ shot, Furnas (São Miguel), 21 Dec 1963; first-winter ♀ ringed in N Carolina, USA, recovered Flores, Jan 1985, three ♂♂, Flores, early Jul 1999; pair, Corvo, 19 Oct 1999, ♂, Caldeirão (Corvo), 17 Oct 2000; ♀, Lajes (Flores), 11 Feb 2001; ♀, Santa Cruz (Graciosa); 3–6 Feb 2003 and a long-staying ♂, Terra Nostra Gardens, Furnas (São Miguel), 12 Oct 2002–27 Jan 2005 at least. **Canaries** ♀ found dead, La Sabinita (Tenerife), 14 Dec 1995; and ♀, Laguna de Barlavento (La Palma), 19 Nov 2000.

EURASIAN WIGEON   *Anas penelope*   Plate 13 (2)
Sp. Silbón Europeo   **P.** Piadeira

Monotypic. **L** 45–51 cm (17½–20″); **WS** 75–86 cm (29½–34″). **IDENTIFICATION** Sexually dimorphic. Like most *Anas*, ♂ unmistakable but ♀ more difficult to identify: ♂ in breeding plumage has chestnut head and upper neck with a yellowish-buff forehead and central crown, the breast and lower neck pinkish-grey, the body predominantly grey and ventral region black and white. In flight the upperwing has a striking white patch on the forewing, lacks any white trailing edge to the secondaries and has a black-bordered green speculum. ♀ has two colour forms, grey and rufous: both are rather uniform and relatively unmarked, but the white belly contrasts strongly with the rest of the plumage; the bill is rather small but overall the species appears rather stocky. In flight appears very white-bellied with a slightly paler forewing. Main confusion risk is with ♀ American Wigeon (which see), but with care they can be separated. Eclipse ♂ resembles ♀ but is richer chestnut and retains upperwing pattern of breeding ♂. Juvenile also similar to ♀ but often has the white belly faintly mottled and the speculum even duller. **VOICE** Best-known call is a clear whistle, *wheeoo*, given by ♂ both in flight and on water. ♀ has a repeated, low, growling *karr-karr-karr*…in flight. **RANGE** Breeds across much of the Palearctic, wintering south to C Africa, India and SE Asia. **STATUS** Varies between archipelagos. **Azores** Vagrant with 23 records, 18 of these since autumn 1998, and possibly more regular than older records suggest (perhaps even an annual visitor). Largest group: 17, Sete Cidades, 2003. **Madeira** Exceptional, fewer than five records in the last 50 years, the most recent, Porto Santo, 8 Nov 2004. **Canaries** Scarce but fairly regular winter visitor with records from all main islands. **Cape Verdes** Extreme vagrant: a ♂, Casas Velhas (Maio), 31 Dec 2004, is the only record.

AMERICAN WIGEON   *Anas americana*   Plate 13 (3)
Sp. Silbón Americano   **P.** Piadeira-americana (Pato-americano)

Monotypic. **L** 45–56 cm (17¾–22″); **WS** 76–89 cm (30–35″). **IDENTIFICATION** A sexually dimorphic species. Breeding-plumaged ♂ easily identified from Eurasian Wigeon but eclipse ♂ and ♀ plumages very similar to latter. In all plumages, the axillaries and median underwing-coverts of American are white, but in Eurasian these areas usually appear greyish. ♂ in breeding plumage unmistakable, with a creamy-white forehead and central crown, broad dark (green-glossed) band extending from eye across head-sides to nape, and rest of head intensely marked grey and white. Breast and flanks pinkish-brown, the upperparts also have a pinkish-brown cast and ventral area black and white. ♀ very similar to ♀ Eurasian but less variable and usually shows more contrast between greyish head and neck and rufous-brown breast and flanks. On the upperwing, American usually has a pale bar in front of the speculum formed by white bases to the greater coverts, which can be present in Eurasian but is greyer and always less pronounced. Eclipse ♂ resembles ♀ but is more chestnut on breast and flanks and retains white forewing of breeding ♂. Juvenile similar to ♀ but generally duller with the dullest speculum. **VOICE** Very similar to Eurasian Wigeon but whistle of ♂ weaker, less piercing and more disyllabic, *wheeoo-wo*. **RANGE** Breeds across most of N North America, wintering south to Panama and N Colombia. **STATUS** Vagrant with records from three archipelagos. **Azores** In excess of 15 records, most from São Miguel but also been recorded once on Terceira, Pico and São Jorge, and twice on Flores. **Canaries** Eight records: adult ♂, Guargacho (Tenerife), 4 Apr 1979; first-winter, Embalse de Los Molinos (Fuerteventura), 23–24 Jan 1990; ♂, Agüimes (Gran Canaria), 22 Feb–29 Mar 1990; (1♂, 3♀♀), Embalse de Los Molinos (Fuerteventura), 30 Jan–18 Feb with one ♀ until 27 Mar 1991; four (2♂♂, 2♀♀), Embalse de Los Molinos (Fuerteventura), 28 Nov 1992–19 Feb 1993; ♀, Embalse de Los Molinos (Fuerteventura), 3 Jan 1995; adult ♀, Embalse de Los Molinos (Fuerteventura), 7 Nov 1997–27 Jan 1998; ♂, Embalse de Valle Molina (Tenerife), 3 Nov 1999–18 Mar 2000; and 2♀♀, Los Silos (Tenerife), 18 Nov. 2002, one until 23 Mar 2003. **Cape Verdes** A pair, Casas Velhas (Maio), 31 Dec 2004.

FALCATED DUCK   *Anas falcata*   Plate 14 (4)
Sp. Cerceta de Alfanjes   **P.** Pato-falcado (Marrequinho-de-foice)

Monotypic. **L** 48–54 cm (19–21″); **WS** 76–82 cm (30–32″). **IDENTIFICATION** Sexually dimorphic. ♂ in breeding plumage unmistakable: large head with long mane glossed green and purple, white throat, breast grey with black scaling, flanks, belly, sides of vent and majority of upperparts grey with fine black scaling, and sides of undertail-coverts buffish bordered black. Elongated and strongly downcurved, sickle-shaped (falcated), tertials are black and grey. In flight has predominantly grey upperwing with green-glossed speculum, bordered in front by a rather diffuse white bar. ♀ has a slender grey bill and rather uniform, greyish-brown head and neck with a hint of a mane. Body is warmer brown with darker upperparts and the paler underparts have prominent, darker V-shaped subterminal marks. In flight upperwing pattern is like that of ♂. Eclipse ♂ resembles ♀ but is darker on crown, nape and upperparts, with a greyer forewing as in upperwing of breeding ♂. Tertials shorter and straighter but still longer than those of ♀. Juvenile also resembles ♀ but is more buff and generally lacks pale subterminal marks on scapulars of adult ♀. **VOICE** Relatively silent when not displaying but ♂ gives a shrill, piercing whistle and a vibrating *rruh-urr*, the ♀ a rather coarse quack. **RANGE** Breeds in NE Asia and winters south to S China and NE India. **STATUS** Extreme vagrant with just one record. **Canaries★** ♂ obtained, Tenerife, although probably of captive origin.

GADWALL   *Anas strepera*   Plate 13 (4)
Sp. Ánade Friso   **P.** Frisada

Monotypic. **L** 46–56 cm (18–22″); **WS** 84–95 cm (33–37½″). **IDENTIFICATION** Sexually dimorphic. Adult ♂ in breeding plumage is dullest *Anas* in the region, with a finely vermiculated grey body, black rear end, browner head and, in flight, a white speculum on inner secondaries. ♀ similar to a small Mallard but has dark bill with orange sides and a partially white speculum. White speculum in both sexes is often visible on the water. Eclipse ♂ resembles ♀ but is greyer and less heavily marked, and retains breeding ♂ wing pattern. Juvenile also resembles ♀ but is brighter overall, with greyer head-sides and yellower sides to bill. **VOICE** Relatively quiet except in display but occasionally gives a Mallard-like *quack*. **RANGE** Breeds over most of the central N Hemisphere. Winters south to S China, N India, Sudan and Mexico. **STATUS** Varies between the two archipelagos on which it has been recorded. **Azores** Vagrant with three records: ten (including three ♂♂), Ribeira Grande (São Miguel), 26 Oct 1991; pair, Lagoa Azul, Sete Cidades (São Miguel), 11–13 Nov 2000; and two, Lagoa das Furnas (São Miguel), 9 Dec 2003 (one until 12 Dec). **Canaries** Scarce winter visitor: usually just odd birds but in winter 1995/96 a flock of up to 13, Embalse de Valle Molina (Tenerife).

EURASIAN TEAL   *Anas crecca*   Plate 14 (2)
Sp. Cerceta Común   **P.** Marrequinha (Marrequinho)

A. c. crecca. **L** 34–38 cm (13½–15″); **WS** 58–64 cm (23–25″). **IDENTIFICATION** Sexually dimorphic. Smallest dabbling duck in the region. ♂ in breeding plumage has a grey body, black and buff undertail, a chestnut head with a broad green band extending back from the eye and buff lines bordering the green, and a white stripe along the side of the upperparts. ♀ has generally brownish, mottled plumage with a relatively indistinct facial pattern, a green speculum and white patch on tail-sides. Very similar to ♀♀ of both Garganey and Blue-winged Teal, but ♀ Garganey has a more pronounced facial pattern, lacks the pale patch on the tail-sides and has a broader white trailing edge to the secondaries, whilst ♀ Blue-winged Teal also lacks a pale patch on the tail-sides, has a more prominent loral spot than Eurasian Teal, has a blue forewing in flight, and lacks any white trailing edge to the secondaries. Eclipse ♂ resembles ♀ but is darker above with a less distinct eye-stripe, and is not so clearly spotted below. The juvenile also resembles ♀ but has a more spotted belly, and occasionally a hint of a dark bar on the cheek, although this is less obvious than on Garganey. As there can be considerable individual variation, ageing can be difficult. **VOICE** ♂ has a clear, high-pitched whistle, *prip-prip*. ♀ fairly silent but does sometimes utter various types of *quack* that are higher pitched and a little faster than those of Mallard. **RANGE** Breeds across the Palearctic and winters south to C Africa, S India and SE Asia. **STATUS** Recorded from all archipelagos but status varies. **Azores** Regular and often fairly numerous winter visitor. **Madeira** Scarce migrant and winter visitor. **Canaries** Regular and fairly common winter visitor, often in small groups and recorded from all the main islands of the group. **Cape Verdes** Vagrant species with seven records: ♀ shot, Boavista, 4 Feb 1898; ♀ obtained, Ribeira Julião (São Vicente), 29 Jan 1924; six, near Fonte Vicente (Boavista), 12 Dec 1993, five ♂♂, four ♀♀, Mindelo sewage farm (São Vicente), 22–27 Jan 1999 (with one ♂ present until 24 Feb); ♀ there, 12 Mar 2000; another juvenile/♀ there, 3 Nov 2001; and another juvenile/♀, 1–15 Mar 2002.

GREEN-WINGED TEAL *Anas carolinensis* Plate 14 (3)
Sp. Cerceta Americana **P.** Marrequinha-americano

Monotypic. **L** 36 cm (14″); **WS** 58 cm (23″). **IDENTIFICATION** Sexually dimorphic. Green-winged Teal is only identified with 100% certainty if it is a ♂ in breeding plumage or a moulting ♂ when it is possible to see the obvious vertical white line on the breast-sides, or traces of this line. Also the buff lines, which border the green band on the head, are much narrower and almost absent compared to Eurasian Teal. ♀ is almost identical to ♀ Eurasian Teal but the pale greater covert bar on the leading edge of the speculum is richer and more cinnamon-buff, although this is not diagnostic and some Eurasian Teal can resemble Green-winged in this feature. It has also been suggested that the head pattern is generally more contrasting than Eurasian Teal but there is much individual variation. Juvenile resembles Eurasian Teal. **VOICE** As that of Eurasian Teal. **RANGE** Breeds across most of N North America and winters south to Honduras and the West Indies. **STATUS** Vagrant with records from three archipelagos. **Azores** A female at Praia da Vitória, 15 Oct 2002 was joined by two more, 28–29 Oct 2002; more recently, on São Miguel, eight ♂♂, Lagoa Azul, Sete Cidades, 23–24 Nov 2003; Lagoa das Furnas, a maximum five ♂♂, 9 Dec 2003–14 Feb 2004 (presumably part of the original eight). **Canaries** About ten records, most from Tenerife but also recorded once on Fuerteventura, at Embalse de Los Molinos, Nov 1997–10 Feb 1998, and once on Lanzarote, at Costa Teguise, 8 Feb 1998. Unsurprisingly, all were ♂♂. **Cape Verdes★** ♂, Mindelo (São Vicente), 18 Dec 2004–13 Jan 2005 at least; then two ♂♂, Pedra Badejo (Santiago), 4 Jan 2005.

MALLARD *Anas platyrhynchos* Plate 13 (5)
Sp. Ánade Azulón **P.** Pato-real

A. p. platyrhynchos. **L** 50–65 cm (20–25½″); **WS** 81–98 (32–38½″). **IDENTIFICATION** Sexually dimorphic. ♂ in breeding plumage is easily recognised by green head and upper neck, purple-brown breast and lower neck, white neck collar, grey body and yellow bill. ♀ generally brown, with darker crown and nape, dark subterminal V-marks and mottles on body, dark eye-stripe, and bill orange-brown with a dusky culmen and tip. Like other ♀♀ of genus, particularly Gadwall and Black Duck, but larger than the former, paler than the latter, and has a white-bordered blue speculum in flight. Eclipse ♂ resembles ♀ but crown and nape even darker, breast rufous-brown and only lightly marked and yellowish-green bill maintained. Juvenile similar to ♀ but is more streaked and flank markings are less V-shaped. **VOICE** Rather vocal, particularly ♀, which gives a rasping *raehb* both in alarm and in contact as a series of spaced, drawn-out notes. Same call but with a faster delivery is used in greeting or as a threat. ♀ gives a varied series of *quack* calls. **RANGE** Breeds over most of central N Hemisphere and winters within, or a little south, of the breeding range. **STATUS** Has bred in the region and its status varies between the three archipelagos on which it has been recorded. **Azores** Has bred on Flores at least four times, in 1903, 1964, 1981 and 1985; otherwise a fairly regular winter visitor, occasionally in large numbers. According to older literature it has been recorded on all main islands. **Madeira** Listed as an occasional visitor, most recently a pair, São Jorge (Madeira), 10 Oct 2002; and a ♀, El Tanque (Porto Santo), 4 Dec 2003. **Canaries** Formerly a fairly common winter visitor, but recent records show it to be scarce and irregular; recorded from all main islands.

(AMERICAN) BLACK DUCK *Anas rubripes* Plate 13 (7)
Sp. Ánade Sombrio **P.** Pato-escuro-americano

Monotypic. **L** 53–61 cm (21–24″); **WS** 85–96 cm (33½–38″). **IDENTIFICATION** Sexes similar. Both sexes look superficially like a very dark ♀ Mallard but are more uniform and show contrasting pale head- and neck-sides. There is no pale in the tail and the purple speculum lacks the white edges of Mallard. Also, in flight, both sexes have very striking white underwing-coverts. ♂ in breeding plumage has entirely dark blackish-brown plumage except pale buffish head- and neck-sides; the bill is yellow and the legs and feet reddish-orange. ♀ similar but has generally less contrasting plumage and an olive or greenish-yellow bill. Eclipse ♂ very similar to ♀. Juvenile also very similar to ♀ but is browner with a greyish bill. Hybrid ♂ tends to show a varying amount of green on the head or some other feature indicative of Mallard. ♀ hybrids are harder to identify but usually show a whitish line at the base of the speculum and are darker than ♀ Mallard. **VOICE** Similar to Mallard. **RANGE** Breeds in E North America, northern populations moving south to winter on the eastern seaboard of the USA south to Florida and the Gulf coast. **STATUS** Has bred in the region but best treated as an accidental visitor, with records from two archipelagos. **Azores** Current status uncertain, but probably best considered accidental. In 1998 many Black Duck × Mallard hybrids were seen on Flores and have been seen in subsequent years in the same area. In autumn 2000, also on Flores, a pure pair of Black Duck was seen with two young, confirming breeding for the first time. Elsewhere, also seen on São Miguel, Terceira and Corvo. **Canaries** One, Punta de Teno (Tenerife), 23 Nov 1993, was also almost almost certainly the individual seen at Puerto de la Cruz and La Orotava (Tenerife), 5 Jan–3 Apr 1994.

(NORTHERN) PINTAIL   *Anas acuta*   Plate 13 (6)
Sp. Ánade Rabudo   **P.** Arrábio

Monotypic. **L** 51–66 cm (20–26″), including 10 cm (4″) tail extension of breeding ♂; **WS** 80–95 cm (31½–37½″). **IDENTIFICATION** Sexually dimorphic. ♂ in breeding plumage unmistakable: head and upper neck chocolate brown, lower neck and breast white, extending as a stripe on head-sides, and body mostly grey with a black and buff ventral region. ♀ similar to congeners in coloration but of more elegant appearance with a thinner bill, plain head and neck contrasting with dark-scalloped upperparts and flanks, and a pointed tail. Eclipse ♂ resembles ♀ but has greyer, more elongated tertials and a more patterned bill. Juvenile also similar to ♀ but upperparts darker and less clearly marked, flanks less scalloped and feathers broadly dark-centred. **VOICE** ♂ gives a mellow *proooop-prooop* similar to Common Teal and ♀ has similar calls to Mallard but quieter. **RANGE** Breeds throughout most of N Hemisphere between *c.* 45° and 70° N, except in Europe where it breeds in a number of isolated locations. Winters south to Panama, N sub-Saharan Africa, India and SE Asia. **STATUS** Accidental on all four archipelagos. **Azores** Vagrant with nine records: ♂, Lagoas das Furnas (São Miguel), pre-1963; eclipse ♂, Sete Cidades (São Miguel), 5–17 Oct 1963; ♂+1, near Lagoa do Congro (São Miguel), 3 Mar 1965; juvenile ♂, Terceira, 1 Nov 1966, 3♂, 1♀, Sete Cidades (São Miguel), 4 Feb 1970; one, São Jorge, 25 Oct 2000; adult ♀, Lagoa Azul, Sete Cidades (São Miguel), 3–9 Dec 2001; ♀, Lagoa Branca (Flores), 30 Oct–2 Nov 2002; and one, Lagoa das Furnas (São Miguel), 10 Dec 2003. **Madeira** Listed as exceptional by Zino *et al.* (1995), the most recent record a ♀, El Tanque (Porto Santo), 16 Oct 2002. **Canaries** Rare and irregular winter visitor, recorded on all main islands except La Gomera. **Cape Verdes** ♀, Ribeira da Madama (Sal), 12 Feb 1996; and a juv/♀, Calheta (Maio), 30 Dec 2004.

GARGANEY   *Anas querquedula*   Plate 14 (5)
Sp. Cerceta Carretona   **P.** Marreco

Monotypic. **L** 37–41 cm (14½–16″); **WS** 60–63 cm (23½–25″). **IDENTIFICATION** Sexually dimorphic. ♂ in breeding plumage is unmistakable: a broad white crescent extends across head-sides from in front of eye to nape, and rest of head, neck and breast are dark pinkish-brown, separated from grey flanks by a vertical white stripe; also has elongated black-and-white scapulars. In flight reveals a large, pale dull blue-grey forewing with a green speculum bordered both in front and behind by a white bar. ♀ is very similar to ♀ plumages of Eurasian Teal, Green-winged Teal and Blue-winged Teal. Differs mainly from ♀♀ of Eurasian and Green-winged Teal in having a more prominent facial pattern and lacking pale stripe on tail-sides. In flight it only has a narrow white line on the greater coverts but a broad white trailing edge to the secondaries and a browner speculum. ♀ Blue-winged Teal has a less well-marked face, but usually a distinct pale loral spot, and in flight a distinctive blue forewing. Eclipse ♂ resembles ♀ but retains upperwing pattern of breeding plumage. Juvenile also resembles ♀ but has central underparts streaked and spotted, not pure white. In flight, bars bordering speculum narrower than those of adult. Juveniles can be sexed and young ♂ has an upperwing similar to adult but not so pale. Young ♀ has a dull grey speculum. **VOICE** When flushed ♂ can give a rattling, knocking call and ♀ a high-pitched *quack*, otherwise relatively silent when not displaying. **RANGE** Breeds across most of Europe and NC Asia, from England to Kamchatka. Winters south to C Africa and S Asia. **STATUS** Recorded from three archipelagos but status varies. **Azores** Vagrant with eight records: immature ♂, Candelária (São Miguel), 8 Jul 1964, captured alive and presented to Ponta Delgada museum, 11 Aug 1964. It was still alive, with clipped wings, when seen by Bannerman & Bannerman, Sep 1964. Also, a ♂ and 2♀♀, Lajes do Pico (Pico), 5 Nov 1990; immature/♀, Sete Cidades (São Miguel), 5–7 Oct 1999; ♂, Cabo da Praia (Terceira), 11 Nov 2001; two, Cabo da Praia (Terceira), 22–30 Sep 2003; immature/eclipse ♂, near Lagoa Seca (Flores), 11 Oct 2003; eclipse ♂, Lagoa Branca (Flores), 2–7 Sep 2004; and one, Lagoa Seca (São Miguel), 3–5 Nov 2004. **Madeira** ♀/juvenile, El Tanque (Porto Santo), 14–19 Aug 2003; and ♀, Lugar de Baixo (Madeira), 26–29 Nov 2003. **Canaries** Currently considered accidental but probably a scarce but regular passage migrant. Recorded from all main islands except Lanzarote and La Gomera. **Cape Verdes** Extreme vagrant: a ♂, Mindelo sewage farm (São Vicente), 14–18 Apr 2001, is the only record.

BLUE-WINGED TEAL *Anas discors* Plate 14 (6)
Sp. Cerceta Aliazul   **P.** Marreca-d'asa-azul (Pato-d'asa-azul)

Monotypic. **L** 37–41 cm (14½–16″); **WS** 60–64 cm (23½–25″). **IDENTIFICATION** Sexually dimorphic. Like others in genus, ♂ unmistakable in breeding plumage with dark blue-grey head, conspicuous white facial crescent in front of eye, black and white vent and a bright blue upper forewing. ♀ is fairly plain grey-brown with a distinct pale loral spot and a blue forewing almost as bright as in ♂, but otherwise very similar to other ♀ teals. Eclipse ♂ resembles ♀ but crown darker, head- and neck-sides more coarsely streaked and upperwing pattern as breeding ♂. Juvenile also similar to ♀ but has darker upperparts and legs and feet are greyish rather than yellowish as in adult. Eclipse ♂, ♀ and immature are all very similar to corresponding plumages of Eurasian Teal, Green-winged Teal and Garganey, and are most easily identified in flight or when preening, when the blue forewing is clearly visible. Also, the face patterns of the various species useful when faced with a lone bird: as mentioned above ♀-type Blue-winged Teal has a distinct whitish oval-shaped area at the bill base, which is not normally shown by Eurasian or Green-winged Teal, and the facial pattern is less pronounced than Garganey. **VOICE** Mainly silent except in display. **RANGE** Breeds over most of C North America and winters from S USA through C America south to Peru and N Brazil. **STATUS** Perhaps best treated as a vagrant, although it is perhaps annual. **Azores** Twenty-two records and recorded annually since 1997, with records from São Miguel (7), Pico (4), Terceira (7), Flores (3) and São Jorge (1). **Canaries** Eight records: one collected, Barranco de la Carnicera (Tenerife), Oct 1901; ♂, Pozo Izqueride (Gran Canaria), 9 Jan 1988; two females/immatures, Embalse de Los Molinos (Fuerteventura), 28 Sep 1989; first-winter ♂, Guargacho (Tenerife), 25 Oct 1992; ♂, Embalse de Los Molinos (Fuerteventura), 15–19 Feb 1993; 1 ♂, 2 ♀♀, Embalse de Los Molinos (Fuerteventura), 23 Oct 1993–31 Jan 1994 (♂ until 3 Mar); ♀, Embalse de Valle Molina (Tenerife), 10–19 Oct 1997; and ♂, Roquito del Fraile (Tenerife) 8 Nov 2002–25 Feb 2003. **Cape Verdes** ♂, Mindelo sewage farm (São Vicente), 12 Mar 2000, is the only record.

(NORTHERN) SHOVELER *Anas clypeata* Plate 14 (1)
Sp. Pato Cuchara (Cuchara Común)   **P.** Pato-colhereiro (Pato-trombeteiro)

Monotypic. **L** 44–52 cm (18–20½″); **WS** 70–84 cm (27½–33″). **IDENTIFICATION** Sexually dimorphic. Both sexes easily distinguished from congeners by the very large, broad bill. ♂ in breeding plumage unmistakable given combination of green head, white breast, chestnut flanks and belly, and prominent blue forewing visible in flight. ♀ resembles a small ♀ Mallard, but has no white trailing edge to green speculum, a darker belly and a pale grey forewing. Eclipse ♂ resembles ♀ but body more rufous, particularly the flanks, head browner and body markings blacker. Juvenile also resembles ♀ but has a darker crown and hindneck, plus paler more spotted underparts. Juveniles can be sexed from an early age by using forewing coloration. **VOICE** ♂ utters a *tuck* note when flushed and, in display, a repeated, rather liquid and hollow *tuck* or *tuck-tuck*. ♀ gives a variety of low quacks including a short descending series of 1–4 long notes followed by 3–9 shorter single quacks. **RANGE** Breeds across much of N Hemisphere. Winters mostly south of the breeding range in northern temperate and tropical zones. **STATUS** Varies between the three archipelagos where recorded. **Azores** Vagrant with 11 records including one recent instance of breeding: one, Paul da Praia (Terceira), summer 1958; ♀ discovered in a collection on Terceira, 1971; 3 ♂♂, 1 ♀, Sete Cidades (São Miguel), 25 Dec 1977; ♂, Lagoa das Furnas (São Miguel), 2 Feb–14 Mar 1983; one, São Miguel, May 1989; ♀, Lagoas das Furnas (São Miguel), 13 Dec 1998; pair with three young, Fajã Grande (Flores), early July 1999; ♀, Sete Cidades (São Miguel), 8–13 Nov 2000; pair there, 4 Dec 2001; ♀, Lagoa Ginjal (Terceira), 21 Oct 2002; and two, Lagoa das Furnas (São Miguel), 9–15 Dec 2003. **Madeira** ♀ shot, Ribeira da Janela (Madeira), 27 Oct 1904; ♀, west of Tabua (Madeira), 1 Jan 1996; and ♂, Lugar de Baixo (Madeira), 28 Nov 2004. **Canaries** Regular winter visitor in very small numbers to all main islands except La Gomera and El Hierro.

MARBLED DUCK (MARBLED TEAL) *Marmaronetta angustirostris* Plate 14 (7)
Sp. Cerceta Pardilla   **P.** Pardilheira

Monotypic. **L** 39–42 cm (15–16½″); **WS** 63–67 cm (25–26″). **IDENTIFICATION** Sexes similar. Adult generally sandy-brown with an obvious dark eye-patch, short shaggy crest on nape (most prominent in ♂), pale-spotted flanks and upperparts, breast and undertail-coverts with darker brown barring, and a slender dark bill. In flight rather pale with very pale secondaries and darker primary-coverts, and lacks a speculum. Juvenile resembles adult but spotting more diffuse and plumage slightly greyer. **VOICE** During display ♂ gives a high-pitched, squeaky, nasal whistle, *vee-veeh* or *jeeeep*, and the ♀ a similar call, but relatively silent at other times. **RANGE** Uncommon with a patchy **DISTRIBUTION** from S Spain and NW Africa, thence in Turkey and Israel east through Asia to extreme W China. Winters mostly in Morocco, Iran and Pakistan. **STATUS** Formerly bred but generally best treated as a vagrant now, although it has recently been regular on Fuerteventura (Canaries), where it bred successfully in 1998, the first confirmed breeding in the archipelago for 140 years. **Madeira** One caught, Funchal harbour (Madeira), 29 Jul 1894. **Canaries** Bred, Maspalomas (Gran Canaria), 1857 (Carl Bolle) and perhaps still bred there until 1914, but thereafter unrecorded in the archipelago until one, Tesejerague (Fuerteventura), 15–19 May 1992, with varying numbers present, on the same island, at Embalse de Los Molinos and Presa de las Peñitas since 1993 until the present. The species bred successfully in 1998, at Presa de Las Peñitas, but this reservoir has since dried out and no further breeding attempts have been recorded, although birds are still occasionally recorded on the island. A recent record, from Maspalomas (Gran Canaria), 14 Feb 2004, gives hope for the future. **Cape Verdes** Three substantiated records, all at unspecified sites in Boavista, where it may have formerly bred: two ♂♂ and a ♀ collected, 10 May 1897; three ♂♂ collected, 5–9 Feb 1898; and two ♂♂ collected, 20 Mar 1924.

RED-CRESTED POCHARD *Netta rufina* Plate 14 (8)
Sp. Pato Colorado   **P.** Pato-de-bico-vermelho

Monotypic. **L** 53–57 cm (21–22½″); **WS** 84–88 cm (33–35″). **IDENTIFICATION** Sexually dimorphic. ♂ in breeding plumage is unmistakable and ♀, although superficially similar to a ♀ Common Scoter, should not be confused with any other species. Breeding ♂ has an orange head, bright red bill, black breast and rear, white flanks and brown upperparts. ♀ has a brown body with paler breast and flanks, crown and nape darker brown with very pale head- and neck-sides, and bill greyish with a pinkish spot near the tip. Eclipse ♂ similar to ♀ but retains red eye and bill from breeding plumage. Juvenile resembles ♀ but bill is all grey and lacks pinkish spot. **VOICE** Usually silent except in display. **RANGE** Breeds in S Spain and France, thence from Turkey across C Asia to NW China and W Mongolia. Winters around Mediterranean, Black and Caspian Seas and across Indian subcontinent to Myanmar. **STATUS** Extreme vagrant. **Canaries** Mentioned for Tenerife by Emmerson *et al.* (1994) but without details.

COMMON POCHARD *Aythya ferina* Plate 15 (1)
Sp. Porrón Común   **P.** Zarro-comum

Monotypic. **L** 42–49 cm (16½–19″); **WS** 72–82 cm (28–32″). **IDENTIFICATION** Sexually dimorphic. ♂ in breeding plumage unmistakable, given chestnut head, black breast, upper mantle and rear, and grey body. The only confusion species are Redhead *A. americana* and Canvasback *A. valisineria*, which reach Europe as vagrants and could appear in our region. ♂ Canvasback has similar plumage but is larger with a longer, blackish bill and paler body. ♂ Redhead also similar but body slightly darker and has a less sloping forehead. ♀ similar to congeners, but structural differences shown by ♂ also apply. It has greyish flanks and back, mottled brown, the head, breast and neck browner, and a pale loral patch, eye-ring and stripe behind eye. Confusion perhaps most likely with ♀ Ring-necked Duck (which see for their separation). Eclipse ♂ resembles breeding ♂ but breast and rear end browner and head and neck duller chestnut. Juvenile similar to ♀ but is browner, more uniform above and lacks pale stripe behind eye. **VOICE** Rather silent except in display. **RANGE** Breeds over most of Europe and C Asia. Winters in W Europe, the Mediterranean basin, the Black and Caspian Seas and thence east across the Indian subcontinent to S China and Japan, also C Africa. **STATUS** Recorded throughout but status varies. **Azores** Accidental with *c.* 10 records, all but one from São Miguel, the other on Terceira in 1966. The most recent sighting was a pair, Lagoa Verde, Sete Cidades (São Miguel), 9–12 Sep 2004. **Madeira** Fewer than five records in the last 50 years. **Canaries** Regular winter visitor, sometimes in small flocks. An unusual record concerns a ♂, Roquito del Fraile (Tenerife), 26–29 Jun 1987. **Cape Verdes** Vagrant with only three records: ♀ collected, São Nicolau, 2 Dec 1898; eight, Pedra Badejo lagoons (Santiago), 6 Feb 1966; and two ♂♂, three ♀♀, Ribeira da Madama (Sal), 12 Feb 1996.

RING-NECKED DUCK *Aythya collaris* Plate 15 (6)
Sp. Porrón Acollarado   **P.** Caturro (Zarro-de-colar)

Monotypic. **L** 37–46 cm (14½–18″); **WS** 61–75 cm (24–29½″). **IDENTIFICATION** Sexually dimorphic. Both sexes share characteristic head shape with an obvious peak to rear crown; the grey, not white, wingbar on upperwing; pale subterminal band on bill (more obvious in ♂); and distinctly S-shaped border between flanks and upperparts. ♂ in breeding plumage superficially resembles Tufted Duck but lacks a crest, has predominantly grey flanks with a white crescent in front and a narrow white band at the bill base. ♀ is more likely to be confused with ♀ Common Pochard but pale loral area whiter, pale subterminal band on bill also whiter and upperparts noticeably darker than flanks. From ♀ Tufted Duck distinguished by distinct whitish eye-ring and, often, a white line behind eye. Eclipse ♂ has blackish parts of plumage browner, whitish lores and undertail-coverts, warm brown flanks and retains yellow irides. Juvenile similar to ♀ but head-sides usually browner and bill darker without a pale subterminal band, although this develops in the first autumn. **VOICE** Relatively silent when not displaying. **RANGE** Breeds across most of S and C Canada and N USA. Winters chiefly on both seaboards of the USA, south to Guatemala and rarely to Panama. **STATUS** Accidental winter visitor to all four archipelagos and now recorded almost annually. **Azores** Though still an accidental visitor in autumn and winter, now over 25 records, including a flock of 12 (2♂ 10♀), São Miguel, winter 90/91. The vast majority are from São Miguel (17) but also recorded on Flores (6), Pico (3) and São Jorge (1). **Madeira** One collected, mouth of Ribeira da Janela (Madeira), 17 Oct 1967. **Canaries** Accidental winter visitor, having been recorded on more than 30 occasions, all but once in Oct–Apr, and all since

1980. Recorded on all main islands except Lanzarote with the majority on Tenerife where almost annual since 1989. **Cape Verdes** The only record is of three ♀♀, Ribeira da Madama (Sal), 16–18 Nov 1999.

FERRUGINOUS DUCK *Aythya nyroca* Plate 15 (2)
Sp. Porrón Pardo **P.** Pêrra (Zarro-castanho)

Monotypic. **L** 38–42 cm (15–16½"); **WS** 63–67 cm (25–26½"). **IDENTIFICATION** Sexes similar. Both resemble a ♀ Tufted Duck but are identified by the richer, more chestnut, plumage, the higher crown reaching a distinct peak in the centre, longer bill and lack of any crest on the hindcrown. In flight the upperwing bar is more extensive than on Tufted Duck. ♂ in breeding plumage has dark chestnut head, neck, breast and flanks, and blackish-brown upperparts. The vent-sides are blackish and the undertail-coverts pure white. The irides are white. ♀ duller with dark irides. Eclipse ♂ resembles ♀ but retains white irides. Juvenile duller than ♀ and undertail-coverts less pure white. **VOICE** Rather silent except in display. **RANGE** Breeds very locally in Spain and France, but commoner from E Europe through C Asia to W China. Winters mainly around Mediterranean, sub-Saharan Africa, Iraq, Iran, Pakistan and N India, with a few in Myanmar and S China. **STATUS** Vagrant with only eight records from three archipelagos. **Azores** ♀, Sete Cidades (São Miguel), 25 Nov 1977. **Canaries** Two shot, Isleta (Gran Canaria), Nov 1829; one collected, Punta de Hidalgo (Tenerife), 17 Nov 1923; ♀, Presa de las Peñitas (Fuerteventura), 8 Dec 1990; ♂, Gran Tarajal (Fuerteventura), 12 Dec 1991–25 Jan 1992; two, Salinas de Janubio (Lanzarote), Jan 1993; and one, Puerto de la Cruz (Tenerife), 13 Nov 1995. **Cape Verdes** ♀ collected, Boavista, 5 Feb 1898.

TUFTED DUCK *Aythya fuligula* Plate 15 (3)
Sp. Porrón Moñudo **P.** Negrinha (Zarro-negrinha)

Monotypic. **L** 40–47 cm (15¾–18½"); **WS** 67–73 cm (26¼–28¾"). **IDENTIFICATION** Sexually dimorphic. ♂ in breeding plumage should not be confused with any other species. Head, breast, upperparts and vent black with a purple sheen to head; flanks white, and there is a long drooping crest from the hindcrown, which is diagnostic. The irides are yellow; the bill bluish-grey with a black nail and tip, and an indistinct pale subterminal band. ♀ similar to other *Aythya* but usually has a short crest on the hindcrown. Generally brown with paler flanks and, like ♂, black bill tip, which helps separate it from either scaup, especially those with extensive white feathering at the bill base. Some in winter have a whitish undertail and can resemble ♀ Ferruginous Duck but do not have such an extensive upperwing bar and, usually, a yellow iris. In flight both sexes have a broad white bar across most of the flight feathers. Eclipse ♂ resembles breeding ♂ but the black is replaced by blackish-brown, the flanks are dusky-brown, and the crest is much reduced. Juvenile is like ♀ but has an indistinct tuft, paler head and upperparts, a small buffish patch on the lores and a darker iris. **VOICE** Generally silent except in display. **RANGE** Breeds over much of Europe and Asia, wintering south to Ethiopia, N India and S China. **STATUS** Recorded from all four archipelagos. **Azores** Thirteen records, mostly on São Miguel, but two from Terceira and one on Pico. **Madeira** Immature ♂ shot, 'Fort S. Tiago' (Madeira), 18 Nov 1906. **Canaries** Regular winter visitor, occasionally in small groups, to Fuerteventura, Gran Canaria, Tenerife, La Palma and El Hierro. **Cape Verdes** Three records: a ♂ and two ♀♀, Pedra Badejo lagoons (Santiago), 6 Feb 1966; ♂, Ribeira de Madama (Sal), 3–17 Nov 1999; and a juvenile/♀, Pedra Badejo (Santiago), 4 Jan 2005.

GREATER SCAUP *Aythya marila* Plate 15 (4)
Sp. Porrón Bastardo **P.** Negrelho (Zarro-bastardo)

A. m. marila. **L** 42–51 cm (16½–20"); **WS** 69–84 cm (27–33"). **IDENTIFICATION** Sexually dimorphic. ♂ in breeding plumage unlikely to be confused except with Lesser Scaup, from which it differs in being larger, with a green sheen to the head, more rounded head shape, less bold upperpart vermiculations, and the white upperwing bar usually extends to the primaries. ♀ also similar to that of Lesser Scaup but could also be confused with a white-faced ♀ Tufted Duck. Compared to Tufted the black on the bill tip of Greater Scaup is restricted to the nail, the head lacks a crest and the white at the bill base is usually more extensive. Best separated from Lesser Scaup by larger size, usually more extensive white facial patches, more rounded head shape and the white wing bar extends onto the primaries. Eclipse ♂ resembles breeding ♂ but is generally duller on the head and breast, occasionally shows some white at the bill base, and the flanks are vermiculated greyish-brown. Juvenile resembles ♀ but is usually paler and more buff below, particularly on the flanks, and the white at the bill base is less extensive. Juvenile ♂ resembles eclipse ♂ by first autumn and attains many adult features in the first winter. **VOICE** Relatively silent except in display. **RANGE** Breeds across most of the northern N Hemisphere. Winters south to S USA, N France, the Black and Caspian Seas, and to Japan and S China. **STATUS** Vagrant with records from three archipelagos. **Azores** Eleven records: ♂ obtained, São Miguel (no other details); ♂, Sete Cidades (São Miguel), 25 Nov 1977; unknown number in a large duck flock, Sete Cidades (São Miguel), Jan–Feb 1981; ♂, Praia da Vitoria, Terceira, 27 Jan 1982, 7 ♀♀, São Miguel, 14–17 Dec 1990; 4 ♀♀/immatures, Lagoa Azul, Sete Cidades (São

Miguel), 19 Nov 1997; first-winter ♂, Praia da Vitoria (Terceira), 23 Oct 1999; pair, Lagoa Azul, Sete Cidades (São Miguel), 8 Nov 2000–18 Feb 2001; immature ♂ there, 3–9 Dec 2001; immature ♂, Lagoa da Lomba (Flores), 30 Oct 2002; and ♀, Lagoa das Furnas and Lagoa Seca (São Miguel), 3–5 Nov 2004. **Madeira** One, São Lázaro (Madeira), late Nov 1985. **Canaries** Ten records from: Fuerteventura (4), Gran Canaria (1), Tenerife (4) and La Palma (1).

LESSER SCAUP *Aythya affinis* Plate 15 (5)
Sp. Porrón Bola **P.** Negrelho-americano

Monotypic. **L** 38–46 cm (15–18″); **WS** 66–74 cm (26–29″). **IDENTIFICATION** Sexually dimorphic. ♂ in breeding plumage differs from similar Greater Scaup in usually having a purple sheen to the black head, a peaked rear crown, more prominent upperpart vermiculations and the white upperwing bar is confined to the secondaries. ♀ can also be identified from ♀ Greater by using head shape and upperwing pattern, and the white facial area is less extensive than Greater Scaup. Great care must be taken with lone females to ensure a correct identification. Eclipse ♂ similar to ♀ but is usually darker on head and breast, greyer on flanks and upperparts with less white at the bill base. Juvenile also very similar to ♀ but often has less white at the bill base and darker irides. First-winter ♂ resembles breeding ♂ but breast and rear end not as black, flanks greyish-brown and upperpart vermiculations less extensive and less well defined. **VOICE** Generally silent except in display. **RANGE** Breeds mostly in interior W and N North America, with isolated pockets further south in the Mid West and Great Lakes region. Winters mainly to the south, on both seaboards of the USA, the West Indies and C America, some even reaching Colombia. **STATUS** Vagrant with records from three archipelagos. **Azores** Seven records all since autumn 1998: adult ♂, Lagoa das Furnas (São Miguel), 31 Oct 1998; 2 ♀♀, Lagoa das Furnas (São Miguel), 29 Nov 1998; first-winter ♂, 6 ♀♀, Lagoa Azul, Sete Cidades (São Miguel), 2 Nov 1999; immature ♂ and a ♀, Lagoa Azul, Sete Cidades (São Miguel), 2–4 Nov 2000; immature ♀, Lagoa das Furnas (São Miguel), 3 Nov 2000; immature ♂, Lagoa Azul, Sete Cidades (São Miguel), 3–9 Dec 2001; and a ♀ there, 25 Dec 2001–2 Jan 2002. **Canaries** Eleven records all but two from Roquito del Fraile, Tenerife: ♀/immature, 28 Nov 1991–10 Mar 1992; ♀, 17 Nov 1994–12 Mar 1995; ♀, 19 Nov 1995–18 Mar 1996; first-winter ♂, 10 Jan–18 Mar 1996; ♀/immature, 12 Nov 1998, joined by 2 ♂♂ and 5 ♀♀, 26 Nov, with the last being 2 ♀♀, 14 Mar 1999; ♀/immature, 2 Nov 1999–1 Apr 2000; and an immature and adult ♀, 15 Nov 2000–2 Feb 2001 at least. The only other records are from Salinas de Castillo del Romeral (Gran Canaria), 5 Jan–6 Feb 2000, and Embalse de Bernardino (Tenerife), Nov 2004–Feb 2005 at least. **Cape Verdes** 3 ♀♀, Mindelo sewage farm (São Vicente), 22 Jan–24 Feb 1999; and a ♀ there, 13 Jan 2005.

COMMON EIDER *Somateria mollissima* Plate 16 (1)
Sp. Eider **P.** Eider (Eider-edredão)

S. m. mollissima. **L** 50–71 cm (20–28″); **WS** 80–108 cm (31½–42½″). **IDENTIFICATION** Sexually dimorphic. ♂ unmistakable: white upperparts, except rear end, and breast (with a pink wash), black crown, underparts and rear end to upperparts, and pale green patches on nape and neck-sides. ♀ is mostly rufous-brown, barred darker, but shares characteristic large head and wedge-shaped bill of ♂. Confusion possible with ♀ King Eider but is larger and less rufous, with feathering on side of bill extending to the nostrils and longer than feathering on culmen. Eclipse ♂ mostly sooty-brown and has white upperwing-coverts and some white feathering on head and upperparts. Juvenile similar to ♀ but is darker and more uniform with an indistinct pale buffish supercilium. Immature ♂♂ obtain white feathering over a period of about two years. **VOICE** Usually silent outside breeding season. **RANGE** Breeds on cooler coasts of N Hemisphere and winters slightly further south. **STATUS** Vagrant with only three records, and all from one archipelago. **Azores** ♀ obtained, Rosta de Cão (São Miguel), prior to 1903; ♂ in breeding plumage obtained, Ponta Delgada (São Miguel), 30 Jan 1952; and ♂ same place, 31 Aug 1996–8 Sep 1998.

KING EIDER *Somateria spectabilis* Plate 16 (2)
Sp. Eider Real **P.** Eider-real

Monotypic. **L** 47–63 cm (18½–25″); **WS** 87–100 cm (34–39″). **IDENTIFICATION** Sexually dimorphic. In breeding plumage ♂ is completely unmistakable given combination of black body, pink breast, pale blue head, large orange frontal shield and red bill. ♀ is similar to ♀ Common Eider but is generally more rufous-brown. The flank markings are angled, not barred as in Common and the feathering along the culmen is longer than on the bill-sides. A small pale area at the bill base is dissected by the distinctly upturned gape line, which gives the species a rather happy facial expression. ♂ in eclipse is darker and more uniform than ♀, usually with some white feathering on the mantle and breast, the frontal shield is reduced, but it retains the white upperwing patches. Juvenile similar to ♀ but is duller and greyer with broader markings on flanks. First-winter ♂ starts to develop pale breast, has greyish patches on head, darker scapulars and the bill is pale orange. **VOICE** Flight call is a low croaking note. Other calls are unlikely to be heard in the region. **RANGE** Breeds along most high-arctic coastlines with most birds wintering at sea just south of the breeding areas. **STATUS** Extreme vagrant with just one, recent record. **Azores** First-winter ♂, Ponta Delgada harbour (São Miguel), 3–4 Nov 2000.

LONG-TAILED DUCK *Clangula hyemalis* Plate 16 (3)
Sp. Havelda **P.** Pato-rabilongo (Pato-de-cauda-afilada)

Monotypic. **L** 36–47 cm (14–18½″) excluding 10–15 cm (4–6″) elongated tail feathers of ♂; **WS** 73–79 cm (29–31″). **IDENTIFICATION** Sexually dimorphic. Both sexes unmistakable in either breeding or non-breeding plumages. ♂ in breeding plumage is predominantly brown-black with greyish-white flanks, white rear flanks and vent, a large pale area around eye and yellowish edges to the scapulars. ♀ in breeding plumage is similar but lacks elongated tail feathers, has duller plumage and a pale collar around neck. In non-breeding plumage, ♂ has head-sides pale grey-brown and a prominent dark patch on the upper-neck-sides. Rest of head and neck are white, as are most of the scapulars and upper mantle. ♀ in non-breeding plumage is similar to breeding plumage but has whiter head-sides. Juvenile is similar to non-breeding ♀ but neck patch less obvious and head is not as white. **VOICE** ♂ gives a nasal, yodelling *ow-ow-owdelee* and both sexes a low, nasal *gak* in flight. **RANGE** Breeds across entire arctic south to S Finland, S Alaska and Labrador. Winters in coastal waters south to Washington, S Carolina, British Isles and Korea. **STATUS** Vagrant with only ten records. **Azores** ♂ obtained, São Miguel, pre-1903, with another ♂ obtained there also pre-1903; ♂ obtained, São Miguel, 1903; one dead, between Praia da Vitoria and Cabo de Praia (Terceira), 27 Dec 1944; one captured, Angra (Terceira), between 1966 and 1968; and one, Praia da Vitoria (Terceira), 14–15 Feb 2004. **Madeira** One recorded by Schmitz prior to 1909, and one, Porto Santo, 8 Nov 2004. **Canaries*** Immature ♂ and immature ♀, Lago de Janubio (Lanzarote), 23–24 Dec 1992; two, Arrecife (Lanzarote), January 1995; and ♂ in breeding plumage, off Punta del Tope (Lanzarote), 9 Jun 2001.

COMMON SCOTER *Melanitta nigra* Plate 16 (4)
Sp. Negrón Común **P.** Negrola (Pato-negro)

Monotypic. **L** 44–54 cm (17–21″); **WS** 79–90 cm (31–35½″). **IDENTIFICATION** Plumage of ♂ is all black except yellow patch on culmen, in front of basal knob of bill. ♀ is brown with paler cheeks and neck-sides and no basal knob on the bill. Juvenile resembles ♀ but has a paler belly. ♂ of American sibling species, Black Scoter *M. americana* (formerly considered a race of Common Scoter) distinguished by its swollen orange-yellow bill with a small black tip. Black Scoter has not been recorded in our region but is a potential vagrant. **VOICE** Generally silent in winter. **RANGE** Nominate breeds in tundra of N Palearctic and winters from Norway and the W Baltic south to Spain and NW Africa. **STATUS** Vagrant with records from three archipelagos. **Azores** Six records but only one in the last 70 years: some, spring 1857, São Miguel; one, May 1865, São Miguel; one, spring 1870, São Miguel; ♀ obtained, Sete Cidades (São Miguel), Oct 1895; immature ringed in Iceland recovered Ponta Delgada (São Miguel), 24 Oct 1927; one, Praia da Vitoria (Terceira), 15 Oct 2000; and one, Cabo da Praia (Terceira), 29 Oct 2004. **Madeira** One shot, Funchal (Madeira), 9 Oct 1895; and a ♀ shot, Madeira, 2 Nov 1936. **Canaries** Accidental visitor, but only three recent records. On Lanzarote, two, Arrecife, Jan 1993, and one there, Jan 1994, and an exhausted immature, El Médano (Tenerife), 13 Dec 1998. Also recorded on Gran Canaria but not within the last 100 years.

SURF SCOTER *Melanitta perspicillata* Plate 16 (5)
Sp. Negrón Careto **P.** Negrola-de-lunetas (Pato-careto)

Monotypic. **L** 45–56 cm (18–22″); **WS** 78–92 cm (31–36″). **IDENTIFICATION** Sexually dimorphic. Adult ♂ in breeding plumage unmistakable: all-black plumage with two white patches on head: a larger one on the nape and a smaller one on the forehead, and a large almost triangular-shaped multicoloured bill are diagnostic. ♀ resembles congeners, being mainly brown but some exhibit a pale nape patch and shares unusual triangular bill shape of ♂. Also ♀♀ have, to a varying degree, two pale patches on head-side (sometimes indistinct or even missing). Juvenile is similar to ♀ but lacks pale nape patch, has paler cheeks and a pale, almost whitish belly. **VOICE** Generally silent except in display. **RANGE** Breeds across N North America from W Alaska to Labrador, wintering on coasts from the Aleutians to Baja California, in the west, and from the Gulf of St Lawrence to N Carolina in the east. **STATUS** Vagrant with only six records from two archipelagos. **Azores** ♀, Ponta Delgada (São Miguel), 21 Nov 1984; immature/♀ at Praia do Populo (São Miguel), 27 Oct 1985; first-winter, Fajã dos Cubres (São Jorge), 9 Oct 1997; first-winter, Lajes (Flores), 7 Nov 1998; another first-winter, Santa Cruz (Flores), same date; one, Porto Pim (Faial), 19 Nov 2001; and first-winter ♂, Lajes (Flores), 14 Nov 2003. **Madeira** One captured alive, Funchal (Madeira), 4 Nov 1988.

BUFFLEHEAD *Bucephala albeola* Plate 17 (1)
Sp. Porrón Albeola **P.** Olho-dourado-de-touca (Pato-olho-de-touca-branca)

Monotypic. **L** 32–39 cm (12½–15"); **WS** 54–61 cm (21–24"). **IDENTIFICATION** Sexually dimorphic. Generally recalls a small Common Goldeneye. ♂ in breeding plumage is unmistakable; the large head is mostly black glossed dark green, purple and bronze, but has a large white patch on the sides and rear. The underparts and neck are white and the upperparts mostly black, except the grey uppertail-coverts and lower rump, and darker grey tail. ♀ has a smaller, brown head with a white patch behind and below the eye, greyish-brown neck, vent, flanks and mantle, with whitish central underparts. The upperwings are a darker brown with a white patch on the inner secondaries. ♂ in eclipse resembles ♀ but has a larger white patch on the head-sides and a greater area of white on the upperwing. Juvenile also similar to ♀ but white patch on head smaller and central underparts mottled grey. ♀ could be confused with Common Goldeneye but, apart from the obvious differences in head pattern, ♀ Bufflehead has a pale grey bill and dark irides. **VOICE** Generally rather silent. **RANGE** Breeds in Alaska, Canada and NW USA. Winters both coastally and inland south to Middle America. **STATUS** Extreme vagrant with just three records. **Azores** ♀, Lajes do Pico (Pico), 10 Nov 1998; and ♀, Santa Cruz (Graciosa), 16 Dec 2000–9 Mar 2001 at least. **Canaries★** One, Salinas de Janubio (Lanzarote), Jan 1994.

COMMON GOLDENEYE *Bucephala clangula* Plate 17 (2)
Sp. Porrón Osculado **P.** Olho-dourado (Pato-olho-d'ouro)

Subspecies unknown. **L** 42–50 cm (16½–21"); **WS** 65–80 cm (25½–31½"). **IDENTIFICATION** ♂ in breeding plumage is black above with green gloss to head, a circular white patch on lores and black-and-white scapulars. The neck and underparts are white. In flight the upperwing is white with black primaries, primary-coverts and leading edge; irides bright yellow and bill black. ♀ has a chocolate-brown head, white collar and grey breast and flanks; irides pale yellow to whitish and bill dark becoming yellow distally (rarely all yellow). Eclipse ♂ similar to ♀ but retains upperwing pattern of breeding ♂. Juvenile also similar to ♀ but is duller, lacks white collar and has a dark bill. Head and bill shape help to avoid confusion with any other diving duck recorded in our region. **VOICE** Usually silent except in display. **RANGE** Breeds across northern parts of the N hemisphere south of the tundra zone, and winters predominantly to the south. **STATUS** Vagrant with records from two archipelagos. **Azores** ♀ obtained, São Miguel, Dec prior to 1888; and 2 ♀♀, Fajã dos Cubres (São Jorge), 13–16 Nov 1998. **Madeira** Recorded less than five times in the last 50 years. In 1936, it was included in Saramento's *Vertebrados da Madeira* but without supporting evidence.

SMEW *Mergellus albellus* Plate 17 (4)
Sp. Serreta Chica **P.** Merganso-pequeño

Monotypic. **L** 38–44 cm (15–17"); **WS** 55–69 cm (22–27"). **IDENTIFICATION** Sexually dimorphic. ♂ in breeding plumage totally unmistakable with its predominantly white plumage, short, loose crest on crown and nape, black mantle, black loral mask and black line either side of nape. Adult ♀ has a predominantly greyish body and dark chestnut head, except white chin, throat and lower cheeks. Eclipse ♂ resembles ♀ but has darker, blacker upperparts and a larger area of white on upperwing-coverts. Juvenile also resembles ♀ but has central underparts greyer and broader white tips to greater coverts and secondaries. **VOICE** Generally silent. **RANGE** Breeds across much of the N Palearctic from N Scandinavia to E Siberia, wintering mainly in the SW Baltic but also from the British Isles south to the W Mediterranean and in Greece, Turkey, Iran and Iraq. Further east, winters in China, Japan and Korea. **STATUS** Extreme vagrant with just one record. **Canaries★** One, Salinas de Janubio (Lanzarote), Jan 1994.

HOODED MERGANSER *Mergus cucullatus* Plate 17 (3)
Sp. Serreta Capuchona **P.** Merganso-capuchinho (Merganso-cabeçudo)

Monotypic. **L** 42–50 cm (16½–20"); **WS** 56–70 cm (22–27½"). **IDENTIFICATION** Sexually dimorphic. A small merganser with a large, bushy crest on the rear crown. ♂ in breeding plumage unmistakable, as the combination of large white patches on the side of the black head, white breast with two black bands, orangey flanks and black upperparts with white shaft-streaks to the tertials are diagnostic. ♀ resembles ♀ Red-breasted Merganser but is much smaller with a smaller, yellowish bill and a much fuller crest. Plumage generally browner and darker than Red-breasted and, in flight, has very little white on upperwing. ♂ in eclipse is like ♀ but has pale irides and a blacker bill. Juvenile also similar to ♀ but has a shorter crest and lacks white streaks on tertials. **VOICE** Usually silent away from breeding grounds. **RANGE** Breeds in C North America and winters mainly in coastal areas south to N Mexico. **STATUS** Extreme vagrant with only four, all recent, records. **Azores** ♀, Lagoa Branca (Flores), 10–11 Feb 2001; another ♀ there, 4 Apr 2002; and ♂ displaying, Caldeirão (Corvo), 8 Apr 2002. **Canaries** ♀, El Pinque (Tenerife), 11 Dec 2001–11 Mar 2002.

RED-BREASTED MERGANSER   *Mergus serrator*   Plate 17 (5)
Sp. Serreta Mediana   **P.** Merganso-de-poupa

Monotypic. **L** 52–58 cm (20½–23″); **WS** 70–86 cm (27½–34″). **IDENTIFICATION** Sexually dimorphic. A medium-sized duck with a long, thin bill and shaggy crest. ♂ in breeding plumage unmistakable with its long, thin, serrated red bill, black head glossed green, rufous-brown breast with black streaks and a white collar above this. Between the breast-sides and the grey flanks is a black area with large white spots, a diagnostic feature. The majority of the upperparts are black and, in flight, it has a large white patch on the inner wing. ♀ mostly greyish with a rusty-brown head and upper neck. Eclipse ♂ resembles ♀ but retains large white patch in the darker wing and the mantle is blacker. Juvenile also similar to ♀ but has central underparts and breast more greyish-brown, less whitish. **VOICE** Usually silent except in display. **RANGE** Breeds over most northern parts of the N Hemisphere. Winters south to N Mexico, the Mediterranean and S China. **STATUS** Vagrant with records from three archipelagos. **Azores** Nine records, the most recent being at Horta (Faial), a ♀ 27–30 Jan 1990 and ♂ 1–17 Feb 2001. **Madeira** Immature ♂ obtained, Machico (Madeira), 10 Nov 1906. **Canaries** One, Puerto de la Cruz and Punta de Hidalgo (Tenerife), 17 Nov 1984–25 Jan 1985. Also reported, Salinas de Janubio, with two adults 2 Oct 1992 and then another two Nov 1993–22 Jan 1994, and at Salinas de Cocoteros and Playa de Cochino, all on Lanzarote, but accompanying details for the last two sites are less than adequate.

(WESTERN) HONEY-BUZZARD   *Pernis apivorus*   Plate 18 (1)
Sp. Abejero Europeo   **P.** Bútio-vespeiro (Falcão-abelheiro)

Monotypic. **L** 52–60 cm (20½–23½″); **WS** 125–145 cm (49–57″). **IDENTIFICATION** Sexes similar. Similar in size to Common Buzzard but distinguished from that species at all ages by its longer, narrower wings, narrower, more prominent, 'cuckoo-like' head and neck, and longer tail with slightly convex sides and rounded corners. Plumage very variable at all ages and can make separation from Common Buzzard problematical. Honey-buzzard is polymorphic, with rufous, pale, medium and dark morphs in adult and juvenile plumages. Adult separated from Common Buzzard by pattern of barring on underwing. On Honey-buzzard there is a broad black bar along the trailing edge and 1–2 narrower bars across base of primaries and outer secondaries, and black of wingtips is restricted to outer tips of primaries. The pattern and spacing of barring on the tail, if visible, is diagnostic. A broad terminal band and then an ample space before the two narrower bands at the base of the tail is totally different to any pattern of Common Buzzard. Juvenile lacks distinctive plumage of adult having more evenly barred underside to flight feathers, a more evenly barred under tail and darker underside to secondaries. **VOICE** Usually silent away from breeding grounds. **RANGE** Breeds throughout much of Europe east to C Asia and winters in Africa, mostly in central equatorial regions but also in smaller numbers in E and S Africa. **STATUS** Vagrant recorded from two archipelagos. **Madeira** Adult, near Monte (Madeira), 12 Aug 1981. **Canaries*** At least seven records, four of them old and lacking details: one seen by Meade-Waldo on Tenerife, one recorded by Cabrera at Laguna (Tenerife) in May, one seen by Polatzek at Treror (Gran Canaria) and another not detailed by Bannerman. Recent records are of one, south of Tefia (Fuerteventura), 16 Apr 1991; one, Teno Massif, 20 Aug 1992; and an adult, San Sebastian (La Gomera), 19 May 1997. Also been recorded from Lanzarote: a specimen found under electric wires between Haria and Ye in the 1980s.

(AMERICAN) SWALLOW-TAILED KITE   *Elanoides forficatus*   Plate 18 (3)
Sp. Elanio Tijereta   **P.** Gavião-tesoura

Subspecies unknown, but the most likely is the nominate *E. f. forficatus*. **L** 52–62 cm (20½–24½″); **WS** 119–136 cm (47–53½″). **IDENTIFICATION** Sexes similar. Unmistakable in all plumages. Adult has head, body and underwing-coverts white, with rest of plumage dark greyish-black glossed purple. Possibly one of the most distinctive raptors, as the long, very deeply forked tail is unique. Juvenile similar to adult but distinguished in early stages by buffy wash to the head and breast, which is quickly lost, making the shorter tail the best feature for ageing. **VOICE** Usually silent outside breeding season. **RANGE** Breeds from coastal SE USA through C America and N South America to NE Argentina. All populations are migratory, spending the non-breeding season in N South America. **STATUS** Exceptional vagrant with just one amazing record. **Canaries** Adult, Costa Calma (Fuerteventura), 19–23 Mar 1993.

BLACK-SHOULDERED KITE   *Elanus caeruleus*   Plate 18 (2)
Sp. Elanio Azul   **P.** Peneireiro-cinzento

E. c. aeruleus. **L** 31–35 cm (12–14″); **WS** 75–85 cm (29½–33½″). **IDENTIFICATION** Sexes similar. Adult unmistakable: grey upperparts, white underparts, black shoulders and black surround to large red eye. In flight, underwing mostly white with dusky primaries. Juvenile resembles adult but browner with white tips to mantle, scapulars, greater coverts and flight feathers, some brownish streaking on breast and pale yellowish eyes. **VOICE** Rather silent, except in display or near nest. **RANGE** Resident in Iberia, over most of Africa and the Indian subcontinent east to the Philippines and south through Indonesia to New Guinea. **STATUS** Extreme vagrant with only two records from the same archipelago. **Canaries**★ Juvenile captured, Guía (Gran Canaria), 1965, and kept in captivity c. 2 years; and one, cliffs at Famara (Lanzarote), 1972.

BLACK KITE *Milvus migrans* Plate 18 (4)
Sp. Milano Negro **P.** Milhafre-preto

M. m. migrans. **L** 55–60 cm (22–23½″); **WS** 135–170 cm (53–67″). **IDENTIFICATION** Sexes similar. A dark, medium-sized raptor with long, broad wings and a shallow-forked tail that is frequently twisted from side to side in flight. Adult has dark brown upperparts with a paler brown panel on the inner wing and dark brown underparts with paler bases to primaries. Juvenile has a paler body with darker streaking on underside, pale tips to greater and primary-coverts, and a paler upperwing panel. Flight action is rather floppy and, in direct flight, the body rises and falls with the wingbeats, rather like a tern. When soaring or gliding it does so on flat or slightly arched wings with the hand angled back and the carpal joint held forward. Beware that, whilst soaring, the tail is often spread and can appear almost straight-ended, losing the characteristic fork. Confusion possible with Red Kite and Marsh Harrier, and with dark-morph Booted Eagle, Common Buzzard and Honey-buzzard. Compared to all of these, except Red Kite, the forked tail is diagnostic. From Red Kite best separated by its darker plumage, lack of rufous tail, shallower tail fork and less prominent window on inner hand of underwing. Marsh Harrier has a less deeply fingered wingtip, no pale bases to the primaries on underwing, a rounded tail and a quite distinct flight silhouette. Booted Eagle also has a rounded and unbarred tail, darker underside to the primaries with a pale wedge on the inner primaries, white 'headlights' on the neck-side and whitish uppertail-coverts. Common Buzzard is smaller, shorter and broader winged, shorter tailed and has a more extensive pale area on the flight feathers of the underwing. Honey-buzzard has a large amount of white on the underwing and should cause the least confusion; although dark juveniles are similar, they always have prominent barring on the flight feathers and tail. **VOICE** Most common is a rather high-pitched mewing *peee-errrrr*. **RANGE** Breeds from Europe through Africa and Asia to Australia, but avoids deserts, heavily forested areas and cooler northern zones. Most of the more northerly populations move south to winter in tropical regions. **DISTRIBUTION** Previously reported as breeding on all main islands of the Cape Verdes, but now probably extinct in the archipelago. **BREEDING** Few data available from our region. Eggs laid mid Mar–early Apr and clutch normally 2–3. **HABITAT** In Cape Verde breeds, or used to breed, on cliffs and rocky ledges but most often seen on coasts, scavenging at harbours and fishing villages. **STATUS** Varies between archipelagos. **Madeira** Vagrant with just three records: one, Salvage Islands, Sep 1990; two, Ponta do São Lourenço (Madeira), Apr 2001; and one, near Ribeiro Frio (Madeira), 8 Sep 2004. **Canaries** Rare passage migrant and very rare winter visitor, sometimes in small flocks, and recorded from all main islands but only regular on the eastern ones. **Cape Verdes** According to Hazevoet (1995), a resident species restricted to Santiago, Santo Antão and Maio, and possibly still extant on Brava and São Vicente; formerly more common but now rare or extinct on many islands. A recent survey, in 1999, failed to locate it on any of the islands mentioned above, and located just one bird, on Boavista.

CAPE VERDE KITE *Milvus fasciicauda* Plate 18 (6)
Sp. Milano de Cabo Verde **P.** Milhafre-vermelho

Monotypic. **L** 50–62 cm (20–24½″); **WS** 134–152 cm (53–60″). **IDENTIFICATION** Slightly smaller than Red Kite with shorter and more rounded wings, and shallower tail fork. The upperparts have less obvious feather fringes and the underparts are paler with narrower shaft-streaks. The tail usually has 8–10 bars on the central rectrices but individual variation is considerable in all of these features. Juvenile similar to adult but is more patterned above and more mottled below, with a more strongly barred tail. **VOICE** If any differences exist in vocalisations, they have never been noted. De Naurois mentioned that this species is 'very silent'. **RANGE** Endemic to Cape Verde. **DISTRIBUTION** Formerly bred on Santiago, Brava, São Nicolau and Santo Antão, but also recorded on São Vicente and Ilhéus do Rombo. Also recorded recently on Boavista and Maio. **BREEDING** In Cape Verde it breeds, or bred, on cliffs and rocky ledges. **HABITAT** Usually associated with rugged montane areas with well-vegetated ravines and some cultivation in surrounding areas, although visits coasts to forage. Formerly more widespread and presumably occupied other habitats, as it was known to scavenge around settlements. **TAXONOMIC NOTE** Formerly considered a subspecies of Red Kite but, following Hazevoet (1995), frequently given specific status in recent years. Recent mitochondrial DNA research suggests that Cape Verde Kite may not be a valid taxon (Johnson *et al.* 2005) **STATUS** Only ever recorded from one archipelago. **Cape Verdes** Now very rare, in the recent past was confined to the interior of Santiago and Santo Antão. On Santo Antão, most often encountered in the vicinity of Tarrafal and Monte Trigo, and on Santiago in the Serra do Pico da Antónia. A recent survey, in 1999, located just two birds on Santo Antão, and its extinction seems imminent. Recent records include two on São Vicente, near the sewage ponds at Mindelo, 12 Mar 2000, and four on Boavista, Jul/Aug 2001, with two, Maio, in the same period. The total population is now considered to be no more than a few individuals, making it one of the most endangered species of raptor in the world.

RED KITE *Milvus milvus* Plate 18 (5)
Sp. Milano Real **P.** Milhafre-real (Milhano)

M. m. milvus. **L** 60–66 cm (23½–26″); **WS** 155–180 cm (61–71″). **IDENTIFICATION** Sexes similar. The only raptor, (except North American vagrant Swallow-tailed Kite) with a deeply forked tail. The rufous underparts, pale head and long, angled wings are also useful features. Adult has a reddish-brown body and an almost white head, both closely streaked darker, and a long deeply forked tail with a bright rufous upperside and black tips to the ends of the fork. Underwing boldly patterned with a distinct, almost white, patch at the base of the primaries, black primary tips and carpal patches, reddish-brown coverts and dark secondaries. Juvenile similar to adult but has a paler body, a paler and more extensive upperwing bar and pale tips to the greater coverts. **VOICE** Generally silent away from the nest. **RANGE** Confined to the W Palearctic. Breeds from S Sweden discontinuously through Europe, south to Morocco and east to the Caucasus and Ukraine. N and C European populations are mostly migratory. **DISTRIBUTION** Nominate race *milvus* formerly bred in the Canaries but became extinct fairly recently. **HABITAT** In the Canaries, used to nest in the endemic Canarian Pine, but only in larger trees and usually at a considerable height above ground. **STATUS** All records from one archipelago. **Canaries** Formerly bred on Gran Canaria, Tenerife, La Gomera and El Hierro, but now extinct on all these islands, the last birds being seen on Tenerife and Gran Canaria in the late 1960s. Since then there have been a few records but the species is now accidental at best.

WHITE-TAILED (SEA) EAGLE *Haliaeetus albicilla* Plate 21 (1)
Sp. Pigargo Europeo **P.** Pigargo (Águia-rabalva)

Monotypic. **L** 70–90 cm (27½–35½″); **WS** 190–250 cm (75–98½″). **IDENTIFICATION** Sexes similar. Large size, broad wings with prominent 'fingers', short wedge-shaped tail and huge bill are features of all ages. Adult unmistakable given its white tail, pale head and huge yellow bill. Juvenile has darker plumage than adult and a dark bill: the tail is less wedge-shaped and the feathers only white in the centres. Immature similar to juvenile but has less uniform plumage. Subadult resembles adult but has dark tips to tail feathers, a rather dark head and dark tip to the bill. Juveniles and immatures are prone to wander. **VOICE** Usually silent outside the breeding season. **RANGE** Breeds across much of the N Palearctic from S Greenland and W Iceland through N and C Eurasia south to Greece and Turkey, the S Caspian, Lake Balkhash and Manchuria. **STATUS** All records are from one archipelago but are old and lack details. **Canaries★** Vagrant with 4–5 very old records from Tenerife, Gran Canaria and Lanzarote. Webb and Berthelot included this species on the strength of a foot from a bird killed on Lanzarote, and Canon Tristram saw one at Arrecife (Lanzarote), Apr 1890.

EGYPTIAN VULTURE *Neophron percnopterus* Plate 19 (1)
Sp. Alimoche Común, Guirre (Canaries) **P.** Britango (Abutre do Egipto)

N. p. percnopterus (Cape Verdes) and *N. p. majorensis* (Canaries). **L** 60–70 cm (23½–27½″); **WS** 155–170 cm (61–67″). **IDENTIFICATION** Sexes similar. Adult unmistakable, being the only large, black-and-white raptor with an obvious wedge-shaped tail. The underparts, tail and underwing-coverts are white, the primaries and secondaries black and the bare head is yellow. The mantle, rump and tail are white, as are the upperwing-coverts except for a dusky area on the inner greater coverts. The white centres to the upper flight feathers are unique to this species. Juvenile mainly blackish-brown with buffish tips to tail, upperwing-coverts, rump, mantle, scapulars and vent, but shares characteristic long wings and wedge-shaped tail of adult. Immature/subadult resembles adult with white areas becoming more extensive with age. Subspecies *majorensis* has only recently been described and, compared to the nominate, is distinctly larger with the white in its plumage impregnated with rufous, particularly on the crown, nape, median coverts, breast and tail. **VOICE** Usually silent but gives varied low whistles, grunts, groans and rattling noises when agitated. **RANGE** Breeds from S Europe east to C Asia and south to S India and Tanzania, with an isolated population in SW Angola and NW Namibia. Northern populations winter south of the breeding range but north of the equator. **DISTRIBUTION** Breeds on the Canaries and Cape Verdes. **BREEDING** On the Canaries, usually from Mar but can start late Feb or as late as mid Apr. Clutch either one or two. On Cape Verde season more protracted, from Nov to Apr. **HABITAT** Breeds on cliffs and rock ledges, but seen anywhere around breeding islands. **STATUS** Recorded from all archipelagos, but only as a vagrant on two. **Azores** Juvenile obtained, Capelas (São Miguel), summer 1932. **Madeira** A few old records without details, except that one was seen on Porto Santo. **Canaries** Formerly a more common and widespread resident, but now restricted to Fuerteventura, Lanzarote, Lobos and Alegranza. Fuerteventura is its last stronghold, but even here it is declining. It is now extinct on Gran Canaria, Tenerife, La Gomera, El Hierro and Montaña Clara. **Cape Verdes** A not uncommon resident on all islands (except Sal) but population definitely declining recently and it is now scarce on Fogo and São Nicolau.

SHORT-TOED (SNAKE) EAGLE *Circaetus gallicus* Plate 19 (2)
Sp. Culebrera Europea **P.** Águia-cobreira

Monotypic. **L** 62-67 cm (24-26"), **WS** 170-190 cm (67-74"). **IDENTIFICATION** Sexes similar. A large raptor with a pale underside. Gives the impression of a huge buzzard rather than a true eagle. The adults can be quite variable but the usual form has a dark hood with white lower breast, belly and underwings marked with coarse dark barring on the belly and underwing-coverts and barred flight feathers and tail. The tail has three, occasionally four, evenly spaced dark bars that can be seen on the upperside as well. The upperparts are fairly uniform with the contrast between the coverts and the flight feathers varying individually. Usually the darkest birds show the least contrast. Some very pale examples, generally subadult birds, lack the dark hood and show hardly any barring on the underside. The most likely confusion species are Honey-buzzard, pale morph Common Buzzard and Osprey, but unlike Short-toed Eagle all of these show some dark carpal markings on the underwing and black or blackish tips to the primaries. Glides with the carpal joints held well forward and with the wings bowed. When hunting often hangs motionless or hovers and frequently dangles its legs. The juvenile is very similar to the adult and is probably inseparable in the field. **VOICE** Rather silent away from the breeding grounds so unlikely to be heard in the region. **RANGE** Breeds from Spain and Morocco east through Turkey and Russia to the region of Lake Balkhash. The western populations winter in sub-Saharan Africa between the Sahel and c. 10°N, from Senegal to Ethiopia. **STATUS** Only recorded from one of the archipelagos. **Canaries** A vagrant with only seven records. One found dead below electricity wires between Haria and Ye (Lanzarote) in the 1980's, a corpse in advanced decomposition found in Barranco de Los Poleos, Macizo de Teno (Tenerife) 6 Sept. 1986, one seen near Guia de Isora (Tenerife) 20 Feb. 1995, one at Toto (Fuerteventura) 4 Mar. 1998, one at Las Galletas (Tenerife) 23 Nov. 1998, one near Monte del Agua (Tenerife) 7 Apr. 1999 and one near the airport (Gran Canaria) 5 Jan. 2001.

(WESTERN) MARSH HARRIER *Circus aeruginosus* Plate 19 (4)
Sp. Aguilucho Lagunero Occidental **P.** Águia-sapeira (Tartaranhão-ruivo-dos-pauis)

C. a. aeruginosus but *C. a. harterti* (North Africa) must be a possibility. **L** 48–56 cm (19–22"); **WS** 120–135 cm (47–53"). **IDENTIFICATION** Sexually dimorphic. The largest of the harriers and similar in size to Common Buzzard. Like congeners, glides and soars with long wings held in a shallow V, which combined with the slender body and long, narrow tail gives harriers a rather distinctive flight silhouette. Adult ♂ has black wingtips, plain blue-grey secondaries and primary bases, brown coverts with a white leading edge, blue-grey tail, yellowish-white head and upper breast, and rest of underparts chestnut-brown. Adult ♀ predominantly dark brown with a creamy-white crown, throat and forewing, often a pale patch on the breast, and a slightly reddish-brown tail. Juvenile similar to ♀ but is darker and has a yellowish tone to creamy areas. In rare dark morph these paler areas are absent and these birds bear a superficial resemblance to Black Kite. However, the flight silhouette is totally different, the tail lacks a shallow fork and there is no pale mid-wing panel on the upperwing. ♂ of the race *harterti* has darker upperparts and paler underparts; the ♀ has the pale areas whiter and more extensive. **VOICE** Usually silent outside breeding season. **RANGE** Breeds from Iberia and NW Africa through Europe east to N Mongolia and Lake Baikal. Some winter close to breeding areas but main wintering area is from Mediterranean Basin, Turkey and the Middle East south to S Africa. **STATUS** Varies between the archipelagos. **Azores** Vagrant with three records: immature, Cabo da Praia (Terceira), 17 Nov 1998; one, Lagoa do Fogo (São Miguel) 31 Jul 1999; and one, Flores, 6 Oct 2002. **Madeira** Fewer than five records in the last 50 years and also recorded from the Salvages. **Canaries** Scarce but regular passage migrant and winter visitor, but no records from La Palma. **Cape Verdes** Accidental winter visitor with more than 20 records, all in Sep–Mar, from Boavista, Sal, Raso, Maio, São Nicolau and Santo Antão.

HEN HARRIER *Circus cyaneus* Plate 19 (3)
Sp. Aguilucho Pálido **P.** Tartaranhão-cinzento (Tartaranhão-azulado)

C. c. cyaneus and *C. c. hudsonicus*. **L** 44–52 cm (17–20½"); **WS** 105–125 cm (41–49"). **IDENTIFICATION** Sexually dimorphic. Larger and broader winged than next two species, with both sexes having a relatively large, square, white rump patch. ♂ is generally pale grey above with black wingtips, and white underparts except grey throat and upper breast. Underwing also whitish with a black tip and black trailing edge to the flight feathers. ♀ has brown upperparts with a conspicuously barred tail and buffish-white underparts streaked brown. Underwing is barred on primaries and secondaries, and underwing-coverts buffish-brown streaked darker. Juvenile similar to ♀ but generally more rufous on underparts with duskier underside to secondaries. N American race *hudsonicus* distinguishable in the field from the nominate *cyaneus*: ♂ has browner upperparts and rufous wedge-shaped spotting on the underparts, particularly prominent on thighs and flanks; ♀ has darker upperparts and has the underparts more warmly tinged cinnamon. Juvenile also darker on upperparts and has rich rufous underparts with much less dark streaking than in *cyaneus*. **VOICE** Usually silent away from breeding grounds. **RANGE** Breeds across much of N Holarctic, wintering south to N South America, S Asia and SE China. **TAXONOMIC NOTE** Recently, subspecies *hudsonicus* has been assigned specific status as Northern Harrier *C. hudsonicus* by some authorities. **STATUS** Recorded from three archipelagos. **Azores** One, Flores, 21 Oct to 2 Nov 2002 (*hudsonicus*). **Madeira** One, Porto Santo, 4 Nov 1999; and immature ♀, Paúl da Serra (Madeira), 12 Oct 2002. **Canaries** Rare and irregular passage migrant and winter visitor recorded from all main islands.

PALLID HARRIER   *Circus macrourus*   Plate 19 (5)
Sp. Aguilucho Papialbo   **P.** Tartaranhão-pálido (Tartaranhão-de-peito-branco)

Monotypic. **L** 40–48 cm (16–19″); **WS** 100–125 cm (39–49″). **IDENTIFICATION** Sexually dimorphic. Adult ♂ identified from both Montagu's and Hen Harriers by diagnostic small wedge of black on wingtip, very pale underwing, body and tail and paler ashy upperparts. Also differs from Hen Harrier in lack of an obvious white rump patch and from Montagu's by lack of black bar on secondaries. ♀ and juvenile plumages are more difficult to separate, but the smaller size, smaller white rump and the slimmer, more pointed wings eliminate Hen Harrier. Also, the pale collar is more pronounced, as is the dark patch on the ear-coverts. Separation from Montagu's is more complicated but ♀ has more uniform secondaries on upperwing, darker secondaries on underwing, lacks barring on greater underwing-coverts and axillaries, the rear pale bar on the underside of the flight feathers is narrower, and the trailing edge of the 'hand' is paler. Juvenile distinguished from ♀ by its unstreaked rufous underparts and underwing-coverts. Compared to Montagu's it has a prominent buffish collar bordered below by a dark brown semi-collar, and paler tips to the primaries. If a collar is present on Montagu's then it is narrower and rather rufous, and the bordering dark collar is less prominent and extensive. **VOICE** Usually silent away from breeding grounds. **RANGE** Breeds in the Ukraine and SW Russia east to Lake Balkash and NW China. Winters predominantly in sub-Saharan Africa and the Indian subcontinent east to S China. **STATUS** Extreme vagrant with just two records from the same island. **Canaries** ♂, Los Rodeos (Tenerife), 26–31 Mar 1990; and an exhausted juvenile there was taken into care, 14 Jan 2005. It was released successfully the following day.

MONTAGU'S HARRIER   *Circus pygargus*   Plate 19 (6)
Sp. Aguilucho Cenizo   **P.** Águia-caçadeira (Tartaranhão-caçador)

Monotypic. **L** 43–47 cm (17–18½″); **WS** 105–125 cm (41–49″). **IDENTIFICATION** Sexually dimorphic. ♂ distinguished from the previous two species by presence of black bars on the secondaries (one above and two below). Otherwise very similar, being predominantly ash-grey above with dark wingtips, but lacks white rump of Hen Harrier, is streaked rufous on the underparts and has a more heavily patterned underwing than either Hen or Pallid Harriers. ♀ is very similar to that of Pallid (which see). Juvenile similar to ♀ but has unstreaked rufous underparts and underwing-coverts, and is also very similar to juvenile Pallid but like ♀ differences are covered under that species. Montagu's has a rare dark morph, which is all dark blackish-brown except the white underside to the primaries and base of the tail feathers. **VOICE** Usually silent away from breeding grounds. **RANGE** Breeds in NW Africa, S and C Europe east to Kazakhstan, and winters in Africa south of the Sahara and in the Indian subcontinent. **STATUS** Varies between the different archipelagos on which it has been recorded. **Madeira** Fewer than five records in the last 50 years, and also recorded on the Salvages. **Canaries** Scarce but regular passage migrant on all main islands, except La Gomera and El Hierro. **Cape Verdes** Two acceptable records: adult ♀, near Rabil airfield (Boavista), 14 Mar 1997; and a juvenile, Morrinho (Maio), 30 Dec 2004.

(NORTHERN) GOSHAWK   *Accipiter gentilis*   Plate 20 (1)
Sp. Azor   **P.** Açor

Subspecies unknown but *A. g. gentilis* is the most likely.   **L** 48–62 cm (19–24½″); **WS** 95–125 cm (37½–49″). **IDENTIFICATION** Sexes similar except in size. Large accipiter with ♀ having wingspan comparable to a Common Buzzard and ♂ to a Carrion Crow. Adult has grey-brown to blue-grey upperparts with a prominent white supercilium and dark cap. Underparts white with extensive barring on breast, belly and underwing-coverts, but undertail-coverts usually plain white and unmarked. ♀ is considerably larger than ♂ and is often browner above, with a less distinct cap and less prominent supercilium, but it is unwise to sex birds on plumage. Juvenile brown above with broad buff fringes to feathers; lacks contrasting dark cap of adult and supercilium is indistinct or lacking; underparts frequently buffish, not white, with dark drop-shaped spots replacing barring of adult. In our region, only likely to be confused with Eurasian Sparrowhawk but the larger size, broader wings, more prominent white undertail-coverts and less distinct barring on the underside of the flight feathers should make recognition relatively straightforward. **VOICE** Silent except near nest. **RANGE** Breeds across much of the Holarctic and is mostly sedentary, except for some post-breeding dispersal and altitudinal movements. Northernmost populations do move south in winter but wintering range is largely within that of breeding. **STATUS** Extreme vagrant with just three records. **Canaries** ♀, Monte del Agua (Tenerife), 7 Jul 1979; ♀, Punta de la Rasca (Tenerife), 10 Jan 2003; and a pair, Anaga Peninsula (Tenerife), 25 Sep 2003.

EURASIAN SPARROWHAWK *Accipiter nisus* Plate 20 (2)
Sp. Gavilán Común **P.** Gavião da Europa (Gavião)

A. n. granti. **L** 28–38 cm (11–15″); **WS** 60–80 cm (23½–31½″). **IDENTIFICATION** Sexually dimorphic. ♂ is smaller than ♀ and is blue-grey above and white below, with rufous cheeks, breast, belly and underwing-coverts barred rufous, underside of flight feathers barred grey-brown and tail has 4–5 dark bands. ♀ is dark brown or grey-brown above with a white supercilium, dark brown underpart barring, and more prominent barring on the underside of the flight feathers than ♂. Juvenile resembles ♀ but is browner above with brown spots and crescents on underparts forming less clear-cut barring. In flight separated from smaller falcons by its shorter, more blunt-tipped wings and square-ended tail. The endemic subspecies *granti* is slightly darker on the upperparts than the nominate *nisus* and more heavily barred below. **VOICE** A loud, shrill *kek-kek-kek-kek* that varies in speed and intensity depending on usage. Also a plaintive *whee-oo* given by begging ♀. **RANGE** Breeds in the Palearctic from W Europe and N Africa east to Kamchatka and Japan, and south to the Himalayas. In Europe, only more northerly populations are migratory but in Asia all winter south to S India and Thailand. **DISTRIBUTION** Breeds only in Madeira and the Canaries. **BREEDING** Nests in various native and introduced trees, with the nest usually placed 6–10 m above ground. Eggs usually laid Apr but in exceptional circumstances delayed until May or even early Jun. Clutch size 2–5, usually three. **HABITAT** Breeds in woodlands but encountered almost anywhere from sea level to the treeline. **TAXONOMIC NOTE** Birds from the Canaries have been subspecifically separated by some authors as *teneriffae,* on account of the barring being less bold than on Madeiran birds, and the barring of the ♂ being more vinaceous-red. However, *teneriffae* is currently considered a synonym of *granti*. **STATUS** Differs between the two archipelagos on which it breeds. **Madeira** Rare and only found on the main island of Madeira. **Canaries** Resident breeder on all main islands, where it is not uncommon in wooded areas, except Fuerteventura and Lanzarote, where it is a rare but regular passage migrant. Whether these involve wandering individuals of the local subspecies is unknown.

COMMON BUZZARD *Buteo buteo* Plate 20 (3)
Sp. Busardo Ratonero **P.** Águia-d'asa-redonda

B. b. rothschildi (Azores), *B. b.. buteo* (Madeira), *B. b. insularum* (Canaries) and *B. b. bannermani* (Cape Verdes). **L** 51–57 cm (20–22½″); **WS** 115–135 cm (45–53″). **IDENTIFICATION** Sexes similar. A medium-sized, polymorphic raptor with a relatively short, rounded tail and fairly long, broad wings with paler patches on the upperwing at the base of the primaries. Although individuals vary considerably, the species is generally brown or rufous-brown above and white below, streaked and barred dark brown. Soars with wings held in a shallow V and the tail well spread. Adult has a blackish terminal tail-band that is broader and more distinct than any of the inner bars, and a paler band across the breast separates the darker upper breast from the belly, whilst the tips and trailing edge of the underwing are black. Juvenile lacks the broad tail-band and the black trailing edge is not as dark or as well defined. All of the island forms are very similar to the nominate *buteo* but they are all slightly smaller and show less individual variation: *rothschildi* is darker above and more densely pigmented belo; *insularum* is browner above and more streaked below, being less densely barred or marked; *bannermani* is slightly paler and less rufous than either of the previous races, its upperparts are rather grey-brown with paler feather fringes. **VOICE** Vocal year-round, the main call being a loud, far-carrying and mewing *peee-yoo*, used mainly in contact between pairs. **RANGE** Breeds from Europe in a broad band across the Palearctic to E Russia and Japan. **DISTRIBUTION** Resident breeder on all four groups. **BREEDING** In the Azores, a rather late breeder with most clutches in May. Likewise in Madeira, where it is also a fairly late breeder. In the Canaries breeding occurs in Mar–May. However, in the Cape Verdes the season starts as early as Jan (or even Dec) and continues until Apr. Clutch size for all races is normally two, but three and even four have been recorded. **HABITAT** Nests on rocky ledges and cliff faces in montane areas and in steep-sided Barrancos, but encountered in almost any habitat from sea level to the high alpine zone. Also breeds in trees but less frequently than in Europe. Martín (1987) notes, that of 60 nests, only nine (15%) were in trees, six different species were used for nesting and the height of the nest varied from 2–20 m above ground. **TAXONOMIC NOTE** Madeiran birds were formerly thought to be a separate subspecies, *harterti,* but this is now considered a synonym of the nominate *buteo*. In the Canaries, the species formerly bred on Lanzarote, as noted by Polatzek in 1908. He described the form *lanzaroteae* from a specimen taken there, although this taxon is no longer recognised. The Cape Verde subspecies, *bannermani,* is considered specifically distinct by Hazevoet (1995) on the grounds that the plumage is less variable than in *buteo,* and the adults are darker rufous brown above; Clouet & Wink (2000) noted that the genetic distance between *bannermani* and Long-legged Buzzard is closer than that between *bannermani* and nominate Common Buzzard, supporting the case for *bannermani* to be accorded specific status. **STATUS** A common resident on all archipelagos except Cape Verde where it is now rare. **Azores** A common breeder on all islands, except Corvo and

Flores. **Madeira** Breeds on Madeira, Porto Santo, and on Ilhéu Chão in the Desertas group until 1996. Regarded as a common resident on the main island but less so on Porto Santo. Also recorded on the Salvage Islands. **Canaries** Breeds on all main islands except Lanzarote, where now extinct as a breeder with no records in recent years. The claim by Bannerman that this species bred on Alegranza in 1913 is unsubstantiated, although he obtained a specimen, and recent surveys of this islet have failed to locate the species. **Cape Verdes** Now exceedingly rare and restricted to Santiago and Santo Antão. Also recorded from Fogo, Brava, São Nicolau, Boavista and São Vicente, from where the type specimen of *bannermani* was obtained in 1913.

LONG-LEGGED BUZZARD *Buteo rufinus* Plate 20 (4)
Sp. Busardo Moro **P.** Bútio-mourisco (Búteo-mouro)

B. r. cirtensis. **L** 50–55 cm (20–22″); **WS** 115–125 cm (45–49″). **IDENTIFICATION** Sexes similar. A polymorphic species with three colour morphs, pale, dark and rufous, but in the race *cirtensis*, which presumably occurs in our region, the dark morph is either absent or very rare. In almost all plumages there are dark carpal patches on both wing surfaces, a dark trailing edge to the underwing and white bases to the underside of the primaries. Also, in all pale morphs and many rufous morphs the head and breast are pale but become progressively darker on the belly. Adult normally has an unbarred, pale rufous tail with an almost white base. Juvenile similar but has the tail barred and a less prominent trailing edge to the underwing. Confusion with Common Buzzard is unlikely in our region because 'Steppe' Buzzard *B. buteo vulpinus* has not been recorded. **VOICE** Similar to Common Buzzard but calls generally shorter and slightly higher. Often rather silent and noticeably much less vocal than Common. **RANGE** Breeds in N Africa, SE Europe, the Middle East, Arabia and Asia, east to Mongolia and India. The N African population is resident and dispersive. **STATUS** Vagrant with a total of 11 records, ten of these from one archipelago. **Madeira** One on the Salvage Islands and then at Ponta do Pargo (Madeira), 18 Sep 2004. **Canaries** One, Alegranza 6 Apr 1995; immature, Punta del Teno (Tenerife), 3 Mar 1996; adult, Tijimiraque (El Hierro), 17 Sep 1996; one, near La Oliva (Fuerteventura), 8 Mar 1998; immature, Gran Valle (Fuerteventura), 23 Nov 1998; adult, Vega Río Palmas (Fuerteventura), 25 & 27 Nov 1998; one, Barranco de Río Cabras (Fuerteventura), 19 Mar 1999; one, Tefia (Fuerteventura), 27 Feb 2001; one, inland of Gran Tarajal (Fuerteventura), 2 Apr 2001; and one, Barranco de La Torre (Fuerteventura), 11 Feb 2003.

ROUGH-LEGGED BUZZARD *Buteo lagopus* Plate 20 (5)
Sp. Busardo Calzado **P.** Bútio-calçado (Búteo-calçado)

Subspecies unknown but *B. l. sanctijohannis* would seem as likely as the nominate. **L** 55–61 cm (22–24″); **WS** 130–150 cm (51–59″). **IDENTIFICATION** Sexes similar. The largest *Buteo* in the region and similar to others in having rather broad wings and a medium-length tail. Best distinguished from resident Common Buzzard by white base to tail which, if present, in Common Buzzard is never as contrasting. Adult has distinctive dark carpal patches on underwings and, often, a dark belly patch. ♂ has at least two black bands on the tail inside the broad terminal band, whereas ♀ usually has a maximum of one visible. On the upperside, ♂ usually has 3–4 narrow bars on the inner tail but ♀ has only 1–2. Juvenile distinguished from adult by the more diffuse terminal tail-band and lacks the distinct trailing edge to the underwing. From above, the pale bases of the primaries form a very distinct patch. N American race *sanctijohannis* has a dark morph, which is not believed to occur in the N European nominate population. Otherwise, the two subspecies are very similar and it would be inadvisable to attempt to separate them in the field. Generally, E North American birds are smaller and darker than the nominate. **VOICE** Usually silent outside the breeding season. **RANGE** Breeds across the N Holarctic and winters south to extreme N Mexico, C and E Europe, east to China, Japan and Korea. **STATUS** Vagrant with just six records from the same archipelago. **Azores*** One, Porto Pim (Faial), 11 Oct–1 Nov 2001 at least; one, Cabo da Praia (Terceira), 1 Nov 2001, and another, Praia da Vitoria, 17 Nov 2001; first-winter, Horta (Faial), 30 Dec 2001; one on Corvo 17 Oct 2002; and one over the airport on Terceira, 2 Dec 2002.

GOLDEN EAGLE *Aquila chrysaetos* Plate 21 (2)
Sp. Águila Real **P.** Águia-real

Subspecies unknown but *A. c. homeyeri* (Iberia, N Africa) seems most likely. **L** 76–93 cm (30–36½″); **WS** 190–240 cm (75–94½″). **IDENTIFICATION** Sexes similar. A large raptor with long wings and tail, the tail is as long as the width of the wing. In all plumages is generally dark brown with a golden-brown nape. Adult fairly uniform dark brown with a paler crown, nape and median upperwing-coverts. Juvenile darker with characteristic white bases to the primaries and outer secondaries, and the tail white-based with a broad dark terminal band. Subadult has less white in wing than juvenile and often less white in the tail. Subspecies *homeyeri* darker than the nominate, with less golden nape patch and less prominent mid-wing panel. Juvenile often has less white in wings and tail. **VOICE** Usually silent. **RANGE** Breeds over much of the N Hemisphere, mainly north of the tropics. Mostly sedentary but northernmost populations move south in winter. **STATUS** Vagrant, only recorded from one archipelago. **Canaries★** Six records: subadult, near Tamaimo (Tenerife), 2 Apr 1980; adult, west of Santiago del Teide (Tenerife), 4 Apr 1980; two, between Bajamar and Punta Hidalgo (Tenerife), 6 Apr 1980; three, between Bajamar and Punta Hidalgo (Tenerife), 22 Apr 1982; two, La Gomera, 23 Apr 1982; and one, north of Vilaflor, early August 1989.

BOOTED EAGLE *Aquila pennata* Plate 21 (3)
Sp. Águila Calzada **P.** Águia-calçada

Monotypic. **L** 50–57 cm (20–22½″); **WS** 115–135 cm (45–53″). **IDENTIFICATION** Sexes similar. A small dimorphic eagle, very similar in size to Common Buzzard. Both pale and dark morphs, and intermediates, have a distinct pale bar on the upperwing, pale uppertail-coverts and a white spot (landing light) on the shoulders. Pale morph has pale underwing-coverts contrasting with dark flight feathers, and an indistinct pale wedge on the inner three primaries. Dark morph has all-dark underwing apart from the same pale wedge on the inner primaries. Pale-morph juvenile resembles adult but usually has a more rufous head and more diffuse underpart streaking. Dark-morph juvenile very similar to adult, usually more uniformly dark below, but only safely identified if pale lines on tips of greater coverts and secondaries visible. Dark phase could be confused with Marsh Harrier or Black Kite but both these species lack the white shoulder spot and conspicuous white uppertail-coverts. **VOICE** Usually silent outside the breeding season. **RANGE** Breeds from Spain and NW Africa through E Europe, Turkey and the Caucasus to Lake Baikal in the northeast, and to N India in the south, with an isolated population in S South Africa. Winters mostly in sub-Saharan Africa and India. **STATUS** Only recorded from two archipelagos. **Madeira** Two records (both pale morphs): Ponta de São Lourenço (Madeira), 21 Mar 1998, and Porto Santo, 18 Apr 1999. **Canaries** Regular but scarce passage migrant and a rare and irregular winter visitor, recorded from all main islands, except La Gomera. Most records are from Tenerife and Fuerteventura.

BONELLI'S EAGLE *Aquila fasciata* Plate 21 (4)
Sp. Águila Perdicera **P.** Águia-perdigueira (Águia de Bonelli)

A. f. fasciatus. **L** 65–72 cm (25½–28″); **WS** 145–175 cm (57–69″). **IDENTIFICATION** Sexes similar. The majority of adults have a whitish patch on the mantle that is often visible at long range and is diagnostic. The underwing has a white leading edge, a broad blackish band across the greater and median coverts, and slightly paler flight feathers. Tail long and indistinctly barred, with a broad black terminal band. Juvenile usually ginger-buff to reddish-brown on underparts and underwing-coverts, lacks the broad tail-band and dark trailing edge, has the tail and flight feathers grey with narrow dark bars, and has dark tips to the outer primaries. The characteristic dark bar on the underwing develops during subadult plumages. **VOICE** Usually silent, except in display. **RANGE** Breeds from Spain and NW Africa discontinuously east to India and S China, also the Lesser Sundas and sub-Saharan Africa. Resident and dispersive but not migratory. **STATUS** Vagrant recorded from one archipelago. **Canaries** Seven records: one captured, Gran Canaria, late 1970s; one found injured; Tenerife, 1978; one, Alegranza, 3 May 1989; subadult, El Golfo (El Hierro), 3 & 18 Mar 1999; juvenile, Pista al Derrabado (El Hierro), 6 Mar 1999; juvenile, near Monte del Agua (Tenerife), 17 Apr 1999; and two, Los Tilos (La Palma), 28 Sep 2002.

OSPREY *Pandion haliaetus* Plate 21 (5)
Sp. Águila Pescadora **P.** Águia-pesqueira

P. h. haliaetus. **L** 55–63 cm (22–25″); **WS** 145–170 cm (57–67″). **IDENTIFICATION** Sexes similar. A medium-large raptor with fairly long, narrow wings and a rather short tail. Always hunts over water, as it feeds exclusively on fish. Adult generally dark-brown above, with a white crown, body (except band of dark streaks on breast) and underwing-coverts, dark carpal patch on underwing, barred flight feathers and a dark eye-stripe extending onto nape. ♀ distinguished from ♂ by its broader breast-band, often on a buffy background. Juvenile similar to ♀ but has marginally paler upperparts with buffy fringes to the upperpart feathers and upperwing-coverts, and a less prominent breast-band. Easily distinguished in flight from only possible confusion species, Short-toed Eagle, by its pure white underparts and underwing-coverts. **VOICE** In display flight gives a mournful whistle *yeelp-yeelp-yeelp*, the alarm call is a sharp, hoarse *kew-kew-kew-…* and the contact call a short, loud, sudden *pyep*. Outside the breeding season usually silent. **RANGE** Breeds around the globe from S Alaska and the USA in the west to Japan and Kamchatka in the east, and also in the Caribbean, the Mediterranean, the Red Sea, the Persian Gulf, N India, SE China and Indonesia to Australia. Northern populations migrate south as far as Peru, Brazil, S Africa, S India and SE Asia. **DISTRIBUTION** In our region, breeds only on the Canaries and Cape Verdes. **BREEDING** In the Canaries breeds from Mar or occasionally mid Feb, but in Cape Verde laying starts Jan and has even been recorded in Dec. Clutch normally 2–3. **HABITAT** On the Canaries breeds only on sea cliffs but in Cape Verde also in palms or on the ground, but mostly near the coast. **STATUS** Varies between the different archipelagos. **Azores** Vagrant with 16 records, seven of these from São Miguel but also recorded on Terceira (5), Pico (2) and Flores (2). With 12 records since 1989, perhaps slightly more regular than records initially suggest. **Madeira** Fewer than five records in the last 50 years. **Canaries** Nearly extinct as a breeder (just 15–20 pairs in total), and now only nests on Lanzarote, Tenerife (1–2 pairs), La Gomera and El Hierro, although formerly bred on the other main islands. Also an uncommon but regular passage migrant, and is most likely to be seen at these periods. **Cape Verdes** Not uncommon resident, breeding on all islands and islets, with an estimated 50 pairs in the entire group.

LESSER KESTREL   *Falco naumanni*   Plate 22 (1)
Sp. Cernícalo Primilla   **P.** Peneireiro-das-torres (Francelho)

Monotypic. **L** 29–32 cm (11½–12½″); **WS** 60–70 cm (23½–27½″). **IDENTIFICATION** Sexually dimorphic. Difficult to distinguish from Common Kestrel, particularly in ♀ and immature plumages. ♂ easier to identify given combination of unspotted chestnut back, blue-grey greater coverts, striking white underwing and relatively unmarked buff body. ♀ and immature best separated from Common Kestrel by claw colour, which varies from white to pale brown, but is never black. Otherwise, close study of a combination of subtle differences can sometimes result in a conclusive identification, but many may have to remain unidentified. On average the underwing of Lesser Kestrel is whiter with less barred flight feathers and usually less pronounced spotting on the underwing-coverts. The streaking on the underparts is finer and the moustachial stripe less pronounced, giving the species a more gentle facial expression than Common Kestrel. First-summer ♂ also very similar to Common Kestrel, as it lacks the blue-grey upperwing panel of adult ♂, and is best identified by the lack of a prominent moustachial stripe and less well-marked upperparts. Juvenile ♀ indistinguishable in the field from adult. **VOICE** Contact call a hoarse, rasping, trisyllabic *chay-chay-chay* unlike any call of Common Kestrel and is diagnostic. Also a *kee-kee-kee-…* similar to that of Common Kestrel but more chattering and faster. **RANGE** Breeds from Iberia and N Africa east discontinuously through S Europe, Turkey and Ukraine to Mongolia and N China. Winters mostly in sub-Saharan Africa and S Arabia. **STATUS** Vagrant with records from three archipelagos. **Azores** ♂ obtained, Fajã da Cima (São Miguel), 4 Apr 1965. **Madeira** Adult ♂, São Gonçalo (Madeira), 3 Aug 1981. **Canaries** Fewer than 15 records, mostly from Tenerife but also reported on Lanzarote, Gran Canaria, Fuerteventura, La Graciosa and on La Gomera, although there are no details for the last record.

COMMON KESTREL   *Falco tinnunculus*   Plate 22 (2)
Sp. Cernícalo Vulgar   **P.** Peneireiro-vulgar (Peneireiro)

F. t. canariensis (C and W Canaries, Madeira and the Salvages), *F. t. dacotiae* (E Canaries and islets), *F. t. neglectus* (N Cape Verdes), *F. t. alexandri* (E and S Cape Verdes), and *F. t. tinnunculus* (accidental in Azores, Madeira, Canaries and Cape Verdes). **L** 30–34 cm (12–13½″); **WS** 60–75 cm (23½–29½″). **IDENTIFICATION** Sexually dimorphic. A small falcon that habitually hovers whilst hunting. Adult ♂ has a grey head, chestnut mantle and wing-coverts that are spotted black, dark flight feathers and a grey tail with a black subterminal band. Head and tail of ♀ dull chestnut-brown; the tail has dark barring, as does the rest of the upperparts. Juvenile resembles ♀ but has broader barring on upperparts, paler underparts and rather than heart-shaped spots on the flanks is streaked. Subspecies *canariensis* differs from nominate *tinnunculus* in being considerably darker with heavier spotting (♂) or barring (♀) on upperparts. ♂ *dacotiae* is paler than both *canariensis* and *tinnunculus* on the upperparts and the spotting is intermediate between those two races. The underparts are paler and much less well marked than in *canariensis* and on average it is slightly smaller. ♀ also averages slightly smaller, is paler and less heavily marked on both upper- and underparts compared to *canariensis*. This is the palest ♀ of all races of Common Kestrel. Subspecies *neglectus* is similar to, but smaller than, *canariensis*, the crown in the adult ♂ is never pure grey, and the tail has little grey and is slightly barred. ♀ is similar to ♂, sexual dimorphism being weak or non-existent in this race, which appears structurally different from the others, being more like a Merlin than a Common Kestrel. Subspecies *alexandri* is noticeably larger than *neglectus* but is on average paler and more spotted. ♂ has a grey crown and the tail is grey but barred darker. **VOICE** The most familiar call is a fast series of short, piercing *kee-kee-kee-kee-…* note. The begging call given by ♀ and young is a vibrant, trilling *vriii* repeated a few times. **RANGE** Breeds throughout much of the Old World. N European and most Asian populations are migratory, wintering in sub-Saharan Africa or S Asia. **DISTRIBUTION** *canariensis* breeds in the Madeiran archipelago, the Salvages, and on El Hierro, La Palma, La Gomera, Tenerife and Gran Canaria in the Canaries; *dacotiae* is confined to Fuerteventura, Lanzarote and the eastern islets of the Canaries; *neglectus* breeds on Santo Antão, São Nicolau, São Vicente, Santa Luzia, Raso and Branco in the Cape Verdes; *alexandri* is also confined to the Cape Verdes, where it breeds on Santiago, Fogo, Brava, Sal, Boavista, Maio and Ilhéus do Rombo. All of these populations are non-migratory. **BREEDING** In Cape Verde, both *alexandri* and *neglectus* breed in Oct–Apr, and both select similar sites in the crown of a palm or on a rocky ledge. One difference between them is clutch size: 4–5 in *alexandri* but only two in *neglectus*. In the Canaries the season depends on altitude, and laying can start as early as Feb on the eastern islands and at lower elevations, but at higher elevations on Tenerife continues until late May. Clutch size is normally four but varies from 2–6. **HABITAT** Occurs throughout the archipelagos. **TAXONOMIC NOTE** Hazevoet (1995) treated both Cape Verdes forms as species,

Neglected Kestrel *F. neglectus* and Alexander's Kestrel *F. alexandri*. **STATUS** That of *tinnunculus* varies between archipelagos, but other races breed on three island groups. **Azores** *tinnunculus* is accidental in autumn and winter with *c.* 25 records. **Madeira** *canariensis* is a common resident and *tinnunculus* a scarce passage migrant; *canariensis* is a resident breeder on the Salvages. **Canaries** Both *canariensis* and *dacotiae* are common and widespread, and *tinnunculus* is a scarce passage migrant recorded from Tenerife and Lanzarote, but undoubtedly overlooked. **Cape Verdes** Both *neglectus* and *alexandri* are common and widespread on most islands on which they occur; *tinnunculus* is a vagrant with only one certain record, an immature ♂ collected on Sal, 10 Mar 1924.

AMERICAN KESTREL   *Falco sparverius*   Plate 22 (3)
Sp. Cernícalo Americano   **P.** Peneireiro-americano

F. s. sparverius. **L** 25–29 cm (10–11½″); **WS** 50–60 cm (20–23½″). **IDENTIFICATION** Sexually dimorphic. A small falcon with a characteristic head pattern consisting of a black moustachial stripe, another black stripe on the rear cheeks, buffy nape with a black central spot, and a grey crown with a chestnut centre. Both sexes are unmistakable. ♂ has orange-brown upperparts, the mantle coarsely barred black whilst the tail has a broad dark subterminal band and white tip. Upperwings have bluish-grey coverts and secondaries, and a row of pale spots on the primaries. ♀ differs from other ♀ kestrels in having more rufous upperparts coarsely marked with black, and a narrower subterminal tail-band. Juvenile ♂ similar to adult but crown centre brown not chestnut, underparts streaked not spotted, and mantle and tail are more heavily barred. Juvenile ♀ also similar to adult and may be indistinguishable in the field, though some show a narrower subterminal tail-band. **VOICE** Common call is a series of five or more shrill *klee* notes. **RANGE** Breeds over most of N, C and S America. Northern populations are migratory and southern ones resident. **STATUS** Vagrant with just three records from the same archipelago. **Azores** One obtained, Terceira, 17 Feb 1968; one obtained, São Miguel, 1970; and one, Arrifes (São Miguel), for 15 days in Mar 1980.

(WESTERN) RED-FOOTED FALCON   *Falco vespertinus*   Plate 22 (4)
Sp. Cernícalo Patirrojo   **P.** Falcão-pés-vermelhos (Falcão-vespertino)

Monotypic. **L** 29–31 cm (11½–12″); **WS** 60–75 cm (23½–29½″). **IDENTIFICATION** Sexually dimorphic. Adult ♂ all dark grey with chestnut thighs, vent and undertail-coverts, and orange-red legs and cere. In flight, distinct silvery-grey flight feathers contrast with darker grey upperwing-coverts. ♀ has a distinctive head pattern with an orangey crown and nape, pale face, dark eye-patches and broad, dark moustachial stripe. The rest of the upperparts are grey, barred black and the underparts orange-buff. Juvenile has rich buff underparts streaked dark, a short, dark moustachial stripe and mark through the eye, and brown upperparts with buffy tips to the body feathers and upperwing-coverts. In flight has a dark trailing edge to the underwing. Confusion possible with Hobby, particularly in juvenile plumage, but juvenile Red-footed Falcon shows greater contrast between paler upperparts and uppertail and darker flight feathers, a prominently barred uppertail, dark trailing edge to the underwing and a smaller and less distinct moustachial stripe. **VOICE** The common call is a high-pitched *kew kew kew* by the ♂ or an ascending *kwee kwee kwee* by the ♀, very similar to the flight call of Hobby. **RANGE** Breeds from E Europe through NC Asia to extreme NW China, and winters in S Africa. **STATUS** Vagrant recorded from three archipelagos. **Azores** Immature ♀ collected, São Miguel, post-1966, as it is not mentioned in Bannerman & Bannerman. **Madeira** Three unconfirmed reports and recently a ♀ at Ponta do Pargo (Madeira), 12 Jun 2003. **Canaries** Seven certain records: unspecified numbers, spring 1890 (a specimen from this influx was seen by Bannerman); one obtained, near La Orotava (Tenerife), prior to 1893; one shot, Tenerife, Feb 1902; ♂, captured on the road to Tejina (Tenerife), May 1907; ♀, picked up exhausted near La Esperanza (Tenerife), 4 Jun 1987; two ♂ ♂, Alegranza, 28 Apr 1994; and ♀, La Caleta (Lanzarote), 9 Feb 1998. Emmerson *et al.* (1994) mention the species for Fuerteventura and Gran Canaria but do not include any details.

MERLIN   *Falco columbarius*   Plate 22 (5)
Sp. Esmerejón   **P.** Esmerilhão-comum (Esmerilhão)

F. c. aesalon has been recorded but *F. c. columbarius* must be possible on the Azores. **L** 25–30 cm (10–12″); **WS** 55–65 cm (22–25½″). **IDENTIFICATION** Sexually dimorphic. A small, compact and dashing falcon. ♂ blue-grey above with almost black primaries, a black subterminal band and narrow white terminal tail-band. The breast and belly are rusty streaked dark, the throat and vent white, and moustachial streak fairly thin and indistinct. ♀ and immature brown above, buffy-white below with bold streaking, and have a prominently barred brown and cream tail. Juvenile almost identical to ♀ and probably inseparable in the field, but the cere has a bluish or greenish tinge, unlike the clear yellow of the adult. **VOICE** Silent away from the nest. **RANGE** Breeds across most of central N Hemisphere. In the Americas it winters south to the equator but in the Palearctic almost all the wintering areas are north of the tropics. **STATUS** Vagrant with at least 11 records on three archipelagos. **Azores** Four records: one obtained, Fajã de Baixo (São Miguel), 8 Nov 1971; first-year ♀, Lomba da Vaca (Flores), 11 Oct 1999; first-year ♂, Ponta Delgada (São Miguel), 25 Oct 1999; and an immature/♀, Cabo da Praia (Terceira), 20 Oct 2002. **Madeira** One old record from the Salvage Islands. **Canaries** Six or seven: ♀/immature, Golf del Sur (Tenerife), 5 Dec 1990, then the same or another, Ciguaña (Tenerife), 3 Mar 1991; one, Los Cristianos (Tenerife), 29 Oct 1993; one, near Arguayo (Tenerife), 28 Jan 1997; one, Melenara (Gran Canaria), 14 Jan 1998; one, El Hierro, 4 & 14 Mar 1999; and one, Tiscamanita (Fuerteventura), 27 Nov 2001.

(EURASIAN) HOBBY   *Falco subbuteo*   Plate 23 (1)
Sp. Alcotán Europeo   **P.** Ógea

F. s. subbuteo. **L** 30–36 cm (12–14″); **WS** 65–85 cm (25½–33½″). **IDENTIFICATION** Sexes similar. Scythe-like wings and relatively short tail give the species a rather swift-like silhouette in flight. Adult dark slate-grey above, slightly paler on the rump and tail, white below but with heavy streaks on breast and belly, black moustachial stripe and chestnut thighs and vent. Immature much browner above with pale buffish fringes to the feathers, yellowish-buff underparts with even heavier streaking on the breast and belly, and lacks chestnut thighs and vent of adult. Confusion possible with Red-footed, Eleonora's and juvenile Peregrine Falcons, with differences between Hobby and Red-footed Falcon dealt with under the latter. Eleonora's Falcon is longer winged, longer tailed, lacks the contrasting chestnut thighs and vent, and has uniform dark underwing-coverts. Juvenile Peregrine has a barred uppertail, more prominent and broader moustachial stripe, and is larger. **VOICE** Usually silent except in courtship and when feeding young. **RANGE** Breeds from Europe and NW Africa east through C Asia to E China, Kamchatka and N Japan. Winters in S Africa, S and SE Asia. **STATUS** Varies between the two archipelagos where recorded. **Madeira** Rare vagrant, the most recent records being one, Porto Moniz (Madeira), 20 Aug 2004, and one, Ponta do Pargo (Madeira), 25 Aug 2004, with another there, 4 Sep 2004. Also recorded on the Salvages. **Canaries** Rare and irregular passage migrant, not recorded on La Gomera or El Hierro.

ELEONORA'S FALCON   *Falco eleonorae*   Plate 23 (2)
Sp. Halcón de Eleonora   **P.** Falcão-da-rainha

Monotypic. **L** 36–40 cm (14–16″); **WS** 85–100 cm (33½–39″). **IDENTIFICATION** Sexes similar. Dimorphic, with both pale and dark forms, the latter accounting for *c.* 25% of the population. Shape in flight, with long, narrow wings and rather long, rounded tail, combined with the all-dark underwing-coverts, should prevent confusion with other adult falcons. Adult pale morph has dark brownish or blue-grey upperparts; the underparts vary from cream to pale chestnut, but are always streaked dark, throat and cheeks white with a black moustachial stripe; the underwing lacks any real pattern but the grey bases to the flight feathers contrast with the dark underwing-coverts and dark trailing edge. Adult dark morph entirely blackish-brown but has same distinctive underwing as pale morph. Juvenile is browner above than pale-morph adult, underparts are rich buff with fine dark streaks and the underside to the flight feathers and underwing-coverts are finely barred, and paler than in adults. Confusion possible with juvenile Hobby, but on Eleonora's the undersides of the flight feathers have paler bases and the underwing-coverts and the wingtips are darker. **VOICE** Main call is a nasal, grating *kyaih-kyaih-kyaih…* but also gives a sharp *ke-ke-ke…* in alarm. Usually silent away from breeding areas. **RANGE** Breeds on small islands and islets in the Mediterranean, the Canaries and NW Africa. Winters mostly in Madagascar but also in E Africa and the Mascarenes. **DISTRIBUTION** Breeds only on the Canaries. **BREEDING** Late summer, with egg laid normally in mid Jul–early Aug, so that young can be fed on migrant birds often caught over the sea. Clutch size 1–4, but 2–3 most common. The nest, which is just a simple depression, is normally located on a rocky ledge or in a small hole in a cliff. **HABITAT** Breeds on islands and rocky coasts. **STATUS** Varies between the three archipelagos on which it has been recorded. **Madeira** Two records, one from the Salvages and two, near Bica da Cana (Madeira), 18 Aug 2002. **Canaries** Summer visitor, breeding on Montaña Clara, Alegranza and Roque del Este, all to the north of Lanzarote. However, often seen in the breeding season in N Lanzarote and has been recorded from all other islands except La Palma. On the other islands it is a rare and irregular passage migrant. **Cape Verdes** Immature ♂ found dead, Praia do Norte (São Vicente), 4 Sep 1998.

LANNER FALCON   *Falco biarmicus*   Plate 23 (3)
Sp. Halcón Borní   **P.** Alfaneque

F. b. erlangeri. **L** 40–50 cm (13½–19½″); **WS** 90–115 cm (35½–45½″). **IDENTIFICATION** Sexes similar. A large falcon for which the only confusion species in the region is either Peregrine or Barbary Falcon. As the only race likely to be seen in the region is *erlangeri* the following details relate solely to that race. Adult has pale greyish-brown upperparts with pale edges to the upperwing-coverts and pale fringes to mantle feathers. Crown and nape pale buff, sometimes with a contrasting dark bar across the top of the forehead, separating the pale forehead from the crown. Moustachial stripe quite long and prominent but generally fairly narrow. Underparts are almost white with dark spotting on the breast and barring on the flanks and thighs. Underwings white with faint streaking on the coverts and barred flight feathers, the wingtips are dark and the tail finely barred with a white tip. Juvenile brown above, more heavily streaked on the breast, belly and underwing-coverts than the adult. Structurally Lanner is longer winged than either Peregrine or Barbary Falcon, and has a longer tail. In all plumages the moustachial stripe is much narrower. Adult has less distinctly marked underparts than Peregrine and lacks the pale peach coloration of barbary Falcon, and the tail is evenly barred throughout its length with no paler area at the base. Juvenile Lanner compared to juvenile Peregrine or Barbary Falcon has darker underwing-coverts, paler bases to the primaries, is warmer brown above and has an evenly barred tail. **VOICE** Usually silent away from the breeding areas. **RANGE** Breeds discontinuously across N Africa, the Middle East, Arabia, and in Italy, SE Europe and Turkey, as well as much of sub-Saharan Africa. Winters in or near breeding range; some are resident but others disperse in the non-breeding season. **STATUS** Only recorded from one island group. **Canaries★** Accidental, having been recorded from all the main islands, except La Palma and El Hierro, with 9–10 records, the vast majority in spring (Feb–May).

PEREGRINE FALCON   *Falco peregrinus*   Plate 23 (5)
Sp. Halcón Peregrino   **P.** Falcão-peregrino

Subspecies unknown on the Canaries or Madeira; *F. p. madens* (Cape Verdes).   **L** 36–48 cm (14–19″); **WS** 85–120 cm (33½–47″). **IDENTIFICATION** Sexes similar. A large falcon most likely to be confused with Barbary Falcon with which it shares bulky build and relatively short tail. Adult has dark blue-grey upperparts; paler lower back, rump and uppertail-coverts, white underparts with fine black barring from the lower breast to undertail-coverts, and fine black spotting on the upper breast. The throat, chin and cheeks are unmarked, pure white, contrasting with the almost black cap, nape, upper cheeks and broad moustachial stripe. Sexes separable by size, ♀ being noticeably bigger than ♂. Also, ♀ tends to be more heavily marked on the underparts than ♂. Juvenile browner on upperparts and has creamy underparts with heavy dark brown streaks from the chest to vent. Adult differs from Barbary Falcon in its darker upperparts, whiter more heavily barred underparts and dark hood. Juvenile is darker than Barbary Falcon and has more heavily streaked underparts. The Cape Verde subspecies *madens* is browner above, with a rufous wash to the head and has a pinkish-buff wash to the underparts. **VOICE** Most common call is a loud, harsh *kek-kek-kek-kek…* given in contact or alarm, the ♀ having a more hoarse quality and lower pitch than ♂. In aerial display often gives a distinctly disyllabic *ee-chip* (♂) or *ee-chup* (♀), sometimes singly but more frequently in short series. **RANGE** Fairly cosmopolitan, breeding on all continents except Antarctica. **DISTRIBUTION** Within our region only breeds on Cape Verdes. **BREEDING** The only published details concerning *madens* were obtained from Ilhéu de Cima in the 1960s, where a fresh egg was found, 26 Jan, and a week-old chick, 6 Mar. **HABITAT** In Cape Verde, normally encountered around sea cliffs or montane regions. **TAXONOMIC NOTE** Cape Verdes *madens* was considered a full species by Hazevoet (1995), on the basis of its morphology. By contrast, some other authors consider that, on morphological grounds, this taxon might be better considered a subspecies of Barbary Falcon. **STATUS** Varies between archipelagos. **Azores★** First-winter ♂, Ponta Ruiva (Flores), 11 Feb 2002; first-winter ♂, Ilhéu da Praia (Graciosa), 1 Dec 2004. **Madeira** One, Quinta da Vista Alegre, Funchal (Madeira), early 1990; and one, Ponta da São Lourenço (Madeira), late Jan or early Feb 2001. Also recorded from the Salvage Islands. **Canaries** Rare and irregular passage migrant and winter visitor, recorded from all main islands except El Hierro. **Cape Verdes** Probably occurs in small numbers on all islands and islets, but is very rare and few data are available.

BARBARY FALCON   *Falco pelegrinoides*   Plate 23 (4)
Sp. Halcón de Berbería   **P.** Falcão-tagarote

F. p. pelegrinoides.   **L** 35–42 cm (14–16½″); **WS** 80–100 cm (31½–39″). **IDENTIFICATION** Sexes similar. The smallest of the large falcons and most likely to be confused with Peregrine, but structurally it is smaller and less bulky than Peregrine and has a shorter tail than Lanner. Adult is blue-grey above with a rusty nape; the underparts are white with fine dark barring and a rufous wash on the breast, belly and underwing-coverts. Tail tends to show a broad subterminal band but that of Peregrine is evenly barred. Juvenile brown above with fine streaking on breast and belly. All ages have a narrower moustachial stripe than that of Peregrine. **VOICE** Very similar to Peregrine but slightly higher in pitch; all call patterns resemble Peregrine. **RANGE** Breeds discontinuously from the Canaries through Morocco, Algeria, Tunisia, Egypt and Israel, and from Iran east to the south of Lake Balkhash. **DISTRIBUTION** Breeds only on the Canaries. **BREEDING** Few data from our region. Egg laying seems to occur second half of Feb or Mar. Usual clutch size 3–4 but instances of five recorded. **HABITAT** Nests on sea cliffs and in montane areas but could be encountered almost anywhere from coastal scrub to alpine areas. **STATUS** Only known from two archipelagos. **Madeira** One, near El Tanque (Porto Santo), 24 Sep 2003. **Canaries** A rare resident that breeds on all the main islands.

RED-LEGGED PARTRIDGE   *Alectoris rufa*   Plate 24 (1)
Sp. Perdiz Común   **P.** Perdiz (Perdiz-comum)

A. r. intercedens (Gran Canaria), *A. r. hispanica* (Azores).   **L** 32–34 cm (12½–13½″). **IDENTIFICATION** Sexes similar and is unlikely to be confused with any other species in our region. Adult has a white throat with a broad black border, which at the lower edge becomes coarse black streaking on the neck and upper breast. Also, a broad white supercilium and flanks coarsely barred white, black and chestnut. Juvenile lacks striking head pattern and flank markings but is usually seen with adults, which aids identification. Subspecies *intercedens* has paler upperparts than *hispanica* with particularly grey central rectrices and rump. **VOICE** Advertising call of ♂ is a harsh *goCHAK-CHAK-CHAK goCHAK goCHAK-CHAK*, and contact call a quiet liquid *tlook* or *tuc*. **RANGE** Resident in Iberia, Germany, France, Switzerland, N Italy and Corsica. **DISTRIBUTION** Breeds on the Azores, Madeira and the Canaries. **BREEDING** Extended season with eggs laid Feb–Jun. Nest a simple scrape on ground, normally well hidden in vegetation. Clutch size is 9–14. **HABITAT** Prefers more open areas and agricultural land in both lowlands and montane regions. **STATUS** Introduced to two and possibly three archipelagos. **Azores** Formerly bred on Pico and Santa Maria, although attempts to introduce the species to São Miguel failed, and the species may no longer survive in a feral state even on the first two islands. **Madeira** Introduced on Madeira (before 1450) and Porto Santo (reintroduced in 1920s) but population only persists due to annual reintroductions for hunting purposes. **Canaries** Breeds only on Gran Canaria, where it is uncommon but widespread. Possibly occurs naturally but deliberate introduction cannot be eliminated. Records from Tenerife and La Palma, but breeding unproven, and on El Hierro *c*. 40 were released in 1985 but whether this population is extant is unclear. Also released on Lanzarote and Fuerteventura, but no proof of its continued presence on either island.

BARBARY PARTRIDGE   *Alectoris barbara*   Plate 24 (2)
Sp. Perdiz Moruna   **P.** Perdiz-moura (Perdiz-mourisca)

A. b. koenigi. **L** 32–34 cm (12½–13½"). **IDENTIFICATION** Sexes similar. Unlikely to be mistaken, as it is the only partridge on those islands where it breeds. Distinctive head pattern with greyish supercilium, greyish throat bordered by a broad, white-spotted, reddish-brown neck-collar and chestnut-brown central crown. Juvenile lacks adult head pattern, and has only a small gorget of chestnut breast spots and less prominent flank bars. **VOICE** Common call is a repeated *kutchuk* or *kutchuk-chuk*, and advertising call of ♂ is a loud, harsh and drawn-out *krrraiik* on a rising pitch. **RANGE** Breeds in the Canaries, Morocco, Algeria, Tunisia, Libya, Egypt and Sardinia. Introduced to Gibraltar. **DISTRIBUTION** Breeds only in the Canaries. Introduced on Porto Santo, in the Madeiran archipelago, but this failed. **BREEDING** Egg laying usually in Mar–Jun, but certainly dependent on rainfall and has been noted in Dec. Clutch size 9–14; exceptionally, 18 reported. Nest a simple scrape on the ground under a bush or amongst grasses. **HABITAT** *Euphorbia* scrub, agricultural land, open pine forest and alpine scrub, from coasts to at least 2,000 m. **STATUS** Less common than formerly, although in certain areas local populations are boosted by annual introductions to permit continued sport hunting. **Canaries** Breeds on all main islands except Gran Canaria, but perhaps introduced on Fuerteventura and certainly so on El Hierro and La Palma. Still relatively common on Lanzarote, but elsewhere uncommon and increasingly scarce.

COMMON QUAIL   *Coturnix coturnix*   Plate 24 (3)
Sp. Codorniz   **P.** Codorniz

C. c. coturnix (Canaries and Madeira; formerly treated separately under name *confisa*), *C. c. conturbans* (Azores) and *C. c. inopinata* (Cape Verdes). **L** 16–18 cm (6–7"). **IDENTIFICATION** Sexes differ. The smallest gamebird of the region and unlikely to be confused. ♂ has rather buffish underparts with dark breast streaking, whitish flank-stripes and a variable black central throat. Upperparts brown with darker markings and white streaks. ♀ is similar to ♂ but has a whitish throat and a less distinct facial pattern. Juvenile similar to ♀ but lacks any defined facial pattern and has the flanks spotted and barred, not streaked. Subspecies *inopinata* is similar to the nominate but is smaller and *conturbans* is smaller and darker with more rufous wing-coverts. **VOICE** Best-known call is the advertising call of the ♂, which is a short series of three liquid, staccato notes *quip-quip-ip*, often transcribed as 'wet-my-lips', followed by a pause of up to several seconds and then the whole phrase is repeated. **RANGE** From the Atlantic islands through Europe and N Africa, W and C Asia to Lake Baikal and N India; also breeds across much of E and S Africa. Most European birds winter in the Sahel between Senegal and Sudan. **DISTRIBUTION** Breeds on all of the archipelagos. **BREEDING** Few data from the Canaries, but egg laying can start in Dec and, with the possibility of a maximum three broods, continue until Aug. Clutch size is 7–12 and the nest is well hidden among weeds, cereal crops or grasses. In Cape Verde, breeds Sep–Dec but, as in the Canaries, this is very dependent on local climatic conditions, and starts earlier at higher altitude. In Madeira can also have 3–4 broods, and the season is extended as on the Canaries. On the Azores season also extended, as eggs have been found in all months Mar–Sep, except August. **HABITAT** Grassland and agricultural areas. **STATUS** Resident and migrant breeder; status varies according to local climate. **Azores** Breeds on all main islands, the population varying annually. **Madeira** A rare breeder on both Madeira and Porto Santo, but formerly much more common. **Canaries** Two populations (one resident, the other migratory) but more common in years following good winter rains. Breeds on all main islands but may vacate eastern islands during longer dry spells. **Cape Verdes** Fluctuating population but sometimes abundant. Resident; breeds on all main islands except Santa Luzia (where occurs occasionally).

HELMETED GUINEAFOWL *Numida meleagris* Plate 24 (4)
Sp. Pintada Común **P.** Pintada

N. m. galeata (Cape Verdes). **L** 60–65 cm (23½–25½"). **IDENTIFICATION** Sexes similar. Dark grey plumage covered with white spots, large body with thin neck, and small bare head with a prominent casque on top are all features of this unmistakable species. Juvenile brownish-buff and paler spotted, as well as lacking casque (helmet) of adult. **VOICE** Common call is a raucous *kik-kik-kik-kik-kaaaaa*, and the contact call is a series of rather metallic but soft *chink* notes. **RANGE** Resident over much of sub-Saharan Africa and in SW Arabia. **DISTRIBUTION** Only certainly breeds on the Cape Verdes but has been recorded breeding wild on the Canaries. **BREEDING** On the Canaries the only breeding data concern a nest with 17 eggs, found near Bajamar, 19 Mar 1981. In Cape Verde clutch size is 8–12 and the breeding season starts Sep and continues into Nov. **HABITAT** Often encountered in scrubby grasslands and lightly wooded valleys, occasionally on more open plains. **STATUS** Introduced on two archipelagos. **Canaries** Current status unclear but possibly extinct, and has been recorded breeding only on Tenerife. **Cape Verdes** Breeds, and is still fairly common, on Santiago, Fogo, São Nicolau and Maio, possibly still extant on Boavista but now believed to be extinct on Brava, Santo Antão and São Vicente.

WATER RAIL *Rallus aquaticus* Plate 25 (1)
Sp. Rascón **P.** Frango-d'água

R. a. aquaticus. **L** 23–28 cm (9–11"). **IDENTIFICATION** Sexes similar. Adult has slate-grey face, foreneck, breast, fore flanks and central belly, the rest of the flanks heavily barred black and white. The undertail-coverts appear white. The crown, nape and upperparts are brown with blackish feather centres, the legs pinkish-red and the bill reddish with a darker tip and culmen. Juvenile lacks pure slate-grey on face and underparts, being browner with a pale throat, the barring on the underparts is less well defined and is sepia and white with broader white bars, the undertail-coverts are buffish and the bill duller and shorter. **VOICE** Typical call is a loud, explosive, screaming series of pig-like squeals and grunts, rising in the middle and then fading. **RANGE** Breeds from Iceland, Ireland, Iberia and NW Africa discontinuously through Europe, the Middle East and Asia to Japan, E China and Sakhalin. Many winter in the breeding range but some disperse in winter. **STATUS** Vagrant with only five records from three archipelagos. **Azores** Specimen with no details; one obtained, near Lagoa de Congro (São Miguel), undated; and one obtained, Angra (Terceira), 2 Sep 1953. **Madeira** One, Porto Santo, 4 Dec 1903. **Canaries★** One heard in dense waterside vegetation, Presa de Las Peñitas (Fuerteventura), 27 Mar 1993.

SPOTTED CRAKE *Porzana porzana* Plate 25 (2)
Sp. Polluela Pintoja **P.** Franga-d'água-grande (Franga-d'água-malhada)

Monotypic. **L** 22–24 cm (9–9½"). **IDENTIFICATION** Sexes similar. Similar in body size to Water Rail but has a shorter neck and a shorter, stubbier bill. Adult is olive-brown above with dark blackish centres to the feathers, and is streaked and spotted white. The chin, throat, head- and upper-breast-sides are blue-grey spotted white, with a broad brown stripe on ear-coverts and neck-sides, with brown flanks heavily barred black and white, and bright buff undertail-coverts. Bill yellowish with a variable amount of orange-red at the base, and legs are greenish. ♀ sometimes separable by the less grey and more heavily spotted face and throat. Juvenile is duller and browner, lacks the grey face, has a whitish chin and throat, and a brownish bill. **VOICE** Relatively silent away from breeding grounds. **RANGE** Breeds from the British Isles and Spain, discontinuously through W Europe to SW Siberia and W China. Most winter in E and S Africa and India, but some winter in W Africa and the Mediterranean Basin. **STATUS** Varies between the various archipelagos on which it has been recorded. **Azores** Six records: one obtained, Santa Maria (no other details); immature ♀ obtained, Ponta Delgada (São Miguel), 13 Oct 1911; adult ♀ obtained, Ponta Delgada (São Miguel), 17 Aug 1953; immature ♂ obtained, São Miguel, 5 Nov 1956; one caught alive, Cabo da Praia (Terceira), 1 Apr 1966; and one obtained, Lagoa Ribeira Seca (São Miguel), 8 Dec 1968. **Madeira** Fewer than five records in the last 50 years. Also recorded on the Salvages. **Canaries** Rare but regular passage migrant, recorded on all main islands. **Cape Verdes** One recent record: one, Mindelo (São Vicente), 13 Jan 2005.

SORA *Porzana carolina* Plate 25 (3)
Sp. Polluela de Carolina **P.** Franga-d'água-americana

Monotypic. **L** 20–23 cm (8–9"). **IDENTIFICATION** Sexes similar. Very similar to Spotted Crake in plumage, structure, size and habits. However, has a distinctly conical bill with a deeper base than that of Spotted Crake, yellowish in colour and lacks any red on the base; crown brown with a dark central stripe. Very small white mark behind eye, if seen, is diagnostic. Adult has black face and central throat, unspotted grey head, neck and breast, lacks V-shaped marks on tertials and inner greater coverts of Spotted Crake, normally exhibits some white, often quite extensive, on undertail-coverts. ♀ sometimes distinguished from ♂ by less extensive black face and more white-speckled upperparts. Juvenile duller than adult, lacks any grey coloration or the black 'mask', has plain buffish-brown underparts and a greenish bill. Undertail-coverts are more extensively buff than in adult. Immature/first-winter similar to adult but less grey, has the 'mask' less intense and black on throat and chin also less extensive and often mottled grey. **VOICE** Characteristic call a descending high-pitched whinnying *whee-hee-hee-hee.....*, also a plaintive ascending whistle *kerwee*. **RANGE** Breeds across much of N America, wintering in S USA through C America to N South America. **STATUS** Vagrant with just one record, although this may have involved breeding. **Azores*** Three, ♂ ♀ and an immature/juvenile, Fajã dos Cubres (São Jorge), 14–16 Nov 1998.

LITTLE CRAKE   *Porzana parva*   Plate 25 (4)
Sp. Polluela Bastarda   **P.** Franga-d'água-bastarda

Monotypic. **L** 18–20 cm (7–8″). **IDENTIFICATION** Sexually dimorphic. In all plumages distinguished structurally from similar Baillon's Crake by longer primary extension beyond tip of the folded tertials. Adult ♂ predominantly blue-grey on face and underparts, with little or no white barring on rear flanks and black-and-white-barred undertail-coverts; upperparts rather dull olive-brown with blackish feather centres and very few, if any, white streaks on mantle and wing-coverts. ♀ has blue-grey replaced mostly by pale buff, except for pale grey supercilium and lores, and whitish chin and throat. In both sexes bill is greenish with a red base, but this is much reduced in ♀; legs yellowish-green to green. Juvenile resembles ♀ but is weakly barred on entire length of flanks, has a whiter supercilium, browner iris and browner legs. **VOICE** Alarm call is a sharp *tyiuck* and contact calls include a low *quec*, a subdued trill and a long series of musical tapping sounds, *dug-dug-dug...* **RANGE** Breeds locally in W and C Europe east to NW China. Wintering areas poorly known, but include sub-Saharan Africa, S Arabia and from Iraq to NW India. **STATUS** Varies between the three archipelagos on which it has been recorded. **Azores** Vagrant with four records: ♀ obtained, Relva (São Miguel), 7 Mar 1894; ♂ obtained, Capellas (São Miguel), Apr 1900; ♀ obtained, Furnas (São Miguel), *c.*1903; and adult ♂, Horta (Faial), 6 Mar 1985. **Madeira** Just one old record from Porto Santo. **Canaries** A scarce and irregular passage migrant, which is unrecorded from the western islands of La Gomera, La Palma and El Hierro.

BAILLON'S CRAKE   *Porzana pusilla*   Plate 25 (5)
Sp. Polluela Chica   **P.** Franga-d'água-pequena

P. p. intermedia. **L** 17–19 cm (7–7½″). **IDENTIFICATION** Sexes similar. In all plumages, the short primary projection can be used to distinguish this species from the similar Little Crake. Adult resembles ♂ Little Crake but has richer brown upperparts with more extensive white streaking extending onto wing-coverts and tertials, flanks heavily barred black and white, and lacks red at the bill base. Juvenile also resembles juvenile Little Crake but has stronger and more extensive barring on the underparts, lacks the pale stripe on the scapulars and has more white markings on the upperparts (which are hollow whirls in Baillon's rather than the complete spots of Little Crake). **VOICE** Characteristic rattling, rasping, frog-like advertising call is unlikely to be heard in our region. Alarm call is a sharp *tak* or *tyiuc* similar to Little Crake. **RANGE** Breeds in isolated pockets in Europe and NW Africa, from Belarus and Ukraine east to Japan and south to Iran, N India and S China; also E and S Africa, Madagascar and Australasia. European birds thought to winter in sub-Saharan Africa. **STATUS** Vagrant to two archipelagos and a rare visitor on a third. **Azores** Two records: one obtained, Ponta Delgada (São Miguel), pre-1905 (no other details); and one obtained, São Roque (São Miguel), 14 Mar 1953. **Madeira** Fewer than five records in the last 50 years. **Canaries** A rare and irregular passage migrant or accidental visitor, recorded only from Lanzarote, Fuerteventura, Tenerife and La Palma.

(AFRICAN) BLACK CRAKE   *Limnocorax flavirostra*   Plate 25 (7)
Sp. Polluela Negra Africana   **P.** Franga-d'água-preta

Monotypic. **L** 20–23 cm (8–9″). **IDENTIFICATION** Sexes similar. Unmistakable; adult all black with bright red legs and a greenish-yellow bill. Juvenile greyish-brown with a whitish throat and chin, and a blackish bill. Immature similar but has the underparts dark grey and the bill greenish. **VOICE** Advertising call unlikely to be heard in our region. Alarm calls include a repeated sharp *chip* or *tyuk* and a harsh *chack*, and the contact call is a soft *pu* or *bup*. **RANGE** Breeds throughout most of sub-Saharan Africa. **STATUS** Only one record, which is also the only record for the W Palearctic. **Madeira** ♂ obtained, São Amaro, near Funchal (Madeira), 26 Jan 1895.

AFRICAN CRAKE   *Crex egregia*   not illustrated
Sp. Guión Africano   **P.** Codornizão-africano

Monotypic. **L** 20–23 cm (8–9″). **IDENTIFICATION** Sexes similar. Adult has olive-brown upperparts with darker blackish-brown centres to the feathers of the mantle, upperwing, scapulars and tertials, but almost no dark streaking on the nape; the head- and neck-sides and breast are grey, the chin and throat white, the flanks, belly and undertail broadly barred black and white. A short white line above the lores, bordered below by a narrow black line; red eye-ring and irides, red bill with a greyish tip and legs greyish-brown. Sexes very similar but ♀ smaller with a less contrasting head pattern, less barring on the central belly and generally has narrower black bars. Juvenile duller and darker than adult, with a dark bill and grey iris. **VOICE** A short, loud *kip* and a loud *tsuck* are the commonest calls. **RANGE** Breeds over much of sub-Saharan Africa. Movements poorly understood but are related to climatic conditions. **STATUS** Extreme vagrant with only one record, which is also the only record for the W Palearctic. **Canaries** Adult recovered, Parque García Sanabria, Santa Cruz (Tenerife), 23 Nov 2001, but died in care.

CORN CRAKE *Crex crex* Plate 25 (6)
Sp. Guión de Codornices **P.** Codornizão

Monotypic. **L** 27–30 cm (10½–12″). **IDENTIFICATION** Sexes similar. Adult has upperparts, except wings, buff-brown with dark centres to all feathers. Underparts are a mixture of buff, grey and white with prominent rufous and white barring on flanks. In flight reveals diagnostic chestnut wing-coverts and chestnut-toned flight feathers. ♀ has less grey on face and underparts than ♂. Juvenile similar to ♀ but has paler upperparts, whiter underparts, less pronounced barring on flanks and pale-spotted wing-coverts. **VOICE** Famous advertising call, the loud, rasping, monotonous *krex-krex* is unlikely to be heard in our region. A short, loud *tsuck* is given in alarm by wintering birds but otherwise rather silent. **RANGE** Breeds from the British Isles, France and Norway east to N China and C Siberia. The winter range is poorly known but most winter in E Africa. **STATUS** Accidental visitor region with records from three archipelagos. **Azores** Twelve records, mostly from São Miguel but others from Faial, Terceira and Santa Maria. **Madeira** Fewer than five records in the last 50 years. **Canaries** Bannerman (1963) believed this species to be a rare passage migrant, but it is now considered accidental, having been recorded from all main islands except La Gomera and La Palma.

(COMMON) MOORHEN *Gallinula chloropus* Plate 26 (1)
Sp. Gallineta Común **P.** Galinha-d'água

G. c. chloropus. **L** 32–35 cm (12½–14″). **IDENTIFICATION** Adult easily identified by combination of red frontal shield, bicoloured (red-and-yellow) bill, and generally dark plumage (brown wings and mantle; slate-grey head, neck and underparts), except a white line on the flanks and white lateral undertail-coverts. Juvenile duller but also has a pale flank line and whitish lateral undertail-coverts. **VOICE** A variety of explosive clucking and chattering calls with a harsh and metallic tone. Most often heard is a low, rolling *krrrruk*, a short *kuk* or a sharp *kittick* given in alarm. **RANGE** Widespread throughout much of the world. **DISTRIBUTION** Formerly bred on the Azores and Cape Verdes but now breeds only in the Canaries. **BREEDING** Commences Feb and continues into Aug, with two or sometimes three broods, but where water levels maintained artificially may breed year-round. Clutch size 4–13, although 6–8 is more normal. Nest usually well hidden under waterside vegetation, although sometimes floating vegetation is used. **HABITAT** Freshwater streams and reservoirs with some dense vegetation near edges. **TAXONOMIC NOTE** The extinct resident Azores population was formerly considered a separate subspecies, *correiana*. However, this taxon was considered invalid by Vaurie (1965). **STATUS** Recorded throughout but status varies between archipelagos. **Azores** Now extinct as a breeder, but is still a regular winter visitor. **Madeira** An occasional visitor and also recorded from the Salvage Islands. **Canaries** A scarce and local breeder on Fuerteventura, Gran Canaria, Tenerife and La Gomera. Migrants boost the resident population in winter, when it has been recorded from all major islands. **Cape Verdes** Formerly bred on Santiago and Boavista, but not reported for 30 years until one, Mindelo sewage farm (São Vicente), 27 Feb–2 Mar 1999.

LESSER MOORHEN *Gallinula angulata* Plate 26 (2)
Sp. Gallineta Chica **P.** Galinha-d'água-pequena

Monotypic. **L** 22–24 cm (8¾–9½″). **IDENTIFICATION** Sexes similar. Smaller than Moorhen with a yellow bill, less distinct flank line and greenish or reddish-brown legs. Adult ♂ has a blackish head, dark grey neck and upper mantle, rest of upperparts and wings dark olive-brown, and underparts dark slate-grey with black central undertail-coverts and white on sides. Bill predominantly yellow with a red shield and culmen, and legs vary greenish to orange or reddish-brown. ♀ has paler brown upperparts, paler grey underparts, especially the belly, with black restricted to around the bill base. Juvenile olive-brown above with buffy-brown head-sides and underparts, greyer flanks and whiter central underparts from the throat to belly. Lacks pale flank line of Moorhen and has tertials and scapulars fringed buff. **VOICE** Calls resemble those of Moorhen including sharp clicking, subdued chuckling and squeaky calls. **RANGE** Breeds over much of sub-Saharan Africa. Well known for its nomadism. **STATUS** Extreme vagrant with just one record. **Canaries★** Immature recovered, Las Palmas (Gran Canaria), 19 Jan 1997, but died in captivity (the first record for the Western Palearctic).

ALLEN'S GALLINULE   *Porphyrula alleni*   Plate 26 (3)
Sp. Calamón de Allen   **P.** Caimão de Allen (Caimão-pequeno)

Monotypic.   **L** 22–24 cm (9–9½″).   **IDENTIFICATION** Sexes similar. Much smaller than both Moorhen and American Purple Gallinule, with adult plumage similar to that of the latter but has an all-red bill, red legs and feet, and dark central undertail-coverts. Juvenile has 'scaly' pattern to brown upperparts, whereas juvenile American Purple Gallinule has rather plain upperparts, and the legs and feet are brownish or red-brown, rather than dull yellowish, and as in adult the undertail-coverts are dark-centred. **VOICE** A variety of harsh, nasal calls including a short *kek* and more drawn-out *kerk*, whilst normal contact call is a subdued *kup*. **RANGE** Breeds in sub-Saharan Africa, with northern populations moving south in the dry season, and considered to be an inveterate wanderer. **STATUS** Vagrant, recorded from three archipelagos. **Azores** Five records: adult obtained, Ribeira Grande (São Miguel), Dec 1879; immature obtained, Furnas (São Miguel), 6 Feb 1902; immature presented to the British Museum, pre-1905; adult female obtained, Vila Franca (São Miguel), Nov 1910; and first-winter ♀ collected, Santa Maria, 17 Dec 1973. **Madeira** Only mentioned by Schmitz (1908) as having been recorded post-1896, with no subsequent reports. **Canaries** Ten records: juvenile, Embalse de los Molinos (Fuerteventura); 12 Jan 1990; juvenile, Las Barranqueras de Valle Guerra (Tenerife), 2 Dec 1991; juvenile, Igueste de San Andrés (Tenerife), 19 Jan 1993; juvenile, Gran Canaria, 31 Dec 1994; juvenile recovered, near Arona (Tenerife), 1 Feb 1996 (released Tejina Ponds next day but not seen subsequently); small influx, Nov/Dec 1999, with an adult, Las Galletas (Tenerife), and juvenile, Puerto de la Cruz (Tenerife), 13 Dec, and juvenile recovered, La Caleta (El Hierro), 19 Dec; adult, Las Palmas (Gran Canaria), 29 Nov–2 Dec 2000; and one, Erjos Ponds (Tenerife), 17 Feb–9 Mar 2003.

AMERICAN PURPLE GALLINULE   *Porphyrula martinica*   Plate 26 (4)
Sp. Calamoncillo Americano   **P.** Caimão-americano

Monotypic.   **L** 30–36 cm (12–14″).   **IDENTIFICATION** Sexes similar. Similar size to Moorhen but in all plumages distinguished from that species by all-white undertail-coverts and lack of a pale flank line. Adult has much more iridescent and colourful plumage than Moorhen, with bright purplish-blue head, neck and underparts, paler hindneck and mantle and breast-sides. Upperparts bright olive-green and frontal shield pale blue-white. Confusion also possible with Allen's Gallinule, but American Purple Gallinule is larger, has bright yellow legs, a yellow bill tip and all-white undertail-coverts. Juvenile has plain olive-brown upperparts with a green tone to the wings, buffy underparts (palest on belly and vent), whitish chin and throat, and all-white undertail-coverts that distinguish it from both Moorhen and Allen's Gallinule, and from the latter the juvenile also differs in having dull yellowish legs and unmarked upperparts. **VOICE** Usual call transcribed as *hiddy-hiddy-hiddy hit-up hit-up hit-up*, the latter section being slower, and the notes are described as being harsh, shrill, rapid and laughing. Also has a sharp and high-pitched *kyik* or *kr-lik*. **RANGE** Breeds from SE USA through C America and the West Indies south to N Argentina. Northern populations are highly migratory and the species is an inveterate wanderer. **STATUS** Vagrant with records from three archipelagos. **Azores** Six records: adult ♂ obtained, São Miguel, 26 Nov 1957; one at sea, off Flores, 8 Apr 1961; immature, Flores, Apr 1966; subadult, Flores, 18 May 1966; immature obtained, Flores, 12 Jul 1969; and adult collected, Vila Franca (São Miguel), 31 Dec 1978. **Madeira** Recorded twice: one found, Funchal harbour (Madeira), 8 Nov 1969; and one caught alive, Parque de Santa Catarina, Funchal (Madeira), 2 Jan 1976. **Canaries** Two records, both of first-years: one, near El Río (Tenerife), 25 Oct 1992; and one found exhausted, Puerto Rico (Gran Canaria), 19 Oct 2004.

COMMON COOT   *Fulica atra*   Plate 24 (5)
Sp. Focha Común   **P.** Galeirão (Galeirão-comum)

F. a. atra.   **L** 36–38 cm (14–15″).   **IDENTIFICATION** Sexes similar. Adult rather plump and all dark, a greyish-black rail with a conspicuous white frontal shield and bill. In flight reveals a narrow whitish trailing edge to the secondaries. Juvenile predominantly greyish-brown with a whitish chin, throat and central breast. First-winter more closely resembles adult but frontal shield is not fully developed, the throat is whitish and the plumage has a brown tint, which is not lost until first spring. Unlikely to be confused with any other species apart from congeners. **VOICE** Varied short, explosive calls, including a short *kow*, *kowk* or *kick*, sometimes two different notes being linked, e.g. *kick-kowk*. ♂ gives a high explosive *pssi* in aggression. **RANGE** Breeds from Europe and N Africa (very local) east to Sakhalin and N China, and south to S India, Sri Lanka, Australia and New Zealand. In Europe, northern and eastern populations migrate as far as Senegal and Sudan. **DISTRIBUTION** Breeds regularly only in the Canaries, but has also bred on the Azores. **BREEDING** Prolonged season, starting Dec and continuing until Aug, usually with a minimum of two broods and clutch size of 4–10, although 6–8 is more normal. Nest constructed on a floating mass of sticks and algae, but variation in water levels can leave these structures stranded on the banks of the reservoirs. **HABITAT** Prefers fresh water but sometimes is encountered around salt works. In the Canaries it breeds on small, man-made, irrigation reservoirs. **STATUS** Breeds in one archipelago and is a regular winter visitor to two others. **Azores** Regular winter visitor but has also bred on a few occasions. **Madeira** Winter visitor only. **Canaries** Breeds locally on Tenerife, Gran Canaria and Fuerteventura, but not always successfully. In winter numbers increase with migrants from Europe and is found throughout the archipelago at this season.

AMERICAN COOT *Fulica americana* Plate 24 (6)
Sp. Focha Americana **P.** Galeirão-americano

F. a. americana. **L** 31–37 cm (12–14½″). **IDENTIFICATION** Sexes similar and much like Common and Red-knobbed Coots, but easily distinguished in adult plumage by the dark (deep red) band on the bill, white sides to the undertail-coverts, dark upper shield, and head and neck distinctly blacker than body. Juvenile, which is unlikely to be seen in our region because this plumage is soon moulted, is similar to Common Coot but has white sides to the undertail-coverts. First-winter plumage very similar to adult but has a slight olive tone to upperparts. **VOICE** Varied harsh, croaking and short, clucking notes with a rather hollow quality, including a harsh *krok*, rather different from Common Coot. **RANGE** Breeds across most of N America, south through C America and the Caribbean to the N and C Andes of S America. Northern populations move south in the winter. **STATUS** Vagrant, only recorded from one archipelago. **Azores** Nine records: one, Lajes (Flores), 25 Oct 1971, had been ringed on Madeleine Island, Canada, 30 Aug 1971; one, São Miguel, 17 Oct 1986; one, 14 Oct 1988; one dead, Pico, 3 Dec 1991; two, Lagoa das Furnas (São Miguel), 9 Nov 1998; one, Flores, 31 Oct 2001; one, Lagoa Azul, Sete Cidades (São Miguel), 3–9 Dec 2001; one, Lagoa das Furnas (São Miguel), 31 Oct–20 Dec 2003 and still present 14 Feb 2004, perhaps even until 14 Oct 2004 at least; and two, Cabo da Praia (Terceira), 20 Nov–14 Dec 2003.

RED-KNOBBED COOT (CRESTED COOT) *Fulica cristata* Plate 24 (7)
Sp. Focha Cornuda **P.** Galeirão-de-crista

Monotypic. **L** 38–42 cm (15–16½″). **IDENTIFICATION** Very similar to Common Coot and best identified at close range, in the breeding season, when the two red knobs atop the frontal shield can be seen. However, identified at all times of year by lack of wedge of feathers between the bill and frontal shield, as on Common Coot. Red-knobbed also lacks the narrow white trailing edge to the secondaries and, in adult plumage, has a distinct bluish tinge to the bill, a useful feature if direct comparison is possible. Juvenile much darker than Common Coot with whitish on underparts restricted to throat. **VOICE** A disyllabic *clukuk* deeper than that of Common Coot, a metallic ringing *croo-oo-k* and varied other croaking, nasal and clucking notes. **RANGE** Predominantly E and S Africa and Madagascar, with relict populations in extreme S Spain and Morocco. Mostly resident but some local movements. **STATUS** Vagrant with all records from one archipelago. **Canaries*** One very old record from Tenerife and, more recently, one, Embalse de Los Molinos (Fuerteventura), 27 Jul 1997, with two there, Mar 2000, another two, 8 Apr 2001; and 2–3, Presa de Las Peñitas (Fuerteventura), Jan 2002.

COMMON CRANE *Grus grus* Plate 27 (1)
Sp. Grulla Común **P.** Grou (Grou-comum)

G. g. grus. **L** 110–120 cm (43–47″); **WS** 220–245 cm (86½–96½″). **IDENTIFICATION** Sexes similar. Very large size should be sufficient to identify this species from all others, except perhaps one of the *Ciconia* storks. In flight, the outstretched neck distinguishes it from any of the larger herons and the lack of a pale belly separates it from the two storks. Adult has grey body with a black-and-white head and neck, red crown and black flight feathers. Juvenile has a brownish head and neck, with a dull brownish tone to the grey body. **VOICE** A loud, trumpeting *krrooah*, a harsher *kraah* and a *kurr*, all are far-carrying. **RANGE** Breeds from England, Scandinavia and E Germany east to the Kolyma River, in Siberia, and N China, and south to Turkey, Iran and Lake Balkhash. Winters mainly in Iberia, N Africa, Sudan, Ethiopia, Turkey, Iraq, the Indian subcontinent and S China. **STATUS** Vagrant with only four records. **Azores** Juvenile ♀ obtained, Arribanas, Arrifes (São Miguel), 27 Nov 1933. **Madeira** One, Quinta do Palheiro Ferreiro, Funchal (Madeira), 15–25 Dec 1987. **Canaries** Three, Fuerteventura, 16 Dec 1999–11 Jan 2000; and one, La Graciosa, 24 Feb–6 Mar 2000, which was possibly one of the same birds.

SANDHILL CRANE *Grus canadensis* Plate 27 (2)
Sp. Grulla Canadiense **P.** Grou-americano

Monotypic. **L** 88–95 cm (35–37½″); **WS** 175–195 cm (69–77″). **IDENTIFICATION** Sexes similar. Smaller than Common Crane but this is difficult to judge on lone birds. Adult is all grey with dull, bare red skin on lores and forecrown. In flight reveals black primaries, but wing pattern is less contrasting than Common Crane. Juvenile has no red on head and is mainly sandy-brown, being more uniform than juvenile Common Crane with browner upperparts. **VOICE** In flight gives a rolling *gar-roooo*, more musical than flight call of Common Crane. Juvenile, before *c.* 10 months, gives a high-pitched, squeaky or trilled *tweer*. **RANGE** Breeds from NE Siberia, N Alaska and arctic Canada south discontinuously to Florida and Cuba. Winters in S USA and in NC Mexico. **STATUS** Extreme vagrant. **Azores** One photographed, Ponta Delgada (Flores), 26 Jun–3 Jul 2000.

LITTLE BUSTARD *Tetrax tetrax* Plate 27 (3)
Sp. Sisón **P.** Sisão

Monotypic. **L** 40–45 cm (16–18″); **WS** 105–115 cm (41–45″). **IDENTIFICATION** Sexually dimorphic. Breeding ♂ is unmistakable with grey face, black neck with white V-shaped collar and white subterminal collar, buff-brown mantle and upperwing-coverts, and white breast and belly. In flight reveals white primaries and secondaries, with black tips to all except inner secondaries, and a black crescent across primary bases. Adult ♀ lacks spectacular head and neck pattern of ♂ and is generally buff-brown with dark vermiculations, except white belly and vent. In flight shows indistinct black barring on secondaries. Non-breeding ♂ is very similar to ♀ but has whiter secondaries and whiter, less well marked, flanks. Juvenile also resembles adult ♀ very closely, but primary- and greater upperwing-coverts buffish and barred darker. **VOICE** Fairly silent but might utter a low grunt if flushed. Also ♂♂ have whistling wingbeat. **RANGE** Breeds discontinuously from Iberia and Morocco through France, Sardinia, Italy and the Balkans, thence Ukraine east to C Asia. Only the more northerly populations are migratory. **STATUS** Vagrant with only three records. **Madeira** One, near Caniço (Madeira), 10 Oct 1897; and one, of 3–4, shot, near Caniço (Madeira), Oct 1940. **Canaries*** Three killed, near Laguna (Tenerife), *c.* 1911.

HOUBARA BUSTARD *Chlamydotis undulata* Plate 27 (4)
Sp. Hubara **P.** Hubara (Abetarda-moura)

C. u. fuertaventurae. **L** 55–65 cm (22–25½″); **WS** 135–170 cm (53–67″). **IDENTIFICATION** Sexes similar. Adult has a striking black-and-white neck frill, rather deep cinnamon basal colour to upperparts, which are heavily vermiculated blackish-brown, white central crown and crest, white underparts and rather turkey-like appearance. ♀ has a less-developed neck frill and crest than ♂. Juvenile resembles ♀ but has no crest and neck frill is reduced to a line of short, black feathers. In flight both sexes have a striking white patch at base of outer primaries; the rest of the flight feathers predominantly blackish. Nominate *undulata* of N Africa is larger than *fuertaventurae* with paler upperparts and dark areas of upperwing are less black. **VOICE** Usually silent but does give a quiet coughing noise *tucq* in alarm (captive birds). **RANGE** *undulata/fuertaventurae* breed from the E Canaries discontinuously across N Africa to Egypt. **DISTRIBUTION** Breeds only on La Graciosa, Lanzarote and Fuerteventura, but also occurs on Lobos. Formerly recorded from S Gran Canaria and probably used to breed there. **BREEDING** Eggs usually laid Jan–Apr but sometimes even in Dec or as late as May–Jun. Clutch normally 2–3 and eggs are rich buff-brown, sparsely spotted and blotched chocolate-brown with pale underlying purplish markings, and quite different from those of *undulata*. Incubation period 28 days and chicks take *c.* 35 days to fledge, and remain with ♀ at least until autumn. **HABITAT** Dry, open desert plains with a sparse scrubby vegetation. **STATUS** Only found on one archipelago where it is an uncommon resident breeder. Elsewhere, has declined dramatically due to excessive hunting. **Canaries** Recent surveys suggest *c.* 240 birds on Fuerteventura, *c.* 270 on Lanzarote and a maximum of 18 on La Graciosa. Apparently fairly stable on Fuerteventura but latest estimates from Lanzarote suggest the population there is much larger than previously suspected.

EURASIAN OYSTERCATCHER *Haematopus ostralegus* Plate 28 (1)
Sp. Ostrero **P.** Ostraceiro

H. o. ostralegus. **L** 40–46 cm (16–18″); **WS** 80–86 cm (31½–34″). **IDENTIFICATION** Sexes similar. Unmistakable, large wader with black-and-white plumage, orange bill and pinkish legs. At rest only belly and vent are white with rest of plumage black, but in flight reveals broad white wingbar and white rump, which extends to a point on back (American Oystercatcher *H. palliatus*, not recorded in our region, only has white on rump). Adult in winter plumage has a variably prominent white half-collar on the throat and neck-sides. Juvenile distinguished by its browner upperparts and duller bill with a dusky tip. **VOICE** Typically a high, piping, shrill *kleep* or more disyllabic *k-peep*. **RANGE** Breeds in Europe east to WC Asia, also occurs in Kamchatka and E China, and winters in Europe, Africa north of the equator, parts of the Middle East, Indian subcontinent and S China. **STATUS** Recorded from all archipelagos but status varies. **Azores** Vagrant with 11 records including 15, Praia da Vitória (Terceira), 27 Sep 1964, and the most recent one, Ponta Delgada (São Miguel), 8 Dec 2003. **Madeira** Fewer than five records in the last 50 years; the most recent one, Funchal (Madeira), 11 Oct 1995; one, Porto Moniz (Madeira), 11 Sep 2004; and one, Lugar de Baixo (Madeira), 15 Oct–7 Nov 2004. **Canaries** A rare and irregular passage migrant/winter visitor recorded from all islands except El Hierro. **Cape Verdes** Nine records, the most recent one, Tarrafal (São Nicolau), 10 Apr 2001.

CANARY ISLANDS (BLACK) OYSTERCATCHER *Haematopus meadewaldoi*
Sp. Ostrero Canario **P.** Ostraceiro das Canárias   Plate 28 (2)

Monotypic. **L** 43 cm (17″); **WS** 83 cm (33″). **IDENTIFICATION** Sexes similar. All sooty-black plumage with orange bill, reddish legs, red eye and red eye-ring. Very little known and juvenile plumage never described. Compared to African Black Oystercatcher smaller (particularly wing length) and has base of inner webs of primaries white, forming patch on open wing, particularly visible on underside. **VOICE** Very similar to Eurasian Oystercatcher but less high-pitched. **RANGE** Confined to eastern islands and islets of the Canaries. **DISTRIBUTION** Formerly bred on Fuerteventura, Lanzarote, Lobos, La Graciosa, Alegranza and Montaña Clara. **BREEDING** Very little is known but presumably similar to African Black Oystercatcher. **HABITAT** Usually on rocky coasts but occasionally on sandy beaches. **TAXONOMIC NOTE** Some authors consider this species to be a form of African Black Oystercatcher *H. moquini*, but morphological differences (and considerable geographical isolation) lend support for specific status. **STATUS** Extinct but was recorded from two archipelagos. **Madeira** Recorded but no details or extant specimens. **Canaries** Last collected, La Graciosa, Jun 1913, but local fishermen reported it on Alegranza until *c.* 1940. Reports of black oystercatchers from Tenerife in 1968 and 1981, but recent coastal surveys of islets off Lanzarote have failed to locate any evidence of its continued existence.

BLACK-WINGED STILT *Himantopus himantopus*
Sp. Cigüeñuela **P.** Pernilongo (Perna-longa)   Plate 28 (3)

Monotypic. **L** 35–40 cm (14–16″); **WS** 67-83 cm (26–32″). **IDENTIFICATION** Sexes similar. Unmistakable given its exceptionally long pink legs, black wings, predominantly white body and fairly long, thin and straight bill. There is often a variable amount of black on the crown and hindneck. Juvenile differs from adult in having black areas with a distinctly brown cast and the feathers (except primaries and secondaries) narrowly fringed pale buff. Also legs duller greyish-pink and crown and hindneck always dusky. **VOICE** Alarm call a monotonous high-pitched *kik-kik-kik-kik-kik…*, also a sharp, nasal *kek*. **RANGE** Breeds discontinuously from Europe east to Mongolia and Vietnam, and south to Sri Lanka, S Africa and Madagascar. European birds winter in sub-Saharan Africa. **DISTRIBUTION** Confined to Sal in the Cape Verdes and recently discovered breeding on Lanzarote in the Canaries. **BREEDING** In Cape Verde breeds Mar–Jul and clutch normally 3–4. On the Canaries it breeds Apr–Jun with a clutch of 3–4. Nests are usually placed close together and are simple depressions in the ground camouflaged with small stones, shell fragments and vegetation. **HABITAT** As a breeder, confined to the Pedra de Lume saltpans on Sal, and the Salinas de Janubio on Lanzarote, where it nests on banks and dykes, but as a passage migrant occurs on any fresh or brackish water, but unlikely to be found on rocky or sandy coasts. **STATUS** Recorded from all archipelagos but status varies. **Azores** Vagrant with 6–8 records: one obtained, São Miguel (no other details); ♂ ♀ obtained, Charco da Madeira (São Miguel), 7 May 1905; five, Lagoa das Impedas, Sete Cidades (São Miguel), 24 Apr 1978, with two, near Lagoa da Congro (São Miguel), 2 May 1978, perhaps part of the same group; two, Vila do Porto (Santa Maria), 17 Jun 1990; one, Lajes do Pico (Pico), 1 Apr 1998; and four, Lajes do Pico (Pico), 8 Apr 2001, with probably the same, Cabo da Praia (Terceira), 22 Apr 2001. **Madeira** Fewer than five records in the last 50 years. **Canaries** Apart from recent breeding records, Laguna de Janubio (Lanzarote), and in 1998 at Vecindario (Gran Canaria), an uncommon, but regular, passage migrant. May also have bred on Fuerteventura on a few occasions in recent years. **Cape Verdes** Latest estimate of the resident breeding population is *c.* 75 birds. Could also occur as an accidental visitor, as it has been recorded from other islands but most of these records probably relate to wandering resident birds.

(PIED) AVOCET *Recurvirostra avosetta*
Sp. Avoceta **P.** Alfaiate (Alfaite)   Plate 28 (4)

Monotypic. **L** 42–45 cm (16½–18″); **WS** 77–80 cm (30–31½″). **IDENTIFICATION** Sexes similar. Unmistakable given combination of striking black-and-white plumage and obvious upcurved bill, which are unique amongst the region's waders, as are the bluish-grey legs. Adult predominantly white with a black cap and nape, black patch either side of mantle and another on median wing-coverts, and black outer primaries. ♀ often distinguished by its shorter and more strongly curved bill. Juvenile has black areas tinged brown; white areas on mantle and wings are initially heavily mottled brown and buff. **VOICE** A loud, piping, oft-repeated liquid *kluip*, and in alarm a similar, but harsher *kloo-eet*. **RANGE** Discontinuous, from Europe across C Asia and in E and S Africa, wintering mainly in Africa, E China, NW India and the Persian Gulf. **STATUS** Varies between the archipelagos. **Azores** Three records: ♂ obtained, Rabo de Peixe (São Miguel), 12 Oct 1932; ♂ obtained, São Miguel, 22 Apr 1950; and one, Charco dos Leimos (São Miguel), 30 Apr–3 May 1979. **Madeira** Exceptional, records include three ♀ ♀, of a small flock, obtained, Madeira, 24 Mar 1950. **Canaries** A rare and irregular passage migrant, and even rarer winter visitor, with records from all main islands except La Gomera. **Cape Verdes** Nine records: one from São Vicente, two from Sal, three from Boavista and three from Maio.

STONE-CURLEW *Burhinus oedicnemus*   Plate 28 (5)
Sp. Alcaraván   **P.** Alcaravão

B. o. oedicnemus (Madeira, Azores), *B. o. insularum* (E Canaries) and *B. o. distinctus* (C and W Canaries, except La Gomera). **L** 40–44 cm (16–17″); **WS** 77–85 cm (30–33½″). **IDENTIFICATION** Sexes similar. A fairly large wader with prominent yellow eyes and long, thick, yellow legs. Adult has pale brown upperparts, streaked darker, and underparts streaked on buffy breast. Diagnostic white bar on the wing-coverts, which is bordered above and below with black. Difficult to observe as it is mostly nocturnal. Juvenile has a less obvious wingbar and lacks prominent whitish supercilium. Endemic Canarian races *distinctus* and *insularum* are smaller than the nominate; *distinctus* has brown coloration similar to *oedicnemus*, but has heavier dark streaking and a paler ground colour; *insularum* is closer to N African race *saharae*, having a sandy-pink aspect, but is more heavily streaked. **VOICE** Typical call a loud, rising, whistling *cur-lee*, and in display gives a series of wailing, mournful whistles rising in pitch, volume and frequency, terminating in loud repeated *cur-lee* calls. **RANGE** Breeds across much of Europe and N Africa, the Middle East and W and S Asia. Northern populations are migratory, wintering in Africa south to N Kenya. **DISTRIBUTION** Breeds only on the Canaries, where the only main island on which it is unknown to breed is La Gomera, and also breeds on islets of La Graciosa and Alegranza, north of Lanzarote. **BREEDING** Mainly Feb–Mar, but season extends from Jan–Jun. Clutch size two and the nest, a small depression in the ground, is usually sited at base of a rock or shrub. **HABITAT** Prefers open areas of stony or sandy semi-desert with scattered bushes. **STATUS** Recorded from three archipelagos but status varies. **Azores** One obtained, São Miguel, Jan 1880. **Madeira** Fewer than five records in the last 50 years. **Canaries** Most frequently encountered on Fuerteventura and Lanzarote.

EGYPTIAN PLOVER *Pluvianus aegyptius*   Plate 28 (6)
Sp. Pluvial Egipcio   **P.** Ave-do-crocodilo (Pluvial do Egipto)

Monotypic. **L** 19–21 cm (7½–8″); **WS** 47–51 cm (18½–20″). **IDENTIFICATION** Sexes similar. Combination of blue-grey, black, white and buff plumage unique and makes the species totally unmistakable. Adult has blue-grey upperwing-coverts, scapulars, tertials and tail, the latter with a broad white terminal band and a narrow black subterminal band. Black crown, mask, mantle, back, V-shaped breast-band and diagonal bar on wings. White supercilium, chin, throat and flight feathers, and buffy-peach breast, belly and vent. Juvenile has a rusty-brown tinge to black areas of head and on the lesser and median wing-coverts. The breast-band is also duller, less distinct, and the flanks and undertail-coverts paler. **VOICE** Most common call is a series of harsh, high-pitched notes, *chersk-chersk-chersk…*; also a softer, single *wheep*. **RANGE** Breeds discontinuously across C Africa, where mainly resident but sometimes undergoes nomadic movements. **STATUS** Vagrant with only one record. **Canaries★** One obtained, near La Laguna (Tenerife), prior to 1893 (held in the Cabrera collection).

CREAM-COLOURED COURSER *Cursorius cursor*   Plate 28 (7)
Sp. Corredor   **P.** Corredeira (Corredor)

C. c. cursor (E Canaries) and *C. c. exsul* (Cape Verdes). **L** 21–24 cm (8–9½″); **WS** 51–57 cm (20–22½″). **IDENTIFICATION** Sexes similar. Unmistakable long-legged wader of desert areas. Adult mainly sandy except black eye-stripe and white supercilia that meet as a V on nape, and grey rear crown. In flight reveals a predominantly black underwing, and black primaries and primary-coverts on upperwing. Short, fine, decurved bill is black and legs whitish. Juvenile has far less distinct head pattern, upperparts marked with narrow, dark subterminal scalloping and underparts less richly coloured. **VOICE** Flight call a repeated piping *quit* or *wit*, also a hoarse, croaking *nhark* or *praak*. When breeding adds a nasal *whow* at the end of the first call. **RANGE** Breeds in N Africa including the Sahel, the Middle East and W Asia. Some populations mostly resident, but others disperse south in winter. **DISTRIBUTION** Breeds on Lanzarote and Fuerteventura, and formerly on Gran Canaria, as well as in Cape Verde (Santiago, São Vicente, Sal, Boavista and Maio). In 2001 recorded breeding in S Tenerife, with two nests (one with two eggs and the other was not examined). **BREEDING** In Cape Verde season lasts Oct–May, whereas in the Canaries lays Feb–May. Clutch normally two and the eggs are laid in a shallow scrape on the ground. **HABITAT** Areas of semi-desert, preferring flat sandy or stony plains. **STATUS** Varies between archipelagos. **Madeira** Extreme vagrant with just one old record. **Canaries** Fairly common and widespread on Lanzarote and Fuerteventura. Formerly bred on Gran Canaria but nowadays accidental. On Tenerife a very scarce passage migrant or accidental to the south of the island, usually in spring. Breeding confirmed there in 2001. **Cape Verdes** Fairly common and widespread on Boavista and Maio, but rarer and more localised on other islands. Accidental visitor to Santo Antão, Santa Luzia and Raso, and status on São Nicolau requires clarification.

COLLARED PRATINCOLE *Glareola pratincola* Plate 28 (8)
Sp. Canastera **P.** Perdiz-do-mar

G. p. pratincola. **L** 23–26 cm (9–10″), **WS** 60–65 cm (23½–25½″). **IDENTIFICATION** Sexes similar. A rather tern-like wader, with long wings, deeply forked tail and small bill. Adult in breeding plumage has very distinct creamy chin and throat with a prominent black border, red bill base and grey-brown upperparts and breast. In flight reveals deeply forked tail, white rump, narrow white trailing edge to secondaries and, in good light, chestnut axillaries and underwing-coverts. Non-breeder similar but facial pattern less pronounced and red bill base duller. Juvenile has scaly upperparts and breast, lacks black border to throat and has paler and less extensive red base to bill. **VOICE** Most likely call to be heard is the alarm call, a harsh, tern-like *kirririk*. **RANGE** Breeds discontinuously in S Europe, N Africa, the Middle East, C Asia and sub-Saharan Africa. Winters exclusively in sub-Saharan Africa. **STATUS** Varies according to archipelago. **Madeira** Listed as an exceptional vagrant by Zino *et al.* (1995). **Canaries** Uncommon passage migrant, mostly in spring, occasionally occurring in small flocks. **Cape Verdes** Five records: ♀ collected, Pedra Badejo lagoons (Santiago), 29 Apr 1898; ♀ collected, Santiago (no location given), 20 Sep 1969; one, Calheta de Baixo (Maio), 27 May 1989; one São Vicente sewage farm, 8–9 Mar 1997; one, Santa Maria saltpans (Sal), 23 May 1997; and one, São Vicente sewage farm, Dec 2003–7 Mar 2004 at least.

LITTLE RINGED PLOVER *Charadrius dubius* Plate 29 (1)
Sp. Chorlitejo Chico **P.** Borrelho-pequeno-de-coleira

C. d. curonicus. **L** 14–17 cm (5½–7″); **WS** 42–48 cm (16½–19″). **IDENTIFICATION** Sexes similar. A small and rather dainty plover. In all plumages, leg colour (pinkish-grey in adults, dull yellowish in juveniles), lack of a prominent wingbar and shorter primary projection help distinguish this species from Ringed Plover. Adult in breeding plumage has an obvious yellow eye-ring, an almost totally blackish bill and a narrow white line above the black band on the forehead. None of these is shown by adult Ringed. Non-breeder has black bands on head and breast turn brown, forehead and supercilium buffish and eye-ring less prominent. Juvenile similar to adult non-breeding but has sandier upperparts with indistinct buff scalloping, a very faint buffish supercilium, duller eye-ring and more yellowish legs. **VOICE** Typical call a whistling *pee-oo* with emphasis on first syllable. Display call given in flight is a repeated harsh *cree-ah*. **RANGE** Breeds across most of Eurasia, also NW Africa and New Guinea. Migratory populations winter south of the breeding range in sub-Saharan Africa, Arabia and Indonesia. **DISTRIBUTION** Only breeds in the Canaries. **BREEDING** Season late Mar–Jul with pairs often having two broods. Clutch 2–4, although latter is most frequent. Nest a shallow depression in ground camouflaged with many small stones and usually fairly near a reservoir, sometimes even on the man-made banks. **HABITAT** Breeds in the vicinity of permanent fresh water and uses many of the dams built for irrigation. **STATUS** Varies between archipelagos. **Azores** Ten records including one obtained, Flores, prior to 1903, and nine records by G. Le Grand in 1977–85, all on São Miguel, except one from Terceira. **Madeira** A very rare breeder that has also been recorded on the Salvages. **Canaries** Breeds in small numbers only on Tenerife, Gran Canaria and Fuerteventura, where presumed resident. On passage also been recorded from Lanzarote, La Palma and El Hierro. **Cape Verdes** Accidental with 16 records, but perhaps more regular, as 14 of these are since 1988.

(COMMON) RINGED PLOVER *Charadrius hiaticula* Plate 29 (2)
Sp. Chorlitejo Grande **P.** Borrelho-grande-de-coleira

C. h. hiaticula and *C. h. tundrae.* **L** 18–20 cm (7–8″); **WS** 48–57 cm (19–22½″). **IDENTIFICATION** Sexes similar. A rather thickset, small plover with a black breast-band, bright orange-yellow legs and bill base, and in flight an obvious white wingbar. Adult in breeding plumage easily separated from Little Ringed Plover by two-tone bill, bright orange-yellow legs, prominent white wingbar and longer primary projection. Non-breeder retains leg colour but black on head and breast replaced with greyish-brown, supercilium and forehead are buffy-brown (joining in front of eye) and yellow-orange base to bill is less obvious and less extensive. Juvenile similar to non-breeding adult but has scaly upperparts, reduced breast-band in centre (sometimes broken), bill mostly blackish and duller legs Subspecies *tundrae* is smaller and darker than the nominate. Racial identification complicated because distinctions unclear due to a north–south cline in size and colour. Hence, *hiaticula* from UK are more distinct from *tundrae* than from nominate birds from Greenland. Identification from its N American counterpart, Semipalmated Plover, can be very difficult and is dealt with under that species. **VOICE** Common call is a whistled, disyllabic *too-lee* rising in pitch and with the emphasis on the second syllable. **RANGE** Breeds from NE Canada and Greenland through N Europe to E Siberia. Winters in Europe, the Middle East and Africa. **STATUS** A regular, and fairly common, winter visitor throughout. **Azores** Regular winter visitor in small numbers, particularly to São Miguel and Terceira. **Madeira** An occasional visitor. **Canaries** Common passage migrant and winter visitor, both subspecies recorded. **Cape Verdes** Fairly common passage migrant and winter visitor, both subspecies recorded.

SEMIPALMATED PLOVER *Charadrius semipalmatus* Plate 29 (3)
Sp. Chorlitejo Semipalmeado   **P.** Batuíra-de-bando

Monotypic. **L** 17–19 cm (7–7½"); **WS** 43–52 cm (17–20½"). **IDENTIFICATION** Sexes similar. Almost identical to Ringed Plover, from which it is most easily identified by call or by detecting the presence of webbing between the middle and inner toes. Ringed has a small amount of webbing between the middle and outer toes, but on Semipalmated this is more extensive. Adult in breeding plumage differs from Ringed in having a stubbier bill, narrower breast-band, usually a narrow yellow orbital ring, less white above and behind eye and averages smaller. A feature for identifying juveniles is that the dark area on the lores is less extensive than on Ringed and characteristically the lower edge meets the bill above the gape, whereas in juvenile Ringed the lower edge of the dark loral area usually meets the bill at the gape. However, there is much variation in these features and extreme care must be exercised in identifying the species. **VOICE** Common calls include a sharp *chuwit* similar to a subdued Spotted Redshank and almost monosyllabic, and a clear, plaintive *chu-wee* rising in pitch with the emphasis on the second syllable, unlike the corresponding call of Ringed Plover and also distinctly less disyllabic. **RANGE** Breeds from Alaska east to Newfoundland. Winters coastally in S USA, C America and most of S America. **STATUS** Vagrant, although probably regular in the Azores. **Azores** Adult ringed in Grindstone, Canada, recovered Santa Maria, 25 Sep 1972. In excess of 17 records since, of which more than 12 were in autumn 1998–autumn 2004. **Madeira** Two records: first-winter, Machico (Madeira), 16 Nov 2000; and one, Tanque (Porto Santo), 20 Oct 2001. **Cape Verdes** Adult, Mindelo sewage farm (São Vicente), 27 Feb 1999; and first-winter, Pedra de Lume saltpans (Sal), 6 Mar 1999, with a claim, from Pedra de Lume, 9 Mar 2004, involving one heard and seen only in flight.

KILLDEER *Charadrius vociferus* Plate 29 (4)
Sp. Chorlitejo Culirrojo   **P.** Borrelho-de-coleira-dupla (Borrelho-de-duas-coleiras)

C. v. vociferus. **L** 23–26 cm (9–10"); **WS** 59–63 cm (23–25"). **IDENTIFICATION** Sexes similar. Easily identified from any other *Charadrius* plover by its larger size and double breast-band. Adult in breeding plumage has brown upperparts, white underparts, double black breast-band, white collar, white forehead bordered above by a black band on front of crown, white supercilium behind eye and black mask. In flight reveals a broad white wingbar, long wedge-shaped tail and diagnostic bright orange-brown lower back, rump and uppertail-coverts. Non-breeder has upperparts with extensive rufous-brown fringes and black areas blackish-brown. Juvenile plumage (quickly moulted) is similar to fresh-plumaged adult but has paler fringes to upperparts. **VOICE** Typical call is a high-pitched, piercing *kill-dee* or *kill-diu* with the second syllable often repeated; also when flushed a thin, clear *diii-di-di*. **RANGE** Breeds over most of N America and the West Indies, also Peru and NW Chile. Northern populations move south in winter to C America and N South America. **STATUS** Vagrant with only ten records, seven of these from one archipelago. **Azores** ♀ collected, Faial, 4 Jan 1928; up to four, Paul da Praia (Terceira), Jan–Feb 1945; one, Flores, 18 May 1966; one, Lagoa (São Miguel), 13 Dec 1977; adult, Ribeira Seca (São Miguel), 8 Feb 1985; one, Santa Cruz (Flores), 6 Nov 1998; and one, Lagoa Azul, Sete Cidades (São Miguel), 1–12 Nov 2001. **Madeira** Included in Johnson's *Handbook of Madeira* (1885) and in subsequent lists on basis of this work with, one, Machico (Madeira), 17 Feb 2004. **Canaries*** One photographed, near Caleta de Fustes (Fuerteventura), 17–20 Mar 1999.

KENTISH PLOVER *Charadrius alexandrinus* Plate 29 (5)
Sp. Chorlitejo Patinegro   **P.** Borrelho-de-coleira-interrompida

C. a. alexandrinus. **L** 15–17 cm (6–7"); **WS** 42–45 cm (16½–18"). **IDENTIFICATION** Sexes differ. A small plover most likely to be confused with either Ringed or Little Ringed Plovers, but in all plumages has small dark patches on breast-sides, blackish or grey-brown legs, thin black bill and all-white tail-sides. Adult ♂ in breeding plumage has white forehead and supercilium, black band on forecrown and line through eye, small black patches on breast-sides, rufous crown and nape, pale grey-brown upperparts and white underparts. ♀ has grey-brown markings on breast-sides and lacks the head pattern of ♂. Non-breeding ♂ is less well marked and colourful than breeding ♂. Juvenile similar to adult ♀ but breast patches paler and has narrow pale fringes to upperparts. **VOICE** Usual call in flight is a soft *pit* or *twit* and alarm call is a hard *prrr* or plaintive *too-eet*. Display-flight call is a repeated, rattling *tjekke-tjekke…* **RANGE** Fairly cosmopolitan, breeding in a broad but incomplete band from the Cape Verdes and Canaries through S Europe, N Africa and C Asia to Japan and south to Hainan, Sri Lanka, Yemen and Somalia. Also, breeds in N America, the West Indies and in S America. More northerly populations are migratory. **DISTRIBUTION** Resident breeder on all four archipelagos: on Canaries breeding not confirmed on western islands of La Gomera, La Palma and El Hierro; in Madeira confined to Porto Santo; on Cape Verdes breeds on Santiago, São Vicente, Sal, Boavista and Maio; and in the Azores on Santa Maria, São Miguel, Terceira, São Jorge and Graciosa, but breeding not confirmed on Pico and Faial. **BREEDING** In the Canaries mainly Mar–May but season extends Feb–Jul, with pairs usually having two broods. Clutch size 2–3, with latter more frequent. Nest a shallow depression in ground camouflaged with small stones and shells. In Cape Verde season more protracted and breeding may occur year-round, but activity peaks Sep–Jun, and clutch size is 1–3. **HABITAT** Most often encountered on sandy shores, but in the Canaries sometimes appears on freshwater reservoirs. **STATUS** A resident breeder on all four archipelagos. **Azores** Scarce on all islands where it breeds. **Madeira** Rare species on those islands where it breeds, and also recorded on the Salvages. **Canaries** Almost extinct as a breeder on Tenerife, still fairly numerous on eastern islands but not common. Also occurs as a scarce passage migrant. **Cape Verdes** Fairly common on all islands where it breeds, but most frequent on the eastern islands of Sal, Boavista and Maio.

LESSER SAND PLOVER (MONGOLIAN PLOVER) *Charadrius mongolus* Plate 29 (6)
Sp. Chorlitejo Mongol Chico **P.** Borrelho-pequeno-de-colar-ruivo (Borrelho-asiático)

Subspecies unknown but *C. m. pamirensis* would seem the most likely. **L** 19–21 cm (7½–8″); **WS** 45–58 cm (18–23″). **IDENTIFICATION** Sexes differ when breeding. A small plover lacking white collar of other *Charadrius* species in the region. ♂ in breeding plumage has deep rufous breast-band with narrow black upper border, white chin and throat, black mask, brown upperparts and white underparts. ♀ in breeding plumage differs in its duller breast-band and has black areas of ♂ replaced by brown. Non-breeding adult loses all rufous and black markings, the forehead is whitish and extends beyond eye as a rather narrow whitish supercilium, and the breast-band is reduced to two large patches on the sides that can almost join in the centre. Juvenile similar to non-breeding adult but has sandy-buff fringes to upperparts and upperwing-coverts. Compared to other species in our region, should be unmistakable, but is very similar to Greater Sand Plover *C. leschenaultii*, which is unrecorded from the region. Generally Greater Sand Plover is larger, longer billed, longer legged and has a more angular head shape with a flatter crown. **VOICE** Most frequent call is a sharp *chitik* but also gives a short *drrit*, both calls rather like those of Turnstone. **RANGE** *pamirensis* breeds in C Asia from Ladakh to the Tien Shan, and winters mostly coastally around W Indian Ocean, from W India and Pakistan through Arabia to E and S Africa. **STATUS** Vagrant with just the one record. **Canaries★** One, Corralejo (Fuerteventura), 26 Apr 1998.

(EURASIAN) **DOTTEREL** *Charadrius morinellus* Plate 29 (7)
Sp. Chorlito Carambolo **P.** Borrelho-ruivo (Tarambola-carambola)

Monotypic. **L** 20–22 cm (8–9″); **WS** 57–64 cm (22½–26″). **IDENTIFICATION** Sexes differ. Adult is unmistakable in breeding plumage given combination of white throat, grey upper breast, white chest-band bordered above and below by narrower black bands, chestnut lower breast and flanks, black belly, and white vent and undertail. Also, long, prominent white supercilia, which meet on nape in a V. ♀ is brighter and has solid dark cap and better-defined supercilium. In non-breeding plumage sexes similar, but long supercilium, scaly upperparts and an indistinct, pale breast-band offer good identification features. Juvenile resembles non-breeding adult but has darker feather centres to upperparts and larger wing-coverts, thus creating a more scaly appearance. **VOICE** Not very vocal, but calls include a soft *pweet-pweet-pweeet*, a *kwip-kwip* or trilling *pyurrr* given on take-off. **RANGE** Breeds from Scotland, across Scandinavia discontinuously to E Siberia, also the Low Countries, the Alps and C Asia, and winters mostly in N Africa and the Middle East. **STATUS** Vagrant with records from three archipelagos. **Azores** Three, Caldeira (Faial), 10 Oct 2003 with one still present 18 Oct 2003. **Madeira** Fewer than five records in the last 50 years, the most recent a first-year, Paúl da Serra (Madeira), 20 Sep 2003. Also recorded on the Salvage Islands. **Canaries** Accidental visitor, or rare passage migrant and winter visitor, to Alegranza, Lanzarote, Fuerteventura, Gran Canaria and Tenerife (almost annual in recent years), most frequently in autumn.

AMERICAN GOLDEN PLOVER *Pluvialis dominica* Plate 30 (1)
Sp. Chorlito Dorado Chico (Chorlito Dorado Americano) **P.** Batuiruçu (Tarambola-dourada-pequena)

Monotypic. **L** 24–28 cm (9½–11″); **WS** 66–72 cm (26–28″). **IDENTIFICATION** Sexes similar. In all plumages could be mistaken for either European Golden Plover or Grey Plover, but distinguished from both by its grey underwing-coverts and axillaries. Structurally, longer winged, longer legged and slimmer bodied than European Golden Plover. In flight lacks the obvious white rump of Grey Plover and has even less of an upper-wingbar than European Golden Plover. Adult in breeding plumage is most similar to European Golden Plover but differs in having large white patches on the breast-sides and a black undertail. Also, normally lacks a white line on the flanks, like on European Golden. Non-breeding adult and juvenile more closely resemble Grey Plover, but American Golden Plover is daintier with a finer bill and more prominent supercilium behind the eye. In flight lacks white rump of Grey Plover. For differences from Pacific Golden Plover see that species. **VOICE** Similar to European Golden Plover but less plaintive and somewhat sharper, a disyllabic *klee-i* or monosyllabic *kleep*. All calls very similar to those of Pacific Golden Plover. **RANGE** Breeds across N North America, from Alaska east to Baffin I. Winters in interior of S South America. **STATUS** Accidental, being almost annual on Canaries in recent years, but still a vagrant to other island groups. **Azores** Twelve records, the first in 1994 and all but one in autumn, with most on Terceira (8), single records on São Miguel and Pico, and two on Flores. **Canaries** About 23 records, most since 1989 and all records post-1983. Recorded on Lanzarote (2), Fuerteventura (4), Gran Canaria (1), Tenerife (15) and El Hierro (1). **Cape Verdes**

Nine records: one obtained, Mindelo (São Vicente), 6 Jan 1924; one, Mindelo (São Vicente), 18 Feb 1966; ♀ collected, Santo Antão, 26 Oct 1972; one, Pedro Badejo lagoons (Santiago), 5 Mar 1997; one, Mindelo sewage ponds (São Vicente), from 22 Jan 1999, joined by a second, 13 Mar, and both present until 20 Apr 1999 at least; adult also there, 14 Apr 2001; another adult there, 14 Nov 2001; adult, same location, 7 Mar 2004; and one, 18–21 Dec 2004.

PACIFIC GOLDEN PLOVER *Pluvialis fulva* Plate 30 (2)
Sp. Chorlito Dorado Siberiano **P.** Tarambola-dourada-siberiana

Monotypic. **L** 23–26 cm (9–10″); **WS** 60–68 cm (23½–27″). **IDENTIFICATION** Sexes similar. Very similar to American Golden Plover and in all plumages structural differences are useful. Tertials are longer in Pacific than in American, thus the primary projection is greater in American Golden Plover. Leg length of Pacific Golden Plover is longer than American, and in flight the toes of Pacific extend beyond the tip of the tail. Also, the rear end of Pacific is rather short, less attenuated than American, and generally it is slightly smaller. Also, in all plumages, the uniform grey-brown underwing differs from the white one of European Golden Plover, but is identical to American Golden Plover. In breeding plumage closely resembles European Golden Plover in that it has a white line on flanks and has inconspicuous white patches on the breast-sides, unlike American which has no white flank line and large white patches on the breast-sides. Differs from European Golden Plover in having the upperparts more coarsely marked, the tertials more coarsely notched and the undertail-coverts heavily marked with black. In non-breeding and juvenile plumages also very similar to European Golden Plover, when structural differences are most useful, along with call and underwing colour. In these plumages, easier to identify from American Golden Plover, as the plumage is more yellowish than the latter. Juvenile separated from adult winter by more prominent gold spotting on upperparts, more obvious supercilium and brighter yellowish-buff breast. **VOICE** Very similar to those of American Golden Plover and include a rapid *tew-ee* and plaintive *kl-ee*, but also has a distinctive whistling *chu-it* similar to Spotted Redshank. **RANGE** Breeds on Siberian tundra from Kara Sea east to the Bering Sea and W Alaska. Winters coastally in Australasia, SW Pacific, Indonesia, SE Asia, the Indian subcontinent and extreme E Africa. **STATUS** Extreme vagrant. **Canaries*** Adult, Playa de Sotavento (Fuerteventura), 8 Nov 1997.

EUROPEAN GOLDEN PLOVER *Pluvialis apricaria* Plate 30 (3)
Sp. Chorlito Dorado Común **P.** Tarambola-dourada

Monotypic. **L** 26–29 cm (10¹–11½″); **WS** 67–76 cm (26½–30″). **IDENTIFICATION** Sexes similar. Slightly smaller and less bulky than Grey Plover. In breeding plumage combination of black underparts bordered with white and golden-spotted upperparts quite distinctive. In non-breeding plumage black is lost and breast is mottled golden-brown. In flight shows white axillaries and underwing-coverts, a narrow bar on upperwing and a dark rump. Most likely to be confused with congeners, particularly American and Pacific Golden Plovers, but careful study of structure and subtle plumage differences assist identification. **VOICE** A plaintive, liquid whistle, *too-ee* or *tloo*, rather monosyllabic with a slight downslur. **RANGE** Breeds from E Greenland and Iceland across Scandinavia and N Europe to W Siberia. Winters mostly in the British Isles, Europe and N Africa. **STATUS** Recorded from three archipelagos. **Azores** Nine records: breeding-plumaged ♂ obtained, Ponta Delgada (São Miguel), prior to 1903; one obtained in winter plumage, 15 Dec 1958; one killed, Flores, 5 May 1968; one, Poço da Luz (Terceira), c.15 May 1968; one, Mosteiros (São Miguel), winter 1978/79; one, Faial da Terra (São Miguel), 15 Sep 1979; one, Praia da Vitória (Terceira), 6 Oct 1979; one, Caldeira (Faial), 22 May 1984; and one, Monte Escuro (São Miguel), 14 May 1985. **Madeira** Fewer than five records in the last 50 years. **Canaries** A scarce and irregular passage migrant and winter visitor, sometimes in small groups and often associated with Northern Lapwing, Grey Plover or Dotterel.

GREY (BLACK-BELLIED) PLOVER *Pluvialis squatarola* Plate 30 (4)
Sp. Chorlito Gris **P.** Tarambola-cinzenta

Monotypic. **L** 27–30 cm (11–12″); **WS** 71–83 cm (28–33″). **IDENTIFICATION** Sexes similar. Unmistakable in breeding plumage with combination of predominantly black underparts, white patches on breast-sides and greyish-white spangled upperparts. Non-breeder could perhaps be confused with other *Pluvialis* but is larger and greyer with a heavier bill. In flight, in all plumages, black axillaries diagnostic but also has a long white wingbar and white rump. Juvenile resembles adult winter, but has darker upperparts with pale yellowish spangling and more obvious streaking on breast and flanks. **VOICE** A loud, melancholy, trisyllabic whistle, *tlee-oo-ee*, with the middle syllable lower. **RANGE** Breeds across arctic N America and Siberia. Winters on coasts from S Canada to S South America and in Europe, Africa, the Middle East, Indian subcontinent, and SE Asia to Australia. **STATUS** Common winter visitor to coasts. **Azores** Regular visitor with records in most months. **Madeira** Listed as exceptional, but a flock of 22, Porto Santo, Jan 2001, and another of similar size, 2003, and one, Machico (Madeira), 1 May 2002, suggest that it is a scarce winter visitor and passage migrant. **Canaries** Recorded from all main islands and in all months, but particularly regular in Nov–Feb. **Cape Verdes** Recorded from all main islands except Brava. Seen in all months and is a regular visitor in Oct–May.

SPUR-WINGED LAPWING (PLOVER) *Vanellus spinosus* Plate 30 (5)
Sp. Avefría Espolada **P.** Tui-tui-ferrão (Abibe-esporado)

Monotypic. **L** 25–28 cm (10–11″); **WS** 69–81 cm (27–32″). **IDENTIFICATION** Sexes similar. Both adult and juvenile recognised by striking black, white and brown plumage. Cap, drooping crest, breast, flanks, upper belly, most of tail and vertical line from chin to breast are black. Rest of face, neck, lower belly, vent, lower rump and uppertail-coverts are white. Upperparts and upperwing-coverts are a uniform, rather pale brown. In flight brown upperwings with a white diagonal bar from the carpal joints to inner secondaries and black flight feathers, white underwing-coverts and black flight feathers. Juvenile resembles adult but has black areas tinged brown and broad buff fringes to upperparts. **VOICE** Typical alarm is a rapidly repeated, sharp, metallic *pitt* or *tick*. **RANGE** Breeds in isolated pockets in Greece, Turkey, the Middle East, the Nile Valley and across C Africa. More northerly populations migrate but mostly resident and dispersive. **STATUS** Extreme vagrant. **Cape Verdes** One, Tarrafal (São Nicolau), 11 Apr 2001.

SOCIABLE LAPWING (PLOVER) *Vanellus gregarius* Plate 30 (6)
Sp. Avefría Sociable **P.** Abibe-sociável (Abibe-gregário)

Monotypic. **L** 27–30 cm (10½–12″); **WS** 70–76 cm (27½–30″). **IDENTIFICATION** Sexes similar. Adult in breeding plumage unmistakable with black crown; white forehead and supercilia meeting in a shallow V on nape, and buffy head-sides and hindneck. Underparts greyish-brown on breast with a black and chestnut belly, and white vent and undertail. Upperparts predominantly grey-brown. In flight reveals black primaries, white secondaries, white uppertail-coverts and tail with black subterminal patch; underwing predominantly white with black primaries. Non-breeding adult loses belly coloration and head pattern more subdued. Juvenile resembles non-breeding adult but has scaly upperparts and dark streaking on breast. **VOICE** Relatively silent away from breeding areas. **RANGE** Breeds in steppes of C Asia. Winters mostly in NE Africa, Pakistan, N India, Middle East and the lower Tigris/Euphrates basin. **STATUS** Extreme vagrant. **Canaries** One, San Andrés, Valverde (El Hierro), 10–12 Feb 1992.

WHITE-TAILED LAPWING (PLOVER) *Vanellus leucurus* Plate 30 (7)
Sp. Avefría Coliblanca **P.** Abibe-de-cauda-branca

Monotypic. **L** 26–29 cm (10–11½″); **WS** 67–70 cm (26½–27½″). **IDENTIFICATION** Sexes similar. Long yellow legs, plain face, striking black-and-white upperwing pattern and all-white tail are diagnostic. Unlikely to be confused, but in flight resembles Sociable Plover; however white on upperwing is more extensive and tail lacks black band of latter. Adult has rather plain brown upperparts, head and upper breast, greyish lower breast, and white belly and vent. Juvenile differs in having breast whitish, mottled grey, and upperparts coarsely marked with dark subterminal bars or dark feather centres. **VOICE** Rather silent away from breeding areas, but gives a *pee-wick* rather similar to Northern Lapwing (quieter and less harsh). **RANGE** Breeds in W Asia, mostly around the Caspian and in Euphrates/Tigris basin. Winters in NE Africa, Pakistan and N India. **STATUS** Extreme vagrant. **Canaries** One, Las Galletas (Tenerife), 18 Nov 1978.

NORTHERN LAPWING *Vanellus vanellus* Plate 30 (8)
Sp. Avefría Europea **P.** Abibe (Abibe-comum)

Monotypic. **L** 28–31 cm (11–12″); **WS** 82–87 cm (32–34″). **IDENTIFICATION** Sexes similar. Adult in breeding plumage has glossy bronze-green upperparts, long black crest, black crown and front of head, black stripe below eye, the rest of head-sides white, broad black band across upper breast and rest of underparts white, except pale orange undertail. ♀ is less well marked on face than ♂. Non-breeding adult has buff head-sides and white chin and throat with some buff fringes to scapulars and coverts. Juvenile resembles non-breeding adult but has a shorter crest and more extensive buff fringes on upperparts. **VOICE** Calls include a rather plaintive, disyllabic *pee-wit* or *wee-ip* and loud, shrill *cheew*. The well-known territorial call is unlikely to be heard in our region. **RANGE** Breeds from Europe across C Asia to E China. Some are sedentary but most winter south of the breeding range, in S Europe and N Africa, the Middle East, Indus basin and S China. **STATUS** A regular and fairly common winter visitor, except to the Cape Verdes. **Azores** Fairly common winter visitor. **Madeira** Fairly common winter visitor to Madeira, but more scarce and irregular on Porto Santo. Also recorded on the Salvages. **Canaries** Regular in small numbers in winter, and recorded from all main islands. **Cape Verdes** Two records: two, Salinas de Pedra de Lume (Sal), 23–24 Dec 1987; and one, between Rabil and Povoação Velha (Boavista), 20 Feb 1999.

(RED) KNOT   *Calidris canutus*   Plate 31 (1)
Sp. Correlimos Gordo   **P.** Seixoeira

C. c. canutus. **L** 23–25 cm (9–10″); **WS** 57–61 cm (22½–24″). **IDENTIFICATION** Sexes similar. In all plumages the large size (for a *Calidris*), fairly short and straight but deep-based bill, and short greenish legs are useful features. Adult in breeding plumage has face and underparts brick red, except whitish vent with some blackish-brown barring. Upperparts mostly blackish-centred with rufous edging and rump pale greyish-white with darker barring. In flight shows a narrow wingbar on upperwing and dull greyish underwing. ♀ has less uniform underparts than ♂. Non-breeder has grey upperparts with narrow white fringes and black shaft-streaks, and rump whitish barred grey. Underparts white with a grey-streaked breast and few grey chevrons on flanks and vent. Juvenile like non-breeding adult but has variable buffish tone to underparts and brownish tinge to upperparts with black subterminal crescents and white fringes to mantle and wing-coverts, creating a rather scaly appearance. **VOICE** A soft, nasal, monosyllabic *knutt* in flight and when feeding in groups. Lone birds are often rather silent. **RANGE** Breeds discontinuously in arctic. Winters in S USA and S South America, the British Isles and parts of Europe, Africa from Tunisia on west coast to S Africa, S Indonesia and Australasia. **STATUS** Recorded from all archipelagos. **Azores** Accidental but possibly regular in small numbers. **Madeira** Rare but regular passage migrant in small numbers. **Canaries** Scarce, but regular, passage migrant and winter visitor, with records from all main islands except La Gomera. **Cape Verdes** Eight records: one reported, near Vila do Maio (Maio), 20 Nov 1897; one, Pedra Badejo lagoons (Santiago), 9 Nov 1988; one same place, 2 Mar 1990; one, Ribeira de Agua (Boavista), 16 Jul 1997; two, Sal Rei (Boavista), 7 Mar 1999; two, Mindelo sewage farm (São Vicente), 12 Mar 2000; seven, Sal Rei (Boavista), 22 Mar 2001; and three, Rabil lagoon (Boavista), 4–17 Mar 2002.

SANDERLING   *Calidris alba*   Plate 31 (2)
Sp. Correlimos Tridáctilo   **P.** Pilrito-das-praias (Pilrito-sanderlingo)

Monotypic. **L** 20–21 cm (8″); **WS** 40–45 cm (16–17″). **IDENTIFICATION** Sexes similar. Found on sandy beaches, where feeds by habitually running back and forth following action of waves. Adult in breeding plumage has chestnut head, upper mantle and breast, with dark streaking on head and upper mantle and dark spotting on breast. Rest of underparts white. Lower mantle and scapulars black-centred with rufous fringes. In flight reveals prominent white wingbar across bases of secondaries and inner primaries, and white sides to rump and uppertail-coverts. In non-breeding plumage basically grey above and white below, with black bill and legs. In flight retains white wingbar, but most of the rest of the wing is black, apart from the grey median coverts. Juvenile has white underparts and black centres to feathers of upperparts, with whitish spots giving a rather spangled appearance. **VOICE** Flight call a sharp but slightly liquid *twick* or *klip*, sometimes repeated as a quick trill. **RANGE** Breeds discontinuously throughout arctic. Winters on coasts throughout much of the world, south of its breeding range, not as far north in east of range. **STATUS** Regular winter visitor to all archipelagos. **Azores** Fairly common winter visitor recorded from all main islands. **Madeira** Regular winter visitor, and also recorded on the Salvages. **Canaries** Common passage migrant and winter visitor recorded from all main islands; some remain most of the year. **Cape Verdes** Regular on passage and in winter when it is fairly common, and recorded from all main islands except Santo Antão.

SEMIPALMATED SANDPIPER   *Calidris pusilla*   Plate 31 (3)
Sp. Correlimos Semipalmeado   **P.** Pilrito-rasteirinho (Pilrito-semipalmado)

Monotypic. **L** 13–15 cm (5–6″); **WS** 34–37 cm (13½–14½″). **IDENTIFICATION** Sexes similar. In all plumages legs and bill black, the bill length is variable but generally fairly short, blunt-tipped and deep-based, and primary projection is short. Adult in breeding plumage is dullest of the four true stints (the others being Little Stint, Red-necked Stint *C. ruficollis* and Western Sandpiper) with little or no rufous on upperparts and mantle, and scapular lines are faint or lacking. Upperparts dark brown with yellowish-buff fringes, scapulars black-centred and wing-coverts and tertials rather plain grey-brown or brown, lacking any obvious rufous margins. Underparts white with a greyish breast-band marked with dark shaft-streaks, the streaking continuing to the fore flanks. In flight white sides to rump and uppertail-coverts, and a narrow white wingbar. In non-breeding plumage grey above and white below, and very difficult to distinguish from other stints, making structural and vocal differences the most important features. Juvenile also rather dull, lacks prominent V on mantle and shows no chestnut in scapulars or on tertial fringes like Little Stint. Very similar to Little Stint and extreme care should be taken in attempting to identify this species. In all plumages, Semipalmated Sandpiper differs from Little Stint in having partial webbing between toes, a blunt-tipped bill and lower pitched, harsher call. In breeding plumage the relatively pale-centred tertials and coverts without rufous fringes always help distinguish a bright Semipalmatd Sandpiper from a dull Little Stint. In juvenile plumage some can approach brightness of juvenile Little Stint but these can be distinguished by the greyer lower two rows of scapulars with darker anchor-shaped marks. **VOICE** Flight call a rather low-pitched *cherk*, *cherp* or *kreet*, distinctly different from *stit-tit* of Little Stint. **RANGE** Breeds across N North America and winters from S Mexico and S Caribbean to N and E South America. **STATUS** On the Azores this species is an accidental visitor, although almost annual since 1993. Vagrant elsewhere with very few records. **Azores** More than 26 records and probably annual, particularly at Cabo da Praia, Terceira. **Madeira** Two, São Lázaro, Funchal (Madeira), 9 Sep 1988. **Canaries** Three records: juvenile, Tejina (Tenerife), 18 Oct 1995; one moulting from first-winter, El Fraile (Tenerife), 3–5 May 1997; and one, Embalse de Los Molinos (Fuerteventura), 11 May 2000. **Cape Verdes** Two records: one, Rabil lagoon (Boavista), 3 Mar 1999; and one, Mindelo sewage farm (São Vicente), 12 Mar 2000.

WESTERN SANDPIPER *Calidris mauri* Plate 31 (4)
Sp. Correlimos de Alaska **P.** Pilrito-miúdo (Pilrito de Maur)

Monotypic. **L** 14–17 cm (5½–7"); **WS** 35–37 cm (14–14½"). **IDENTIFICATION** Sexes similar. Usually has a rather strikingly long bill but there is some overlap between longest-billed ♀ Semipalmated Sandpiper and shortest-billed ♂ Western Sandpipers, and also has partially webbed toes as in Semipalmated. Adult in breeding plumage has rufous ear-coverts, crown-sides and both upper and lower scapulars. Underparts white, but heavily streaked on breast with arrowhead marks on sides extending to flanks. Non-breeder as other stints, grey above and white below, making structure and call important. However, Western tends to have a streaked breast not shown by Semipalmated Sandpiper. Primary projection of Western is even shorter than Semipalmated. As mentioned, the bill is usually obviously longer, slightly decurved, thinner overall, finer tipped and with a narrower base compared to Semipalmated. Juvenile fairly readily identified as rufous on upper scapulars forms a V bordering the mantle that contrasts with the grey lower scapulars and wing-coverts, and is not shown by any other stints. **VOICE** Flight call higher pitched than in Semipalmated Sandpiper, a rather drawn-out and shrill *jeet* or *cheet*. **RANGE** Breeds in N and W Alaska and extreme NE Siberia, wintering on both coasts of S USA through C America and the Caribbean to N South America. **STATUS** Vagrant with 12 records. **Azores** One, Madalena (Pico), 1 Oct 1978; juvenile, Lajes de Pico (Pico), 21–26 Sep 1988; adult, Cabo da Praia (Terceira), 3–4 Sep 1998, joined by a second on 4th; juvenile, Cabo da Praia (Terceira), 14 Oct 1998; one, Cabo da Praia (Terceira), 18 Oct 2000; one there, 22–23 Oct 2001, and a juvenile there, 16 Oct 2002. **Madeira** One obtained, São Lázaro, Funchal (Madeira), 1 Nov 1979; and one, Funchal (Madeira), 28 Sep–5 Oct 1988. **Canaries★** One, Puerto de la Cruz (Tenerife), 26–27 Dec 1960; and one, El Médano (Tenerife), 27–28 Oct 1991.

LITTLE STINT *Calidris minuta* Plate 31 (5)
Sp. Correlimos Menudo **P.** Pilrito-pequeno

Monotypic. **L** 12–14 cm (5–5½"); **WS** 34–37 cm (13½–14½"). **IDENTIFICATION** Sexes similar. Presence of a prominent V on mantle/scapulars in both breeding and juvenile plumages helps distinguish the species from its American counterparts. Structurally, Little Stint has the longest primary projection of all the dark-legged stints, unwebbed toes and the bill shorter than in Western Sandpiper but with a finer tip than Semipalmated Sandpiper. Adult in breeding plumage has face, cheeks, neck and upper breast rufous with brown streaking, rufous fringes to scapulars and most wing-coverts, and an obvious pale V on sides of mantle. Rest of underparts white with no markings on flanks. In non-breeding plumage almost identical to other dark-legged stints. Juvenile has upperparts with obvious rufous fringes to dark feather centres, a prominent pale V on the mantle-sides and rufous fringes to tertials. Also, on the crown-sides there is a narrow white line that joins the supercilium in front of the eye, giving the impression of a 'split' supercilium, a feature only rarely shown by Semipalmated Sandpiper and never by Western. **VOICE** Flight call is a short *stit* or *stit-it*, and is one of the most useful identification features, particularly for non-breeders. **RANGE** Breeds in the arctic, from Scandinavia to C Siberia and winters mainly in Africa, the Middle East and Indian subcontinent. **STATUS** Recorded from all archipelagos but status varies. **Azores** Accidental visitor with over 20 records. **Madeira** Fewer than five records in the last 50 years (Zino *et al.*1995), with recent records, El Tanque (Porto Santo), 16 Oct 2002 and 14–19 Aug 2003, and Machico (Madeira), 25–27 Sep 2003. **Canaries** Regular passage migrant in small numbers and a scarce winter visitor, but no records on La Gomera. **Cape Verdes** Formerly considered rare but recent records show it to be a regular passage migrant and winter visitor in small numbers.

TEMMINCK'S STINT *Calidris temminckii* Plate 31 (6)
Sp. Correlimos de Temminck **P.** Pilrito de Temminck

Monotypic. **L** 13–15 cm (5–6"); **WS** 34–37 cm (13½–14½"). **IDENTIFICATION** Sexes similar. The only stint to have white tail-sides. Rather dull plumage and usually has pale greenish or yellowish legs. In non-breeding plumage looks similar to a minute Common Sandpiper, and in all plumages tail projects noticeably beyond folded wings, exaggerating this similarity. Never shows pale mantle V of other stints. Adult in breeding plumage has variable number of black-centred mantle feathers and scapulars fringed pale rufous, and rest upperparts rather plain grey-brown. Underparts white with a grey wash on upper breast and gorget of brown streaks. Non-breeder has upperparts uniform grey-brown and underparts white with large patches on breast-sides that sometimes join in the middle. Juvenile resembles adult winter but has upperparts with buff tips and dark subterminal lines, and is generally browner. **VOICE** Flight call a distinctive high-pitched trill, *tirrr*, or more drawn-out *tirrr-tirr-tirr*. **RANGE** Breeds across N Palearctic from Scandinavia to E Siberia and winters mostly in C Africa, the Indian subcontinent and SE Asia. **STATUS** Varies within the region. **Azores** Two records: one obtained, prior to 1888; and six, Mosteiros (São Miguel), Oct 1978. **Madeira** One, El Tanque (Porto Santo), 3 Sep 2002. **Canaries** A rare and irregular passage migrant and winter visitor, most frequent on Lanzarote, Fuerteventura and Tenerife, but also noted on Gran Canaria and La Gomera. **Cape Verdes** Two records: one, Santa Maria saltpans (Sal), 24 Nov 1989; and one, Rabil lagoon (Boavista), 16 Mar 1999.

LEAST SANDPIPER *Calidris minutilla* Plate 32 (1)
Sp. Correlimos Menudillo **P.** Pilrito-anão (Pilrito-minúsculo)

Monotypic. **L** 11–12 cm (4–5″); **WS** 32–34 cm (12½–13½″). **IDENTIFICATION** Sexes similar. The smallest of all the stints. Shares pale legs of Temminck's Stint but differs from latter in having better patterned plumage and grey tail-sides. Non-breeder distinctly browner than dark-legged stints but not as plain as Temminck's. Adult in breeding plumage has dark brown upperparts with rufous and grey fringes, darker than any other stint. Underparts white except brown wash on breast and complete gorget of streaks recalling minute Pectoral Sandpiper. Non-breeding adult grey-brown above with darker but ill-defined feather centres. Streaking on breast is not as heavy as in breeding plumage. Juvenile has narrow mantle V, split supercilium, dark mantle and scapulars fringed with chestnut, and browner tertials and wing-coverts with pale chestnut. Underparts white with breast washed buff and streaked brown. **VOICE** Common call is a soft, high-pitched, shrill, rising *trreeep* often delivered in short series; also gives a lower *prrrt*. **RANGE** Breeds across N North America and winters from S USA to S America north of the Tropic of Capricorn. **STATUS** Recorded from three archipelagos and probably best treated as a vagrant. **Azores** Accidental but probably annual in small numbers, with more than 25 records and most frequently reported from the quarry at Cabo da Praia (Terceira). **Canaries★** Four records: two, Salinas de Janubio (Lanzarote), Jan 1993, and two also there, Jan 1994; one, Salinas de Los Cocoteros (Lanzarote), Jan 1994; and a moulting adult, 2 km west of Tarajalejo (Fuerteventura), 12–22 Aug 1994. **Cape Verdes** Two records: one, Mindelo sewage farm (São Vicente), 9–10 Mar 1996, and another there, 12 Mar 2000.

WHITE-RUMPED SANDPIPER *Calidris fuscicollis* Plate 32 (2)
Sp. Correlimos Culiblanco **P.** Pilrito-de-sobre-branco (Pilrito de Bonaparte)

Monotypic. **L** 15–18 cm (6–7″); **WS** 40–45 cm (16–18″). **IDENTIFICATION** Sexes similar. In all plumages, white uppertail-coverts, long wings projecting well beyond the tail when folded (a feature only shared with Baird's Sandpiper), a relatively short, straight bill; quite distinct, whitish supercilium and paler base to lower mandible. Adult in breeding plumage has pale chestnut wash to crown, ear-coverts and scapular fringes. Underparts white with neat, dark streaks on breast and flanks. Mantle, scapulars and tertials black-centred, fringed paler with chestnut, grey and buff. Coverts grey-brown and edged whitish. Non-breeding adult obviously greyer and plainer than either breeding adult or juvenile. Basically grey above and white below with dark shaft-streaks on upperparts, greyish suffused breast and fine streaks on breast and upper flanks. Juvenile usually has fairly prominent white lines on sides of mantle, chestnut cap and chestnut fringes to dark-centred upper scapulars. Lower scapulars much greyer and have whitish fringes. **VOICE** Flight call very distinctive, a mouse- or bat-like, high-pitched squeak. **RANGE** Breeds in N North America and winters in S South America. **STATUS** Accidental with records from three archipelagos. **Azores** Probably a regular autumn visitor with *c.* 50 records and records from all main islands except Santa Maria and Graciosa. Sometimes seen in small flocks, with 48, 34 and 25+ among those recorded. **Madeira** Juveniles, Funchal (Madeira), 5–6 Oct 1993, and, El Tanque (Porto Santo), 16 Oct 2002. **Canaries** Vagrant/accidental with at least 13 records, most on Tenerife but also recorded twice on La Gomera and Fuerteventura, and once on La Palma and El Hierro. The most recent record was one, Roquito del Fraile (Tenerife), 24–25 Oct 1999.

BAIRD'S SANDPIPER *Calidris bairdii* Plate 32 (3)
Sp. Correlimos de Baird **P.** Pilrito-de-bico-fino (Pilrito de Baird)

Monotypic. **L** 14–17 cm (5½–7″); **WS** 40–46 cm (16–18″). **IDENTIFICATION** Sexes similar. Similar in size and shape to White-rumped Sandpiper but slightly more attenuated, with a finer bill and in all plumages has broad dark central rump and uppertail-coverts, and an all-black bill. Adult in breeding plumage lacks rufous tones of White-rumped and has unmarked flanks. In non-breeding plumage browner on upperparts than White-rumped and has a buffier breast with less distinct streaking. Also supercilium less distinct, shorter and more buff than White-rumped. Juvenile lacks pale mantle lines and is also duller with more scaly upperparts, a less distinct supercilium and buffish tinge to breast. **VOICE** Flight call is a soft, rolling trill, *krreeet*. **RANGE** Breeds from extreme NE Siberia across N

North America to NW Greenland, and winters in W and S South America. **STATUS** Possibly regular in autumn, in very small numbers, on Azores, but certainly a vagrant elsewhere. **Azores** Sixteen records, most from Cabo da Praia (Terceira), but four records from São Miguel and one from São Jorge. **Canaries** Three records: juvenile, Amarilla Golf, then Golf del Sur (Tenerife), 18–21 Oct 1990; one, El Médano (Tenerife), 24 Apr 1992; and adult, Roquito del Fraile (Tenerife), 23 Nov–6 Dec 1992.

PECTORAL SANDPIPER *Calidris melanotos*   Plate 32 (4)
Sp. Correlimos Pectoral   P. Pilrito-de-colete (Pilrito-peitoral)

Monotypic. **L** 19–23 cm (7½–9″); **WS** 42–49 cm (16½–19″). **IDENTIFICATION** Sexes similar. In all plumages heavily streaked breast ends sharply at white belly and has a point in centre. A medium-sized *Calidris* with a short, slightly decurved, bicoloured bill and usually yellowish legs. Looks quite long-winged in flight, with a rather weak upperwing bar and white rump with a broad blackish centre. In breeding plumage upperparts dark brownish with contrasting fringes varying from chestnut to buffish. Pale supercilium is inconspicuous, giving it a slightly capped appearance. ♂ tends to show a darker breast than ♀ and is noticeably larger, but this can only be judged when seen together. In non-breeding plumage adult similarly patterned but distinctly duller and less rufous. Juvenile brighter, the most obvious differences being a prominent white V on the mantle-sides and the more distinct supercilium. Could be confused with a small ♀ Ruff but latter has a much shorter primary projection, the feet extend beyond the tip of the tail in flight and it lacks heavy breast streaking. **VOICE** Flight call a short, trilled *krrrt* resembling Curlew Sandpiper but somewhat deeper. **RANGE** Breeds across N North America and Siberia, wintering in small numbers in Australia and New Zealand, but mostly in S South America. **STATUS** Accidental visitor with records from all archipelagos. **Azores** More than 40 records; noted on all main islands except Graciosa. **Madeira** One, Machico (Madeira), 7 Feb 1962; one, Funchal (Madeira), 27–30 Aug 1994, and at El Tanque (Porto Santo), one 20 Oct 2001, juvenile 19 Sep 2002, and one 24 Sep 2003. **Canaries** More than 25 records but none from La Gomera or La Palma. **Cape Verdes** Two at Tarrafal (Santiago), 16-17 Oct 2001

SHARP-TAILED SANDPIPER   *Calidris acuminata*   Plate 32 (5)
Sp. Correlimos Acuminado   P. Pilrito-acuminado

Monotypic. **L** 17–21 cm (7–8″); **WS** 42–48 cm (16½–19″). **IDENTIFICATION** Sexes similar. Very similar to the preceding species but in all plumages lacks heavy streaking on neck and breast that stops abruptly on upper belly, a feature characteristic of Pectoral Sandpiper. Similar in size and shape to Pectoral but slightly smaller, slightly longer legged, with a flatter crown, shorter neck and shorter, uniformly dark, bill. Pattern and coloration of underparts, unlike Pectoral, highly variable between seasons and ages. Adult in breeding plumage unlike any other *Calidris*, being streaked and spotted on neck and upper breast, becoming chevrons on lower breast and flanks. Head, neck and upper breast suffused buff but rest of underparts white. Upperparts dark-centred with chestnut and buffish fringes adding to overall scaly appearance. Supercilium is white, more distinct than on Pectoral, the rufous cap is brighter and more distinct, and it has a much more prominent white eye-ring. Non-breeding adult much duller and less well marked on underparts, the markings being reduced to some fine streaking on foreneck and breast, the latter with a rather greyish suffusion. Supercilium still distinct as is rufous cap, despite being much duller. Juvenile probably the most spectacular *Calidris*, with a rich chestnut crown bordered below by a prominent white supercilium that broadens behind eye, upperparts dark brown with bright chestnut, white or buff fringes to all feathers, and underparts white with a strong orange-buff wash on breast, a gorget of fine streaks across upper breast, a few fine streaks on the breast-sides and often some fine streaking on the rear flanks and undertail-coverts. **VOICE** Flight call a rather subdued *wheeep*, sometimes run together to give a twittering swallow-like call. Calls higher pitched and less reedy than those of Pectoral Sandpiper. **RANGE** Breeds in NE Siberia from the Lena River delta east to Chaun Gulf. Winters mostly in Australia, on New Guinea and in New Zealand. **STATUS** Vagrant with just four records from two archipelagos. **Azores** Adult, Cabo da Praia (Terceira), 23 Aug 2000. **Canaries*** Three records, the first two involving one, Salinas de Janubio (Lanzarote), Jan 1993 and Jan 1994, and then a juvenile, Roquito del Fraile (Tenerife), 9 Nov 1996.

CURLEW SANDPIPER   *Calidris ferruginea*   Plate 32 (6)
Sp. Correlimos Zarapitín   P. Pilrito-de-bico-comprido

Monotypic. **L** 18–23 cm (7–9″); **WS** 42–46 cm (16½–18″). **IDENTIFICATION** Sexes similar. Larger than Dunlin and in flight shows distinctive white rump. Bill also longer and usually more decurved. Adult in breeding plumage has head, neck and most underparts striking brick red. Non-breeder has grey upperparts and underparts white with greyish wash on breast-sides, but size and bill length help distinguish it at this season. Juvenile like non-breeding adult but has browner, scalier upperparts and buffish suffusion to breast. **VOICE** Flight call is a short, soft, trilling *chirrip* or *kirrip*. **RANGE** Breeds in NW and N Siberia, wintering in Africa, Arabia and from

Indian subcontinent through SE Asia and Indonesia to Australia. **STATUS** Recorded from all archipelagos. **Azores** Formerly considered a vagrant but recent records suggest that it is a scarce passage migrant, particularly in autumn. **Madeira** Listed by Zino *et al.* (1995) as occasional: one, El Tanque (Porto Santo), 16 Oct 2002, and one, Machico (Madeira), 18 Sep 2003. **Canaries** Regular passage migrant in small numbers and a rare winter visitor. **Cape Verdes** Fairly common passage migrant and winter visitor, sometimes in flocks numbering in the hundreds.

DUNLIN   *Calidris alpina*   Plate 32 (7)
Sp. Correlimos Común   **P.** Pilrito-de-peito-preta (Pilrito-comum)

C. a. schinzii, *C. a. alpina* and *C. a. arctica*.   **L** 16–22 cm (6–9″); **WS** 38–45 cm (15–18″). **IDENTIFICATION** Sexes similar. Black belly of breeding-plumaged adult is diagnostic. In non-breeding plumage grey upperparts and breast with rest of underparts white. Bill is variable in length but is usually fairly long and distinctly downcurved at the tip. Juvenile has diagnostic spotting on flanks and breast-sides. Races sometimes separable, but only in breeding plumage; *alpina* has rusty-red fringes to upperparts, in *schinzii* these are more yellowish-red and in *arctica* they are paler reddish-yellow and less extensive. Of these three, *alpina* has the longest bill and is the largest, with *schinzii* smaller and shorter billed, and *arctica* has the shortest bill and is smallest. **VOICE** Flight call is a distinctive shrill, reedy *chrrreep*. **RANGE** Breeds across N Holarctic and winters mainly on coasts in N Hemisphere. **STATUS** Regular passage migrant and winter visitor. **Azores** Scarce or accidental visitor with most records on São Miguel and Terceira. **Madeira** Scarce but regular visitor and also recorded on the Salvages. **Canaries** Regular and fairly common passage migrant, and a scarce winter visitor. **Cape Verdes** Rare and irregular passage migrant and winter visitor.

PURPLE SANDPIPER   *Calidris maritima*   Plate 33 (1)
Sp. Correlimos Oscuro   **P.** Pilrito-escuro

Monotypic.   **L** 20–22 cm (8–9″); **WS** 42–46 cm (16½–18″). **IDENTIFICATION** Sexes similar. Slightly larger than Dunlin and easily identified by yellow base to bill and yellowish legs. Very unlikely to be seen in breeding plumage, when it has dark brownish-black upperparts with extensive white and chestnut fringes, a white supercilium and white underparts with heavy dark streaking on breast and flanks. Non-breeding adult has dark slate-grey head, breast and upperparts with greyish fringes to wing-coverts, and white underparts below breast with dark streaking on flanks. Juvenile resembles breeding adult but is less rufous above and has rather neater, scaly appearance to wing feathers. In flight looks very dark apart from narrow white wingbar and white sides to rump. **VOICE** Flight call a short, sharp, liquid *quit* or disyllabic *quit-it*. **RANGE** Breeds discontinuously in the arctic from NE North America and Greenland to W Siberia, and breeds further south in Scandinavia. Winters just south of breeding range in N America, the British Isles, N Europe and Scandinavia. **STATUS** Accidental with records from three archipelagos. **Azores** Twenty-four records, with the majority from São Miguel (16), but also recorded on Terceira (4), Flores (3) and Faial (1). **Madeira** Listed as exceptional by Zino *et al.* (1995). **Canaries★** About ten records with birds seen on Lanzarote, Tenerife, La Gomera and La Palma.

STILT SANDPIPER   *Calidris himantopus*   Plate 37 (3)
Sp. Correlimos Zancolín   **P.** Pilrito-pernilongo (Maçarico-de-perna-longa)

Monotypic.   **L** 18–23 cm (7–9″); **WS** 43–47 cm (17–18½″). **IDENTIFICATION** Sexes similar. Adult in breeding plumage unmistakable given combination of barred underparts, rather dark upperparts with pale fringes to feathers, rusty ear-coverts and crown separated by whitish supercilium; long greenish or yellowish legs, and long, slightly decurved, blunt-tipped bill. Non-breeding adult has rather plain grey-brown upperparts with white fringes to larger wing-coverts and tertials, white below with rather indistinct streaking on breast extending to flanks and undertail-coverts, and a distinct white supercilium. Juvenile has mantle, scapulars and tertials dark brownish-black with cinnamon fringes, but these soon fade to off-white, the underparts whitish with indistinct streaks on breast, flanks and undertail-coverts, the breast washed buff, and a whitish supercilium. In flight and in all plumages has a white rump, no obvious wingbars and the feet project well beyond the tail. **VOICE** Flight call is a rather soft trilling *drrrp* somewhat reminiscent of Curlew Sandpiper, but generally rather silent. **RANGE** Breeds from NE Alaska to N Canada. Winters predominantly in C South America south to Uruguay and NE Argentina. **STATUS** Extreme vagrant. **Azores** Adult, Ribeira Grande (São Miguel), 28 Jan 1984.

BUFF-BREASTED SANDPIPER   *Tryngites subruficollis*   Plate 33 (2)
Sp. Correlimos Canelo   **P.** Pilrito-acanelado (Pilrito-canela)

Monotypic.   **L** 18–20 cm (7–8″); **WS** 43–47 cm (17–18½″). **IDENTIFICATION** Sexes similar. Similar to Ruff and shares small-headed appearance, but is even smaller than ♀ of latter. Easily identified by combination of buff of face and most underparts, short black bill, obvious dark eye, bright yellow legs and scaly brown-and-buff upperparts. Adults in breeding and non-breeding plumages are very similar, but in non-breeding plumage buffy-brown edges to upperpart feathers usually broader with less buff, more brown. Juvenile very similar to adult but underparts usually have less extensive and paler buff on belly and coverts appear scalier. In flight lacks wingbar and white sides to rump (both features of Ruff). **VOICE** Often rather silent but sometimes gives a low, growling call in flight. **RANGE** Breeds across N North America and winters in S South American grasslands. **STATUS** Vagrant/accidental with records from two archipelagos. **Azores** Ten records: six, Horta (Faial), 9 Oct 1979; one, Vila Franca (São Miguel), Sep 1985; one, Praia da Vitória (Terceira), 28–29 Sep 1994; adult, near Lajes do Pico (Pico), 3 Oct 1997; one, Cabo da Praia (Terceira), 9 Oct 1998; juvenile, Cabo da Praia (Terceira), 11–12 Sep 2001; one, Santa Cruz airport (Flores), 16–18 Sep 2001; one, near Serra do Cume (Terceira), 26 Oct 2002; one, Cabo da Praia (Terceira), 22–30 Sep 2003; and juvenile, Cabo da Praia (Terceira), 1 Sep 2004. **Canaries** Thirteen records: one collected, S Tenerife (no other details); two, Embalse de Cigüeña (Tenerife), 29 Sep 1984; one, Costa Teguise Golf Course (Lanzarote), for two weeks in Sep 1985; adult, Golf del Sur (Tenerife), 15–18 Sep 1989; adult, Amarilla Golf (Tenerife), 5–13 Oct 1990; juvenile, Golf del Sur (Tenerife), 21–23 Sep 1991; juvenile, Embalse de Cigüeña (Tenerife), 13–26 Sep 1992; juvenile, Golf del Sur (Tenerife), 12–21 Oct 1995; juvenile, Roquito del Fraile (Tenerife), 9 Nov 1996; one, same place, 19 Sep–23 Oct 1998, with another there, 14–15 Sep 1999; one, Amarilla Golf (Tenerife), 17 Sep 1999, and another there, 6–16 May 2002.

RUFF *Philomachus pugnax*   Plate 33 (3)
Sp. Combatiente   P. Combatente

Monotypic. **L** ♂ 26–32 cm (10–12½"), ♀ 20–25 cm (8–10"); **WS** ♂ 54–58 cm (21–23"), ♀ 48–52 cm (19–20½"). **IDENTIFICATION** Sexually dimorphic. Large *Calidris*-like wader with much variation in plumage, both individually and sexually. Very noticeable size difference between sexes. In all plumages smallish head, fairly long neck, long legs, relatively short, slightly decurved bill and slightly humpbacked appearance are useful pointers to identification. ♂ in breeding plumage has variably coloured head and neck plumes, ruffs, which make the species unmistakable. In non-breeding plumage ♂ resembles ♀ but is distinctly larger. ♀, otherwise known as a reeve, generally dark blackish-brown on wings and mantle with all feathers fringed buff, breast buff of variable intensity, and belly and undertail white. In flight looks quite long-winged with a narrow white wingbar and white oval patches on sides of rump formed by unusually long uppertail-coverts. Juvenile differs from adult in having more scaly upperparts, a stronger buff coloration on head, neck and breast, and legs greenish or brownish. **VOICE** Usually silent but occasionally utters low grunting sounds. **RANGE** Breeds across N Palearctic, wintering mainly in sub-Saharan Africa but also in the British Isles, discontinuously around Mediterranean, in Arabia and the Indian subcontinent. **STATUS** Regular passage migrant and rare winter visitor with records from all archipelagos. **Azores** Accidental with *c.* 25 records and has been almost annual since 1997. **Madeira** An occasional visitor. **Canaries** Regular, but uncommon, passage migrant and a rare and irregular winter visitor. **Cape Verdes** More than 20 records, and considered to be an uncommon, but regular, passage migrant and winter visitor.

JACK SNIPE *Lymnocryptes minimus*   Plate 33 (4)
Sp. Agachadiza Chica   P. Narceja-galega

Monotypic. **L** 17–19 cm (6½–7½"); **WS** 38–42 cm (15–16½"). **IDENTIFICATION** Sexes similar. The smallest snipe, with a noticeably short bill; lacks central crown-stripe of other snipes and bobs continually whilst feeding. Flight weaker than Common Snipe and if flushed take-off is less explosive, much closer to observer and bird usually returns to a point near the take-off. Tail lacks any white on sides and, uniquely in snipes, is wedge-shaped. In all plumages most obvious features are the two broad yellowish-buff stripes on mantle-sides, the prominent split supercilium, the dark crescent below the eye and the streaked (not barred) flanks and breast. Juvenile distinguished from adult by slightly more pointed tail feathers and less distinct streaking on undertail, but both features are difficult to assess correctly. **VOICE** Rather silent away from breeding grounds, but rarely gives an almost inaudible *gah* or *scah* when flushed. **RANGE** Breeds across NC Palearctic from Scandinavia to Siberia, and winters discontinuously across British Isles, Europe, Africa north of the equator, the Indian subcontinent and SE Asia. **STATUS** Recorded from all archipelagos but rare. **Azores** Only two definite records: one obtained, Vila Franca do Campo (São Miguel), undated; and one, Poço da Queiré, Sete Cidades (São Miguel), Aug 1961, with a probable, Lagoa de Peixe, Arrifes (São Miguel), 24 Mar 1962. **Madeira** Vagrant: records include one shot, Machico (Madeira), 15 Mar 1889, and a ♂ obtained, San Antonio (Madeira), 6 Nov 1956. **Canaries** Rare and irregular passage migrant and winter visitor, unrecorded from La Gomera and unconfirmed on Lanzarote. **Cape Verdes** One, west of Vila do Maio (Maio), 26 Aug 1986.

COMMON SNIPE *Gallinago gallinago*   Plate 33 (5)
Sp. Agachadiza Común   **P.** Narceja (Narceja-comum)

G. g. gallinago. **L** 25–27 cm (10–10½″); **WS** 44–47 cm (17–18½″). **IDENTIFICATION** Sexes and all plumages similar. Easily recognised by long, straight bill, striped head pattern, rather dumpy shape, fairly short legs and complicated brown, white and blackish plumage. Crown dark brown with distinct buff central stripe, prominent supercilium, buffy stripes on mantle, streaked brownish breast and white belly and flanks with dark blackish-brown barring on latter. Juvenile has more neatly fringed wing-coverts, broken only by dark shafts, whereas adult has broad shaft-streaks. **VOICE** When flushed gives an abrupt harsh *scaaap*. In display a monotonous repeated *chip-per* and a short, sharp *chick*. When diving during display flight produces a hollow, throbbing sound created by air passing through tail feathers. **RANGE** Breeds across Palearctic from W Europe to E Siberia, Kamchatka and Sakhalin. Winters in S Europe, C Africa, the Middle East, Indian subcontinent and SE Asia. **DISTRIBUTION** Breeds only in the Azores. **BREEDING** Very little known in our region. **HABITAT** Prefers wet, marshy ground but also areas of vegetation near open water. **STATUS** Recorded from all four archipelagos but status varies from resident breeder to vagrant. **Azores** Resident breeder, but European birds also visit the islands regularly in winter. Breeds on all main islands except Corvo, Graciosa and Santa Maria, but population small and endangered due to hunting and habitat loss. **Madeira** Uncommon winter visitor. **Canaries** Regular and not uncommon on passage and in winter with records from all main islands. **Cape Verdes** Vagrant with 12 records. First recorded, Mindelo sewage farm (São Vicente), 1996, with six further records at this site, as well as records from Sal, Boavista (2) and Santiago (2).

WILSON'S SNIPE *Gallinago delicata*   Plate 33 (6)
Sp. Agachadiza Americana   **P.** Narceja-americana

Monotypic. **L** 25–27 cm (10–11″); **WS** 44–47 cm (17–18½″). **IDENTIFICATION** Sexes and all plumages similar and very similar to corresponding plumages of Common Snipe. Wilson's Snipe always has 16 tail feathers but Common Snipe usually has 14 (sometimes 18). This species lacks warm tones of Common Snipe and is much colder, with the dark areas blacker and pale areas whiter. Supercilium can show a distinct bulge in front of eye, whereas that of Common Snipe is more parallel-sided. Axillaries and underwing-coverts have dark barring equal to or broader than white barring, producing a darker looking underwing. White on trailing edge of secondaries on upperwing much narrower, and fringes of scapulars lack contrast shown between inner and outer edges in Common Snipe; they can even appear scalloped when held relaxed. The upper breast is streaked on a brown ground colour, whereas the lower breast and flanks are heavily barred on an off-white background. Juvenile almost identical to adult but has complete white fringes to coverts broken only by dark shafts. **VOICE** Very similar to Common Snipe but possibly averages slightly lower in pitch. **RANGE** Breeds across N America, mostly north of 40°. Winters mostly from C USA to N South America. **STATUS** Vagrant with records from two archipelagos. **Azores★** Six recent records: one, Corvo, 2 Oct 1988; one, near Lagoa Comprida (Flores), 8 Nov 1998; one, near Lagoa Comprida (Flores), 12 Oct 1999; one, Lagoa Seca (Flores), 13 Sep 2001; one, Cabo da Praia (Terceira), 28 Oct 2001; and one, Lagoa do Ginjal (Terceira), 26 Oct 2002. **Canaries** Two records: one, Embalse de Valle Molina (Tenerife), 9 Dec 1998; and one, Roquito del Fraile (Tenerife), 20–22 Sep 2002.

GREAT SNIPE *Gallinago media*   Plate 33 (7)
Sp. Agachadiza Real   **P.** Narceja-real

Monotypic. **L** 27–29 cm (10½–11½″); **WS** 47–50 cm (18½–20″). **IDENTIFICATION** Sexes and all plumages similar. A large, bulky snipe that lacks prominent white trailing edge on secondaries of Common Snipe and has a relatively short bill. At rest, the breast and flanks are more intensely barred than on Common Snipe, often extending onto belly, and white tips to wing-coverts form obvious white lines on folded wing. In flight underwing more heavily barred, the white tips to wing-coverts form narrow white wingbars not present in Common Snipe, and the tail has prominent white sides, but the latter are only truly visible when the tail is spread on take-off or landing. Juvenile, although similar to adult, has less prominent pale tips to upperwing-coverts and white outer tail feathers barred dark brown. **VOICE** When flushed sometimes gives a quiet, single or double, croaking *etch-etch*. **RANGE** Breeds from Scandinavia and E Europe east to Kazakhstan and C Russia. Winters predominantly in sub-Saharan Africa. **STATUS** Vagrant with two records. **Madeira** One obtained, Madeira, by Padre Schmitz *c.* 1900, but exact date unknown. Identified by Guillemard, the specimen no longer exists. **Canaries★** One, Laguna (Tenerife), prior to 1893.

SHORT-BILLED DOWITCHER *Limnodromus griseus*   Plate 34 (1)
Sp. Agujeta Gris   **P.** Maçarico-de-bico-curto (Maçarico-escolopáceo-americano)

Subspecies unknown, but *L. g. hendersoni* and *L. g. griseus* are the most likely to occur. **L** 25–29 cm (10–11½″); **WS** 45–51 cm (18–20″). **IDENTIFICATION** Rather snipe-like in size but actions and plumage suggest cross between snipe and a godwit. Unlikely to be mistaken for any other species except Long-billed Dowitcher. Adult *hendersoni* in breeding plumage has rather washed-out orangey underparts and only a faintly spotted breast, little or no barring on flanks and spotted undertail-coverts. Subspecies *griseus* has stronger markings on breast and undertail-coverts, but belly whitish. Non-breeding adult basically grey above and whitish below with throat and upper breast grey and finely speckled, and flanks barred grey. Juvenile dark blackish-brown above with broad rusty-buff fringes to feathers, particularly scapulars and tertials, which usually have conspicuously pale internal markings. Underparts have an almost orange wash to foreneck and breast, with distinct dark spotting on breast-sides. Although distinctions between juvenile Short- and Long-billed Dowitchers can be fairly obvious, some birds can show characters of the other species and great care is needed in identifying such individuals. **VOICE** A rather rapid mellow *tu-tu-tu* or *chu-du-du*, similar to flight call of Turnstone and unlike equivalent calls of Long-billed Dowitcher. **RANGE** Breeds in S Alaska and in C and E Canada. Winters from S USA to N South America. **STATUS** Vagrant with only five records, all from one archipelago. **Azores** One, Lagoa de Furnas (São Miguel), 2 Nov 1978, and presumably the same, Ribeira Quente, Dec 1978; juvenile, Lagoa Branca (Flores), 5–6 Sep 1998; juvenile, Cabo da Praia (Terceira), 13–17 Sep 1999; another juvenile there, 1–17 Oct 2000; and a non-breeding adult there, 3 Sep 2003–18 Feb 2004 at least.

LONG-BILLED DOWITCHER   *Limnodromus scolopaceus*   Plate 34 (2)
Sp. Agujeta Escolopácea   **P.** Maçarico-de-bico-comprido (Maçarico-escolopáceo-europeu)

Monotypic. **L** 24–26 cm (9½–10″); **WS** 46–52 cm (18–20½″). **IDENTIFICATION** Sexes similar but ♀ averages larger and longer billed. Unlikely to be confused with any other species except very similar Short-billed Dowitcher, from which most easily distinguished by call. Adult in breeding plumage differs from Short-billed in generally having more extensive orange-red underparts, more densely spotted foreneck and barred upper breast, and coloration of underparts darker red. Non-breeding adults even harder to separate, but usually grey of upperparts and breast slightly darker, with no fine speckling on breast. Juvenile easiest to separate from Short-billed, as scapulars, inner wing-coverts and tertials have narrower reddish-buff fringes and any internal markings are usually lacking or at most very inconspicuous. In all plumages pale bars on tail are never broader than the dark bars, and are usually much narrower. However, the two species can prove impossible to separate unless the call is heard. **VOICE** A high-pitched strident *keek* often repeated in a short accelerating series, *keek-keek-keek-keek-keek*. **RANGE** Breeds in NW North America and NE Siberia, wintering mainly in S USA and N Central America. **STATUS** Vagrant with two records. **Azores** A juvenile dowitcher calling like this species, flushed Cabo da Praia (Terceira), 1 Oct 2000, but not relocated. **Madeira** A dowitcher sp. recorded on the Salvages. **Canaries*** First-winter, Roquito del Fraile (Tenerife), 4 Jan–3 May 1986; and juvenile, Embalse de Cigüeña (Tenerife), 21–24 Sep 1996.

(EURASIAN) WOODCOCK   *Scolopax rusticola*   Plate 33 (8)
Sp. Chocha Perdiz   **P.** Galinhola

Monotypic. **L** 33–35 cm (13–14″); **WS** 56–60 cm (22–23½″). **IDENTIFICATION** A stocky, pigeon-sized snipe of wooded habitats. All plumages similar; an intricate mix of black, white and brown (of various shades) bars, vermiculations and spots. When flushed flies away with zigzag action, revealing much rufous on rump and tail. Display flight (roding) has rather jerky, flicking wingbeats and is performed just above treetops at dusk. **VOICE** When roding gives series of 2–5 guttural croaks followed by a high-pitched sneeze, *quorr-quorr-quoro pissp*, repeated constantly but broken occasionally for brief intervals. When flushed often silent but sometimes gives a harsh snipe-like *schaap*. **RANGE** Breeds from W Europe across the Palearctic to Sakhalin and Japan. Populations on Atlantic Islands resident but throughout rest of range mostly migratory. In the west, some reach N Africa in winter, although it is rare there. **DISTRIBUTION** Resident breeder in the Azores, where breeds on all islands, in Madeira where restricted to the main island of Madeira, and in the Canaries, where absent from Lanzarote and Fuerteventura. **BREEDING** Prolonged season, with display starting Jan and young reported Feb–Jul, suggesting the species is double-brooded. Clutch normally 3–4 but five recorded. Nest a shallow depression among leaf litter, often placed close to an old tree trunk. **HABITAT** Confined to well-vegetated areas where it feeds in leaf litter, and is difficult to observe except when roding. **STATUS** Breeds on three archipelagos. **Azores** Reported to be declining and breeding is only sporadic on Santa Maria and Graciosa. **Madeira** Also appears to be declining. **Canaries** Appears to be declining on many islands, except for La Gomera where it is still fairly common. Migrants reported from Alegranza and Fuerteventura.

BLACK-TAILED GODWIT   *Limosa limosa*   Plate 34 (3)
Sp. Aguja Colinegra   **P.** Milherango (Maçarico-de-bico-direito)

L. l. limosa and *L. l. islandica*. **L** 36–44 cm (14–17½″); **WS** 70–82 cm (27½–32″). **IDENTIFICATION** Sexes similar but ♀ larger and longer billed, and in breeding plumage the underparts coloration of the ♀ is paler and less extensive. In all plumages distinguished from Bar-tailed Godwit by longer tibia, straighter bill, prominent white rump, black tail and white wingbar on upperwing. Adult in breeding plumage has throat to upper belly brick red, rest of underparts white barred black on flanks, and upperparts dark blackish-brown with chestnut and cinnamon markings. Non-breeders have the upperparts grey-brown, the foreneck and breast grey and rest of underparts white with some mottling on flanks. Juvenile resembles non-breeding adult but has darker upperparts with buffish fringes to scapulars and tertials, and a more buffy-orange neck and breast. **VOICE** Most common call is a sharp *kip* sometimes repeated in series. **RANGE** Breeds in Iceland and W Europe to W Asia, with isolated populations in E Asia, and winters in the British Isles, W Europe, around the Mediterranean, sub-Saharan Africa, India, SE Asia, Indonesia and Australia. **STATUS** Recorded from all archipelagos but the status varies. **Azores** Scarce passage migrant and winter visitor. **Madeira** An occasional visitor, the most recent records being one, Machico (Madeira), 25 Sep 2003, and one, Lugar de Baixo (Madeira), 27 Nov–1 Dec 2003. Also recorded on the Salvages. **Canaries** A regular passage migrant in small numbers, but a rare winter visitor, and recorded from all main islands. **Cape Verdes** Five records: two, Baia do Curralinho (Boavista), 29 Oct 1897; one, Baia das Gatas (São Vicente), 16 Jan 1966; one, west of Povoação Velha (Boavista), 2 Dec 1986; two, Pedra de Lume saltpans (Sal), 30 Aug & 30 Oct 1988; and 16, near Vila de Maio (Maio), 22 Oct 2000.

BAR-TAILED GODWIT *Limosa lapponica* Plate 34 (4)
Sp. Aguja Colipinta **P.** Fuselo

L. l. lapponica. **L** 37–41 cm (14½–16″); **WS** 70–80 cm (27½–31½″). **IDENTIFICATION** Sexes similar in non-breeding plumage but ♀ has a longer bill. Similar to Black-tailed Godwit but easily distinguished in all plumages by the barred tail and lack of white wingbar. Other differences include a more upcurved bill, shorter tibia and white back. Adult ♂ in breeding plumage has entire underparts rich rufous, whereas ♀ is much paler with the colouring often confined to the foreneck and breast. Upperparts dark with chestnut fringes to mantle and scapulars, whilst the brightest ♀♀ may show some chestnut fringing, but it is more restricted and often less intense. Adult in non-breeding plumage has brownish-grey upperparts with darker feather centres creating a streaked appearance. Breast and neck have a greyish-brown wash and some fine dark streaking, but otherwise rest of underparts whitish. Juvenile resembles non-breeding adult but upperparts browner with bright buff fringes, and breast and neck more brownish-buff. **VOICE** In flight a sharp *ke-vick* or *kik* and a more nasal *kurruc*. **RANGE** Breeds from N Scandinavia discontinuously across N Russia and in W Alaska. Winters in the British Isles, parts of Europe, Africa, Arabia and India, east through Indonesia to Australia and New Zealand. **STATUS** Recorded from all archipelagos and is a regular visitor. **Azores** Scarce and irregular passage migrant and winter visitor. **Madeira** Scarce passage migrant, listed by Zino *et al.* (1995) as occasional. **Canaries** Regular passage migrant and winter visitor with records from all main islands. **Cape Verdes** Regular winter visitor, in small numbers, to eastern islands of Sal, Boavista and Maio, but also recorded on Santiago, Fogo and São Vicente.

WHIMBREL *Numenius phaeopus* Plate 34 (6)
Sp. Zarapito Trinador **P.** Maçarico-galego

N. p. phaeopus and *N. p. hudsonicus*. **L** 40–46 cm (16–18″); **WS** 76–89 cm (30–35″). **IDENTIFICATION** Sexes similar but ♀ larger and longer billed. A medium-sized curlew with a dark crown, obvious pale central crown-stripe and prominent supercilium. Adult in breeding plumage dark olive-brown above with most feathers notched, spotted or fringed whitish or buffish. Underparts whitish with a buff suffusion on breast and neck, and dark streaking on much of the neck, breast and flanks. In flight reveals a white back and rump, with a dark-barred tail. Underwing-coverts and axillaries white with variable narrow brown barring. Non-breeding adult similar and juvenile resembles adult but is usually more buff on breast and has more extensive buff notches, spots and fringes on upperparts. American race *hudsonicus* is similar, but most easily identified in flight, when brown rump and lower back clearly visible. Other features that help distinguish *hudsonicus* from *phaeopus* include a bolder head pattern, the generally buffier tone (particularly of the underparts), the buff rather than white basal colour to the underwing with well-barred axillaries, and the less distinct breast streaking. Juvenile *hudsonicus* has browner upperparts than *phaeopus*, and more buff on underparts, with a similar underwing to the adult. **VOICE** Common call is a far-carrying series of fast, rippling, mellow whistles *bibibibibibi…* or *pupupupupupupu*. **RANGE** Nominate breeds discontinuously across N Palearctic and winters on coasts throughout most of Europe, Africa, Asia and Australia, mainly south of 30°N; *hudsonicus* breeds in Alaska and N Canada, wintering from S North America to Tierra del Fuego and S Brazil. **STATUS** Recorded from all four archipelagos. **Azores** Nominate regular, in small numbers, as a passage migrant and winter visitor. Records of *hudsonicus* include one, Oct 1985; one, Cabo da Praia (Terceira), 1 Nov 2001; another there, 9 Oct 2002; another there, 6 Sep–18 Nov 2003; and two, Ponta Delgada (São Miguel), 20 Dec 2003. Le Grand stated that he saw some *hudsonicus* during his extended stay in the islands (late 1970s until mid 1980s), but failed to provide any further details. **Madeira** Encountered in all months but is more common on passage and most numerous in winter. Also recorded from the Salvages. **Canaries** Regular passage migrant and winter visitor, but encountered all year. **Cape Verdes** Regular passage migrant and winter visitor. There has been a single record of *hudsonicus* in the archipelago; one, Preguiça (São Nicolau), 15 Feb 1991.

SLENDER-BILLED CURLEW *Numenius tenuirostris* Plate 34 (5)
Sp. Zarapito Fino **P.** Maçarico-de-bico-fino

Monotypic. **L** 36–41 cm (14–16″); **WS** 80–92 cm (31½–36″). **IDENTIFICATION** Sexes similar but ♀ is slightly larger and longer billed. Difficult to separate from congeners, but in all plumages is marginally smaller than Whimbrel, the foreneck and breast are whiter than either Whimbrel or Eurasian Curlew, with the dark streaks being more sharply defined. Also, marks on the flanks are more rounded or heart-shaped than in either confusion species. Compared to both Whimbrel and Eurasian Curlew, the bill is shorter, less decurved, thinner overall and has a finer, more pointed tip. Adult in breeding plumage has a white ground colour to the head, neck and underparts, with streaks on the head, neck and breast. Facial pattern more pronounced than that of Eurasian Curlew but not as bold as in Whimbrel. The lower-breast-sides and fore flanks are marked with distinct blackish spots. In flight reveals white from back to tail, with some dark streaking on rump and uppertail, and tail barred dark brown. Underwing-coverts and axillaries almost pure white with very little evidence of dark barring. Non-breeding adult very similar to breeding-plumaged birds, but has reduced flank spotting. Juvenile resembles adult but spots on flanks replaced by streaks and tertials have more distinct triangular pale notches. **VOICE** A short *cour-lee*, similar to Eurasian Curlew but weaker, faster and higher pitched with a less guttural quality. Also a shriller *cue-ee* given in alarm and unlike any call of Eurasian Curlew. **RANGE** Breeding area unknown, but is somewhere in Russia, and wintering areas also unknown, although tiny numbers wintered in Morocco in the early 1990s. Very close to extinction. **STATUS** Vagrant with five uncertain records. **Azores** One seen with three Whimbrel, Ilhéu de Vila Franca (São Miguel), Mar 1979; and one with two Eurasian Curlew, Mosteiros (São Miguel), 28 Jan 1981. Also, a bird seen by Sr Álvares Cabral, in early Oct 1964, at Mosteiros (São Miguel), is believed to have been this species (Bannerman & Bannerman 1966). **Canaries★** Three unconfirmed records: one reported, Lanzarote, early 1970s; one, Janubio (Lanzarote), 5 Sep 1976; and with four Eurasian Curlew, Corralejo (Fuerteventura), 9 Nov 1976.

EURASIAN CURLEW *Numenius arquata* Plate 34 (7)
Sp. Zarapito Real **P.** Maçarico-real

N. a. arquata. **L** 50–60 cm (20–23½″); **WS** 80–100 cm (31½–39″). **IDENTIFICATION** Sexes similar but ♀ is larger and longer billed. Distinctive and easily recognised by its large size (the largest wader in the region) and long, decurved bill. Could be confused with Whimbrel but is larger with a plainer face and lacks the obvious crown-stripe of the latter. Adult in breeding plumage has rather variable ground colour and intensity of streaking, but is generally grey-brown, obviously streaked or blotched darker, with a buffy-brown head, neck, breast and upperparts, and rather plain face. In flight reveals white rump and lower back, and predominantly white underwing and axillaries with some dark brown barring. Non-breeder similar but less bright, the buff tones being replaced by duller brown. Juvenile initially identifiable by combination of shorter bill, strong buff notches and fringes to upperparts, buffier head and neck, and more lightly streaked flanks. First-winter difficult to age unless have some retained juvenile scapulars or tertials. **VOICE** Most common and familiar call is far-carrying rising *cour-lee*, from which species acquires its common name. Also a rapid *tyuyuyuyu* in alarm. **RANGE** Breeds from the British Isles, discontinuously across C and N Europe to Siberia and C Asia. Winters (mainly on coasts) in Europe, Africa, Arabia, islands in Indian Ocean, and S and E Asia. **STATUS** Recorded from all archipelagos but **STATUS** varies. **Azores** Accidental, with *c.* 20 records. **Madeira** Listed as occasional and also recorded from the Salvages. **Canaries** Regular but scarce passage migrant and winter visitor, more frequent on the eastern islands. **Cape Verdes** Vagrant with only seven records: one, near Santa Maria (Sal), 23 Nov 1989; one, south of Vila do Maio (Maio), 2 Apr 1991; one, Costa de Boa Esperança (Boavista), 29 Jan 1995; one, Ponta do Sol (Santo Antão), 1 Feb 1999; two, near Curral Velho (Boavista), 20 Feb 1999; one, Raso, 21 Apr 1999; and one, Curral Velho (Boavista), 18 Mar 2002.

UPLAND SANDPIPER *Bartramia longicauda* Plate 30 (9)
Sp. Correlimos Batitú **P.** Maçarico-do campo

Monotypic. **L** 28–32 cm (11–12″); **WS** 64–68 cm (25–27″). **IDENTIFICATION** Sexes similar. Quite distinctive with a proportionately small head, long tail (which extends well beyond folded wingtips) and a rather short, almost straight, bill. The yellow legs remove any confusion with small *Numenius* species as yet unrecorded in our region, and unlikely to be confused with any other species. In all plumages generally brown above and white below, with heavily barred flanks and a plain face; the relatively thin neck is also heavily streaked darker. Juvenile distinguished from adult by scaly upperparts, the wing-coverts brown-centred with pale fringes, flanks less well marked and tertials dark brown with buff notches, quite unlike the olive-brown with dark bars of adult. **VOICE** Calls include a piping, whistled *quip-ip-ip-ip* with the last syllable on a lower pitch, and a liquid *pulip-pulip*, both given in flight.

RANGE Breeds over most of C North America and winters in S America, mostly between S Brazil and S Argentina. **STATUS** Vagrant with only eight records from three archipelagos. **Azores** Immature obtained, Ribeira das Tainhas (São Miguel), 25 Jan 1966; adult, Populo (São Miguel), 2 Feb 1979; immature, Fajã Grande (Flores), 6 Nov 1998; immature, Ribeira Seca (São Miguel), 3 Oct 1999; and immature, Ponta Delgada (Flores), 15–18 Oct 2000. **Madeira** Two records of birds shot, Caniço de Baixo (Madeira), 15 Oct 1973 and 12 Nov 1973. **Canaries** Adult Amarilla Golf (Tenerife), 20–22 Sep 2002.

SPOTTED REDSHANK   *Tringa erythropus*   Plate 35 (1)
Sp. Archibebe Oscuro   **P.** Perna-vermelha-bastardo (Perna-vermelha-escuro)

Monotypic. **L** 29–32 cm (11½–12½"); **WS** 61–67 cm (24–26½"). **IDENTIFICATION** Sexes similar. Sooty-black breeding plumage with fine white spotting, particularly on scapulars, tertials and wing-coverts, unique among waders. In flight reveals striking white back and upper rump. In non-breeding plumage could be mistaken for Common Redshank, but Spotted Redshank is slightly larger with a much longer, slimmer bill and has a fairly prominent, but short, supercilium. In flight does not show obvious white secondaries like Common Redshank, instead upperwing is mainly grey with some pale notches and bars on secondaries and larger coverts. Juvenile distinguished from non-breeding adult in being darker above with heavier white markings, and underparts more extensively and heavily barred. **VOICE** Common flight call a rather sharp *chewit* that falls then rises in pitch. **RANGE** Breeds in a narrow band across N Scandinavia and N Russia. Winters discontinuously from W Europe and W Africa across Asia to Vietnam and E China. **STATUS** Recorded from all four archipelagos but status varies. **Azores** At least eight records, six from São Miguel and two from Terceira. **Madeira** Listed as exceptional. **Canaries** Regular but scarce passage migrant and irregular winter visitor, but no records on La Gomera and El Hierro. **Cape Verdes** Eight records, with half on São Vicente (4), and also recorded on Boavista (3) and Santiago (1).

COMMON REDSHANK   *Tringa totanus*   Plate 35 (2)
Sp. Archibebe Común   **P.** Perna-vermelha (Perna-vermelha-comum)

T. t. totanus and *T. t. robusta*. **L** 27–29 cm (10½–11½"); **WS** 59–66 cm (23–26"). **IDENTIFICATION** Sexes similar. In all plumages unlikely to be mistaken given combination of red legs, red base to bill, mainly grey-brown plumage and, in flight, the white lower back and secondaries. Adult in breeding plumage has irregular coarse dark markings on both upper- and underparts. Non-breeding adult has much plainer grey-brown upperparts and breast with some pale barring on flanks and undertail. Juvenile has paler, more yellow-orange legs, a duller red base to bill, much browner upperparts, heavily spotted and notched buff, and underparts quite heavily streaked and barred dark brown. Subspecies *robusta* is larger than nominate and has larger dark spotting on breast and belly in breeding plumage. **VOICE** Common flight call is a piping, double-syllabic downslurred *teu-hu* or trisyllabic *teu-huhu*. Also a long, mournful *tyuuuu* from the ground in alarm. **RANGE** Breeds from Iceland discontinuously across Europe and, in a fairly narrow belt, across S Russia and C Asia to N and E China, also N India and Tibet. Winters in W Europe, N and C Africa to S Asia. **STATUS** *totanus* recorded from all the archipelagos and is a regular visitor; *robusta* reported from Madeira and is also believed to have occurred on the Canaries, but is probably only a vagrant. **Azores** At least 16 records from: Santa Maria (1), São Miguel (6), Terceira (7), Pico (1) and Faial (1). **Madeira** Fairly regular passage migrant and also recorded on the Salvages. **Canaries** Regular passage migrant and scarcer winter visitor, recorded from all major islands. **Cape Verdes** Recent observations suggest that it is regular but uncommon visitor. Recorded in all months Aug–Apr, with most records from the saltpans on Sal and sewage farm on São Vicente.

MARSH SANDPIPER   *Tringa stagnatilis*   Plate 35 (3)
Sp. Archibebe Fino   **P.** Perna-verde-fino

Monotypic. **L** 22–25 cm (9–10"); **WS** 55–59 cm (22–23"). **IDENTIFICATION** Similar to Greenshank but separated in all plumages by its obviously longer legs, much daintier appearance and very fine, straight bill. In flight the area of white on the back and rump is narrower than in Greenshank and the full length of the toes extends beyond the tail tip. Adult in breeding plumage greyer above than Greenshank with more regularly spaced dark markings, and breast and flanks more finely marked. Non-breeding adult is rather plain grey above and white below, with a more distinct supercilium behind the eye than Greenshank, a paler forehead and less distinct eye-stripe. Juvenile also similar to Greenshank but supercilium more distinct behind the eye and central breast whiter. Distinguished from adult by browner upperparts, finely marked with buff notches giving a rather speckled appearance. **VOICE** Flight call is a mellow *keeuw* reminiscent of Greenshank but somewhat thinner, quicker and higher pitched. Also a loud *chip*, sometimes repeated, in alarm when flushed. **RANGE** Breeds in a rather broad belt from the Ukraine to east of Lake Baikal. Winters in Africa and through S Asia to Australasia. **STATUS** Vagrant with records from three archipelagos. **Azores** Four records: one obtained, Santa Maria (no other details); one, Pico do Negro (São Miguel), 25 Dec 1977; one, Cabo da Praia (Terceira), 11 Oct 1982; and one, Lajes do Pico (Pico), 25 Aug 1998, with two probable records, São Miguel, Aug/Sep 1960 and Sep1978. **Canaries** Nine records: one, salinas north of Arrecife (Lanzarote), 9 Jan 1982; adult, Guargacho and then Ciguaña (Tenerife), 14–24 Mar 1990; one, Barranco de Río Cabras (Fuerteventura), 21 Mar 1991; one, Embalse de Los Molinos (Fuerteventura), 21 Apr 1994; one, Charca de Maspalomas (Gran Canaria), 8 Apr1994, with another there, 21 Oct 1994; one, Barranco de La Torre (Fuerteventura), 7 Dec 1998; one, Caleta de Fustes (Fuerteventura), 31 Jan 1999; and one, reservoir north of Triquivijate (Fuerteventura), 18 Apr 2001. **Cape Verdes** Three records: one, Baia das Gatas (São Vicente), 16 Jan 1966; one, Pedra Badejo lagoons (Santiago), 6 Feb 1966; and one, near Santa Maria (Sal), 23 Nov 1989; Hazevoet (1995) treated the first two as unconfirmed, but in view of Bannerman's usual care I believe should be accepted.

(COMMON) GREENSHANK *Tringa nebularia*   Plate 35 (4)
Sp. Archibebe Claro **P.** Perna-verde (Perna-verde-comum)

Monotypic. **L** 30–34 cm (12–13½"); **WS** 68–70 cm (27–27½"). **IDENTIFICATION** Sexes and all plumages similar. A large *Tringa* with greenish or yellowish legs and a relatively long, rather heavy, slightly upturned bill. Plumage generally brownish-grey above and white below, with a streaked breast. Adult in breeding plumage heavily streaked and barred on head, neck and breast, upperparts brownish-grey with a variable number of dark-centred scapulars, underparts white, bill dark grey and legs usually olive- or grey-green. Non-breeding adult similar but has more uniform, paler grey upperparts with an indistinct scaly appearance, streaking on head and neck less extensive and bold, and legs duller. Juvenile rather dark grey-brown above with fine buffy-white feather fringes and tertial edges more heavily marked than in adult. Underparts more extensively streaked than non-breeding adult and legs often yellowish-green. This species can display bright yellow legs, in which case great care must be taken to avoid confusion with Greater Yellowlegs. In flight both adult and juvenile have rather plain dark upperwings, a white rump extending to a point high on back and a pale tail with some barring in centre. **VOICE** Usually very vocal, the most common call being a three-note *teu-teu-teu*. **RANGE** Breeds across the C Palearctic from the N British Isles to Kamchatka and winters in a few places in the W Palearctic but mostly in Africa, Arabia, the Indian subcontinent, SE Asia, Indonesia and Australasia. **STATUS** A regular passage migrant and winter visitor. **Azores** Prior to 1990 only 12 records but it is now clearly a regular passage migrant, at least in autumn. **Madeira** Regular passage migrant in small numbers and also recorded on the Salvages. **Canaries** Regular passage migrant, in small numbers, and scarce winter visitor with records from all main islands. **Cape Verdes** Regular passage migrant and winter visitor with records from all main islands except São Nicolau, Fogo and Brava.

GREATER YELLOWLEGS *Tringa melanoleuca*   Plate 35 (5)
Sp. Archibebe Patigualdo Grande **P.** Perna-amarela-grande

Monotypic. **L** 29–33 cm (11½–13"); **WS** 70–74 cm (27½–29"). **IDENTIFICATION** Sexes similar. Larger and bulkier than similar Lesser Yellowlegs, with a longer, thicker and slightly upturned bill. Differs from Greenshank, mainly in its proportionately longer, orange-yellow legs (though Greenshank can show yellowish legs) and absence of white wedge extending up rump and back in Greenshank. Adult in breeding plumage has head, neck and breast very heavily streaked, forming spots and chevrons on flanks, lower breast and even upper belly. Lesser Yellowlegs in this plumage has much less coarse markings on flanks. Non-breeding adult has browner upperparts with fine pale spotting, notches and fringes. On underparts streaking less pronounced than in breeding plumage. Lesser Yellowlegs in non-breeding plumage is slightly paler with less extensive underparts streaking. Juvenile has dark brownish upperparts with many pale whitish spots and notches, and head and neck strongly streaked brown. In Lesser Yellowlegs the breast streaking in juveniles is much more diffuse and less prominent. **VOICE** Call is a trisyllabic *chew-chew-chew* very like Greenshank but the third note invariably drops in pitch. **RANGE** Breeds from S Alaska, across SC Canada to Newfoundland, and winters from S USA to S South America. **STATUS** Vagrant with only seven records. **Azores** Adult winter obtained, São Miguel (no other details); one, Lagoa de São Braz (São Miguel), Aug 1964; two observations, S Pico, late Sep 1978; one, Lagoa Branca (Flores), 14–17 Oct 2000; and a juvenile at Cabo da Praia (Terceira), 18 Sept 2003. **Canaries** One, near Playa Blanca (Lanzarote), 18–19 Nov 1989. **Cape Verdes** One, near Tarrafal (Santiago), 16–17 Oct 2001, and another there, 19 Oct 2003.

LESSER YELLOWLEGS *Tringa flavipes*   Plate 35 (6)
Sp. Archibebe Patigualdo Chico **P.** Perna-amarela-pequeno

Monotypic. **L** 23–25 cm (9–10"); **WS** 59–64 cm (23–25"). **IDENTIFICATION** Sexes similar. A slim, attenuated, medium-sized, grey-and-white sandpiper with a straight, fine bill and long, orange-yellow legs. In flight has an obvious square white rump patch. Breeding adult has head, neck and breast heavily streaked dark brown, with relatively little streaking on flanks, and rest of underparts white. Mantle, scapulars and tertials dark blackish-brown with some whitish spots and notches, wing-coverts mostly dull grey-brown with fine, pale fringes, a short white supercilium and white eye-ring. Non-breeding adult is paler with less intense and more diffuse streaking. Juvenile resembles adult winter but is darker and browner above with more contrasting pale spots. Most likely to be confused with either Greater Yellowlegs or Wood Sandpiper, the differences being mentioned under those species. **VOICE** Call is a high-pitched *tiu* given singly or 2–4 on same pitch in rapid succession. **RANGE** Breeds in Alaska, NW and C Canada and winters from S USA to S South America. **STATUS** Accidental with records from all archipelagos. **Azores** About 25 records, mostly from Terceira and São Miguel, but also noted on Faial, Pico, São Jorge, Flores and Corvo. **Madeira** One, Lugar de Baixo (Madeira), 11 Sep 1985. **Canaries** Eight records, all since 1989: one, El Médano (Tenerife), 15 Aug 1989; one, Presa de Bernadino (Tenerife), 3 Nov 1990; one, Fuerteventura, winter 1990–91, at Las Salinas, 18 Nov 1990 & 24 Mar 1991, and at Barranco de la Torre, 13–16 Jan 1991; one, Charco de Maspalomas (Gran Canaria), 12–20 Sep 1995; one, Los Silos (Tenerife), 4–11 Oct 1998; one, Neuvo Horizonte (Fuerteventura), 5 Feb 2000; one, Charca de Maspalomas (Gran Canaria), 14–16 May 2000; and juvenile, Roquito del Fraile (Tenerife), 5–26 Oct 2001. **Cape Verdes** Seven very recent records: one, Ribeira do Ervatão (Boavista), 17–21 Mar 1999; first-winter, Rabil lagoon (Boavista), 13 Mar 2000; one, Pedra de Lume saltpans (Sal), 26 Sep 2001–20 Mar 2002, with two there, 20 Apr 2003; one, near Tarrafal (Santiago), 19 Oct 2003; one, near Santa Maria (Sal), 24 Oct 2003; and one, Pedra da Lume (Sal), 24 Oct 2003, increasing to four, late Dec, with three still, 9 Mar 2004 and one, 17 Apr 2004.

SOLITARY SANDPIPER *Tringa solitaria* not illustrated
Sp. Andarríos Solitario **P.** Maçarico-solitário (Perna-verde-solitário)

Monotypic. **L** 18–21 cm (7–8″); **WS** 55–59 cm (22–23″). **IDENTIFICATION** Sexes and all plumages similar. At all ages best separated from Green Sandpiper by all-dark rump and central tail. Usually has a longer primary projection, the folded primaries extending further beyond the tip of the tail, a slightly paler underwing and narrower wings. Adult in breeding plumage has more distinctly spotted mantle and scapulars, and streaking on head, neck and breast is less heavy than Green Sandpiper. Non-breeding adult similar but has less spotting and head, neck and breast are plainer and slightly paler grey-brown. Chin and throat almost white. Juvenile resembles non-breeding adult but initially has bright buff spots, the upperparts slightly browner and neck and breast less plain with a distinct brownish tinge. Autumn plumage is noticeably fresher than that of adult. **VOICE** Similar to Green Sandpiper but normally slightly softer, giving an excited *peet* or *peet-weet-weet* when flushed. **RANGE** Breeds across N North America from Alaska to Newfoundland. Winters from SE USA through West Indies and C America to Argentina and Uruguay. **STATUS** Very rare vagrant with records from three archipelagos. **Azores** Three recent records: adult, Sete Cidades (São Miguel), 9 Sep 1997: one, 2 km east of Lagoa Branca (Flores), 6 Sep 1998; and juvenile, Cabo da Praia (Terceira), 12 Sep 1999. **Canaries*** One, Embalse de Valle Molina (Tenerife), 18 Mar 2004. **Cape Verdes** Two records: adult in winter plumage, Rabil lagoon (Boavista), 12 Mar 1997; and one, Mindelo (São Vicente), 17 Dec 2004–13 Jan 2005 at least.

GREEN SANDPIPER *Tringa ochropus* Plate 36 (1)
Sp. Andarríos Grande **P.** Maçarico-bique-bique (Pássaro-bique-bique)

Monotypic. **L** 21–24 cm (8–9½″); **WS** 57–61 cm (22½–24″). **IDENTIFICATION** Sexes and all plumages similar. A very dark *Tringa*, in flight with conspicuous white rump and almost black underwings, and larger and bulkier than Wood Sandpiper, with distinctive dull green legs and densely barred tail. Adult in breeding plumage has heavy dark streaking on head and neck, very dark olive-brown upperparts with small whitish spots, a prominent white supercilium in front of eye and narrow white eye-ring. Non-breeding adult has even smaller upperparts spots and head and breast paler grey-brown and less heavily streaked. Juvenile like non-breeding adult but has spotting on upperparts more buff. Most likely to be confused with Solitary Sandpiper, but this has dark rump and central tail with more finely barred sides, a slightly decurved bill, a bolder eye-ring and noticeably slighter build. **VOICE** Flight call is a loud, whistling *tlueet-wit-wit* with the second two notes higher than the first. **RANGE** Breeds from Scandinavia across Russia generally north of 50°N, and winters mostly in Africa, India, SE Asia and Indonesia, but also as far north as S England. **STATUS** Recorded from all archipelagos but status varies. **Azores** Only five definite records: one obtained, Lagoa des Empadadas (São Miguel), 17 Aug 1964; one, Madalena (Pico), Sep 1977; two, Populo (São Miguel), 10 Oct 1977; one, Santa Cruz (Flores), Aug 1978; and four, Ribeira Grande (São Miguel), Nov 1978. **Madeira** Fewer than five records in the last 50 years, more recently one (El Tanque (Porto Santo), 24 Sep 2003, and also recorded on the Salvages. **Canaries** Regular passage migrant and winter visitor in small numbers. **Cape Verdes** Only eight records, five from Santiago, two from São Vicente and one from Boavista.

WOOD SANDPIPER *Tringa glareola* Plate 36 (2)
Sp. Andarríos Bastardo **P.** Maçarico-de-dorso-malhado (Maçarico-bastardo)

Monotypic. **L** 19–21 cm (7½–8″); **WS** 56–57 cm (22–22½″). **IDENTIFICATION** Sexes similar. Adult in breeding plumage has coarse pale-spangled dark brown upperparts, white supercilium, white throat contrasting with the dark-streaked head and neck, and dark-barred flanks. Legs greenish-yellow. Non-breeder has less boldly spangled upperparts and breast washed grey and less boldly streaked. Legs more yellow. Juvenile like non-breeding adult but has buff-spotted upperparts, brownish mottling on neck and breast, and brighter yellow legs. Most likely to be confused with Green Sandpiper or Lesser Yellowlegs. Compared to Green Sandpiper, Wood Sandpiper is paler, slimmer, longer legged and has an obvious supercilium behind eye, as well as yellowish legs that extend well beyond the tail tip in flight, a pale greyish-white underwing and less contrasting white rump. Lesser Yellowlegs differs in being larger and greyer with proportionately longer, brighter yellow legs, and at rest the primaries clearly project beyond the tip of the tail. **VOICE** Flight call is a rapid, high-pitched *chiff-if-if* sometimes repeated in a longer sequence that rises in pitch. **RANGE** Breeds across most of the Palearctic, mainly north of 50°N, and winters mostly in the S Old World, but is more common in sub-Saharan Africa and India than in Australasia. **STATUS** Recorded from all archipelagos but status varies. **Azores** Ten records: immature ♂ obtained, Lagoa Rasa (São Miguel), 5 Sep 1956; one, Sete Cidades (São Miguel), 9 Sep 1966; two, Pico do Negro (São Miguel), 25 Dec 1977; five, Charco dos Leimos (São Miguel), 30 Apr 1979; five, Sete Cidades (São Miguel), 19 Nov 1979; adult in breeding plumage, Cabo da Praia (Terceira), 26 Aug 1991; adult, same place, 13 Aug 1996, another there, 8 Oct 2000, and 14 Dec 2003; and one, Lagoa Azul, Sete Cidades (São Miguel), 11 Sep 2004. **Madeira** Fewer than five records in the last 50 years. **Canaries** Scarce but regular passage migrant and a rare and irregular winter visitor; no records from La Gomera. **Cape Verdes** Uncommon but regular passage migrant and winter visitor, with most records on São Vicente and Santiago.

TEREK SANDPIPER   *Xenus cinereus*   Plate 36 (3)
Sp. Andarríos del Terek   **P.** Maçarico-sovela

Monotypic. **L** 22–25 cm (9–10″); **WS** 57–59 cm (22½–23″). **IDENTIFICATION** Sexes and all plumages similar. Combination of fairly long but clearly upturned and often yellowish-based bill, and short yellow-orange legs should make it unmistakable. Adult in breeding plumage has grey upperparts and breast with a dark carpal area, dark line on scapulars and an obscure white supercilium. Non-breeder plainer, almost unmarked, grey on upperparts. Juvenile darker grey-brown above with greyer tertials and coverts, with pale buff fringes and rather faint dark subterminal bar. In flight most obvious feature in all plumages is the distinct white trailing edge to secondaries. **VOICE** Flight call is a rippling trill, *du-du-du-du-du*, somewhat reminiscent of Whimbrel but softer. **RANGE** Breeds in Finland and Russia thence east through Siberia. Winters discontinuously from Africa through Arabia and S Asia to Australia. **STATUS** Vagrant with only two records. **Canaries** Adult in breeding plumage, Arrecife (Lanzarote), 21 Aug 1985; and one, Cigüeña (Tenerife), 10 Apr 1993.

COMMON SANDPIPER   *Actitis hypoleucos*   Plate 37 (1)
Sp. Andarríos Chico   **P.** Maçarico-das-rochas

Monotypic. **L** 19–21 cm (7½–8″); **WS** 38–41 cm (15–16″). **IDENTIFICATION** Sexes and all plumages similar. A small shorebird with characteristic bobbing walk and rather spasmodic flight action. Generally olive-brown above and white below, with obvious dark breast-side patches, and legs usually greenish-grey. Adult in breeding plumage has slight bronzy gloss to upperparts with darker streaks and dark subterminal mark on some larger feathers. Also, patches on breast-sides form an almost complete breast-band. Non-breeder has plainer upperparts and smaller breast patches. Juvenile has pale buff and dark brown barring on wing-coverts, pale fringes to most upperparts and distinct dark and buff notches on tertials. In flight, in all plumages, a long wingbar and two blackish bars on underwing-coverts, although latter is difficult to see. Very similar to the next species and great care should be taken on the Azores, where both are accidental. **VOICE** Common call is a high-pitched trisyllabic *tsee-wee-wee* distinct from calls of Spotted Sandpiper, but some overlap is known. **RANGE** Breeds across most of Europe and N Asia to Kamchatka, Sakhalin and Japan. Winters in Africa (mostly south of the Sahara), coastal Arabia, the Indian subcontinent, through SE Asia and Indonesia to Australia. **STATUS** Recorded from all archipelagos and is a regular winter visitor. **Azores** Accidental visitor. **Madeira** Fairly common passage migrant and winter visitor. **Canaries** common passage migrant and winter visitor. **Cape Verdes** Regular passage migrant and winter visitor.

SPOTTED SANDPIPER   *Actitis macularius*   Plate 37 (2)
Sp. Andarríos Maculado   **P.** Maçarico-pintado (Maçarico-maculado)

Monotypic. **L** 18–20 cm (7–8″); **WS** 37–40 cm (14½–16″). **IDENTIFICATION** Sexes similar. Very similar to Common Sandpiper in non-breeding and juvenile plumages. Adult in breeding plumage easily distinguished from Common Sandpiper by black spotting on white underparts and pinkish bill with a dark tip. Non-breeder perhaps best identified by its shorter tail, pinkish bill base and by the pinker or yellower legs, but is also rather greyer on upperparts and has smaller breast-side patches. Juvenile separated from Common Sandpiper by the shorter tail and plainer upperparts, more strongly marked coverts and lack of pale- and dark-spotted fringes to tertials. The bill also tends to be bicoloured, pinkish at the base with a dark tip. In flight wing bar is less extensive than in Common Sandpiper, as it does not reach the inner wing. Underwing of Spotted Sandpiper only has one diagonal dark bar on underwing-coverts and the dark trailing edge extends across all of the inner secondaries, but these last two features are difficult to see. **VOICE** Vagrants often rather silent but calls include a single *peet* or double *peet-weet* recalling Green Sandpiper. Can also give calls indistinguishable from Common Sandpiper. **RANGE** Breeds across most of USA and Canada, wintering mostly in C and S America and the West Indies. **STATUS** Recorded from all archipelagos. **Azores** Regular in autumn and may overwinter rather frequently. There are records for all months except May–Jul, with *c.* 70 records. **Madeira** One, Funchal (Madeira), 26 Apr 1960; juvenile, Machico (Madeira), 3 Oct 1993; summer-plumaged adult, Machico (Madeira), 27 Aug–6 Sep 2002; one, El Tanque (Porto Santo), 4–6 Dec 2003; and one, Machico (Madeira), 9–10 Jan 2004. **Canaries** One, Puerto de la Cruz (Tenerife), 1 Sep 1975; one, Arrecife (Lanzarote), 24 Mar 1980; adult winter, Puerto del Carmen (Lanzarote), 2 Nov 1985; juvenile, Valle Gran Rey (La Gomera), 16 Nov 1991; juvenile, Las Galletas (Tenerife), 11 Nov–1 Dec 1995; adult winter, Las Salinas (La Palma), 17 Oct 1996; juvenile, Corralejo (Fuerteventura), 7 Nov 1997; first-year, Roquito del Fraile (Tenerife), 3 Sep 1997–25 Feb 1998, with it or another, Las Galletas, 22 Mar 1998: first-winter moulting to adult breeding, Las Galletas (Tenerife), 17 Feb–21 Apr 1999; juvenile, Charco del Cieno, Valle Gran Rey (La Gomera), 25 Sep 1999; and juvenile, Las Salinas del Carmen (Fuerteventura), 26 Oct–5 Nov 1999. **Cape Verdes** Three records, two of them, Mindelo sewage farm (São Vicente), 24 Feb–2 Mar 1999 (a second bird reported 28 Feb was not substantiated) and 15 Mar 2002; and one, Pedra de Lume (Sal), 24 Mar 2004.

WILLET *Catoptrophorus semipalmatus* Plate 36 (4)
Sp. Playero Aliblanco   **P.** Maçarico-d'asa-branca (Maçarico-semipalmado)

Subspecies unknown, but *C. s. semipalmatus* is the most likely. **L** 33–41 cm (13–16″); **WS** 70–80 cm (27½–31½″). **IDENTIFICATION** Sexes similar. A godwit-sized, *Tringa*-like shorebird with a very characteristic wing pattern. The white bar across the entire upperwing contrasts with the black primaries and grey coverts, whilst the underwing is patterned black/white/black, and contrasts with the pale body. Legs thick, longish and blue-grey, bill straight and fairly deep-based but relatively short compared to godwits. Adult of race *semipalmatus* in breeding plumage has heavily streaked and barred neck, breast and flanks, with grey-brown upperparts barred whitish-buff and dark brown. Non-breeder rather plain grey-brown above and white below, with narrow white fringes to upperparts and grey wash on breast and flanks. Juvenile similar but slightly darker above with buff fringes and dark subterminal line to most feathers, the tertials and rear scapulars notched with buff. **VOICE** Flight call is a loud, clear *kyaah-yah* or *kleee-lii*, or a simple descending *haaaa*. When flushed gives loud *week* or shorter *kip*. **RANGE** Breeds on the eastern coast of USA, West Indies and inland in NW USA and S Canada. Winters along coasts to N South America. **STATUS** Vagrant with only three records. **Azores** One long dead, Populo (São Miguel), 12 Mar 1979; and two records, Topo, off São Jorge, 5 Jul 1984, and 6 Jun 1985, which may refer to the same individual.

(RUDDY) TURNSTONE *Arenaria interpres* Plate 37 (4)
Sp. Vuelvepiedras   **P.** Rola-do-mar

A. i. interpres. **L** 21–26 cm (8–10″); **WS** 50–57 cm (20–22½″). **IDENTIFICATION** Sexes similar. Very distinctive wader with fairly short, orange legs, short, wedge-shaped, slightly upturned bill and distinctive dark-and-white upperparts in flight. In breeding plumage has strongly patterned black-and-white head and breast, mostly chestnut-orange upperparts and pure white rear underparts. Non-breeder and juvenile much duller but easily identified by combination of short, orange-red, legs and short, slightly upturned, black bill. Both have mottled blackish and dull brown upperparts and head, dark grey-brown breast with pale patches on sides. Juvenile has wing and mantle feathers fringed buff giving scaly appearance to upperparts, and a paler brown head, duller black breast and more obvious pale patches on head. **VOICE** Flight calls include a nasal, rattling *tuk-a-tuk-tuk* and sharp *teu*. **RANGE** Breeds across N Holarctic and winters on coasts throughout most of world. **STATUS** Recorded from all four archipelagos. **Azores** Recorded in all months but is mainly a winter visitor and has been recorded from all islands. **Madeira** Regular winter visitor to all main islands and also recorded on the Salvage Islands. **Canaries** Seen in all months but most common as a winter visitor and passage migrant, when it is the most abundant wader in the archipelago. Recorded from all islands and islets. **Cape Verdes** Recorded in all months but most common as a passage migrant and winter visitor, and recorded from all islands and islets.

WILSON'S PHALAROPE *Phalaropus tricolor* Plate 38 (1)
Sp. Falaropo de Wilson   **P.** Pisa-n'água (Falaropo de Wilson)

Monotypic. **L** 22–24 cm (9–9½″); **WS** 39–43 cm (15–17″). **IDENTIFICATION** Sexually dimorphic. ♀ in breeding plumage is brighter than ♂. The largest, longest-legged and longest-billed of the phalaropes. Unmistakable in breeding plumage when patterning on head and neck is diagnostic. Throat and supercilium white, crown and hindneck pale grey, with broad black eye-stripe reaching onto the neck-sides where it becomes deep chestnut, and rufous foreneck. ♂ also has darker, browner upperparts. Both sexes at this season have black legs, and in all plumages bill quite long and very fine, almost needle-like. Non-breeder basically pale grey above and white below, with a white supercilium that can extend well down grey neck-sides, and yellow legs. Juvenile plumage is lost very early and is unlikely to be seen in our region but has darker brown upperparts than non-breeding adult, with pale buff fringes to feathers. Legs yellowish-flesh. In flight, in all plumages, plain unmarked upperwings, square white rump and the toes project beyond the tail tip. Most likely confusion is with Lesser Yellowlegs and Stilt Sandpiper, but unusual pot-bellied appearance of Wilson's Phalarope helps to distinguish it. **VOICE** Usually rather silent but can give a rather soft nasal *aangh*. **RANGE** Breeds in C North America and winters south of the equator in S America. **STATUS** Vagrant with only 11 records, eight from one archipelago. **Azores** Adult in breeding plumage, Mosteiros (São Miguel), 18 Aug 1979; juvenile, Praia da Vitoria (Terceira), 19 Aug 1992; juvenile, Cabo da Praia (Terceira), 30 Jul 1997; first-winter, Cabo da Praia (Terceira) 3–8 Sep 1998, then again (or another), 14 Oct 1998; juvenile, Lagoa do Fogo (São Miguel), 30 Sep 2000; juvenile, Lagoa Verde, Sete Cidades (São Miguel), 16 Sep 2001; one, Cabo da Praia (Terceira), 28–29 Sep 2002; and one, Lajes do Pico (Pico), 11–16 Aug 2003. **Madeira** Just one old record. **Canaries** One, Los Cristianos (Tenerife), 16 Dec 1976; and one, Los Silos, autumn 2004.

RED-NECKED PHALAROPE *Phalaropus lobatus* Plate 38 (2)
Sp. Falaropo Picofino **P.** Falaropo-de-bico-fino

Monotypic. **L** 18–19 cm (7–7½″); **WS** 32–36 cm (12½–14″). **IDENTIFICATION** Sexually dimorphic. ♀ in breeding plumage is brighter than ♂. Adult unmistakable in breeding plumage but non-breeder could easily be confused with Grey Phalarope. In all plumages, bill is much finer and more pointed than Grey Phalarope, and Red-necked Phalarope is marginally smaller and daintier. Breeding adult has crown, back of neck, mantle and cheeks dark grey, chin and throat white, variable amount of orange-red on neck, and dark upperparts with buff lines on sides of mantle and scapulars. In non-breeding plumage adult is white below and grey above with an obvious black mask. The whitish edges to larger feathers form indistinct lines along sides of mantle and the scapulars, not shown by Grey Phalarope. Also centres to feathers of upperparts are slightly darker than in Grey Phalarope. Juvenile blackish-brown above with buff fringes to feathers forming stripes on sides of mantle and on scapulars. Initially, neck and breast washed greyish-buff, darker than in Grey Phalarope, but this soon fades to whitish. **VOICE** Usual flight call is a short *chep* or *cherp*, lower and throatier than Grey Phalarope. **RANGE** Breeds continuously across N Holarctic and winters mainly off Peru, N Indonesia and in the Arabian Sea. **STATUS** Vagrant with 4–5 records. **Azores** One obtained, Arrifes (São Miguel), prior to 1903; and ♀ obtained, 16 May 1906 (no details). **Madeira** Just one old record. **Canaries★** One, Arrecife (Lanzarote), 7 Feb 1968 and possibly two there, 6 Sep 1976. **Cape Verdes** One in winter plumage, Pedra de Lume saltpans (Sal), 17 Jan–4 Feb 1995.

GREY (RED) PHALAROPE *Phalaropus fulicarius* Plate 38 (3)
Sp. Falaropo Picogrueso **P.** Falaropo-de-bico-grosso

Monotypic. **L** 20–22 cm (8–9″); **WS** 40–44 cm (16–17″). **IDENTIFICATION** Sexually dimorphic. ♀ in breeding plumage brighter than ♂. Adult in breeding plumage easily recognised by its white face, chestnut-red underparts, dark blackish-brown upperparts with rufous and buff fringes, and yellow bill with a dark tip. Non-breeder has head and underparts white except dark greyish-black crown and nape and black mask. Upperparts grey with narrow white fringes to feathers. In flight reveals dark grey wings with distinct white wingbar on upperside, and white underwings. In this plumage, very similar to Red-necked Phalarope, from which best be identified by combination of plainer, paler grey mantle and shorter, thicker, more blunt-tipped bill. Juvenile resembles juvenile Red-necked Phalarope but has narrower mantle V, scapular V is usually lacking and breast and neck are washed paler pinkish-buff. **VOICE** Flight call is a short, sharp, high-pitched *pit* or *wit*. **RANGE** Breeds discontinuously across N Holarctic and winters mainly off coasts of Chile, W Africa and SW Africa. **STATUS** Recorded from all archipelagos. **Azores** Only *c*. 12 records but groups of *c*. 15 and *c*. 20 northeast of Faial, Sep 2001, suggest that it is more common in Azorean waters than records suggest. **Madeira** Zino *et al*. (1995) listed one old record, with one at sea, southeast of Madeira, 31 Oct 1999. **Canaries** Martín & Lorenzo (2001) list the species as accidental, but it is probably a scarce migrant only recorded rarely due to the lack of observers at sea. **Cape Verdes** Fewer than ten records but again probably overlooked due to its pelagic habits.

POMARINE SKUA *Stercorarius pomarinus* Plate 38 (4)
Sp. Págalo Pomarino **P.** Moleiro do Árctico (Moleiro-pomarino)

Monotypic. **L** 46–51 cm (18–20″) includes tail of up to 19 cm; **WS** 125–138 cm (49–54″). **IDENTIFICATION** Sexes similar. Full-adult breeding plumage attained in fifth summer, when usually displays spoon-shaped extensions to central tail feathers but occasionally one is broken and, rarely, both are missing; they are moulted twice a year, in

Nov/Dec and Mar/Apr. This feature makes adults of both morphs readily identifiable from similar Arctic Skua; in all other plumages these species can be difficult to identify, and the structure of Pomarine Skua is a very important aid. Larger and bulkier than Arctic Skua and often appears rather pot-bellied in flight. Wings broad based, the width at the body being greater than the length of the body from the wings to the tail tip, but excluding any extensions. **Juvenile** has upperparts dark brown with narrow pale tips to feathers, the rectrices dark with paler bases, the bill pale grey with a black tip and legs greyish. Three colour types. Pale type extremely rare and has head, neck and underparts creamy buff; dark type uncommon and is uniform dark brown on head and body, and almost lacks any pale barring on underwing. The intermediate type is the most numerous and has a greyish-brown head and body, with heavily barred underwings (a feature shared with the pale type). All three have a double white flash on the underwing, formed by pale bases to the primaries and primary-coverts, which is shown very rarely in Arctic Skua and is never as pronounced as in Pomarine. Also, all juveniles, except the very darkest individuals, have prominently pale-barred uppertail-coverts and none has any sign of a pale neckband, which is a feature of many juvenile Arctic and Long-tailed Skuas. In subsequent plumages, dark types are very difficult to age because of the similarity of all plumages, and are best judged on bare-part coloration and the projection of the central rectrices. **First-winter/first-summer** resembles juvenile but has paler hindneck, sometimes a paler belly and a hint of a dark cap. The tail-tip projection is 10–41 mm, the bill often has darker cutting edges and is more greyish-yellow, and the feet are all black. **Second-winter** similar to non-breeding adult but retains pale barring on axillaries and underwing-coverts. The tail-tip projection is similar to that of first-summers. **Second-summer** has a tail-tip projection of 20–70 mm, which is similar to non-breeding adult, and occasionally displays a twisted tip. The pale morph has head and body similar to adult but less well defined, and underwing-coverts and undertail-coverts barred like those of first-summer. **Third-winter** also resembles non-breeding adult but has a less clear-cut dark cap and displays some barring on underwing, although this is less extensive than in second-winter. **Third-summer** has tail-tip projection averaging shorter than adult and bare parts similar to breeding adult, but some show some pale spotting on tarsus. Pale morph has head and body like breeding adult, but underparts not as clean, with some barring on belly. Uppertail-coverts still partially barred and underwing-coverts not as uniformly dark as in adult. **Fourth-winter** is like non-breeding adult but usually has a few barred underwing-coverts or pale tips to dark underwing-coverts and axillaries; bare parts like adult. **Fourth-summer** closely resembles adult breeding but some have a few barred underwing-coverts and uppertail-coverts. The tail-tip projection is slightly shorter than in breeding adult. **Adult breeding** is dimorphic but pale morph outnumbers dark morph by 9:1; any intermediate morphs are extremely rare and constitute less than 1% of population. Pale morph in breeding plumage has dark cap and yellowish neck-sides, some ♂♂ and all ♀♀ have a dark band on upper breast, predominantly white lower breast and belly, dark vent, and dark brown upperparts with a pale flash on base of primaries. Underwing mostly dark but also has pale flash at base of primaries. Dark morph in breeding plumage is all dark blackish-brown, most having smaller pale area on base of primaries compared to pale morph. **Adult non-breeding** very variable and can resemble both breeding adult, but with pale barring on undertail-coverts, or juvenile but with uniformly dark underwing-coverts. This plumage is usually obtained in winter quarters and is unlikely to be seen in our region. **VOICE** Usually silent away from the colonies. **RANGE** Breeds discontinuously in the tundra within Arctic Circle. Winters south to C Chile, S Africa and Australasia. **STATUS** Recorded from all archipelagos but status varies. **Azores** At least 14 records, most in recent years, suggesting that this species is probably regular in Azorean waters on autumn migration. **Madeira** Fewer than five records in the last 50 years (Zino *et al.* 1995). However, since publication of the checklist, the number of observers dedicated to seawatching has increased, and the species is clearly a regular passage migrant in small numbers in autumn. **Canaries** Scarce passage migrant noted in both spring and autumn, with records from all main islands except Lanzarote. **Cape Verdes** Regular visitor in small numbers, recorded in Oct and Feb–May.

ARCTIC SKUA *Stercorarius parasiticus* Plate 38 (6)
Sp. Págalo Parásito P. Moleiro-pequeno (Moleiro-parasítico)

Monotypic. **L** 41–46 cm (16–18″) includes tail of up to 18 cm; **WS** 110–125 cm (43–49″). **IDENTIFICATION** Sexes similar. Full-adult plumage breeding plumage obtained in fifth summer. In all plumages, similar to Pomarine Skua and structural differences are important for identification. Arctic Skua is smaller and slimmer with a smallish head and bill, a narrower wing base and pointed projections to central rectrices, a useful feature in all but juveniles and moulting birds. In flight, the deepest part of the underbody is at the breast, and flight action more buoyant than Pomarine, the long pointed wings and powerful flight often create a rather falcon-like appearance. **Juvenile** has three colour types: pale, intermediate and dark. The rare pale type has an almost whitish belly, a pale creamy to almost orangey head, rusty-tinged brown upperparts, a coarsely vermiculated rump, barred underwing-coverts and a pale crescent on base of primaries, which if present on upperwing is a diagnostic feature. The second pale underwing crescent at the base of the primary-coverts, if present at all, is less conspicuous than Pomarine. Intermediate type typically warm brown with distinct rusty tinge to entire underparts, particularly prominent in barring on underwing and undertail-coverts. A rusty neckband is present, which is not shown by Pomarine in similar plumage. Most juveniles are of this type, but individual variation is considerable. Dark type rare and varies from coffee-brown to almost blackish-brown. It appears almost uniformly dark, usually with some indistinct rusty barring on upperparts and a faint rusty neckband. **First-winter/first-summer** similar to juvenile but loses rusty tone and starts to obtain black on tibia and scattered pale patches on head and body. Tail more pointed than juvenile and is elongated 30–46 mm. **Second-winter** similar to both first- and second-summer. **Second-summer** has greater tail elongation than first-summer. Pale morph basically resembles adult on head and body, and juvenile on underwing. Dark morph distinguished in much same way as pale morph, but is generally more difficult to age. **Third-winter** similar to adult in both morphs, but has varying amount of barring on underwing-coverts and still some pale on bill and tarsus. **Third-summer** tail elongation is longer than second-summer but still shorter than adult, and plumages of both morphs resemble adult but are less clearly defined. Usually shows some barring on greater and median underwing-coverts. Bill is similar to adult but legs still show some pale tarsus spots. **Fourth-winter/summer** is like adult but can sometimes be aged by presence of a few barred feathers on otherwise uniformly dark underwing. **Adult breeding** very variable and has three colour morphs. Pale morph has dark cap (less dark or extensive as in Pomarine), often a pale band above bill, yellowish-tinged neck, whitish underbody, all dark underwing and variable dark breast-band. Dark morph all sooty brown except for white wing flashes and has slight capped appearance. Intermediate morph (which is commonest) resembles dark morph but has paler yellow head-sides and hindneck, and paler greyish-brown underparts. Paler intermediates resemble pale morph but have dirtier underparts. **Adult non-breeding** resembles breeder but cap less well defined and uppertail- and undertail-coverts are barred like juvenile. **VOICE** Rather silent away from the colonies. **RANGE** Breeding range circumpolar in N Hemisphere, wintering south to S South America, S Africa, Australia and New Zealand. **STATUS** Recorded from all archipelagos but status varies. **Azores** Previously considered accidental but recent records suggest that it is a scarce passage migrant, most frequent in autumn. **Madeira** Listed by Zino *et al.* as exceptional, but recent reports reveal that it as a rare passage migrant, most frequent in autumn. **Canaries** Scarce and irregular passage migrant, and rare winter visitor, with records from all main islands except La Palma. **Cape Verdes** Rare passage migrant and winter visitor.

LONG-TAILED SKUA   *Stercorarius longicaudus*   Plate 38 (5)
Sp. Págalo Rabero   **P.** Moleiro-rabilongo (Moleiro-de-cauda-comprida)

S. l. longicaudus. **L** 48–53 cm (19–21″) includes tail of up to 29 cm; **WS** 105–115 cm (41–45″). **IDENTIFICATION** Sexes similar. Full-adult breeding plumage attained in fifth summer. The smallest and slightest of the skuas but has the longest tail projection. Most similar to Arctic Skua, particularly in dark-morph juvenile but structural differences consistent at all ages. Long-tailed Skua is a similar size to a Black-headed Gull, although larger individuals overlap with small Arctic Skua. Head shorter and more rounded, bill shorter and proportionately heavier, wing narrower (especially the hand) and it has a longer and narrower body and tail. **Juvenile** as variable as juvenile Arctic Skua and has three colour morphs. Pale types have pale whitish or yellowish head and belly, bicoloured upperparts with fine but distinct barring on mantle and upperwing-coverts, evenly barred black and white uppertail- and undertail-coverts, and prominently barred pale and dark underwing. Intermediate types have a darker, greyer head with greyish-brown neckband, less extensive white underparts and darker upperparts. Most numerous dark type is all sooty-brown with heavy barring on underwing and uppertail- and undertail-coverts, but in extreme cases almost all barring is absent, except narrow fringes to mantle feathers. **First-winter/first-summer** similar to juvenile but underparts whiter and upperwing-coverts plain grey-brown. Darker individuals develop a paler hindneck. **Second-winter** similar to adult non-breeding but have barred underwing like juvenile. A few develop adult-like underwing and are almost indistinguishable from adult non-breeding. **Second-summer** has head and body similar to adult breeding but cap duller and less well defined. Underwing and uppertail-and undertail-coverts all still predominantly barred but in some parts of the underwing-coverts and axillaries are uniformly dark. Tail projection is shorter than breeding adult but is often distinctly longer than any Arctic Skua. **Third-winter/third-summer** similar to corresponding adult plumage and most are probably indistinguishable, but a few retain some barred underwing-coverts and in summer some have remnants of a grey breast-band and pale belly. **Fourth-winter/fourth-summer** almost identical to adult but some show combination of shorter tail projection, a few barred underwing-coverts and some dark on underparts. **Adult breeding** has distinct black cap, prominent contrast between pale grey to brownish-grey upperparts and black flight feathers and tail; underparts pale on breast becoming grey on belly, with no white wing flashes on either surface, but the outer 2–3 primaries have white shafts on upperside (only the outermost shows a white shaft on underwing). The existence of a dark morph is disputed but some have more extensively dark underparts and these are considered intermediate morphs. **Adult non-breeding** combines dark underwing of adult with head and body pattern of immature. **VOICE** Rather silent away from breeding grounds. **RANGE** Breeds around Arctic, from Scandinavia through Siberia, Alaska and Canada to Greenland, and winters in Southern Ocean. **STATUS** Rare visitor recorded from all archipelagos. **Azores** Accidental, but a record of 53 in Azorean waters, 12 Aug 1995, shows that it may be more common than land-based records suggest. **Madeira** One, near Baixa de Nordeste (Porto Santo), May 1990; juvenile, from the Porto Santo ferry, 21 Sep 1999; and one, off Ponta da Cruz (Madeira), 23 Sep 1999. All other records from Porto Moniz (Madeira): adult, 10 Aug 1998; adult, 25 Sep 1999; 11 adults, 23 Aug 2003; 6 adults and 1 juvenile, 5 Sep 2004; and 2 adults next day. **Canaries** Accidental visitor first recorded 1988 with records from La Gomera, Tenerife and Fuerteventura, but may prove more regular. **Cape Verdes** One, at sea south of Boavista, 25 Apr 1976.

GREAT SKUA (BONXIE)   *Stercorarius skua*   Plate 38 (7)
Sp. Págalo Grande   **P.** Alcaide (Moleiro-grande)

Monotypic. **L** 53–58 cm (21–23″); **WS** 132–140 cm (52–55″). **IDENTIFICATION** Sexes similar. Full-adult breeding plumage attained after second summer. The largest and bulkiest of the skuas with a rather barrel-shaped body, large head, broad wings and very short tail. Juvenile similar to adult but generally darker and more uniform with narrower white primary flashes. First-winter/first-summer also similar to non-breeding adult but moult cycle differs, although ageing is difficult as plumages are so similar. Head more uniformly dark than adult and lacks strong yellow hackles on neck and breast-sides. All subsequent immature plumages resemble adult and ageing almost impossible. If present, pale tarsus spots are probably evidence of immaturity. Adult breeding generally rufous-brown with variable dark cap, yellow streaking on neck- and breast-sides, blotchy underparts, wing-coverts paler than blackish flight feathers and very prominent white wing flashes at bases of primaries visible on both wing surfaces. Both darker and paler types exist, the darker birds being more uniform blackish-brown and paler birds blotchier brownish-buff. With wear some develop a pale frontal blaze, although this is never as obvious as on some South Polar Skuas. Non-breeder resembles breeding adult but has darker head with much-reduced yellow streaking on neck. **VOICE** Usually rather silent away from breeding grounds. **RANGE** Breeds in Scotland, Iceland, Jan Mayen I., Spitsbergen, Norway, the Kola Peninsula and Novaya Zemlya. Main wintering range is from W British Isles south to Senegal and Gambia, but also ranges widely through N Atlantic reaching coasts of Greenland, N and S America, and the Mediterranean. **STATUS** A scarce but regular visitor. **Azores** More than 20 records and appears to be fairly regular autumn/winter visitor. **Madeira** Scarce passage migrant/winter visitor, with increasing records in recent years. **Canaries** Scarce but regular passage migrant and winter visitor, with records from all main islands. **Cape Verdes** Nine records but, like other migrant seabirds, probably more common than records suggest.

SOUTH POLAR SKUA   *Stercorarius maccormicki*   Plate 38 (8)
Sp. Págalo Polar (Págalo de McCormick)   **P.** Alcaide do Antárctico

Monotypic. **L** 51–54 cm (20–21″); **WS** 125–135 cm (49–53″). **IDENTIFICATION** Sexes similar. Full-adult plumage attained some time after second summer; one bird aged at 36 months was indistinguishable from adult. At all ages, dark morph and intermediate morph (dark type) very similar to Great Skua and great care should be taken in attempting to identify South Polar Skua. Generally, South Polar is less powerful than Great Skua, the head is proportionately smaller, the bill appears shorter and slimmer, the wings slightly narrower and the overall plumage colder and greyer, lacking any warm brown tones. Juvenile much less variable than adult, with most having medium grey head and underparts, but a few are blackish-grey whilst some have head paler than body. Mantle and coverts dark greyish-black, the bill distinctly bicoloured and prominent white flashes at base of primaries. Immature resembles adult following post-juvenile moult, but streaking on neck usually weaker or lacking until three years old. Breeding adult has three colour morphs. Pale morph is pale greyish to pale brown on head and underparts, and this coloration sometimes extends onto mantle, which is diagnostic. Both wing surfaces are uniformly dark, except the pale flashes on the primary bases, and tail and rump also dark, creating a rather two-toned appearance to the bird. **VOICE** Usually silent away from breeding areas. **RANGE** Breeds around Antarctic continent and on S Shetlands. Adult movements poorly understood but juveniles and immatures migrate through Atlantic and Pacific oceans, and into Indian Ocean. **STATUS** Vagrant with only four records. **Azores★** Three records from Mosteiros (São Miguel): one, 3 Sep 1996; immature, 16 Sep 1998; and one, 13 Nov 2000. **Canaries★** Immature, from ferry between La Gomera and Tenerife, 17 Jul 2002.

PALLAS'S GULL (GREAT BLACK-HEADED GULL)   *Larus ichthyaetus*   Plate 40 (1)
Sp. Gavión Cabecinegro   **P.** Gaivotão-de-cabeça-preta

Monotypic. **L** 57–61 cm (22½–24″); **WS** 149–170 cm (59–67″). **IDENTIFICATION** Sexes similar. Adult breeding plumage obtained in fourth summer and in this plumage is totally unmistakable, as combination of large size and black hood are unique. In other plumages structural characteristics are useful. The long, sloping forehead that peaks behind eye, combined with the long, heavy bill produces a very distinctive head shape. In flight wings are, in proportion, longer, slimmer and more pointed than other large gulls. Immatures usually acquire a grey mantle and white underparts by their first winter, unlike any other similar-sized gull. Although adult plumage is obtained by fourth year, from second year plumage closely resembles that of adult. Like all hooded gulls, Pallas's Gull loses black head in winter, this being reduced to a dark patch behind eye, which can extend indistinctly over crown. **First-winter** has white head and body with dark marks around eyes and dusky patch behind eye that extends diffusely over rear crown, white eye-crescents above and below eye, and dark patch on lower hindneck contrasting with grey mantle. Blackish-brown outer primaries, pale grey-brown panel in mid-wing, neat black tail-band, dark tipped pale bill and greyish legs. **First-summer** may develop partial hood and pale mid-wing panel becomes increasingly greyer, legs acquire greenish tone and bill is yellowish and can show distinct subterminal band. **Second-winter** has narrower tail-band and more extensive grey upperwing. By **second-summer** hood almost complete and white primary tips and terminal tail-band are reduced or lacking through wear. **Third-winter** has tail-band very faint or reduced to spots, and primaries show more black and less extensive white spots than non-breeding adult. **Third-summer** similar to third-winter but acquires full hood and has reduced white primary tips. **Adult non-breeding/fourth-winter** similar to adult breeding but has winter head pattern and bill as second-winter. **Adult breeding** has complete black hood, distinct white crescents above and below rear of eye, yellow bill with black subterminal band and a reddish tip, and greenish-yellow legs. **VOICE** Rather quiet away from colonies but flight call is a deep, nasal *naagh*. **RANGE** Breeds discontinuously from Black Sea east to Mongolia and winters mainly on coasts of E Mediterranean south to Kenya, Yemen, S India and Myanmar. **STATUS** Extreme vagrant. **Madeira** One, Funchal harbour (Madeira), 4 Dec 1933. **Canaries*** Adult in almost complete breeding plumage, Los Cristianos (Tenerife), 31 Mar 1995.

MEDITERRANEAN GULL   *Larus melanocephalus*   Plate 39 (1)
Sp. Gaviota Cabecinegra   **P.** Gaivota-de-cabeça-preta

Monotypic. **L** 36–38 cm (14–15″); **WS** 92–100 cm (36–39½″). **IDENTIFICATION** Sexes similar. Adult breeding plumage attained in third summer. Most likely to be confused with Black-headed Gull but has a black not brown head, deeper-based bill with black subterminal band and no black on primaries, except on outer web of the outermost, which feature is particularly noticeable in flight. First-year more likely to be confused with Common Gull, from which it differs in its more contrasting upperwing pattern, shorter deeper bill and narrower tail-band. In first-winter, and all subsequent plumages, mantle is much paler grey than in Common Gull. Like all hooded gulls, loses hood in winter when head only has dark patch behind eye that often extends diffusely over nape or rear crown. **First-winter** has plain grey mantle, blackish tail-band and across secondaries, blackish outer primaries, white head with black mask behind eye, blackish bill with pale base, and legs varying from grey to reddish. **First-summer** has less prominent wing and tail markings, incomplete black hood, bare parts highly variable and can resemble either first-winter or adult. **Second-winter** has head pattern as that of first-winter, but wings pale grey with variable black on outer primaries and tail shows no sign of band. **Second-summer** similar to second-winter but black hood usually full and white primary tips reduced or lacking. **Third-winter/non-breeding adult** resemble adult breeding but head pattern is still similar to first-winter. **VOICE** Relatively silent away from breeding grounds, but occasionally gives distinctive *kee-ow*, which is rather quiet and not harsh. **RANGE** Breeds at isolated colonies around Mediterranean, Black Sea and, more recently, English Channel, and inland in Hungary, Turkey and Russia. Winter range includes British and N European coasts, Mediterranean and Black Sea coasts, and the Atlantic coast of NW Africa south to Mauritania. **STATUS** Varies between the three archipelagos where it has been recorded. **Azores** Adult in winter plumage, Ponta Delgada (São Miguel), Aug 1979, then first-winters, Praia da Vitoria (Terceira), 19 Feb 2003, Ponta Delgada (São Miguel), 11 Dec 2003, Praia da Vitoria/Cabo da Praia (Terceira), 14–16 Feb 2004, and Ponta Delgada (São Miguel), 7 Nov 2004. **Madeira** Occasional according to Zino *et al.* (1995) but recent records suggest it is perhaps more regular, with first-winters, Funchal, 16 Nov–23 Dec 2001, 8–16 Feb 2002, and 22 Nov 2004. **Canaries** Rare and irregular winter visitor to coasts and recorded from all main islands except La Palma. One, ringed in Rhine Valley *c.* 20 km north of Strasburg, 4 Jun 2001, seen Bocabarranco (Gran Canaria), 2 Feb 2002.

LAUGHING GULL   *Larus atricilla*   Plate 39 (2)
Sp. Gaviota Guanaguanare   **P.** Gaivota-alegre (Guincho-americano)

Monotypic. **L** 36–41 cm (14–16″); **WS** 100–125 cm (39½–49″). **IDENTIFICATION** Sexes similar. Adult breeding plumage attained in third summer. Larger and darker on mantle than Black-headed Gull and intermediate in size between that and Common Gull. Most likely confusion is with Franklin's Gull (also a vagrant to our region), but similar to Black-headed and Mediterranean Gulls in certain plumages. A structural feature of Laughing Gull is the rather long bill, which sometimes appear downcurved, and is often sufficient to distinguish it from all confusion species. **Adult in breeding plumage** has dark grey mantle and upperwings with blackish primaries often lacking any white tips except when fresh. Hood black with prominent white crescents above and below eye. Bill dark red with a black subterminal band and legs also dark red. **Juvenile** unlikely to be seen in our region. **First-winter** has uniform dark grey mantle and scapulars, greyish hindneck, breast and flanks, contrasting with white forehead, throat and belly. Dark patch on rear of ear-coverts often extends over crown, and has white crescents above and below eye. Outer wing and secondaries blackish, the latter with white tips, and a broad blackish tail-band. In this plumage, Laughing Gull superficially resembles Mediterranean Gull but is much darker on mantle and has dusky markings on breast and flanks. **First-summer** resembles first-winter but inner wing-coverts very faded and tail-band less pronounced. Grey on flanks, breast and hindneck often less extensive. **Second-winter** resembles adult winter but separated by more extensive black on outer wing plus presence of some black on tail and secondaries. Bill can have some hint of red at tip. **Second-summer** is like second-winter but has full or almost complete hood and predominantly dull red bill. **Adult non-breeding** lacks blackish hood and has only a greyish patch of variable intensity on ear-coverts that usually extends across nape and rear crown. Bare parts are more blackish but bill has red tip. White tips to outer primaries are more obvious. In flight shows very little white on upperwing and, at rest, pale tips to primaries are very small or lacking. **VOICE** Usual call is a high-pitched, laughing *ha-ha-ha*. **RANGE** Breeds on east coast of N America from Nova Scotia to Florida, Mexico and through Caribbean to Venezuela. Winters as far south as NE Brazil and N Chile. **STATUS** Accidental with just three records. **Azores★** Adult in almost complete breeding plumage, Fajã Grande (Flores), 6 Nov 1998; and first-winter with a broken wing, Corvo, 9 Aug 2004. **Canaries** Second-winter, San Sebastian (La Gomera), 12 Jan & 12 Feb 1996, and possibly present since 19 Dec 1995.

FRANKLIN'S GULL *Larus pipixcan* Plate 39 (3)
Sp. Gaviota de Franklin **P.** Gaivota-das-pradarias (Gaivota de Franklin)

Monotypic. **L** 32–36 cm (12½–14″); **WS** 85–95 cm (33½–37½″). **IDENTIFICATION** Sexes similar. Adult breeding plumage attained in second summer. Most likely confusion species is Laughing Gull. Franklin's separated from other hooded gulls by much darker grey mantle, which approaches that of paler form of *graellsii* Lesser Black-backed Gull. Also, identified easily in winter by presence of at least a half-hood, unlike any other related species. Distinguished from Laughing Gull by shorter bill, very broad white crescents both above and below eye (in all plumages), paler mantle and upperwing, and grey centre to tail, which is diagnostic and unique in adult gulls, but it can be difficult to see. **Juvenile** unlikely to be seen in our region. **First-winter** has half-hood, grey mantle and scapulars, blackish primaries, blackish bar on secondaries, and black band on tip of greyish tail does not extend onto outer pair of rectrices, which are white. **First-summer** has half-hood, grey upperparts and upperwing, except black tip and white trailing edge to inner primaries and secondaries; outer primaries have small white tips and there is sometimes a partial secondary bar or tail-band. **Second-winter** resembles first-summer but often has pink flush to underparts, black on primaries is less extensive, white tips larger and white line separating black primary tips from grey bases is present on middle primaries. **Adult summer** has full black hood, pinkish flush to underparts, less extensive black on outer primaries than second-winter, and white line separating black from grey bases extends across all primaries. White tips to primaries more extensive than in second-winter. **Adult non-breeding** resembles adult summer but only has half-hood. **VOICE** A soft *krruk* or a shrill, repeated *kuk*. **RANGE** Breeds in C North America from Canada south to Minnesota, and winters mainly on Pacific coast from Guatemala to Chile. **STATUS** Vagrant with only three records. **Azores** One, Santa Cruz (Flores), 15 Jun 1984. **Madeira** One, Funchal (Madeira), Jul 1979. **Canaries** Winter-plumaged adult, Playa de los Charcos (Lanzarote), 14 Jan 1998.

LITTLE GULL *Larus minutus* Plate 39 (5)
Sp. Gaviota Enana **P.** Gaivota-pequena

Monotypic. **L** 25–27 cm (10–10½″); **WS** 75–80 cm (29½–31½″). **IDENTIFICATION** Sexes similar. Adult breeding plumage attained in third summer. The smallest gull in the world. Juvenile resembles juvenile Kittiwake and adult could be confused with Bonaparte's Gull or, perhaps, Black-headed Gull. In all plumages, small size, rather rounded wingtips (except first-years) and buoyant flight action are useful identification pointers. **Adult breeding** very distinctive with black hood extending onto upper neck, pinkish flush to underparts, pale grey mantle and upperwings, except white tips to all flight feathers, forming complete white trailing edge to both wing surfaces, and underwing otherwise uniformly blackish. **Juvenile** has predominantly dark blackish-brown crown, hindneck and mantle (extending onto breast-sides), prominent dark W pattern on wings, faint dark secondary bar and dark tail-band. Underwing predominantly white. **First-winter** similar to juvenile but mantle is grey and darker areas faded and less prominent. **First-summer** resembles first-winter but has variable (but never complete) hood, dark W and tail-band even more faded and usually broken in middle. **Second-winter** has dusky undersides to primaries and secondaries, white head except dusky patch on crown and spot on ear-coverts, white body and tail, pale grey upperwing and mantle, with blackish subterminal marks on outer primaries and white trailing edge. **Second-summer** resembles adult breeding but has underwing with whiter axillaries and coverts. **Adult non-breeding** has dusky patch on the crown, dark ear-spot and eye-crescents, grey hindneck and dark blackish underwing. **VOICE** A short, harsh, hard, nasal *kik-kik* somewhat reminiscent of a *Chlidonias* marsh tern. **RANGE** Breeds mainly in Siberia and the Baltic region. Winters at sea, from the English Channel, North Sea and Irish Sea south through the Mediterranean to NW Africa and the Caspian. **STATUS** Rare or accidental visitor recorded on three archipelagos. **Azores** Three records of first-winters: one dead, salt works (Santa Maria), 1 Jan 1983; Ribeira Grande (São Miguel), 28 Dec 2001; and, Porto Pim (Faial), Feb 2002. **Madeira** One, Funchal (Madeira), 13 Jan 1981, was present a few days. **Canaries** Rare winter visitor and passage migrant, recorded on Lanzarote, Gran Canaria and Tenerife.

SABINE'S GULL *Larus sabini* Plate 39 (4)
Sp. Gaviota de Sabine **P.** Gaivota de Sabine

Monotypic. **L** 27–32 cm (10½–12½″); **WS** 90–100 cm (35½–39½″). **IDENTIFICATION** Sexes similar. **Adult breeding** plumage attained in second summer; in this plumage, unmistakable with a dark grey hood and black border, and black bill with yellow tip. In all plumages has a striking upperwing pattern and notched tail. **Juvenile** has extensive grey-brown head and breast-sides, scaly grey-brown upperparts and inner wing-coverts, white inner primaries and secondaries, black outer primaries, black tail-band and black bill. **First-winter** has wings, tail and bill of juvenile combined with head and body plumage of adult winter. **First-summer** resembles adult but hood incomplete and retains some juvenile features, such as smaller white primary tips and a few subterminal marks on tail. **Second-winter** resembles adult non-breeding and only those birds with some retained immature feathers are separable. **Adult non-breeding** has hood reduced to dark eye-crescents and variable area on nape and hindneck. White primary tips reduced or lacking through fading and wear. VOICE A harsh tern-like *prree* most frequently given by juveniles. RANGE Breeds in arctic and winters off S Africa and W South America. STATUS Accidental but perhaps more regular than records suggest. **Azores** Five records: one, Flores, Jun 1950; one, Ponta Delgada (São Miguel), Oct 1977; adult in breeding plumage, between São Miguel and Terceira, 5 Jun 1984; one, Horta (Faial), 12 Jun 1984; and one, off Mosteiros (São Miguel), 28 Sep 1998. **Madeira** Immature, Funchal Harbour (Madeira), 9 Jan 1994; and first-winter, Porto Moniz, 26 Jan 2003. **Canaries** Considered accidental by Martín & Lorenzo (2001) but probably occurs regularly in Canarian waters. **Cape Verdes** Two records, both adults: 9 km south of Santo Antão, 14 Apr 1976, and southeast of Boavista, 24 Apr 1976. May be more regular than records suggest.

BONAPARTE'S GULL *Larus philadelphia* Plate 42 (5)
Sp. Gaviota de Bonaparte **P.** Garrincho-americano (Gaivota de Bonaparte)

Monotypic. **L** 28–30 cm (11–12″); **WS** 90–100 cm (35½–39½″). IDENTIFICATION Sexes similar. Adult breeding plumage attained in second summer. In all plumages resembles slightly larger Black-headed Gull. **Juvenile** unlikely to be seen in our region but differs from Black-headed in the predominantly black bill, lack of ginger tones to mantle and breast-sides, and lack of dusky underside to inner primaries. **First-winter** has blacker and more well-defined ear-spot, grey mantle a shade darker than in Black-headed, underwing lacks dusky inner primaries as in juvenile, and bill shorter and blacker. **First-summer** has variable black hood but is otherwise similar to first-winter except for dark carpal bar, secondary bar and tail-band more faded. **Adult non-breeding** has a narrow black trailing edge to primaries, white tail, head pattern as first-winter and a grey hindneck and breast-sides. **Breeding adult** has dark greyish-black hood, white eye-crescents and underparts often with a pink flush. VOICE Feeding call is a harsh, rasping tern-like *tee-er*. RANGE Breeds across Alaska and N Canada and winters in USA, C America and West Indies. STATUS Vagrant with only nine records. **Azores** Six records; first three at Point Delgada (São Miguel), Nov 1977, Sep 1979, and Dec 1982 (all of 2-3 birds), with first-winters, Lajes do Pico (Pico), 11 Nov 1998, Horta (Faial), 20 Oct 2002.and Praia da Vitoria (Terceira), 5 Nov 2001 and 19–21 Feb 2003. **Canaries** Three: adult, Lanzarote, 22 Dec 1995; first-winter, Punta del Camisón (Tenerife), 27 Nov 2000; and second-winter, Puerto Naos (La Palma), also 27 Nov 2000.

BLACK-HEADED GULL *Larus ridibundus* Plate 39 (6)
Sp. Gaviota Reidora **P.** Garrincho (Guincho-comum)

Monotypic. **L** 34–37 cm (13½–14½″); **WS** 100–110 cm (39½–43½″). IDENTIFICATION Sexes similar. Adult breeding plumage attained, by vast majority, in second summer. **Juvenile** has extensive ginger brown on head, mantle, scapulars and breast-sides, dark brown carpal bar, dark subterminal bar across secondaries, dark tips to primaries, fairly extensive white on outer wing and blackish subterminal band on otherwise white tail. Legs fleshy coloured and bill fleshy with a dark tip. **First-winter** resembles juvenile but mantle and scapulars grey; carpal bar, secondary bar and tail-band slightly faded and head has characteristic winter pattern of a dark spot behind eye, often indistinctly extending across rear crown, and dusky eye-crescents with faint greyish band over crown. **First-summer** is like first-winter but has an incomplete blackish-brown hood of variable extent and bars on wings and tail very worn and faded. **Adult non-breeding** loses all markings on tail, which is now all white, wings are grey with white wedge on outer wing and black tips to outer primaries, underwing has extensive blackish on outer primaries and head pattern is like first-winter. **Adult breeding** is like adult non-breeding but has full brown hood and white eye-crescents, and some have a pinkish flush on underparts. VOICE Common call is a loud, downslurred, rolling *karrr* or *krreearrr*, also a sharp *kek* when feeding. RANGE Breeds from Iceland and W Europe east to Kamchatka, Sakhalin and NE China, and winters south to the Gulf of Guinea. STATUS Regular winter visitor with records from all archipelagos. **Azores** Recorded from all islands. **Madeira** Regular winter visitor in small numbers. **Canaries** Regular winter visitor and passage migrant to all islands; most frequent in Nov–Feb. **Cape Verdes** Regular but uncommon winter visitor with most records from Santiago, São Vicente, Sal and Boavista.

SLENDER-BILLED GULL *Larus genei* Plate 40 (3)
Sp. Gaviota Picofina **P.** Gaivota-de-bico-fino

Monotypic. **L** 42–44 cm (16½–17½″); **WS** 100–110 cm (39½–43½″). IDENTIFICATION Sexes similar. Adult breeding plumage attained in second summer, when should be unmistakable, but other plumages closely resemble slightly smaller and much commoner Black-headed Gull. In breeding plumage differs from Black-headed primarily in lack of dark hood and having a distinct pinkish wash to white underparts, although some Black-headed can show the latter. At other ages structural differences probably of greater use, as plumages of the two are rather similar. Perhaps the most important feature of Slender-billed Gull is its distinctive head shape: the more sloping forehead and distinctly longer, but not thinner, bill give it an elongated profile. First-year Slender-billed differs from Black-headed in having paler brown wing markings, paler bill and legs and, if present, a smaller and less prominent ear-spot, a feature also shared in adult non-breeding plumage. At all ages, except early juvenile, iris pale yellowish, whereas in Black-headed the iris is always dark. **Juvenile** has extensive grey-brown on head and mantle. **First-winter** develops pale grey mantle, bars across carpals and secondaries (darker on latter), dark tail-band, pale grey ear-spot, and pale bill and legs. **First-summer** resembles first-winter but bars on wings and tail are faded. **Adult non-breeding** has pale grey mantle, scapulars and upperwing, with dark tips to white outer primaries, pale grey ear-spot, pale pinkish flush to underparts and dark red bill and legs. **Adult breeding** similar but has all-white head and stronger pinkish flush to underparts, particularly on breast and belly. VOICE A rolling *krerrr* similar to Black-headed Gull but lower and more nasal. RANGE Breeds in isolated populations from Mediterranean and NW Africa east to Pakistan and Kazakhstan. Winters mostly in Mediterranean, Caspian and Red Seas, and in Persian Gulf. STATUS Only recorded from two archipelagos and status varies considerably. **Canaries** Breeding attempted 1976 but despite eggs being laid no young were seen. On Fuerteventura probably a regular visitor, but on other islands it is a vagrant recorded only from Lanzarote (twice) and Tenerife (once). **Cape Verdes** Two records: immature, Santa Maria (Sal), 16 Feb 1994; and first-winter, Feijoal Lagoa (Sal), 12 Jan 2005.

AUDOUIN'S GULL   *Larus audouinii*   Plate 40 (2)
Sp. Gaviota de Audouin   **P.** Gaivota de Audouin

Monotypic. **L** 48–52 cm (19–20½″); **WS** 115–140 cm (45½–55″). **IDENTIFICATION** Sexes similar. Adult breeding plumage attained in fourth summer. Quite distinctive and appears more elegant than Yellow-legged or Lesser Black-backed Gulls, with which it is most likely to be confused. Adult unique in having a rather stubby red-based bill with a yellow tip and black band on gonys. Upperparts paler than either Yellow-legged or Lesser Black-backed and upperwing lacks prominent white trailing edge. Wingtips more predominantly dark, as white primary tips are smaller, and only a very small 'mirror'. Seen perched grey-green legs also diagnostic. Juvenile and first-winter more difficult to identify and structural differences are among the best features, namely the generally shorter and stubbier bill, more elongated forehead, longer neck when alert and longer, narrower wings. Also, the more strongly patterned underwing, prominent white U formed by uppertail-coverts and two-toned bill are useful at these ages. **Juvenile** has smooth-looking greyish-brown head and underparts with darker patch on rear flanks, very scaly upperparts, brownish-black primaries, secondaries and tail, and blackish soft parts. In flight prominent white U on uppertail-coverts and whitish central panel on underwing, across greater coverts and inner greater primary-coverts. **First-winter** resembles juvenile but head and underparts whiter and less smooth grey-brown, upperparts less distinctly scaly and scapulars grey with brown centres (diagnostic). **First-summer** resembles first-winter but head and underparts whiter with dark half-collar at base of neck, wings and tail worn and faded, mantle has much grey and bill paler (sometimes reddish) with dark tip or subterminal band. **Second-winter** more like adult but has prominent blackish bar on secondaries, black subterminal band on tail, extensively dark outer wing, predominantly white underwing, dark eye-crescents and brown streaking on nape. Bill often has reddish base. **Second-summer** resembles second-winter but head white and dark areas of upperwing faded and browner. Bill base usually reddish and there is a red orbital ring. **Third-winter** similar to adult but has extensive black on outer greater primary-coverts and tail sometimes has dark subterminal marks. **Third-summer** is like third-winter but white primary tips reduced or lacking through wear. **Adult non-breeding** has black wingtip confined to primaries, all-white tail, dark red bill with black subterminal band and yellow tip (often appearing all black at long range), and greyish or greenish legs. **Adult breeding** resembles adult non-breeding but has white primary tips reduced or lacking through wear. VOICE All calls harsher and deeper than those of Yellow-legged Gull. RANGE As a breeder confined to the Mediterranean, where it occurs in isolated populations. Winters in the Mediterranean and in coastal NW Africa. STATUS Accidental, recorded from two archipelagos. **Madeira** First-winter, Funchal harbour (Madeira), 16 Nov 2000. **Canaries** Recorded from Lanzarote (twice), Fuerteventura (three times), Tenerife (nine times) and La Gomera (once). One, Fuerteventura, Jan 2003, had been ringed as a chick in Jun 1999 in the Ebro Delta.

RING-BILLED GULL   *Larus delawarensis*   Plate 40 (4)
Sp. Gaviota de Delaware   **P.** Gaivota-de-bico-riscado (Gaivota do Delaware)

Monotypic. **L** 43–47 cm (17–18½″); **WS** 120–155 cm (47–61″). **IDENTIFICATION** Sexes similar. Adult breeding plumage attained in third summer. In all plumages confusion possible with Common Gull but Ring-billed is slightly larger and bulkier with a heavier bill, more aggressive facial expression and paler grey upperparts. Juvenile unlikely to be seen in our region as this plumage is usually quickly lost. **First-winter** has mantle and scapulars pale grey, head and underparts white with distinct brownish spotting on nape/hindneck and crescents, spots and bars on breast-sides and flanks. Tail-band rather variable but rarely as clear-cut as Common Gull and uppertail-coverts usually rather strongly barred brownish. Upperwing dark on outer primaries with a dark bar on secondaries and predominantly dark coverts. Pale grey mid-wing panel formed by greater coverts and inner primaries. Bill pink with a broad dark tip, and a small pale spot develops on extreme tip. **First-summer** resembles first-winter but head and body whiter, dark areas on wings and tail paler and pale bill tip more extensive. **Second-winter** has more extensively grey upperwing with just traces of a tail-band and sometimes a hint of a bar on secondaries. If any, there is only one small mirror on the outer primary, often visible only on underwing. Iris pale and band on bill well developed. **Second-summer** resembles second-winter but dark areas of wings faded, head and body whiter with just a few pale spots on hindneck, and eye, bill and legs obtain yellow coloration of adult. **Adult non-breeding** has distinct dark spotting on head and hindneck, black wingtip has white mirror on two outer primaries and tail white. Legs and bill yellow, the latter with broad black band near tip. **Adult breeding** similar but head white and bill and legs brighter. VOICE Similar to Yellow-legged Gull but higher pitched and more nasal. The long-call, which is delivered with head thrown back, differs from Common Gull in being higher pitched and harsher but not as nasal. Usually rather silent away from breeding grounds. RANGE Breeds in S and C Canada and N USA. Winters mainly in E USA, West Indies and C America. STATUS A scarce but regular visitor recorded from all archipelagos but status varies between them. **Azores** Probably annual, sometimes occurring in flocks, as in Jan 1982, when 48 were present on Terceira, and in Feb 2001 56 were on Terceira, at Praia da Vitoria with 34 there 19 Feb 2003. **Madeira** Accidental or scarce visitor, sometimes in small groups. **Canaries** Accidental visitor mostly in winter, with more than 20 records and recorded from all main islands except La Gomera. **Cape Verdes** Two records: first-winter moulting to second-summer, Rabil lagoon (Boavista), 22 Apr 2001; and first-winter, Mindelo (São Vicente), 17–28 Dec 2004.

COMMON GULL (MEW GULL) *Larus canus* Plate 40 (5)
Sp. Gaviota Cana   P. Famego (Gaivota-parda)

L. c. canus and *L. c. brachyrhynchus*. **L** 40–42 cm (15½–16½"); **WS** 110–130 cm (43½–51"). IDENTIFICATION Sexes similar. Adult breeding plumage attained in third summer. In all plumages most likely to be confused with Ring-billed Gull but Common Gull is daintier with a distinctly smaller bill and darker grey upperparts. **Juvenile** unlikely to be seen in our region. **First-winter** closely resembles Ring-billed Gull but has a more solid and clear-cut black tail-band, less contrast between grey mid-wing panel and dark coverts and secondary bar, darker grey mantle and less distinct markings on head and underparts. **First-summer** is similar but head and body whiter, upperwing pattern faded and pale mid-wing contrasts with darker mantle. **Second-winter** has head and body like first-winter but tail usually white, sometimes with a few dark subterminal marks. Upperwing predominantly grey with black tip and two white mirrors. Bill typically grey-green with a dark tip or subterminal band. **Second-summer** is like second-winter but has head whiter and black on wings browner. **Adult non-breeding** has white head with fine streaking and spotting, densest on hindneck, and mantle and wings blue-grey with dark tip on outer six primaries. Outer two primaries have a large white mirror and all have prominent white tips. Bill yellowish with narrow band or dark subterminal mark. **Adult breeding** is similar but has a white head, yellow or yellowish-green bill and smaller white primary tips. VOICE Most frequent is a laughing, high-pitched and rather nasal *kleeoo*. RANGE Breeds from Iceland and the British Isles east to Kamchatka and Sakhalin, and south to Lake Balkhash, and in NW North America. Winters in and south of breeding range to C Europe and Turkey east to S China, and south to N Mexico. STATUS Rare and irregular winter visitor with records from three archipelagos. **Azores** Accidental with 12 records: São Miguel (9), Terceira (2) and Flores (1). The only record of the N American subspecies *brachyrhynchus* for the region is from the Azores★: a non-breeding adult, Praia da Vitoria (Terceira), 18 Feb–24 Mar 2003. **Madeira** Listed by Zino *et al.* (1995) as exceptional. **Canaries** Scarce and irregular winter visitor recorded from Lanzarote, Fuerteventura, Gran Canaria and Tenerife. Some of these, particularly some of the older records, may relate to Ring-billed rather than Common Gull.

LESSER BLACK-BACKED GULL *Larus fuscus* Plate 41 (1)
Sp. Gaviota Sombría   P. Gaivota-d'asa-escura

L. f. graellsii, *L. f. intermedius* and *L. f. fuscus*. **L** 52–67 cm (20½–26½"); **WS** 135–155 cm (54–61"). IDENTIFICATION Sexes similar. Adult breeding plumage attained in fourth summer. **Juvenile** has head and underparts streaked dark grey-brown, neatly scaled mantle and scapulars, dark primaries and primary-coverts, dark bar on secondaries, whitish rump and solid blackish subterminal band on tail. **First-winter** closely resembles juvenile but dark patch on ear-coverts less well defined and scapulars more uniform and less scaly. **First-summer** similar but head and underparts whiter and upperparts more uniform through wear. **Second-winter** has some adult-coloured feathers on mantle and scapulars, but is otherwise like first-winter, except whiter tail with narrower subterminal band, brownish (rather than barred) inner wing-coverts and bill often extensively pale with a dark subterminal mark.

Second-summer resembles second-winter but head and underparts whiter and mantle and scapulars like adult, a dark ashy grey. Bill and legs yellowish or bright yellow, the former with a dark subterminal mark. **Third-winter** resembles non-breeding adult, but black wingtip less well defined, mirror often lacking, tail has an indistinct band and there is a dark subterminal mark on bill. **Third-summer** resembles third-winter but head is white and wings show brownish patches through fading and wear. **Adult non-breeding** has streaking on head, grey or black mantle and upperwing, with a small white mirror on outer primary and, often, an even smaller one on second. Underwing predominantly white with a dark subterminal trailing edge and blackish outer primaries. Bill deep yellow with a red spot near gonys. **Adult breeding** is similar but with white head and body, and the upperparts show some brownish tones through fading; *graellsii* is palest of the three subspecies and therefore the closest to *atlantis* Yellow-legged Gull. Mantle and upperparts dark slate-grey contrasting with black wingtip, and very heavily streaked on head in winter. Subspecies *intermedius* has darker upperparts but still shows some contrast with black wingtip, and the winter head streaking is usually less heavy than *graellsii*. Nominate *fuscus* is the darkest and shows hardly any contrast with the black wingtip, it usually has only one mirror (unlike *graellsii*), the head shows very little streaking in winter, and it is generally smaller and slimmer. Great care should be taken if attempting to identify the various subspecies, as intergrades occur. **VOICE** All calls similar to those of Yellow-legged Gull. **RANGE** Breeds from Iceland east to the W Kola Peninsula and White Sea, and south to NW Spain and Portugal. Winters within the breeding range and south to w and E Africa. **STATUS** A regular visitor particularly in winter and recorded on all archipelagos. **Azores** Regular winter visitor. **Madeira** Fairly common winter visitor and passage migrant. **Canaries** Common winter visitor and passage migrant, but recorded in all months with records from all islands. Most are *graellsii* and *intermedius* but *fuscus* has also been reported. In 2001 the first breeding record was reported from Alegranza, with another pair on Montaña Clara, where breeding was suspected but not confirmed. **Cape Verdes** *c.* 12 records from Santo Antão, São Nicolau, Sal and Boavista. Both *graellsii* and *fuscus* reported.

AMERICAN HERRING GULL *Larus smithsonianus* Plate 41 (3)
Sp. Gaviota Argéntea Americana **P.** Gaivota-prateada-americana

Monotypic. **L** 53–65 cm (21–25½″); **WS** 120–150 cm (47–59″). **IDENTIFICATION** Sexes similar. Adult breeding plumage attained in fourth summer. Resembles Herring Gull at all times but structurally slightly larger and heavier with a flatter forehead and stronger bill. Most easily distinguished in first-winter and juvenile plumages. **Juvenile** has head and underparts dark chocolate- or greyish-brown with an almost velvety appearance, and generally more uniform than Herring Gull. Rump heavily barred dark brown to greyish-brown, whilst undertail-coverts distinctly but narrowly barred and tail uniform blackish-brown. Underwing-coverts darker and more uniform than Herring Gull thus contrasting more with paler flight feathers. **First-winter** similar but head usually paler, even almost white, and contrasts considerably with dark marks on hindneck/upper mantle and fairly uniform brown underparts. Most of plumage is like juvenile but mantle and, to a lesser degree, scapulars obtain first-winter feathers similar to Herring Gull, grey with dark subterminal markings of variable extent. In some, bill is pink with a black tip as in Glaucous Gull. The easiest to identify at this age are those with a combination of pale head, pink bill with black tip and velvety brown underparts. **First-summer** is similar to first-winter but through wear and bleaching; head and underparts become paler and flight feathers and tail bleach brownish. Best separated from same-age Herring Gull by the mainly dark tail and dark-barred rump. Possibly the best feature in **second-winter** plumage for separating this species from Herring Gull is the contrast between white rump and predominantly black tail, but upperwing is more contrasting with slightly darker secondaries and no small mirror on p10, as shown by many Herring Gull. By **second-summer** head and underparts mainly white and mantle and scapulars predominantly grey, and very similar to Herring Gull but tail markings broader and darker. **Third-winter** and **third-summer** plumages almost identical to corresponding plumages of Herring Gull and only the darker, and occasionally more extensive, tail-band may enable identification. Subsequent plumages of American Herring Gull are all very similar to Herring Gull. **Adults** have long grey tongues on p6–8, unlike nominate Herring Gull, and solid black markings on p5–6, which are not shown by nominate Herring Gull. However, most birds in fourth-year or adult plumages may have to remain unidentified. **VOICE** Common call is a *ki-auwww* often combined with a cackling *gag-ag-ag* similar to Herring Gull but lower and faster. Calls of American Herring Gull when played back to Herring Gull do not evoke a response. **RANGE** Breeds across much of N North America. Winters in the breeding range and south to S Central America. **STATUS** Vagrant with records from just two archipelagos. **Azores** Five records: first-winter, Ponta Delgada (Flores), 17 Sep 2001; 3–4 first-winters, Praia da Vitória (Terceira), 19–21 Feb 2003; 3 first-winters, Ponta Delgada (São Miguel), 21 Feb 2003; adult, Horta (Faial), 6 Apr 2004; and first-winter, Cabo da Praia (Terceira), 31 Oct 2004. **Madeira** First-year, Machico (Madeira), 18 Sep 2003.

HERRING GULL *Larus argentatus* Plate 41 (4)
Sp. Gaviota Argéntea **P.** Gaivota-prateada (Gaivota-argêntea)

Subspecies unknown. **L** 56–64 cm (22–25″); **WS** 138–150 cm (54–59″). **IDENTIFICATION** Sexes similar. Adult breeding plumage attained in fourth summer. **Juvenile** rather greyish-brown overall with darker flight feathers and dark subterminal band on tail, but compared to other large juvenile gulls rather featureless. In flight has a conspicuous pale window on inner primaries, a feature much reduced on Yellow-legged Gull and lacking on Lesser Black-backed. Rump darker than on Yellow-legged or Lesser Black-backed Gulls, and underwing-coverts paler than Lesser Black-backed. Legs pink and bill black but often has an extensive pale base. **First-winter** resembles juvenile but head and sometimes underparts whiter, the new scapulars and mantle feathers paler and greyer with dark subterminal bars, and retained juvenile wing feathers worn and paler. Much individual variation in plumage and extremes could be mistaken for Glaucous or American Herring Gulls, but Herring Gull differs from Yellow-legged Gull in having a darker head, darker underparts and usually a paler bill base. **First-summer** is like first-winter but head and underparts paler with wear and darker feathers on wing and tail bleach brownish. **Second-winter** also shows much individual variation but is generally similar to first-winter with plainer grey feathering appearing on mantle, greater coverts more vermiculated, paler head and underparts, eye becomes paler, obvious pale tip to bill, paler and whiter underwing, whiter rump and less extensive tail-band. Some show a pale mirror on p10. It is unusual for any second-winter Herring Gull to show combination of grey saddle and solid black tail-band, features that distinguish Yellow-legged Gull of a similar age. **Second-summer** resembles second-winter but head and underparts predominantly white, mantle and scapulars uniform grey, wings and tail very faded with wear, most have pale eyes and bill may start to attain adult coloration but retains extensive subterminal band. By **third-winter** overall appearance is between second-winter and adult winter. Compared to adult dark wingtip is more extensive and less clear-cut, inner wing has brown markings, particularly on greater and lesser coverts and tertials, highly variable in extent, and pattern of dark subterminal markings on tail and bill, although often like adult, usually paler and still shows dark subterminal mark. Yellow-legged Gull at same age resembles adult even more closely than Herring Gull does, as it has many fewer brown markings on upperwing. **Third-summer** is like third-winter but head and underparts white, brown markings on inner wing faded to whitish. **Fourth-winter** very similar to adult winter and many are unidentifiable, but dark markings on primary-coverts and on both mandibles are features of this age. **Adult breeding** has pure white head, underparts, rump and tail; the upperwings are grey (with a slight bluish tone), with black tips, white trailing edges and large white mirrors on p9–10, the eye is yellow or whitish with a fleshy orbital ring, the bill yellow with a red spot on gonys and legs flesh-coloured. **Adult winter** resembles adult breeding but has strong brown streaking and spotting on head. Adult Yellow-legged Gull has less white on wingtip than Herring Gull with smaller mirrors, yellow legs and, in nominate *michahellis* Yellow-legged Gull, reduced dark streaking on head in winter. **VOICE** Common call is a loud, wailing *keeaahh*, combined with a rather anxious, cackling *gag-ag-ag*. **RANGE** Breeds in Iceland, Scandinavia, the White Sea and Baltic regions, the British Isles and in coastal N Europe to W France. Some winter in breeding range but others reach south to N Spain. **STATUS** Vagrant with records from just two archipelagos. **Madeira** A couple of records, the most recent a third-winter, Funchal harbour (Madeira), 26 Oct–8 Nov 2002. **Canaries★** Five records: one collected, Las Palmas (Gran Canaria), Oct 1912; adult, Cigüeña (Tenerife), Jan 1990; adult, El Médano (Tenerife), 1 Dec 1991; third-winter, Punta de Hidalgo (Tenerife); and first-winter, Santa Cruz (La Palma), 27 Nov 2004.

YELLOW-LEGGED GULL   *Larus michahellis*   Plate 41 (2)
Sp. Gaviota Patiamarilla   **P.** Gaivota-de-patas-amarelas

L. m. atlantis and *L. m. michahellis*. **L** 55–67 cm (22–26″); **WS** 138–155 cm (54–61″). **IDENTIFICATION** Sexes similar. Adult breeding plumage attained in fourth summer. The following refers to race *atlantis* except where stated. **Juvenile** basically intermediate between juveniles of Herring Gull and Lesser Black-backed Gull, with often a pale panel on inner primaries, although this is never as distinct as Herring Gull, the rump whiter than Herring Gull and tail-band blacker and thus more contrasting, the lower belly, forehead and throat paler than Herring Gull, and underwing-coverts dark brown with a gingery tone in some, similar to some Lesser Black-backed Gulls. **First-winter** resembles juvenile and many appear to retain juvenile plumage through the winter. **First-summer** similar but head and underparts whiter with some developing grey feathers on mantle and wing-coverts. **Second-winter** usually has a grey saddle and sometimes grey median coverts, a paler head than first-winter with a more contrasting dark mask around eye, rump usually white but black subterminal tail-band like first-winter, barring on wing-coverts weaker than first-winter giving an almost uniformly dark appearance, which is also less contrasting than first-years. **Second-summer** resembles second-winter but head whiter, eye often yellow and legs distinctly yellow. By **third-winter** resembles adult, unlike most third-winter Herring Gulls. Streaking on head is frequently as strong as adult, giving a hooded appearance, the bill has more extensive dark markings, the pale primary tips are indistinct and the tail still has an extensive dark band. **Third-summer** resembles third-winter but head white or has just a faint greyish area around eye. **Fourth-winter** is like adult non-breeding but some have dark markings on primary-coverts, tail and bill. Legs paler than adult breeding. **Adult breeding** is darkest mantled of Yellow-legged Gull plumages and some, particularly those in the Azores, closely approach the coloration of *graellsii* Lesser Black-backed Gull. The head, underparts, rump and tail are white, the black of wingtips more extensive than Herring Gull with a narrow white mirror on p10, and on some individuals p9, bill yellow with a red gonys-spot and legs bright yellow. **Adult non-breeding** similar but has variable fine streaking on head, particularly on crown and ear-coverts, and in birds from the Azores this streaking is very extensive, creating a hooded appearance. Legs paler and duller yellow. Subspecies *michahellis* is, on average, larger and bulkier than *atlantis*, with a more rounded head, steeper forehead and more slender bill. Upperparts are paler grey, the eye generally darker yellow and bill is also darker yellow. **VOICE** All calls are deeper and more nasal than Herring Gull, closely resembling those of Lesser Black-backed Gull, and varying from a guttural *gágagaga*, a powerful, hollow, disyllabic *gulo* or *gao* to a pitiful, smothered *kiau-kiau-kiau-kiau*. **RANGE** Breeds from W France discontinuously south along Atlantic coast, around Mediterranean and on the Atlantic Islands. Some winter as far north as Britain, Denmark and S Sweden, and south to Senegambia and the Gulf of Guinea. **DISTRIBUTION** A resident breeder on the Azores, Madeira and Canaries. In winter the local population in the Canaries is augmented by a small number of *michahellis*. **BREEDING** Clutch normally 2–3, although 1–4 known. Nests less than elaborate, mostly of sticks, seaweed and feathers, and placed near a rock or plant, but are used habitually by the same birds each year. Eggs are usually laid in first half of April and hatching begins early May. **HABITAT** Can be seen almost anywhere, although the coastal zone is the main habitat. For example, on the eastern Canaries it is often encountered among sand dunes at Jandía, Fuerteventura, and in Madeira they have been seen on the windswept, high-altitude grassy plain on Paul da Serra. **TAXONOMIC NOTE** Some authorities consider the race *atlantis* should be considered specifically distinct, as Atlantic Gull *L. atlantis*. Others believe that birds on the Azores are the only true representatives of this taxon and that more study of those on Madeira, the Canaries and, in particular, on the Atlantic coast from N Iberia to Morocco is required to discover their true affinities to *L. m. atlantis*. **STATUS** A common resident breeder on three archipelagos but only an irregular visitor to the fourth. **Azores** Breeds on all main islands with an estimated population of 2,800 pairs. The largest colonies are those at Ilhéus das Cabras, off Terceira (400 pairs), Ilhéu do Topo, off Sao Jorge (300 pairs), and Lagoa do Fogo, on São Miguel (250 pairs). **Madeira** Breeds on all islands including the Salvages, and the population is estimated at 6,000 pairs. **Canaries** Breeds on all islands and islets except La Graciosa. Population estimated at 4,000–4,700 pairs, with La Gomera (1,000) and Gran Canaria (900) having the largest numbers. **Cape Verdes** True status unknown due to identification problems between Yellow-legged and Lesser Black-backed Gulls in immature plumages, but probably a scarce winter visitor, with records from São Vicente, São Nicolau, Raso and Boavista.

ICELAND GULL   *Larus glaucoides*   Plate 42 (1)
Sp. Gaviota Polar   **P.** Gaivota-branca (Gaivota-polar)

L. g. glaucoides. **L** 52–60 cm (20½–23½″); **WS** 140–150 cm (55–59″). **IDENTIFICATION** Sexes similar. Adult breeding plumage attained in fourth summer. In all plumages very similar to Glaucous Gull and structural differences are important, including the smaller bill and more rounded head of Glaucous, in addition to the longer primary projection giving this species a more attenuated rear end at rest. **Juvenile/first-winter** generally rather milky coffee-coloured, with pale primaries and barred upperwing-coverts, mantle, rump and tail, and usually a predominantly dark bill. **First-summer** similar but head and underparts usually very pale, almost white, as are wing-coverts. Mantle and scapulars are also often very pale but with coarse dark barring. **Second-winter** is very whitish with less barring than first-winter and head and underparts often have some coarse streaking. Diagnostic first-winter bill pattern is still evident, but usually the base is paler, resulting in a pattern similar to Glaucous Gull. **Second-summer** similar but head and underparts lack coarse streaking and distinctive bill pattern lost. Some grey usually evident on mantle and scapulars. **Third-winter** is similar to adult, with grey appearing on wings, and rump is white. Head similar to adult and often has dense streaking. **Third-summer** is like third-winter but lacks streaking (or is only lightly streaked) on head, and remaining brown areas on wings and tail are more faded. **Adult non-breeding** has pale grey upperwings and mantle with a white trailing edge to wings. Head, body and tail white with extensive brownish streaking on head and bill often has greenish tinge (not shown by Glaucous). **Adult breeding** similar but head white and bill is brighter. The orbital ring is usually red unlike Glaucous. Note that although as yet unrecorded from our region, the possibility that some birds are of the race *kumlieni* must be considered. **VOICE** Similar to Herring Gull but slightly higher pitched. **RANGE** Nominate *glaucoides* breeds on coasts of S Greenland and winters in this range as well as in Iceland, NE North America and, rarely, in NW Europe. Race *kumlieni* breeds in arctic NE Canada and winters in NE North America but has been recorded in NW Europe. **STATUS** Varies between archipelagos on which it has been recorded. **Azores** Previously considered accidental but recent observations suggest that it occurs regularly in small numbers in winter. **Madeira** Accidental with the most recent records in Dec 1999, 12 Jan 2002, 12 Mar 2002 and again 7 Apr 2002. **Canaries★** Two records: one, Puerto de Santiago (Tenerife), 14 Feb 1996; and first-winter, various locations S Tenerife, 1–9 Dec 1999.

GLAUCOUS GULL   *Larus hyperboreus*   Plate 42 (2)
Sp. Gaviota Hiperbórea   **P.** Gaivotão-branco (Gaivota-hiperbórea)

L. h. hyperboreus. **L** 62–68 cm (24½–26½″); **WS** 150–165 cm (59–65″). **IDENTIFICATION** Sexes similar. Adult breeding plumage attained in fourth summer. On average, noticeably larger than Iceland Gull, being closer to Great Black-backed Gull rather than Yellow-legged Gull in size. Also has a much larger bill and more angular head shape, which afford a much more aggressive expression. Primary projection much shorter than in Iceland Gull, giving a blunter appearance to the rear body. In juvenile and first-winter plumages bill colour diagnostic and makes identification easier than in some other plumages, being very obviously bicoloured with a dark tip and basal two-thirds pinkish. In all other plumages it is very similar to Iceland Gull and structural differences are the best means of separating the two. **VOICE** All calls are hoarser than Iceland Gull. **RANGE** Breeds in the arctic and in Europe winters regularly to *c.* 50°N. **STATUS** Accidental visitor to three archipelagos. **Azores** About 20 records, most from São Miguel but also on Faial, Terceira, Graciosa and Flores. **Madeira** Listed by Zino *et al.* (1995) as exceptional. **Canaries** Eight records: one, Arrecife (Lanzarote), 29 Mar & 17 Apr 1984; one, Corralejo (Fuerteventura), 4 Apr 1984; second-winter, Las Palmas (Gran Canaria), 13 Mar 1998; adult, Puerto de la Cruz (Tenerife), 18 Feb–5 Mar 1999; one, El Médano (Tenerife), 16 Nov 1999; immature, Santa Cruz (La Palma), 25 Oct 2000; immature, Tazacorte (La Palma), 20 Nov 2000; and second-winter, Punta de Jandía (Fuerteventura), 18 Jan 2003.

GLAUCOUS-WINGED GULL   *Larus glaucescens*   Plate 42 (3)
Sp. Gaviota de Bering (Gaviota Glauca)   **P.** Gaivota do Mar de Bering

Monotypic. **L** 66 cm (26"); **WS** 147 cm (58"). **IDENTIFICATION** Sexes similar. Adult breeding plumage attained in fourth summer. Juvenile and first-year resemble Glaucous Gull but bill is all black or has just a small pale base, primaries more uniform with rest of upperwing, whereas on Glaucous Gull they are obviously paler, and greater coverts more uniform with less distinct barring. Second-winter and second-summer differ in similar ways from same-age Glaucous Gull as first-years but are even more uniform and lack any barring on wing-coverts. Adult coloration appears on mantle and scapulars but bill still mostly black in second-winter and is more bicoloured in second-summer, with dark tip still being more extensive than Glaucous Gull. Third-winter and third-summer resemble adult but have wing-coverts washed brown, the tail has variable pale brown colouring, the primary pattern is less distinct and there is often a dark subterminal mark on bill. Adult non-breeding resembles Glaucous Gull but has extensively grey-patterned primaries, which feature also distinguishes Glaucous-winged from Glaucous in adult breeding plumage. At a distance wingtip of Glaucous-winged appears grey, whereas that of Glaucous is distinctly white. **VOICE** Similar to Herring Gull but slightly more hollow and flatter. **RANGE** Breeds from the Commander Is. through the Aleutians, S Alaska, W Canada and NW Washington. Winters south to Japan and Baja California. **STATUS** Extreme vagrant. **Canaries** Third-year, photographed at La Restinga (El Hierro), 7–10 Feb 1992.

GREAT BLACK-BACKED GULL   *Larus marinus*   Plate 41 (5)
Sp. Gavión   **P.** Gaivotão-real

Monotypic. **L** 64–78 cm (25–30¾"); **WS** 150–165 cm (59–65"). **IDENTIFICATION** Sexes similar. Adult breeding plumage attained in fourth summer. In all plumages large body size and heavy bill should be sufficient to separate it from commoner species. **Juvenile** similar to juvenile Yellow-legged Gull but darker coverts contrast more with overall paler ground colour of body, and pale barring on greater coverts is broader. **First-winter** is similar but head and underparts whiter and mantle and scapulars have more complex barring. **First-summer** resembles first-winter but head and underparts even whiter, pattern on mantle and scapulars less contrasting, darker feathering on wings and tail worn and faded, and bill has small pale tip and often a paler base. **Second-winter** is like first-winter but usually has a few blackish-grey feathers on mantle and scapulars, the innerwing-coverts (particularly the greater coverts) are more uniform grey-brown with finer barring than in first-years, and bill although highly variable has a broader pale tip than in first-years with a diffuse pale base. **Second-summer** resembles second-winter but has less streaking on white head and underparts, mantle and scapulars mostly blackish with a few brown feathers and bill pale whitish or yellowish with a variable black subterminal band or black patches, and some even show traces of a red gonys-spot. **Third-winter** has white head and underparts with some dark streaking on hindneck and around eye, and upperparts start to resemble adult, but mantle and scapulars less black and many brown-tinged feathers among lesser and greater coverts, whilst white tail shows remnants of dark band and bill is dull yellow with black subterminal mark and signs of a red gonys-spot. **Third-summer** similar but head and underparts white, upperwing appears patchier through wear and white primary tips reduced or lacking. **Fourth-winter** is like adult winter but white tip to p10 is broken by a subterminal black mark, the bill can show blackish subterminal marks and upperparts sometimes retain an indistinct brownish tinge. **Adult breeding** has white head, rump, tail and underparts with blackish upperparts and black primaries, a white trailing edge to inner primaries and secondaries, and white tips to the outer primaries especially large on p9–10. Bill yellow with a red gonys-spot and legs pink. **Adult non-breeding** resembles adult summer but head has some fine dark streaking around eye and on hindneck. **VOICE** All notes deeper and harsher than other large gulls. A deep, echoing *oow-oow-oow* is distinctive and laughing display call is slower and shorter than Herring or Lesser Black-backed Gulls. **RANGE** Breeds in NE North America,

Greenland and N Europe to NW Russia. Many winter in the breeding range but also some southward movement, with European birds recorded on Iberian and Moroccan coastlines. **STATUS** Scarce and irregular winter visitor with records from three archipelagos. **Azores** Regular visitor in small numbers, most frequent Sep–May but recorded all months and from all islands. **Madeira** Accidental visitor with the most recent records a first-winter, Funchal (Madeira), 14–15 Mar 2001, and adult, Lagoa near Santa da Serra (Madeira), 19–21 Mar 2001. **Canaries** Scarce and irregular winter visitor recorded from all main islands except La Gomera and El Hierro.

(BLACK-LEGGED) KITTIWAKE *Rissa tridactyla* Plate 42 (4)
Sp. Gaviota Tridáctila **P.** Gaivota-tridáctila

R. t. tridactyla. **L** 38–40 cm (15–15½″); **WS** 95–120 cm (37½–47″). **IDENTIFICATION** Sexes similar. Adult breeding plumage attained in second or third summer. **Juvenile** unlikely to be seen in our region, it resembles first-winter but always has black half-collar on lower hindneck and bill is all black. **First-winter** has white head with a dark ear-spot, pale grey rear crown and hindneck, and black half-collar often reduced or replaced by dark grey. Upperwing tricoloured, with grey coverts, black carpal bar and outer wing forming W pattern, and whitish trailing edge to inner wing. Mantle grey and slightly forked tail is white with a black bar on tip. Bill black and legs dark with a brownish tinge. **First-summer** similar but plumage heavily worn and faded, and bill often develops pale base. **Second-winter** resembles adult non-breeding and some are indistinguishable, but others show a more extensive black wingtip with black subterminal marks on up to seven (rather than five) outer primaries, some dark markings on outer wing-coverts and alula and bill still retains some dark. **Second-summer** birds that can be identified from adult breeding show one or more of the characters described for second-winter, and retain partial or full winter head pattern throughout the summer. **Adult non-breeding** has white head with a dark ear-spot that can extend over rear crown, ill-defined dusky eye-crescents and pale grey rear crown and hindneck. Underparts, rump and tail white, upperparts grey and upperwings predominantly grey with a white trailing edge and solid black tip except small white tips on p4–6. Bill yellow or greenish-yellow, iris blackish-brown and legs blackish or dark grey, rarely yellow, orange, pink and even red. Kittiwake is obviously darker grey in adult plumage than to Common Gull. **Adult breeding** is like adult non-breeding but head white. **VOICE** On breeding grounds common call is an excited, nasal *kitti-waaak*, also a gruff *kek-kek-kek* and in flight a short, nasal *kia*. **RANGE** Breeds discontinuously along northern coasts of much of the N Hemisphere. Most European birds winter at sea south to *c*. 32°N. **STATUS** Regular visitor with records from all archipelagos, but status varies between them. **Azores** Regular winter visitor with most records in Dec–Feb and noted on all. **Madeira** Rare winter visitor and passage migrant. **Canaries** Irregular winter visitor with records from all main islands. **Cape Verdes** Uncommon winter visitor.

GULL-BILLED TERN *Gelochelidon nilotica* Plate 43 (1)
Sp. Pagaza Piconegra **P.** Tagáz (Gaivina-de-bico-preto)

G. n. nilotica, although Azores records could relate to *G. n. aranea*. **L** 35–38 cm (14–15″); **WS** 100–115 cm (39–45″). **IDENTIFICATION** Sexes similar. Generally resembles Sandwich Tern and best distinguished from latter by longer and broader wings, shorter and thicker all-black bill, uncrested black cap, grey rump and central tail, shallower tail fork and, often distinct, dark trailing edge to primaries. Perched, the longer legs of Gull-billed are quite noticeable. Adult in breeding plumage has uncrested black cap, stout black bill, white underparts and grey upperparts. In non-breeding plumage black cap is reduced to a dark patch behind eye. Juvenile has brownish wash to crown and brown tips to feathers of mantle and upperwing, and is much plainer than juvenile Sandwich. First-winter differs from same-age Sandwich in plainer upperwing and the eye patch that does not continue on nape. **VOICE** Common flight call is a nasal *ger-vik* or *kay-vek*. **RANGE** Quite widespread, breeding from Europe, discontinuously, to Australia and China; also in the Americas. Northern populations are the most migratory, reaching as far south as Botswana and Peru. **STATUS** Recorded from all four archipelagos and probably an accidental visitor. **Azores** Three records: one, Povoação (São Miguel), 4 Oct 1967; adult in breeding plumage, Ponta Delgada (São Miguel), 4 Sep 1996; and another in breeding plumage, Madalena (Pico), 25 Jul 1997. **Madeira** An old record mentioned by Schmitz in 1900 and an adult, Funchal (Madeira), 13 Apr 2002. **Canaries** Accidental visitor or rare and irregular passage migrant, recorded from all main islands except La Palma. **Cape Verdes** Two records: two, south of Sal Rei (Boavista), 31 Jan 1995; and two, Calheta (Maio), 30 Dec 2004–2 Jan 2005.

CASPIAN TERN *Hydroprogne caspia* Plate 43 (2)
Sp. Pagaza Piquirroja **P.** Garajau-grande (Gaivina-de-bico-vermelho)

Monotypic. **L** 47–54 cm (18½–21″); **WS** 130–145 cm (51–57″). **IDENTIFICATION** Sexes similar. The largest European tern and similar in size to Lesser Black-backed Gull. Breeding adult has very thick red bill with a dark tip, extensive black cap, grey upperparts and white underparts. Wings broad but rather blunt and tail short. Flight action recalls a *Larus* gull with slow heavy wingbeats. Extensive dark underside to primaries, which is shown by all ages, is useful feature to separate this species from Royal Tern. Non-breeding adult loses very little of black cap and lacks masked pattern of other terns in this plumage. Juvenile also has extensive black cap and lacks dark carpal bar of other large terns. First-winter also has rather plain upperwings compared to other large terns. **VOICE** Flight call is a harsh, hoarse, croaking, heron-like *kraa* or disyllabic *kra-ahk*. **RANGE** Breeds discontinuously from Europe across Asia to Australia, and in Africa and N America, and in winter disperses within same regions. **STATUS** Vagrant with only seven records. **Madeira** One, Porto Santo, 30 Jul 1987. **Canaries** One, Las Salinas (Fuerteventura), 2 Aug 1995. **Cape Verdes** Seven records: one, Sal Rei (Boavista), 24 Mar 1993; one, Rabil lagoon (Boavista), 27–31 Jan 1995, another there, 21 Feb–9 Mar 1999; one, between Raso and Branco, 5 Mar 1999; one, Porto Grande (São Vicente), 13 Mar–21 Apr 1999; one, Rabil lagoon (Boavista), 22 Apr 2001; and one, Casas Velhas (Maio), 30–31 Dec 2004.

ROYAL TERN   *Sterna maxima*   Plate 43 (3)
Sp. Charrán Real   **P.** Garajau-real

S. m. albididorsalis; *S. m. maxima* could occur in the Azores.   **L** 45–50 cm (18–20″); **WS** 125–135 cm (49–53″).
IDENTIFICATION Sexes similar. A large tern, only slightly smaller than Caspian Tern and most likely to be confused with latter. Royal Tern is less heavily built than Caspian and looks much more like a large Sandwich Tern, with narrower wings, a finer more orange bill with no dark tip, paler undersides to primaries, a longer and deeper forked tail and, at rest, a more obvious crest. In non-breeding plumage has a white forehead and crown with black reduced to a shaggy nape patch, quite unlike the black cap streaked white of Caspian. Juvenile has a dark carpal bar, distinctive pale mid-wing panel and the lack of an almost complete dark cap should prevent confusion with juvenile Caspian. **VOICE** Flight call is a *kirrrup* similar to Sandwich Tern but lower pitched. **RANGE** Breeds on Pacific and Atlantic coasts of S USA south to the West Indies and Mexico, also equatorial W Africa. Winters south to Peru and Argentina and, in Africa, from Morocco to Angola. **STATUS** Vagrant with five records from three archipelagos. **Azores** Adult in breeding plumage, Terceira, 1 Jun 2003, and presumably the same, Praia Islet (Graciosa), 24–27 Jun 2003. **Canaries★** Adult, west of Puerto Rico (Gran Canaria), 1 Aug 1989; and three adults, Punta de Abona (Tenerife), 28 Mar 1997. **Cape Verdes** Immature, Santa Maria (Sal), 5 Feb 1995; and three, Rabil lagoon (Boavista), 22 Apr 2002.

LESSER CRESTED TERN   *Sterna bengalensis*   Plate 43 (4)
Sp. Charrán Bengalí   **P.** Garajau-pequeno (Garajau-bengalense)

S. b. torresii.   **L** 35–37 cm (14–14½″); **WS** 92–105 cm (36–41″). **IDENTIFICATION** Sexes similar and similar in size to Sandwich Tern but slightly darker grey and has a deeper based orange bill. The only other obvious difference is in adult plumage, when the rump, central tail and uppertail-coverts of Lesser-crested Tern are grey rather than white. At all ages, best separated from Royal Tern by smaller size and thinner, more finely tipped bill. First-winter upperwing pattern less contrasting adults darker than Royal, with grey rump and tail-centre. **VOICE** Typical call is a harsh *kirrrik-kirrrik* like Sandwich Tern. **RANGE** Breeds mainly SW Pacific and Indian oceans, but extends to Red Sea and into Mediterranean. In winter Mediterranean population migrates to W Africa. **STATUS** Very rare vagrant. **Madeira** Adult, off Selvagem Pequena, 8 Aug 1989. **Canaries** Adult in winter plumage, Puerto del Carmen (Lanzarote), 27 Jan 1987.

SANDWICH TERN   *Sterna sandvicensis*   Plate 43 (5)
Sp. Charrán Patinegro   **P.** Garajau (Garajau-comum)

S. s. sandvicensis but *S. s. acuflavida* could occur in the Azores.   **L** 36–41 cm (14–16″); **WS** 95–105 cm (37½–41″).
IDENTIFICATION Sexes similar. The largest tern regularly recorded in the region. Adult in breeding plumage has black cap and shaggy crest, long slender black bill with a yellow tip, pale grey upperparts with dusky outer primaries and white trailing edge to upperwing, white rump and tail, and white underparts. Non-breeder has forehead and crown white. Juvenile has scaly upperparts, a complete dusky cap and shorter bill lacking yellow tip. First-year like adult but retains juvenile feathering on coverts, secondaries and tail, and yellow bill tip tiny or lacking. **VOICE** Most frequent call is the distinctive flight call, a loud, grating *kirrik*. **RANGE** In the New World breeds in coastal E USA and the Caribbean; in Eurasia in coastal W Europe, NW Mediterranean, the Black and Caspian Seas. Eurasian birds winter as far south as S Africa. **STATUS** Recorded throughout but status varies. **Azores** Accidental with only 13 records but increased observer coverage may change this. **Madeira** Listed by Zino *et al.* (1995) as exceptional, but recent records suggest it may be regular in winter, with a maximum five, Caniçal (Madeira), 19 Oct 2002. **Canaries** Fairly common passage migrant and winter visitor, with large numbers on Fuerteventura, e.g. in March 1992 when 1,600 at Playa de Sotavento. **Cape Verdes** Only *c.* 12 records but perhaps commoner than records suggest.

ROSEATE TERN *Sterna dougallii* Plate 43 (6)
Sp. Charrán Rosado **P.** Gaivina-rosada (Andorinha-do-mar-rósea)

S. d. dougallii. **L** 33–38 cm (13–15″); **WS** 72–80 cm (28–31½″). **IDENTIFICATION** Sexes similar. Resembles both Common and Arctic Terns but shorter winged and has longer tail streamers. In breeding plumage identified by its red-based black bill, very pale grey upperparts, almost white underparts, often with pinkish suffusion to breast, and lack of any obvious dark trailing edge to underside of primaries. In winter overall paleness of plumage is perhaps the best feature. Juvenile resembles that of Sandwich Tern due to boldly scaled upperparts and dusky cap, but upperwing closer in pattern to Common Tern. **VOICE** Call is a distinctive *chir-vick* not unlike Sandwich Tern. Also a rasping *krraahk* and on breeding grounds a variety of similar calls. **RANGE** Breeds eastern seaboard of N America and in Caribbean, NW Europe and the Azores, S Africa, the Indian Ocean and in the W and SW Pacific. European population winters exclusively in W Africa. **DISTRIBUTION** Breeds on the Azores and Madeira, where rare, and also noted breeding on Los Roques de la Sal, off El Hierro in the Canaries. **BREEDING** Only a sporadic breeder away from the Azores, thus all details are from latter. Season is Apr–Jun with the earlier dates from the western islands and later dates in the east. Clutch size is 1–3 but one is the most frequent and three unusual. Breeds in mixed colonies with Common Tern on offshore rocks and islets, usually in fairly inaccessible areas. Nest is a simple scrape or depression in the ground and there is usually just one brood, although a replacement clutch is sometimes laid. **HABITAT** Breeds on rocky coasts, sand-spits and shingle beaches. Winters in similar habitats on warm tropical coasts, avoiding fresh water. **STATUS** Varies between the archipelagos. **Azores** One of the last European strongholds with an estimated 547 pairs in 1994. **Madeira** Very rare breeder (probably less than annual). Formerly bred on Ilhéu de Fora in the Salvages but not since 1994. **Canaries** Accidental but has bred at least once, a pair on Los Roques de la Sal (El Hierro) in 1987. Recorded from all main islands except La Gomera. **Cape Verdes** One caught exhausted, Santa Maria (Sal), 14 Apr 1998, but probably under-recorded due to its pelagic habits on migration.

COMMON TERN *Sterna hirundo* Plate 44 (1)
Sp. Charrán Común **P.** Gaivina (Andorinha-do-mar-comum)

S. h. hirundo. **L** 31–35 cm (12–14″); **WS** 77–98 cm (30–38½″). **IDENTIFICATION** Sexes similar. Medium-sized tern very similar to Arctic Tern and care must be taken in separating them. Common Tern has a longer head and bill, with the highest point of the head being behind the eye. Legs longer than Arctic and perched adults have the tail streamers equal to or shorter than the wingtip. In flight, the wings of Common Tern are broader and appear more central on the body than in Arctic. Adult in breeding plumage has dark tip to the red bill, grey belly paler than upperparts and broad white wedge between black cap and bill. The key features, however, are visible in flight. Common Tern has only the inner primaries translucent, the dark trailing edge to the primaries is broader, shorter and more diffuse, and there is a distinct dark wedge on the outer primaries of the upperwing, which becomes more obvious in summer due to wear. In juveniles the upperwing pattern is quite distinctive. On Common Tern there is an obvious dark carpal bar and narrow grey bar on the secondaries, with a distinct pale mid-wing panel. The upperside of the primaries is duskier on Common Tern and the trailing edge much more diffuse. Adult winter has white forehead and underparts, dark carpal bar and an all-black bill. **VOICE** Alarm call is a screechy *kreee-arrr* or *kreeeahh* slightly lower than that of Arctic Tern. Also a short and oft-repeated *kit* or characteristic *kierri-kierri-kierri*. **RANGE** Breeds in C and E North America and from Europe through much of C Asia to Sakhalin and the Kuril Is. Winters south to S South American, S Africa and S Australia. **DISTRIBUTION** Breeds on Madeira, the Salvage Islands, Canaries (except Lanzarote where perhaps bred in the past), and the Azores. **BREEDING** Season in the Canaries is late Mar–May but laying usually takes place Apr. Clutch size is 2–3 and the nest is a simple ground depression on isolated coasts or offshore islets, but also recorded nesting on a small fishing boat. **HABITAT** Breeds both on coastal and inland waters, but in winter is mostly marine. **STATUS** On the Canaries the breeding population is very small and in danger of extinction. In Madeira numbers fluctuate, but on average the population appears stable. On the Azores the species is common and seems likely to remain so. **Azores** Population estimated at 4,015 pairs in 1989. **Madeira** Breeds on all islands in the archipelago. **Canaries** Probably no more than *c.* 50 pairs with the majority on the western islands of La Gomera, La Palma and El Hierro. On migration sometimes fairly numerous, e.g. *c.* 1,000, Playa de Sotavento (Fuerteventura), 18 Mar 1992. **Cape Verdes** Eleven records but probably more regular than sightings would suggest.

ARCTIC TERN *Sterna paradisaea* Plate 44 (2)
Sp. Charrán Ártico **P.** Gaivina-ártica (Andorinha-do-mar-ártica)

Monotypic. **L** 33–35 cm (13–14″); **WS** 75–85 cm (29½–33½″). **IDENTIFICATION** Sexes similar. Very similar to Common Tern but can usually be identified with care. The head of Arctic Tern is shorter and more rounded, with the highest point immediately above the eye. Perched, the legs are noticeably shorter than Common Tern, indeed Arctic Tern can appear to be sitting. In adults, if the tail projects beyond the primaries this is a good feature, but on some the wingtip and tail can fall level, as in Common Tern. In flight Arctic Tern has narrower wings that appear to be positioned ahead of the central body. Adult in breeding plumage shares black cap of Common Tern but white wedge between this and bill is much narrower. Underparts darker than Common Tern, the breast greyer and belly darker and exhibits greater contrast with underwings, and white cheeks well demarcated from breast. Upperparts are similar but on Arctic Tern there is only a narrow but well-defined trailing edge to primaries (all of which are translucent), a feature also visible on the underwing. Bill shorter than Common Tern and is coral-red without an obvious black tip. Adult winter plumage is not obtained fully until birds reach Antarctic wintering grounds. Juvenile and first-winter Arctic Tern lack dark secondaries of Common Tern and carpal bar is less pronounced, whilst Arctic Tern does not show a pale mid-wing panel like that of Common Tern. **VOICE** Similar to Common Tern, a screechy *prree-ah* in alarm, but more frequently heard from migrants is a short and oft-repeated *kik*. **RANGE** Breeds in arctic and subarctic south to S Britain, and winters throughout Southern Ocean, some even to the pack ice of Antarctica. **STATUS** Recorded from all archipelagos but is an accidental visitor. **Azores** Six records but two of these were of small flocks in late autumn. **Madeira** First recorded 1981 and still fewer than five records. **Canaries** Accidental but possibly more frequent than records suggest. Recorded from all main islands except El Hierro but many records are from inter-island ferries. **Cape Verdes** Only recorded on a few occasions and true status uncertain.

FORSTER'S TERN *Sterna forsteri* Plate 44 (7)
Sp. Charrán de Forster **P.** Gaivina de Forster

Monotypic. **L** 33–36 cm (13–14″); **WS** 73–82 cm (29–32″). **IDENTIFICATION** Sexes similar and very similar to the three preceding species. Adult in breeding plumage whiter below than Common Tern and has paler primaries, lacking dark wedge of that species. Bill slightly deeper than Common Tern and more orange-red in colour. Legs distinctly longer and, like bill, more orange in colour. The grey-centred tail is also clearly different to that of Common Tern. Adult in non-breeding plumage has white crown and prominent black mask from eye to ear-coverts, and some black streaking on nape, which facial pattern is more like Gull-billed Tern than any medium-sized tern. Upperwing-coverts all grey and lack dark carpal bar of Common Tern. First-winter similar to non-breeding adult but has shorter tail streamers and dark tips to tail feathers. Seen at rest, the dark-centred tertials are often obvious and useful for identification. **VOICE** Rather more nasal and lower pitched than Common Tern, including a rapidly repeated *kek*, harsh *kerr* and nasal *kyarr*. **RANGE** Breeds in N America inland from the prairie provinces of Canada south to California, and on the east coast from New Jersey Texas. Winters from C California to Guatemala in the west, and in the east from Virginia to the C Caribbean. **STATUS** Vagrant with only three certain records. **Azores** One photographed, Praia da Vitoria (Terceira), 18–21 Feb 2003; and first-winter, Cabo da Praia (Terceira), 31 Oct 2004. **Madeira** One, probably this species, Madeira, Aug 1981. **Canaries*** Adult, Los Gigantes (Tenerife), 19 May 2000.

WHITE-CHEEKED TERN *Sterna repressa* Plate 44 (3)
Sp. Charrán Cariblanco **P.** Gaivina-arábica

Monotypic. **L** 32–34 cm (12½–13½″); **WS** 75–83 cm (29½–33″). **IDENTIFICATION** Sexes similar. Similar to Common Tern and Arctic Tern but adult breeding plumage is noticeably darker grey and more uniform, with an obvious white facial streak separating the black cap from the grey underparts. Long tail streamers and grey vent make separation from Whiskered Tern relatively straightforward. In non-breeding plumage grey rump and tail prevents confusion with any other medium-sized *Sterna*. First-winter similar to Common Tern but upperwing more uniform and darker grey, carpal bar often broader, secondary bar less obvious and rump and tail grey. **VOICE** Both advertising and alarm calls are like those of Common Tern. **RANGE** Breeds in the Red and Arabian Seas, and off Kenya and W India. Winter range poorly known, although probably off E African and W Indian coasts. **STATUS** Extreme vagrant. **Canaries*** One claimed, near Famara (Lanzarote), 11–25 May 1977.

LITTLE TERN *Sternula albifrons* Plate 43 (7)
Sp. Charrancito **P.** Chilreta (Andorinha-do-mar-anã)

S. a. albifrons. **L** 22–24 cm (9–9½″); **WS** 48–55 cm (19–22″). **IDENTIFICATION** Sexes similar. Small size and rather hurried flight usually sufficient to confirm identification. Adult in breeding plumage has black crown, nape and eye-stripe, white forehead and shallow V above eye, along with rest of head and underparts. Upperparts predominantly pale grey with black outer primaries and whiter rump and tail. Bill yellow with a black tip and legs yellow-orange. Non-breeding adult has white lores, more extensive white forehead and a darker, pale-based, bill. Juvenile/first-winter similar but crown browner, feathers of upperparts tipped brown and there is a dark carpal bar on upperwing. Note that Least Tern *S. antillarum* of N America could occur in our region, particularly in the Azores. It differs from Little Tern in its grey rump and tail, and in its nasal, trumpet-like call. **VOICE** A sharp *kik-kik* and a rasping *kriet*. **RANGE** Breeds on coasts and rivers of Europe and W Asia discontinuously through India, SE and E Asia, and Indonesia to Australia. Northern populations winter as far south as S Africa. **STATUS** Varies between the archipelagos where it has been recorded. **Azores** Three records: one, Santa Cruz (Flores), 16 Jun 1984, with another there, 3–23 Oct 2002; and juvenile, Ponta Delgada (São Miguel), 12–13 Sep 2004. **Madeira** Fewer than five records in the last 50 years, the most recent ten, São Vicente (Madeira), 26 Mar 2004. **Canaries** Rare and irregular passage migrant recorded from all main islands except El Hierro. **Cape Verdes** Accidental visitor with 14 records, but possibly a regular passage migrant and winter visitor in small numbers. All records have been from Rabil lagoon/Sal Rei (Boavista), apart from one, Maio, 7 Oct 1988, which was the first record for the archipelago.

BRIDLED TERN *Onychoprion anaethetus* Plate 44 (5)
Sp. Charrán Embridado **P.** Gaivina-de-dorso-castanho (Garajau-embridado)

O. a. melanoptera. **L** 35–38 cm (14–15″); **WS** 77–81 cm (30½–32″). **IDENTIFICATION** Sexes similar. Dark upperparts including rump and deeply forked tail are features shown by two only species, Bridled Tern and Sooty Tern. Adult Bridled Tern has dark grey-brown upperparts and white underparts with a faint greyish wash. Crown and nape black and hindneck greyish-white forming a collar, which can be hard to see. White patch on forehead narrower than in Sooty Tern and extends behind eye as a short supercilium. Dark line on lores does not taper towards the bill as in Sooty Tern. Juvenile recalls adult but has a much paler crown and generally scaly appearance to paler upperparts and upperwing-coverts, although in some upperparts much darker and scaly appearance less obvious. **VOICE** Vagrants usually silent but may give varied rather short, harsh calls. **RANGE** Breeds in the tropics but disperses outside breeding season. Breeding range and pelagic habits less extensive than Sooty Tern. **STATUS** All records are from the same archipelago. **Azores** Vagrant but has been recorded breeding. One, Ilhéu das Cabras (Terceira), 6 Jun 1984; nest with two eggs, Topo (São Jorge), 2 Jun 1985, and in 1985–86, a pair, Caloura (São Miguel); one, Ilhéu do Vila (Santa Maria), 1–4 July 1997; one, Ilhéu da Praia (Graciosa), 14–15 Aug 2000; and three records in summer 2004, from Ilhéu do Vila (Santa Maria), 7–8 Jul, Mosteiros (São Miguel), 11 Jul, and , Ilhéu da Praia (Graciosa), 31 Jul, which perhaps involved just one individual.

SOOTY TERN *Onychoprion fuscata* Plate 44 (4)
Sp. Charrán Sombrío **P.** Gaivina-de-dorso-preto (Andorinha-do-mar-escura)

O. f. fuscata. **L** 43–45 cm (17–17½″); **WS** 82–94 cm (32½–37″). **IDENTIFICATION** Sexes similar. Adult very similar to Bridled Tern. However, Sooty Tern is larger, with distinctly more blackish upperparts and a broader white forehead. It lacks the thin, white supercilium and pale hindneck collar of Bridled Tern and has a narrower, tapering loral stripe. In flight the underside of the primaries is darker than Bridled Tern. Juvenile entirely dark sooty brown, except pale lower belly and undertail-coverts, with whitish tips to upperwing-coverts and on mantle, forming long narrow bars. **VOICE** Usually only vocal around breeding islands where gives characteristic *ker-wacki-wak*, often transcribed as *wide-awake*. **RANGE** Breeds mostly on islands in the tropics. Disperses outside breeding season but still mainly found in tropical waters. **DISTRIBUTION** Recently discovered breeding on Vila Islet, off Santa Maria in the Azores. **HABITAT** Mostly pelagic, coming to land only to breed, generally on smaller islands and islets. **STATUS** Recently recorded breeding but still best treated as a vagrant. **Azores** Recently recorded breeding. One, Vila Islet (off Santa Maria), 1 Jun 1903, had been present previous year; one, Santa Maria, 23 May 1990; one, Topo (São Jorge), Jun 1992; one, Capelinhos (Faial), 4 Jul 1992, and in 1999–2000 one summered on Ilhéu da Praia (Graciosa), where also present, 5–28 Jun & 11 Sep 2003, and 5 Sep 2004. In recent years reported breeding on Vila Islet, where up to four individuals seen regularly since 1989. Breeding was proven when a single chick fledged, 8 Aug 1994. At least one pair has attempted to breed subsequently, with two pairs in 1998, but in 2000 a fledgling was taken by a Common Buzzard. On Ilhéu da Praia (Graciosa), proof of breeding came in the form of one incubating a single egg, 12–19 May 2004. **Madeira** Possibly bred on the Salvages in 1982, and an adult seen on there, 17 Aug 1992. **Canaries** Six records: one, La Tejita (Tenerife), Feb 1984; one, between Tenerife and La Gomera, 3 Sep 1988; at least one, Playa de Sotavento (Fuerteventura), 18 Mar 1992; one, Caleta de Fustes (Fuerteventura), 1 Mar 1993; another there, 9 Feb 1998; and juvenile, between Tenerife and La Gomera, 3 Oct 2001.

WHISKERED TERN *Chlidonias hybrida* Plate 44 (6)
Sp. Fumarel Cariblanco **P.** Gaivina-dos-pauis (Gaivina-de-faces-branca)

C. h. hybrida. **L** 23–25 cm (9–10″); **WS** 74–78 cm (29–31″). **IDENTIFICATION** Sexes similar. Adult in breeding plumage has predominantly dark grey body and upperwings, with black cap, white cheeks and red bill. In flight white underwings and undertail conspicuous. In non-breeding plumage generally very pale grey above and white below, with black-streaked rear crown and black patch behind eye. Lacks white-collared effect of other *Chlidonias*. First-winter similar to non-breeding adult but retains juvenile wing and tail pattern; upperwing usually lacks dark carpal bar of other *Chlidonias* at same age, and tail has narrow, dusky subterminal band. At this age can be difficult to separate from White-winged Tern but structural differences such as thicker bill, broader wings, stockier body, less rounded crown and distinctly notched tail are useful features. Also has a more extensively dark hindcrown and rump is only slightly paler than tail, whereas in White-winged Tern it is obviously whiter. **VOICE** Calls include a short *kek* often repeated in series and a harsh, rasping *cherk*. **RANGE** Breeds from W Europe discontinuously east through the Middle East and C Asia to Ussuriland, China and Assam, also S and E Africa, Madagascar and Australia. Most European birds winter in sub-Saharan Africa. **STATUS** Recorded from three archipelagos. **Azores** Juvenile, Lagoa Ginjal (Terceira), 21–26 Oct 2002; 7–8 *Chlidonias* terns, Lagoa das Furnas (São Miguel), 3–5 Nov 2004, of which four (and probably all) were this species, and at least five still present, 29 Jan 2005. **Madeira** First recorded, Porto Santo, 23 May 1954 and still fewer than five records. **Canaries** Scarce and irregular passage migrant recorded from the central and eastern islands and possibly La Palma.

BLACK TERN *Chlidonias niger*   Plate 45 (1)
Sp. Fumarel Común   **P.** Gaivina-preta

C. n. niger and possibly *C. n. surinamensis* on the Azores. **L** 22–24 cm (9–9½"); **WS** 64–68 cm (25–27"). **IDENTIFICATION** Sexes similar. Adult in breeding plumage has black head and underparts, except white vent, mantle and wings dark grey and rump and tail paler grey-brown. Upperwings grey with darker outer primaries and, unlike White-winged Tern, underwing-coverts pale. Adult in non-breeding plumage has solid black cap, grey rump and distinctive dark breast-side patches, none of which are shown by White-winged Tern at similar age. Juvenile has extensive black cap, dark breast-side patches, grey rump and, usually, a less contrasting dark saddle compared to juvenile White-winged Tern. Juvenile of American race *surinamensis* distinguished from European nominate by combination of features. The dark patches on the breast-sides are larger and merge with the dark grey flanks. The rump is very dark grey, almost concolorous with the tail and mantle. The head pattern is more reminiscent of White-winged Tern in having a paler crown, which may appear streaked. Upperwing-coverts very dark grey and almost the same colour as the mantle, and the pale fringes to the mantle, wing-coverts and primaries are very narrow. **VOICE** Flight calls include a harsh, nasal *klii-a*, a shrill *kyeh* and sharp *kik*. **RANGE** Breeds across C North America and from Europe east to the Caspian, Kazakhstan and C Asia. European populations winter predominantly in coastal W Africa. **STATUS** Accidental visitor with records from three archipelagos. **Azores★** Three records: moulting adult, Cabo da Praia (Terceira), 3–8 Sep 1998; adult in non-breeding plumage, Santa Cruz airport (Flores), 12–13 Oct 1999; and juvenile/first-winter, Praia da Vitória (Terceira), 22 Oct 1999. **Madeira** Fewer than five records in the last 50 years. **Canaries** Scarce and irregular passage migrant and winter visitor, recorded from all main islands except La Gomera and El Hierro.

WHITE-WINGED (BLACK) TERN *Chlidonias leucopterus*   Plate 45 (2)
Sp. Fumarel Aliblanco   **P.** Gaivina-d'asa-branca

Monotypic. **L** 20–23 cm (8–9"); **WS** 63–67 cm (25–26"). **IDENTIFICATION** Sexes similar. Adult in breeding plumage unmistakable given combination of black body and underwing-coverts, white upperwing-coverts and white rump and tail. Non-breeding adult more likely to be confused with Black Tern but crown is whiter, it lacks dark patches on the breast-sides and rump is whitish. Juvenile has saddle darker than Black Tern, which contrasts more with the pale grey inner wing, lacks dark patches on the breast-sides and has a white rump. Differences between White-winged and Whiskered Terns in juvenile/first-winter plumage dealt with under that species. **VOICE** A loud, harsh *krek* somewhat harsher than that of Black Tern. Also a harsh, creaking *kesch* likened to the call of Grey Partridge *Perdix perdix*. **RANGE** Breeds from C and E Europe east through C Asia to S China. Winters south to S Africa, Sri Lanka, Indonesia and Australia. **STATUS** Accidental visitor. **Madeira** Adult in breeding plumage, Funchal (Madeira), 24 May 1963. **Canaries★** Various records from Lanzarote, where it is apparently regular in winter offshore, with two records from La Palma: one, Santa Cruz, 31 Aug 1979; and immature, La Restinga (El Hierro), 23 Sep 2001.

COMMON GUILLEMOT *Uria aalge*   Plate 45 (4)
Sp. Arao Común   **P.** Airo (Arau-comum)

Subspecies unknown but *U. a. albionis* is the most likely.   **L** 38–43 cm (15–17"). **IDENTIFICATION** Sexes similar. In breeding plumage has head, upper breast and upperparts brownish and underparts white. Non-breeding adult and first-winter generally dark brownish above and white below, with head-sides predominantly white except a dark line reaching back from and below eye. Most likely confusion species are Brünnich's Guillemot and Razorbill, and differences are dealt with under those species. **VOICE** Unlikely to be heard in our region. **RANGE** Breeds discontinuously on coasts in N Hemisphere, from *c.* 40°N to 75°N, and winters at same latitudes but northern populations move south. **STATUS** Extreme vagrant. **Canaries** Two records: specimen identified by C. Bolle, in the Léon collection in Gran Canaria, and one, Arrecife (Lanzarote), 6 Dec 2002.

BRÜNNICH'S GUILLEMOT   *Uria lomvia*   Plate 45 (5)
Sp. Arao de Brünnich   **P.** Airo-de-freio (Arau de Brünnich)

U. l. lomvia. **L** 39–43 cm (15–17″). **IDENTIFICATION** Sexes similar. Very similar to Common Guillemot and great care is needed to identify this species. Brünnich's has a thicker and stubbier bill with a pale stripe along the base of the cutting edge on upper mandible (beware Common Guillemot carrying fish), and has darker upperparts than most Common Guillemot with no obvious dark streaking on flanks. First-winter and adult non-breeding, the most likely plumages to be seen in the region, have the dark area on head-sides more extensive and finishing well below eye. Also Brünnich's lacks dark eye-stripe of Common Guillemot. In flight underwing-coverts and axillaries whiter than Common Guillemot and feet project beyond tail, unlike Razorbill. **VOICE** Usually silent away from breeding grounds. **RANGE** Circumpolar, breeding discontinuously through arctic and winters to the south, but usually not as far south as Common Guillemot. **STATUS** Vagrant with only four records, all from the Azores and three from the same island. **Azores** One killed, Santa Cruz, Flores, 15 Dec 1965, and another captured there, Feb 1966, with a third apparently captured, S Flores, Mar 1965; one at Porto do Pocinho (Pico), 7 Feb 2001

RAZORBILL   *Alca torda*   Plate 45 (6)
Sp. Alca Común   **P.** Torda-mergulheira

A. t. islandica. **L** 37–39 cm (14½–15″). **IDENTIFICATION** Sexes similar. Resembles the two *Uria* species but has a larger, broader bill and pointed tail (rounded in *Uria*). Adult breeding plumage has blackish head and upperparts (darker than Common Guillemot but similar to Brünnich's). Bill broad and blunt-tipped with a white vertical stripe near tip, and an obvious white loral line, which is diagnostic but only visible at close range. Non-breeder lacks dark stripe behind eye of Common Guillemot and head-sides not as extensively dark as Brünnich's Guillemot. In this plumage, the loral line is much reduced and even harder to see. In flight shows more contrast on underwing than either of the *Uria* species. **VOICE** Usually silent away from breeding grounds. **RANGE** Confined to N Atlantic where breeds in NE North America, Greenland, Iceland, Scandinavia, the British Isles and N France. Winters south to the Mediterranean and NW Morocco. **STATUS** Accidental visitor. **Azores** Two records, one obtained at Rosto de Cão (São Miguel) prior to 1888 and one at Porto Pim (Faial), 5 Jan 1985, which had been ringed in Ireland. **Canaries** Winter vagrant that has been recorded from Lanzarote, Fuerteventura, Gran Canaria, Tenerife and La Palma; most recent record one, Puerto de la Cruz (Tenerife), 5 Apr 1980.

LITTLE AUK   *Alle alle*   Plate 45 (3)
Sp. Mérgulo Marino   **P.** Torda-miúda (Torda-anã)

A. a. alle. **L** 17–19 cm (7–7½″). **IDENTIFICATION** Sexes similar. Small size, very short stubby bill and dark underwing should prove unmistakable. Confusion could occur with lone juveniles of other species, but dark underwing only shared with Puffin, from which easily identified by smaller size, narrower wings and very small bill. Breeding adult has dark blackish upperparts, head and upper breast, with white rest of underparts, white tips to inner secondaries and some white markings on scapulars. Non-breeding adult and first-winter have white head-sides, throat and breast with a partial dark band on the breast-sides. **VOICE** Silent away from breeding grounds. **RANGE** Breeds in high arctic, on Greenland, Iceland, Novaya Zemlya, Spitsbergen, Jan Mayen and Franz Josef Land. In winter moves south to *c.* 42°N. **STATUS** Accidental winter visitor with records from three archipelagos. **Azores** At least 15 records prior to 2004, all Oct–Apr, suggesting that the species is regular in small numbers. **Madeira** Two records: one collected, 20 Nov 1940; and one, Funchal harbour (Madeira), 20 Dec 1958. **Canaries*** Records from Alegranza, La Graciosa, Lanzarote and La Palma, with an influx to the eastern islands in Jan 1980 accounting for most records; the La Palma bird is a specimen with no details other than location, Santa Cruz de La Palma.

(ATLANTIC) PUFFIN   *Fratercula arctica*   Plate 45 (7)
Sp. Frailecillo Común   **P.** Papagaio-do-mar

F. a. grabae. **L** 26–30 cm (10–12″). **IDENTIFICATION** Sexes similar. Totally unmistakable in adult breeding plumage with its black-and-white plumage and huge, bright red, blue and yellow bill. In non-breeding and juvenile plumage easily distinguished from other auks by the black band on the upper breast and massive bill, although this is smaller in juveniles. In flight has a grey underwing, unlike any of the *Alca* or *Uria* species, and is larger and bigger headed than Little Auk. **VOICE** Generally silent away from breeding grounds. **RANGE** Confined to N Atlantic where breeds in E North America, Greenland, Iceland, N Europe, Spitsbergen and Novaya Zemlya. Winters in this range, but also south to the Canaries in the east and to *c.* 39°N in the west. **STATUS** Probably regular in small numbers and recorded from three archipelagos. **Azores** Regular winter visitor in small numbers. **Madeira** Accidental visitor but perhaps more common than records suggest. **Canaries** Scarce and irregular winter visitor, and a vagrant at other seasons. Recorded from all the main islands except El Hierro.

BLACK-BELLIED SANDGROUSE *Pterocles orientalis* Plate 46 (1)
Sp. Ganga Ortega **P.** Cortiçol-de-barriga-preta

P. o. orientalis. **L** 33–35 cm (13–14″). **IDENTIFICATION** Sexually dimorphic. In our region unmistakable as it is only resident sandgrouse and both sexes have an obvious large black patch on belly. ♂ has grey head and breast with dark orangey patch on throat (with black centre) and upperparts heavily marked with dark golden-brown spots. ♀ has finely spotted yellow-brown breast and vermiculated upperparts. Juvenile resembles adult ♀ but upperparts and breast more cinnamon-pink with less distinct, finer markings. In flight both sexes show conspicuous white underwing-coverts and axillaries, and prominent black belly. At certain times they can appear rather pigeon-like, although confusion is more or less impossible at close quarters. **VOICE** Flight call is a far-carrying bubbling *churr-urr-urr-g…* **RANGE** Breeds in Spain and NW Africa, thence from Turkey east to Chinese Turkestan and south to Afghanistan. **DISTRIBUTION** Resident breeder on Fuerteventura in the Canaries. Also recorded from Lanzarote and may breed there. **BREEDING** Main season is Mar–Jun, although a nest with one egg was found late Jul. Clutch size 2–3 and nest is a simple depression in ground. **HABITAT** Sandy and stony plains with little vegetation. **STATUS** Only recorded from one archipelago. **Canaries** Not uncommon on Fuerteventura; highest population density is in south, in Parque Natural de Jandía. Records from Lanzarote scarce: a few pairs possibly breed. May have bred on Gran Canaria prior to mid-19th century. More recently, recorded from La Graciosa.

PIN-TAILED SANDGROUSE *Pterocles alchata* Plate 46 (2)
Sp. Ganga Ibérica **P.** Ganga (Cortiçol-de-barriga-branca)

Subspecies unknown. **L** 31–39 cm (12–15″). **IDENTIFICATION** Sexually dimorphic. Easily separated in flight from Black-bellied Sandgrouse by white belly and long tail projection. adult ♂ breeding has olive-green mantle, nape, crown and neckband with yellowish spotting on mantle, reddish-brown breast bordered above and below by a narrow black band, black-barred green wing-coverts, black bib and eye-stripe, and heavily vermiculated lower back, rump and tail. ♂ in non-breeding plumage heavily vermiculated on mantle and scapulars, with white throat and paler reddish-brown breast. ♀ has white bars on wing-coverts, is heavily vermiculated over entire upperparts, with paler brown face and breast-band, and double black band on neck. Juvenile resembles ♀ but markings on head less distinct, patterning on underparts less pronounced and lacks elongated central rectrices. **VOICE** Common flight call is a loud, nasal, ringing *catar-catar, ga-ga* or *ga-ng*. **RANGE** Breeds in Iberia, NW Africa, the Middle East and SW Asia east to Lake Balkash in Kazakhstan, and Lake Issyk-Ku in Kyrgyzstan. Western populations are resident and nomadic. **STATUS** Extreme vagrant with just one recent record and a few unsubstantiated older references. **Canaries★** Two flying over Barranco de la Costilla (Fuerteventura), 15 Dec 1998.

ROCK DOVE (FERAL PIGEON) *Columba livia* Plate 47 (4)
Sp. Paloma Bravía **P.** Pombo-das-rochas

C. l. livia (including *C. l. canariensis* and *C. l. atlantis*). **L** 31–34 cm (12–13½″); **WS** 63–70 cm (25–27″). **IDENTIFICATION** One of the most familiar species and its domestic form, Feral Pigeon, is widespread in urban areas worldwide. Wild-type birds have pale grey upperwings with two black bars and a white rump patch that can be very indistinct or even lacking on the birds in Canaries (previously treated as a distinct subspecies, *canariensis*). Birds from the Azores and Cape Verdes were originally named *atlantis* due to their overall dark appearance. Those from Madeira were treated as *atlantis* but might have been placed in *canariensis*. Feral Pigeon comes in all sorts of combinations of colours, ranging from all dark to all white, as well as piebald, variegated and even reddish-brown. **VOICE** Call is a moaning *oorh* or *oh-oo-oor* rising slightly in pitch in the middle. **RANGE** Difficult to state with certainty, given confusion over original and current ranges. **DISTRIBUTION** Resident breeder on all of the archipelagos. **BREEDING** Breeds Mar–Jul but possibly year-round, in caves and on rock ledges on cliffs, mainly on coasts but also in mountains. In urban areas breeds on any form of ledge it can find such as windows of buildings, bridges etc. Clutch size two but single eggs reported and several broods are raised. **HABITAT** Wild-type birds best looked for in more remote coastal areas or in equally remote and inaccessible montane regions. Feral Pigeons are ubiquitous. **STATUS** The feral form is common on all archipelagos but the pure form is increasingly rare. **Azores** Common species throughout but pure wild birds increasingly hard to find. **Madeira** Common throughout and in the Salvages. **Canaries** Common and widespread, with wild birds still present locally, but the vast majority show at least some signs of hybridisation with Feral Pigeons. **Cape Verdes** The wild form is still found on Santiago, Santo Antão and São Nicolau, with feral populations existing on these islands as well as on Fogo, Brava and São Vicente.

(COMMON) WOODPIGEON *Columba palumbus* Plate 47 (5)
Sp. Paloma Torcaz **P.** Pombo-torcaz

C. p. azorica (Azores), *C. p. maderensis* (formerly on Madeira, now considered extinct) and *C. p. palumbus*. **L** 40–42 cm (16–16½″); **WS** 75–80 cm (29½–31½″). **IDENTIFICATION** Sexes similar. A large bulky pigeon, clearly bigger than Rock Dove. Best identification features are the large white patches on the neck-sides and, in flight, the white transverse bar on the upperwing. Also in flight, black wingtips and single broad black terminal band on tail are useful features. Adult predominantly grey with iridescent green feathers on nape and neck-sides, below which are the obvious white neck patches; breast vinous-pink becoming paler pink on belly and pale grey on vent. Iris pale yellow and bill yellowish with a reddish base and paler tip. Juvenile duller and lacks iridescent feathering and white patches on neck. The iris is darker and greyer and the bill is also darker. Extinct race *maderensis* was darker than nominate and had greyish-wine colour on underparts more pronounced. That confined to the Azores, *azorica*, also darker than nominate, particularly on head and rump, with a deeper, more vinous colour on breast. **VOICE** Common call is a variable, far-carrying *croo COOOO cooo coo-coo*, with the first syllable audible only at close range. **RANGE** Breeds across most of Europe and in NW Africa east to WC Asia, Afghanistan and SE Iran. Some northern populations migrate but is mostly resident in the south. **DISTRIBUTION** Breeds on the Azores and formerly on Madeira. **BREEDING** Details from Madeira are scant but apparently bred year-round and a clutch of two eggs was recorded. Details from the Azores also limited but breeding proven May–Sep. Clutch size is two and the nest is sited in a tree 3–7.5 m above ground. Incubation and fledging periods for both subspecies presumed similar to nominate. **HABITAT** On Madeira formally inhabited forests in north of the island. In the Azores also prefers well-vegetated areas, although native plants are not essential. **STATUS** Varies between the three archipelagos where it has been recorded. **Azores** Fairly common on all main islands except Flores and Corvo. **Madeira** Endemic *maderensis* now extinct and was apparently very scarce even in the second half of the 19th century; *palumbus* is an exceptional visitor and has been recorded from the Salvages. **Canaries** Scarce and irregular visitor, recorded from all main islands except La Gomera and La Palma.

TROCAZ PIGEON (LONG-TOED PIGEON)   *Columba trocaz*   Plate 47 (1)
Sp. Paloma Torqueza   **P.** Pombo de Madeira

Monotypic. **L** 38–40 cm (15–16″); **WS** 72–76 cm (28–30″). **IDENTIFICATION** Sexes similar. A large pigeon, about size of Woodpigeon but with shorter wings. Darker grey upperparts than Woodpigeon, and lacks white bars on upperwing and neck patches, but does have a diagnostic silvery-grey neck patch that can be difficult to see. Adult has dark slate-grey upperparts with some green iridescence on hindneck and violet iridescence on upper mantle. Flight feathers and tail almost black, the latter with a broad, paler grey subterminal band. Breast vinaceous but rest of underparts grey. Bill red with a yellow tip and legs and feet bright coral red. Iris yellow with a red eye-ring. Juvenile duller and browner, lacks glossy iridescence, with a browner breast, grey bill and buff fringes to scapulars, tertials and wing-coverts. **VOICE** Similar to that of Woodpigeon but generally a quieter *coo-coo cooo cook*. Display also similar to Woodpigeon but again quieter and more muffled. **RANGE** Found only in the Madeiran archipelago. **DISTRIBUTION** Endemic to Madeira. **BREEDING** Season apparently year-round but with an obvious peak in Feb–Jun. Clutch normally one but two eggs have been recorded. Nest an untidy pile of dry sticks, mostly *Erica*, in trees, on the ground in a rock crevice, or more rarely in a low bush. Incubation period seems to be 19–20 days and chick fledges at 28 days. **HABITAT** Restricted to remnant native laurel forest, most of which is in the north of the island, although a few small pockets persist in the south. Occurs in agricultural areas when food is short but is then avidly hunted. **STATUS** Endemic to one archipelago. **Madeira** Recent surveys suggest a population of 10,000 birds, although past surveys have estimated as few as 500 individuals. The species' shyness and difficulties of access to its habitat could have been factors in the apparent under-estimation.

BOLLE'S PIGEON   *Columba bollii*   Plate 47 (2)
Sp. Paloma Turqué   **P.** Pombo-turqueza

Monotypic. **L** 35–37 cm (14–14½″); **WS** 65–68 cm (25½–27″). **IDENTIFICATION** Sexes similar. Very similar to Trocaz Pigeon but with no overlap in ranges confusion is impossible, and is more likely with Laurel Pigeon. Although these species do overlap identification is relatively straightforward in most cases. Bolle's slightly smaller, lacks the silvery neck patch and is generally slightly darker. The grey tail-band is more contrasting than on Trocaz, because on Bolle's the base of the tail is much darker. Compared to Laurel Pigeon, the upperwing of Bolle's Pigeon is bicoloured, with dark primaries and secondaries, and grey coverts. Wings more pointed and flight more rapid and direct than Laurel Pigeon, and more often flies low over canopy. The most obvious feature is the tail pattern, which on Bolle's Pigeon is clearly banded, black–grey–black, whereas on Laurel Pigeon the tail is grey with a broad off-white band at the tip. Adult dark blue-grey with darker flight feathers, and a darker tail with a broad, pale grey subterminal band. Some iridescent green feathering on the sides and hindneck, a purple sheen to the upper mantle and breast deep vinaceous. At close range the bill appears predominantly red with a slightly more orange tip. Legs and feet are red. Juvenile rather different, having upperparts sepia-brown and underparts a more rufous-brown, and shows very little if any iridescence. Wing-coverts and scapulars have buff fringes, the bill is black and the feet and legs dull brownish-red. **VOICE** Territorial call is a quiet cooing *truu-truuu-tru* and advertising call a *ruor rur ruor* *rup*. **RANGE** Endemic to the Canaries. **DISTRIBUTION** Breeds on western islands of Tenerife, La Gomera, La Palma and El Hierro. Formerly bred on Gran Canaria but not been recorded there since 1889. **BREEDING** Season apparently year-round but peaks in Nov–May. Clutch is just one egg. Nest a rather untidy platform of sticks placed in various tree species within laurel forest, especially *Erica arborea* but also *Laurus azorica*, *Myrica faya*, *Ilex canariensis* and *Erica scoparia*. Height varies at 1–15 m but 6–8 m most frequent. Rather surprisingly (but very rarely), has been reported as nesting on ground. Incubation is 18–19 days and fledging period 30–35 days. **HABITAT** Confined to montane laurel forests where closely associated with the densest areas. **STATUS** Only known from one archipelago. **Canaries** Fairly common in its habitat, except on El Hierro where it is uncommon. Recent surveys suggest that the population may be significantly larger than previously thought, with more than 2,000 birds on Tenerife, over 1,000 on La Gomera and possibly more than 3,000 on La Palma.

(WHITE-TAILED) LAUREL PIGEON *Columba junoniae* Plate 47 (3)
Sp. Paloma Rabiche **P.** Pombo-rabil

Monotypic. **L** 37–38 cm (14½–15″); **WS** 64–67 cm (25–26½″). **IDENTIFICATION** Sexes similar. Slightly larger and more slender in build than Bolle's Pigeon, with even shorter and more rounded wings. The slower, more floppy flight, often well above the canopy, can also be distinctive. Upperwing concolorous sooty-brown, rump and uppertail grey with an off-white band on the distal third on all except the central rectrices, which is even more obvious from below, as the band is whiter and contrasts greatly with the dark underside of the tail. At close range, the head of Laurel Pigeon is darker than Bolle's Pigeon with a more extensive metallic green sheen. The vinous underparts are darker and the coloration more extensive than in Bolle's Pigeon and the bill is mostly off-white with a vinaceous-red base. Adult has rather uniform dark brown upperwings and mantle, greenish gloss to crown and nape, grey back and rump and dark vinaceous maroon underparts. Juvenile similar but has indistinct pale fringes to inner wing feathers and reduced iridescence on neck. Best separated from juvenile Bolle's Pigeon by tail pattern and lack of any blue-grey on underparts. **VOICE** Common call of ♂ is a deep, hoarse *huUUU-uhuUU-uhuUU-uhuUU*. **RANGE** Endemic to the Canaries. **DISTRIBUTION** Breeds on western islands of Tenerife, La Gomera and La Palma, and recently discovered on El Hierro where presumed to breed in small numbers. **BREEDING** Possibly year-round but certainly prefers Apr–Aug. Like Bolle's Pigeon clutch is just one egg. Nest, unlike that of Bolle's, is located on rocky ledges, in crevices and small caves, and rarely under boulders or fallen tree trunks. Incubation period unknown but presumed to be 18–20 days; young leave the nest prematurely when still unable to fly. Suffers low reproductive success and predators take many young birds. **HABITAT** Prefers steep valleys just below the high laurel forest zone, although these areas are still extensively clad in tree heath and laurel, and other dense vegetation. **STATUS** Only known from one archipelago. **Canaries** In 1985 Emmerson estimated the population to be a minimum of 1,200–1,480 individuals, but recent surveys suggest the population on La Palma alone exceeds this.

AFRICAN COLLARED DOVE (BARBARY DOVE) *Streptopelia roseogrisea* Plate 46 (3)
Sp. Tórtola de Cabeza Rosa **P.** Rola-rosada

S. r. 'risoria'. **L** 29–30 cm (11½–12″); **WS** 45–50 cm (18–20″). **IDENTIFICATION** Sexes similar. Smaller than Eurasian Collared Dove with a shorter tail. Most '*risoria*' (Barbary Dove – the domesticated form of African Collared Dove) are small with very pale sandy upperparts and underparts, making them unmistakable, but others are much closer to Eurasian Collared Dove in plumage, and certain care must be taken with the identification. Adult has pale sandy-brown upperparts with a narrow black semi-collar on hindneck bordered above and below with white. Outerwing-coverts pale grey and flight feathers brown with buff fringes. Head and underparts pinkish-grey becoming white on vent and undertail-coverts. In flight tail appears considerably darker than Eurasian Collared with more contrasting white tips. Iris dark red and feet red. Juvenile resembles adult but semi-collar indistinct, iris pale yellowish and feet dull greyish-brown. **VOICE** Very different from Eurasian Collared Dove, a distinctly two-part call, with a short *cook* followed by a long, rolling *rrrrrrrrrrooooooooo*. **RANGE** Breeds across Africa in the Sahel from Senegambia and Mauritania east to E Sudan, N Ethiopia, N Somalia and W Arabia. Resident and partially migratory. **DISTRIBUTION** Introduced on the Canaries. **BREEDING** Little documented in the Canaries but season appears prolonged, probably year-round but prefers Mar–April and Oct–Nov. Normal clutch is two and nest usually located in an ornamental tree such as *Casuarina* or *Eucalyptus*. **HABITAT** Most frequently encountered around habitation where it frequents parks and gardens. **STATUS** Introduced on one archipelago. **Canaries** Breeds on all main islands except El Hierro.

EURASIAN COLLARED DOVE *Streptopelia decaocto* Plate 46 (4)
Sp. Tórtola Turca **P.** Rola-turca

S. d. decaocto. **L** 31–33 cm (12–13″); **WS** 47–55 cm (18½–22″). **IDENTIFICATION** Sexes similar. Very similar to African Collared Dove and some are indistinguishable using plumage alone. Upperparts pale greyish-brown (greyer than African Collared) with blackish-brown primaries. Head and underparts pinkish-grey with greyish belly and undertail. Black semi-collar on hindneck bordered above and below with white. Uppertail more extensively pale than African Collared and central tail feathers concolorous with rump. Undertail is black based with a contrasting white tip. ♀ slightly browner on head and nape and less pink on underparts. Juvenile recalls adult but lacks black half-collar and has more prominent pale fringes to upperparts and wing-coverts. **VOICE** Advertising call is a loud, low-pitched, repeated cooing *kroo-kooo ku*. **RANGE** Breeds from W Europe and NW Africa discontinuously east through Middle East and Indian subcontinent to China and Korea. **DISTRIBUTION** Only known to breed on the Canaries. **BREEDING** Few data from the Canaries but season appears protracted and variable according to location. Clutch size two and nest located in a tree, in species such as *Ficus nitida*, *Coussapoa dealbata* and *Eucalyptus globulus*. **HABITAT** Concentrated in urban areas where frequents parks and gardens. **STATUS** Recorded from two archipelagos. **Madeira** One, Salvages, Sep 1990 and two, Caniçal (Madeira), 1 Sep 2004, with up to four, Prainha, just east of Caniçal, 10–12 Apr 2005. **Canaries** Breeds on all main islands except El Hierro, where birds have been seen but breeding is unconfirmed.

(EUROPEAN) TURTLE DOVE *Streptopelia turtur* Plate 46 (5)
Sp. Tórtola Común **P.** Rola-brava (Rola-comum)

S. t. turtur and *S. t. arenicola*. **L** 26–28 cm (10–11″); **WS** 47–53 cm (18½–21″). **IDENTIFICATION** Sexes similar. Adult has dark-centred scapulars, wing-coverts and tertials with broad rufous fringes. Forehead to nape pale grey with black-and-white striped patch on neck-sides and mantle grey-brown. Underparts vinous on breast but belly white. Tail dark with broad white tips to all but central feathers and narrow white fringes to outer feathers. Iris orange and surrounding bare skin red. ♀ resembles ♂ but usually paler and duller. Juvenile much duller than adult with head and breast dull brown. Lacks striped neck patch, has thinner and duller fringes to upperparts with less distinct dark centres. Eye-ring pinkish. **VOICE** Most characteristic is the advertising call, a deep purring *tuoorrr tuoorrr tuoorrr*. **RANGE** Breeds from W Europe and N Africa east to W China. Winters in Sahel zone of sub-Saharan Africa, from Senegal and Mauritania east to the Red Sea. **DISTRIBUTION** Breeds only in Madeira and the Canaries. **BREEDING** Season late Mar–Jul. Clutch size two. Nest, a rather flimsy structure, is located in a wide variety of vegetation at heights of 0.5–7 m. **HABITAT** Wide range, from pine and laurel forests to drier coastal areas and the alpine zone, thus also occupying a wide range of altitudes. **STATUS** Recorded throughout but status varies. **Azores** Accidental visitor with just three records. **Madeira** Listed as a common breeder but first confirmed as recently as 1964. **Canaries** Common breeder on all islands. The subspecies *S. t. arenicola* is an accidental visitor with just two records, one, near Las Palmas (Gran Canaria), 5 May 1913, and one, Las Mercedes (Tenerife), (date unknown). **Cape Verdes** Ten records, nine in Aug–Oct, the other, 28 Feb 1999, Mindelo sewage farm (São Vicente).

LAUGHING DOVE (PALM DOVE) *Streptopelia senegalensis* Plate 46 (6)
Sp. Tortola Senegalesa **P.** Rola-dos-palmares (Rola do Senegal)

Subspecies unknown. **L** 25–27 cm (10–11″); **WS** 40–45 cm (16–18″). **IDENTIFICATION** Sexes similar. The smallest of the *Streptopelia* in the region. More colourful than Eurasian or African Collared Doves, with which it shares some structural characters, and compared to Turtle Dove is longer tailed and shorter winged with much plainer upperparts. Adult has display feathers on foreneck, these being rufous-brown with black centres. Head, neck and breast distinctly vinous-pink, grading to white on belly and vent. Upperparts reddish-brown with extensive blue-grey on upperwing across outer wing-coverts and secondaries, with dark brown primaries and grey-brown central tail, the outer feathers having dark blackish bases and prominent white tips. Juvenile similar but duller and lacks necklace. **VOICE** Common call is a series of 4–8 bubbling coos, *coo-coo, cooroocoo-coo-coo* with a laughing quality. **RANGE** Breeds over a large portion of Africa, S Arabia, the Middle East and Indian subcontinent. Mostly resident but some seasonal movements occur and is expanding in Morocco. **DISTRIBUTION** Has probably bred on Fuerteventura in the Canaries. **BREEDING** As this species has only recently started to colonise Fuerteventura, very few data are available. **HABITAT** Areas of ornamental vegetation, particularly large gardens around major hotels. **STATUS** Only recorded from one archipelago. **Canaries** Recorded from Fuerteventura and Tenerife, but is certainly a vagrant (if records do not refer to escapes) on latter. On Fuerteventura, where it may be a natural coloniser, observed regularly around La Lajita where juveniles have been reported with adults. There are also records from other parts of Fuerteventura with some evidence for a small breeding population on the island.

NAMAQUA DOVE *Oena capensis* Plate 46 (7)
Sp. Tortolita Rabilarga **P.** Rolinha-rabilonga

O. c. capensis. **L** 26–28 cm (10–11″) including tail of 11–12 cm, **WS** 28–33 cm (11–13″). **IDENTIFICATION** Sexually dimorphic. Adult ♂ unmistakable and easily identified by combination of very small size, long graduated tail, black face and breast, rufous underwing and red-and-yellow bill. ♀ shares structure of ♂ but lacks black on face and breast, and bill is less brightly coloured. Juvenile similar to ♀ but initially has much shorter tail. Upperparts highly variegated with much buff spotting on wings. **VOICE** Advertising call is a quiet low-pitched *coo hoo woooooo*, but otherwise rather silent. **RANGE** Breeds sub-Saharan Africa, Madagascar, parts of Arabia and in S Israel. Mainly resident but some local movements. **STATUS** Extreme vagrant. **Canaries★** Adult ♂, of unknown origin, Golf del Sur (Tenerife), 19–22 May 1997. **Cape Verdes** Adult ♂ photographed, Praia Preta (Maio), 21 Jul 1995.

BUDGERIGAR   *Melopsittacus undulatus*   not illustrated
Sp. Periquito Común   **P.** Periquito

Monotypic. **L** 18 cm (7″). **IDENTIFICATION** Sexes similar. Small, long-tailed parrot with black-and-yellow barred upperparts, green rump and underparts, yellow face, forehead and throat, with blue spots on cheeks continuing as line of black spots on throat, blue cere and white iris. ♀ has a brownish cere when breeding. Immature resembles adult but is duller and has barring on forehead, spots on throat ill defined or lacking and iris greyish. **VOICE** A pleasant warble frequently given in flight and a subdued, disyllabic screech; also often utters a quiet chatter when feeding. **RANGE** Australia but absent from Tasmania and coastal regions. **DISTRIBUTION** Confirmed to breed on Tenerife but records of escapees exist from other islands in the archipelago. **BREEDING** Details are scant but in Australia nests in holes in trees and lays 4–6 eggs. **HABITAT** Parks and gardens. **STATUS** A small feral population may still exist. **Canaries** On Tenerife a population of *c.* 10 pairs might still occur in Santa Cruz, particularly around La Granja park.

SENEGAL PARROT   *Poicephalus senegalus*   not illustrated
Sp. Lorito Senegalés   **P.** Periquito-massarongo

P. s. senegalus. **L** 23 cm (9″). **IDENTIFICATION** Sexes similar. Adult has green upperparts with a yellowish rump, grey head, green upper breast and thighs, yellow lower breast and belly with orange tinge, and staring yellow iris. In flight reveals bright yellow underwing- and undertail-coverts. Immature resembles adult but is duller with brownish head and dark brown iris. **VOICE** Series of short screeches and rather high-pitched whistles. **RANGE** Confined to W Africa, from Senegal and Guinea to N Cameroon and S Chad. **DISTRIBUTION** Confirmed to breed on Tenerife. **BREEDING** Details are very scarce but in natural range nests in holes in trees, with a clutch size of 2–4. **HABITAT** Most frequently observed in parks and gardens. **STATUS** Only recorded from one archipelago where records refer to escapes. **Canaries** A few seen regularly on Tenerife in Santa Cruz, Puerto de la Cruz/La Orotava and Las Americas/Los Cristianos. The only confirmed breeding was by a pair in Santa Cruz in May 1991. Also recorded on La Gomera.

MONK PARAKEET   *Myiopsitta monachus*   Plate 48 (2)
Sp. Cotorra Argentina   **P.** Caturrita

Subspecies unknown. **L** 33 cm (13″). **IDENTIFICATION** Sexes similar. Adult mostly green with a grey face and breast, the latter with rather indistinct barring, and a yellow band on belly. Tail long and graduated, and bill horn-coloured. Juvenile similar but forehead has green tinge. **VOICE** Highly vocal, with a wide vocabulary including various chatters, screeches and squawks. **RANGE** In natural range breeds in Bolivia, Brazil and Argentina. **DISTRIBUTION** Breeds only on the Canaries. **HABITAT** Frequents parks and gardens within several populated areas. **STATUS** Only recorded from one archipelago. **Canaries** Introduced breeder on Fuerteventura, Gran Canaria, Tenerife and La Palma, but no breeding records since 1994 on the latter. Encountered on other islands but most records refer to escapees.

ROSE-RINGED (RING-NECKED) PARAKEET   *Psittacula krameri*   Plate 48 (1)
Sp. Cotorra de Kramer   **P.** Periquito-rabijunco

P. k. krameri. **L** 38–42 cm (15–16½″) including a 22–26 cm tail. **IDENTIFICATION** Sexes differ. Almost completely bright grass green with a very long, narrow, pointed tail with blue central feathers. ♂ has narrow black line on neck that broadens on lower cheek and forms a completely black chin. Also has a pinkish collar and pale grey nape and neck-sides. ♀ lacks head and neck markings, the central tail feathers are green, and the tail is shorter. Juvenile like ♀ but has a slightly paler bill and greyish iris (yellowish-white in adult). **VOICE** Common call a harsh, shrieking *kee-ak* often repeated in groups of 3–6. Highly vocal with a wide range of shrieks, squawks and whistles. **RANGE** In natural range breeds from Senegal across Africa to Somalia and Sudan, and in Indian subcontinent and Myanmar. **DISTRIBUTION** Breeds only in the Canaries. **HABITAT** Parks and gardens in several populated areas. **STATUS** Recorded from all four archipelagos. **Azores★** Recent records from Ponta Delgada (São Miguel) suggest a population is becoming established. **Madeira** Recent records, from the Funchal area, suggest the presence of a small feral population. **Canaries** Introduced breeder on Lanzarote, Gran Canaria and Tenerife. Recorded on Fuerteventura but breeding unconfirmed. **Cape Verdes** One obtained, Praia (Santiago), 5 Apr 1909, was the nominate subspecies and may have been a genuine vagrant. Also records from Mindelo (São Vicente), Jan–Mar 1992, and Praia (Santiago), May–Sep 1993. In 1995 birds were still present in Praia and although breeding was unproven it appears that the species is establishing itself there.

GREAT SPOTTED CUCKOO *Clamator glandarius* Plate 48 (3)
Sp. Críalo **P.** Cuco-rabilongo

C. g. glandarius. **L** 38–40 cm (15–15½″) including the 14–18-cm tail. **IDENTIFICATION** Sexes similar. Adult and juvenile plumages both unmistakable. Adult has silvery-grey crown and ear-coverts, with a small crest at rear and a blackish nape. Upperparts brownish-grey with large white spotting on tips of wing-coverts, scapulars and tertials. Underparts predominantly white with a yellowish tinge to throat and upper breast. Juvenile has black crown and ear-coverts, with a shorter crest, blacker upperparts with smaller white spotting and bright rufous primaries. **VOICE** Often silent outside breeding season. **RANGE** Breeds in Iberia, S France, Italy, the Middle East, and extensively in sub-Saharan Africa. Most European birds winter in sub-Saharan Africa, although move to W Morocco and the Nile Valley. **STATUS** Recorded from three archipelagos but is an extreme vagrant on two. **Madeira** Two records: ♂ shot, Madeira, 6 Nov 1939; and juvenile, Ponta do Pargo (Madeira), 23 Aug 2003. **Canaries** Scarce, irregular passage migrant and even scarcer winter visitor. Recorded from all main islands except La Gomera and La Palma. **Cape Verdes** Juvenile photographed, Boavista, at an undisclosed site and on an unknown date, probably in 2000.

COMMON CUCKOO *Cuculus canorus* Plate 48 (4)
Sp. Cuco **P.** Cuco (Cuco-canoro)

C. c. bangsi (Azores and Canaries) and *C. c. canorus* (all archipelagos). **L** 32–34 cm (12½–13½″). **IDENTIFICATION** Can be sexually dimorphic as ♀ has two colour morphs. Adult ♂ is a familiar bird, with grey upperparts, head and breast. Rest of underparts white with continuous fine dark barring on belly and less extensive barring on vent and undertail-coverts. Iris, eye-ring, bill base and feet yellow. Adult ♀ grey morph similar to adult ♂ but has a more rusty-buff breast with some dark barring; brown morph is rusty-brown above and on breast, and has most of plumage barred dark. Juvenile also has two colour morphs: grey one similar to adult grey morph but has white patch on nape and completely barred underparts; brown morph resembles brown morph ♀ but also has white nape patch and completely barred underparts, less rufous upperparts and some white feather tips. Subspecies *bangsi* from Iberia, the Balearics and NW Africa is smaller, especially in wing length, and ♀ of grey morph has more extensive and richer rufous on breast-sides. **VOICE** Possibly the most familiar of all European bird calls: advertising call of ♂ is a far-carrying *cuc-koo* with variations. ♀ has a loud, bubbling trill, likened to either Little Grebe or Whimbrel. **RANGE** Breeds from Europe and NW Africa east to Kamchatka, Sakhalin, Japan and E China. All W Palearctic populations winter in E and S Africa. **STATUS** Varies in the different archipelagos. **Azores** Nine records, six of these from São Miguel, two Santa Maria and one São Jorge. Most are old, but the most recent was an adult ♀ brown morph, Parc Forestier (Santa Maria), 7 Jun 1986. Measurements suggest that both *bangsi* and *canorus* have occurred. **Madeira** Fewer than five in the last 50 years, and also recorded from the Salvages. **Canaries** Fairly regular passage migrant recorded on all main islands. Both *bangsi* and *canorus* have been recorded. **Cape Verdes** Four records: immature collected, Brava, 5 Sep 1898; one, inland of Praia (Santiago), 16 Aug 1951; one, near Chão Bom (Santiago), 16 Jan 1986; and one, near Vila do Maio (Maio), 26 Aug 1986.

BLACK-BILLED CUCKOO *Coccyzus erythrophthalmus* Plate 48 (5)
Sp. Cuco Piquinegro **P.** Papa-lagarta-de-bico-preto (Cuco-de-bico-negro)

Monotypic. **L** 27–31 cm (10½–12″). **IDENTIFICATION** Sexes similar. Only likely to be confused with similar Yellow-billed Cuckoo, but as its name suggests has a mostly black bill with a blue-grey base to lower mandible. Plumage rather dull brown above and white below. Adult has underside of tail grey with white tips and black subterminal bands on median and outer feathers. Upperwing relatively uniform dark brown and eye-ring red. Juveniles and first-winters have eye-ring yellow, with a hint of rufous in the flight feathers, and tail feathers lack subterminal black bars and have smaller, more diffuse white tips. **VOICE** Usually silent away from the breeding grounds. **RANGE** Breeds in N America east of the Rockies, from S Canada south to Arkansas and N Carolina. Winters in NW South America south to Peru and Bolivia. **STATUS** Extreme vagrant. **Azores** One obtained, São Miguel, pre-1903; and one, Santa Cruz (Flores), 15 Oct 1969.

YELLOW-BILLED CUCKOO *Coccyzus americanus* Plate 48 (6)
Sp. Cuco Piquigualdo **P.** Papa-lagarta-norte-americano (Cuco-de-bico-amarelo)

C. a. americanus. **L** 28–32 cm (11–12½″). **IDENTIFICATION** Sexes similar. Smaller than Common Cuckoo and only likely to be confused with Black-billed Cuckoo, but has most of the lower mandible and cutting edge of the upper mandible yellow, hence its name. Adult has upperparts paler than in Black-billed with a slightly greyer tone and very distinct bright rusty primaries. Undertail black with broad white tips to median and outer feathers, and eye-ring yellow. Juveniles and first-winters have undertail dark grey with less distinct pale tips, and eye-ring yellow in first-winter and grey juvenile. Post-juvenile moult commences early, so unlikely to be seen in complete juvenile plumage in our region. **VOICE** Usually silent outside breeding season. **RANGE** Breeds from S Canada south to Florida Keys and Mexico. Winters mainly in S America from Colombia and Venezuela south to Argentina. **STATUS** Vagrant with at least 11 records from one archipelago. **Azores** One, Praia da Vitória (Terceira), 19 Nov 1932, with another there, Nov 1953, and in Oct/Nov 1965 there were birds on at least five islands (Santa Maria, Terceira, Faial, Flores and Corvo); one, Ponta dos Rosais (São Jorge), 13 May 1966; one obtained, São Miguel, 17 Apr 1977; one found dead, Sete Cidades (São Miguel), 6 Nov 1985; and one, between Relva and Rabo de Asno (São Miguel), 23 Oct1990.

BARN OWL *Tyto alba* Plate 49 (1)
Sp. Lechuza Común **P.** Coruja-das-torres

T. a. alba (central and western Canary Islands), *T. a. detorta* (Cape Verdes), *T. a. gracilirostris* (Fuerteventura and Lanzarote) and *T. a. schmitzi* (Madeiran archipelago). **L** 33–35 cm (13–14″); **WS** 85–93 cm (33½–36½″). **IDENTIFICATION** Sexes similar. A medium-sized owl with a distinctive heart-shaped facial disk and dark eyes. In *alba* underparts and face almost pure white, except a few dark spots on chest and flanks. Upperparts golden-buff, finely streaked, mottled and vermiculated grey and brown. Legs feathered almost to base of toes. ♀ differs slightly from ♂ in having more spots on underparts, but this is highly variable and trying to sex birds in the field should only be attempted on a known pair. Juvenile down is lost when bird fledges, by which time it resembles adult. Subspecies *gracilirostris* differs from *alba* in having underparts more orange-buff and more heavily spotted. Central facial disk is more rufous than *alba* and upperparts clearly darker. This subspecies is also smaller and, as its scientific name implies, has a smaller and slimmer bill than all subspecies of the region. Subspecies *detorta* is the darkest form in the region and is browner above than *gracilirostris*, from which it also differs in its much larger and longer bill; *schmitzi* of Madeira is intermediate between *alba* and *gracilirostris* and has generally darker upperparts than *alba*. Underparts paler than either *gracilirostris* or *detorta*, and can range from white to buff, but are more heavily spotted than *alba*. However, the facial disk is much closer in coloration to *alba* than either of the other forms. **RANGE** Fairly cosmopolitan, breeding in N, C and S America, Europe, Africa, Arabia, the Indian subcontinent, SE Asia, Indonesia and Australia. Mainly resident but northernmost populations in N America move south in winter. **DISTRIBUTION** Breeds throughout, except the Azores. In the Canaries, endemic *gracilirostris* is confined to the eastern islands of Fuerteventura, Lanzarote and Alegranza, but *alba* breeds on Gran Canaria, Tenerife and El Hierro, and has been seen recently on La Palma and La Gomera; *schmitzi* is widespread in Madeira and *detorta* is found on the main islands of Santiago, Brava, Santo Antão, São Nicolau and probably Fogo, plus the islets of Branco and Raso, in the Cape Verdes. **HABITAT** On Madeira, occurs in various habitats such as urban and agricultural areas, forests and rocky mountains. On the Canaries, breeds in small caves on coastal cliffs or in barrancos, usually near cultivated areas where it hunts. Occasionally seen around large urban areas. On Cape Verde *detorta* is reported to be widespread in mountains and cultivations, but rarer in more open country, and not uncommon around towns and villages. **TAXONOMIC NOTE** Hazevoet (1995) treated those on Cape Verde as a separate species, the Cape Verde Barn Owl *T. detorta*. **STATUS** Recorded from all four archipelagos but only breeds on three. **Azores** Five records all from the same island. One obtained, Sete Cidades (São Miguel), *c*. 1903; one, hunting near Ponta Delgada (São Miguel), 1 Sep 1996; one, Solar do Conde, Capelas (São Miguel), 11 Nov 2000; one, near Lagoa (São Miguel), 11 Jul 2004; and one, Hotel Bahia Palace (São Miguel), same date. **Madeira** *schmitzi* is a common breeder on Madeira, Porto Santo and the Desertas. Also recorded from the Salvages. **Canaries** *alba* on Tenerife is estimated to number at least 161 pairs, but on Gran Canaria it breeds at much lower densities. On the western islands, it breeds on El Hierro, but on La Palma and La Gomera, although recorded, there is no proof of breeding; *gracilirostris* is a scarce resident on Fuerteventura and the eastern islets, but is more common on Lanzarote, where an estimated 50–100 pairs breed. **Cape Verdes** *detorta* is a not uncommon and fairly widespread resident on the islands where it occurs.

(EURASIAN) SCOPS OWL *Otus scops* Plate 49 (2)
Sp. Autillo **P.** Mocho-d'orelhas

O. s. mallorcae. **L** 19–20 cm (7½–8″); **WS** 53–63 cm (21–25″). **IDENTIFICATION** Sexes similar. A very small owl with obvious short ear-tufts, which when relaxed can be difficult to see. At a distance appears either greyish or rufous-brown, depending on colour morph, with a pale shoulder line and slightly paler face and underparts. At close range the intricate pattern of black streaks, white spots and dark vermiculations is obvious. Juvenile closely recalls adult but is paler with shorter ear-tufts, smaller white spots on lower scapulars, less distinct streaking on upperparts and finer dark streaking on underparts. **VOICE** A plaintive whistle *tyuu*, repeated every few seconds, often for long periods. **RANGE** Breeds from Iberia and NW Africa through Europe and the Middle East to C Asia. Most European birds winter in the savanna belt of sub-Saharan Africa. **STATUS** Recorded from three archipelagos and generally a rare and irregular visitor. **Madeira** Occasional, occurring seasonally. **Canaries** Rare and irregular passage migrant recorded from all main islands except La Gomera and El Hierro. **Cape Verdes** A juvenile, photographed on a ship 52 km northwest of Sal, 7 Sep 1985.

(EURASIAN) **EAGLE OWL** *Bubo bubo*   not illustrated
Sp. Búho Real   **P.** Bufo-real (Bufo)

Subspecies unknown but *B. b. ascalaphus* would seem the most likely.   **L** 47–59 cm (18½–23¼"); **WS** 126–149 cm (49½–59"). **IDENTIFICATION** Sexes similar. Measurements given are those of race *ascalaphus*; the nominate race *bubo* is noticeably larger. Rather pale but large owl with prominent ear-tufts, smaller than those of *bubo*. Highly variable coloration, from pale sandy to almost as dark as *bubo*, the latter having quite heavily marked underparts, but paler birds show just a few triangular or drop-shaped dark marks on breast. Facial disk much paler than *bubo* with dark borders at sides and yellow eyes. **VOICE** Advertising call of ♂ is a hooting *waha* or *hoowaha* with the stress on the *wah*, but this call is unlikely to be heard from vagrants. **RANGE** *ascalaphus* is resident across N Africa, and in the Middle East and Arabia. **TAXONOMIC NOTE** Some authors regard *ascalaphus* as a species, the Pharaoh Eagle Owl *B. ascalaphus*. Anyone fortunate enough to encounter this species in our region should attempt to differentiate between *ascalaphus* and *bubo*. **STATUS** Extreme vagrant with just two records. **Canaries★** One found dead under high-tension wires, Lanzarote, 1980s, and one, Barranco de Río Cabras (Fuerteventura), Dec 1993. Both records may relate to birds escaped from captivity, and neither was identified to subspecies.

SNOWY OWL *Bubo scandiacus*   Plate 49 (3)
Sp. Búho Nival   **P.** Coruja-das-neves (Bufo-branco)

Monotypic. **L** 53–66 cm (21–26"); **WS** 142–166 cm (56–65"). **IDENTIFICATION** Sexually dimorphic. Both sexes unmistakable due to size and predominantly white plumage. Adult ♂ pure white except for a few dark subterminal marks on primaries, secondaries and tail feathers, large yellow eyes, and black claws and bill. ♀ larger but also has many dark marks on wings and tail, and extensive barring on upperparts and underparts. Immature ♀ even more heavily barred and can even look rather grey at distance. Immature ♂ similar to adult ♀ but dark barring is narrower. **VOICE** Rather silent away from nest. **RANGE** Circumpolar from Greenland across N Europe and Siberia to Alaska and N Canada. Winters south of breeding range, in N Europe, C Asia to Japan and N USA. **STATUS** Extreme vagrant. **Azores** One obtained, Faial, 14 Jan 1928, was presumably ship-assisted, although natural vagrancy is possible.

(NORTHERN) **HAWK OWL** *Surnia ulula*   Plate 49 (4)
Sp. Lechuza Gavilana   **P.** Mocho-rabilongo (Coruja-gavião)

S. u. caparoch. **L** 36–39 cm (14–15"); **WS** 74–81 cm (29–32"). **IDENTIFICATION** Sexes similar. Medium-sized owl with a long, graduated tail and rather large, flat-topped head. In flight reveals blunt-tipped wings more like those of a hawk than another owl. Dark grey-brown upperparts with fairly extensive white spotting, underparts whitish with broad dark brownish bars and a brownish tinge to flanks and belly. Facial disk mostly white with thick blackish borders on foreneck and some dark streaking, particularly around eyes and below bill. Feet and tarsus feathered, eyes yellow. Juvenile plumage unlikely to be seen in the region. **VOICE** An excited, shrill *ki-ki-kikikikikiki* in alarm, rather reminiscent of Merlin. **RANGE** *caparoch* breeds in N North America from Alaska to Labrador, and winters in S Canada and the N USA. **STATUS** Extreme vagrant. **Canaries★** One obtained in autumn, but no other details.

TAWNY OWL *Strix aluco*   Plate 49 (5)
Sp. Cárabo Común   **P.** Coruja-do-mato

Subspecies unknown, but *S. a. mauretanica* would seem the most likely. **L** 37–39 cm (14½–15"); **WS** 94–104 cm (37–41"). **IDENTIFICATION** Sexes similar. A medium-sized owl with no ear-tufts and a rather compact appearance. Plumage varies from rufous-brown to greyish-brown, paler on underparts. Two pale false eyebrows, whitish outer webs to scapulars forming a pale line on shoulder and pale tips to secondary-coverts creating an indistinct wingbar. Entire plumage streaked, mottled and vermiculated darker. Head quite large and distinctly rounded, and in flight wings fairly short but also distinctly rounded. **VOICE** Most common call is the shrill *ke-vick* given by both sexes, the male advertising call is the familiar hooting *hoooooo.........hoo, hoohoohoo hoooooo* with a *c.* 4-second pause after the initial hoot. **RANGE** Resident across Europe and in NW Africa, east discontinuously to W Siberia, E China and Taiwan. **STATUS** Vagrant with just seven records from one archipelago. **Canaries★** Two reported by Cabrera as being killed Tenerife, pre-1893; von Thanner, who resided on Tenerife 1902–18 reported one that had been brought to him; one,

Puerto de la Cruz (Tenerife), May 1975; one, Playa de las Americas (Tenerife), 1 Jan 1980; one heard, El Cedro (La Gomera), 16–17 Aug 1985; and one, Costa Calma (Fuerteventura), 3 Feb 2004.

LONG-EARED OWL   *Asio otus*   Plate 49 (6)
Sp. Búho Chico   **P.** Bufo-pequeno

A. o. canariensis (Canaries) and *A. o. otus* (Azores).   **L** 35–37 cm (14–14½″); **WS** 90–100 cm (35½–39″). **IDENTIFICATION** Sexes similar but ♀ often has more richly coloured underparts and more heavily streaked upperparts. A slim, medium-sized owl with long ear-tufts that are often visible, otherwise very similar to Short-eared Owl. Obvious facial disc is pale ochraceous-tawny with dark blackish rim. Upperparts rich buffy-brown with dark streaking, spotting and mottling, and row of whitish spots on scapulars. Underparts paler than upperparts and streaked throughout, heavier on foreneck and upper breast, with fine crossbars on lower breast and belly. Eyes yellowish-orange. In flight, large yellowish-buff patch at base of primaries and 4–5 narrow black bands on primary tips and secondaries. Tail also finely barred. Legs and toes feathered. The mesoptile (fluffy juvenile) has tail, wings and facial disk as adult but ear-tufts not fully developed. Retains downy whitish-buff feathering, barred dusky, on head, mantle, underparts and wing-coverts. Downy chick is whitish with bare skin pinkish. Canarian subspecies *canariensis* distinctly smaller and darker than nominate and has darker reddish-orange eyes. **VOICE** Advertising call of ♂ is a far carrying and deep *whoop* repeated every few seconds. ♀ has much weaker and more nasal note only audible at close range. Fledged young give a loud, drawn-out, high-pitched *peeeeee-ee* often likened to the noise made by a squeaky gate. **RANGE** Breeds from Canaries and the Azores across Europe and C Asia to Japan, and much of North America. More northern populations are largely migratory. **DISTRIBUTION** Breeds on the Azores, on São Miguel, Graciosa, Terceira, São Jorge, Pico and Faial, and on all of the Canaries except Lanzarote. **HABITAT** Widespread on the Canaries occurring in all habitats from the arid southern coastal zone to over 2,000 m, where recorded in the alpine zone. Also frequently recorded in towns where often found in parks or gardens. On the Azores has also been recorded in a wide range of habitats. **STATUS** A common breeder on two archipelagos. **Azores** Common resident but like all nocturnal species population difficult to ascertain. **Canaries** Common resident on most islands on which it is found but rare on Fuerteventura where only recently discovered to breed.

SHORT-EARED OWL   *Asio flammeus*   Plate 49 (7)
Sp. Lechuza Campestre   **P.** Coruja-galhofa (Coruja-do-nabal)

A. f. flammeus.   **L** 37–39 cm (14½–15″); **WS** 95–110 cm (37½–43″). **IDENTIFICATION** Sexes similar but ♀ distinctly buffier. Overall appearance is like that of Long-eared Owl, but lacks obvious ear-tufts and is partially diurnal. Upperparts yellow-brown to pale ochraceous-buff, heavily streaked and spotted dusky, and tail has 4–5 dark bars (Long-eared has 6–8 much narrower bars). Underparts paler, ranging from pale yellow-tawny to ochre-whitish, distinctly streaked brown. In flight best distinguished from Long-eared Owl by white trailing edge to inner primaries and secondaries, and almost all-black wingtip. Juvenile unlikely to be seen in our region. **VOICE** Most likely to be heard in our region is the alarm call, which is a harsh *chef-chef-chef*. **RANGE** Breeds from Iceland and the British Isles east to Kamchatka and Sakhalin, and across N America, Andes of Peru and Ecuador, and in S South America. Northern populations are migratory, wintering well south of breeding range. **STATUS Azores** Five records: one collected, Sete Cidades (São Miguel), prior to 1903; another collected, same island, presented to British Museum, 1903; one, Caldeira do Guilherme (Terceira), 13 Mar 1971; four, Lagoa do Fogo (São Miguel), 13 Mar 1980; and one, near São Sebastião (Terceira), 20 Sep 1994. **Madeira** Considered frequent by Zino *et al.* (1995). **Canaries** Scarce and irregular winter visitor recorded from all main islands except El Hierro. **Cape Verdes** Seven records; a wing, found near Santa Maria (Sal), 23 Feb 1983; one, Raso, 13 Mar 1990; one flew in off the sea, Santa Maria (Sal), 15 Oct 1990, and another there, 4 Dec 2001; one, Raso, 18 Oct 2001; one, *c.* 15 km off Santiago, 5 Nov 2001; and one at sea, northeast of Fogo, 16 Oct 2002.

MARSH OWL   *Asio capensis*   Plate 49 (8)
Sp. Lechuza Mora   **P.** Coruja-galhofa-mourisca (Coruja-moura)

A. c. tingitanus.   **L** 35–37 cm (14–14½″); **WS** 90–95 cm (35½–37½″). **IDENTIFICATION** Sexes similar. Resembles other *Asio* species but differs in having almost uniform dark brown breast, upperparts and upperwing-coverts, and dark eyes. Lower breast and belly pale but finely barred darker. Generally, rather like a small but very dark Short-eared Owl, but patch on upperside of primaries is pale orange-brown. Juvenile plumage moulted shortly after fledging and is unlikely to be seen in our region. **VOICE** Common call is a harsh, croaking, frog-like *kaaa*. **RANGE** Confined to Africa where mostly resident. Predominantly sub-Saharan, breeding discontinuously from Senegambia and Ethiopia south to the Cape, and in Madagascar and NW Morocco. **STATUS** Extreme vagrant. **Canaries*** One, between Bajamar and Punta Hidalgo (Tenerife), 30 Mar–13 Apr 1983. Habitat in this area is atypical of the species, as noted by the observer.

EUROPEAN NIGHTJAR *Caprimulgus europaeus* Plate 48 (7)
Sp. Chotacabras Gris **P.** Noitibó-cinzento (Noitibó da Europa)

C. e. meridionalis. **L** 26–28 cm (10–11″); **WS** 57–64 cm (22½–25″). **IDENTIFICATION** Sexes similar. Long-winged, long-tailed nocturnal species that may recall either a small falcon or a cuckoo. Flight action somewhat reminiscent of an *Oceanodroma* petrel. Plumage a cryptic mix of browns, greys and black, with a narrow white throat-band and pale spots on shoulders. In flight ♂ reveals white spots on wing and tail, which are much reduced or lacking in ♀. ♀ also has white on throat indistinct or lacking. Juvenile resembles adult ♀ but is generally paler and throat patch even smaller. **VOICE** Territorial song is a long churring trill and the call a soft *cu-ik*. Unlikely to be heard in our region. **RANGE** Breeds from W Europe and N Africa east to around Lake Baikal and W China, and south to Iran. Entire population winters in sub-Saharan Africa mostly in the east and south. **STATUS** Accidental visitor with records from three archipelagos. **Azores** One obtained, Ginetes (São Miguel), *c.* 1953. **Madeira** One mentioned by Harcourt as having been reported by a Mr Hinton, and one shot, Salvages, Apr 1895. **Canaries** Accidental visitor with a handful of records from Montaña Clara, Fuerteventura, Gran Canaria and Tenerife, most recently one, Inagua (Gran Canaria), 2 May 1991, and one, Monte de El Raso (Gran Canaria), 12 Jun 2000.

RED-NECKED NIGHTJAR *Caprimulgus ruficollis* Plate 48 (8)
Sp. Chotacabras Pardo (Chotacabras Cuellirrojo) **P.** Noitibó-de-nuca-vermelha

Subspecies unknown but *C. r. ruficollis* is most likely. **L** 30–32 cm (12–12½″); **WS** 65–68 cm (25½–27″). **IDENTIFICATION** Sexes similar but white areas on throat, tail and wings are generally larger and more prominent in ♂. Larger and longer tailed than European Nightjar, with a distinct rusty half-collar on hindneck, and throat and upper breast also distinctly rustier. Also lesser coverts paler and less contrasting with Red-necked. All wing-coverts have broad, pale tips and there is a large white patch on the breast-sides. White wing spots are slightly more extensive than on European Nightjar, as are the white spots on the undertail. Juvenile resembles adult ♀ but is paler and greyer, with a less distinct rufous collar and less distinct white throat patch. **VOICE** Quite different from European Nightjar, being a louder but low-pitched *kutuk* often repeated for several minutes, but is unlikely to be heard in our region. **RANGE** Breeds in Spain, Portugal, Morocco, Algeria and Tunisia. Some in Morocco appear to remain in the breeding range in winter, but most move south to W Africa. **STATUS** Extreme vagrant. **Madeira** Old record, Porto Santo (Noronha & Schmitz 1902). **Canaries*** One dead, Tenerife (possibly 1971); one photographed, Barranco de Esquinzo (Fuerteventura), 12–20 May 1987, and probably the same bird there, 10 Jun 1987.

COMMON NIGHTHAWK *Chordeiles minor* Plate 48 (9)
Sp. Añapero Yanqui **P.** Bacurau-norte-americano (Noitibó-americano)

C. m. minor. **L** 23–25 cm (9–10″); **WS** 59–68 cm (23–26″). **IDENTIFICATION** Sexes differ. Similar to European Nightjar but has slightly longer and more pointed wings, with a noticeably shorter tail and shallowly forked tip. White patches on wing are more proximally positioned and more prominent than in European Nightjar, and white on throat is much more extensive. ♂ has a white subterminal tail-band, which is lacking in ♀. Also, ♀ has a narrower primary bar, which is often buffy, as is the throat patch. Juvenile resembles adult ♀ but is paler and greyer, and has a narrow white trailing edge to wing. **VOICE** On breeding grounds ♂ gives an insect-like, buzzing *peeent* repeated every few seconds, but unlikely to be heard in our region. **RANGE** Breeds over most of N and C America, and mostly winters in S America south to C Argentina. **STATUS** Extreme vagrant. **Azores** One, on board a ship 75 miles southwest of Pico, 28 Sep 1966; female obtained, Flores, 27 Oct 1968; and one, Fajã dos Cubres (São Jorge), 5–7 Oct 2001. **Canaries** Juvenile found dead, Las Mercedes (Tenerife), Dec 1972.

CHIMNEY SWIFT *Chaetura pelagica* Plate 50 (1)
Sp. Vencejo de Chimenea **P.** Rabo-espinhoso

Monotypic. **L** 12–13 cm (5″); **WS** 31–32 cm (12–12½″). **IDENTIFICATION** Sexes similar. A small, dark swift with a short, cigar-shaped body, rather broad wings and square-ended tail, which is unlikely to be confused in our region. Dark greyish-brown with a slightly paler throat and upper breast, and needle-like points to tail feathers (not usually visible in the field). Flight can be rather fluttering, almost bat-like, when feeding, and more typically swift-like. Lack of white throat and rump should make separation from Little Swift, the only likely confusion species, fairly straightforward. **VOICE** Unlikely to be heard in our region but typical call is an uneven and chattering series of chip notes. **RANGE** Breeds extensively throughout N America east of the Rockies and winters in W South America. **STATUS** Vagrant with only 4–6 records. **Azores*** One, probably this species, Horta (Faial), 31 Oct 1990; three, Santa Cruz (Flores), 2–3 Sep 1998; maximum four, Lagoa Azul, Sete Cidades (São Miguel), 26–27 Oct 1999; and one, in north of São Miguel, 26 Oct 2002. **Canaries*** One, reported from Tenbel (Tenerife), four times between 2 Oct and 19 Dec 1997, and one, Matagorda (Lanzarote), 2 Jan 2001.

CAPE VERDE SWIFT *Apus alexandri* Plate 50 (2)
Sp. Vencejo de Cabo Verde **P.** Andorinhão de Cabo Verde

Monotypic. **L** 13 cm (5"); **WS** 34–35 cm (13½–14"). **IDENTIFICATION** Sexes similar. Smaller than Plain Swift, with shorter wings and the shallowest tail fork of breeding *Apus* species in the region. A broad, but ill-defined pale throat patch and palest underparts of any W Palearctic *Apus*. Flight is weak and fluttering. **VOICE** Call higher pitched than Common Swift and is more abrupt and harsher, with a distinctive reeling quality. **RANGE** Confined to Cape Verde. **DISTRIBUTION** Recorded from all islands (except Santa Luzia), but probably confined to Santiago, Fogo, Brava, Santo Antão and São Nicolau as a breeder. Resident but inter-island movements occur. **BREEDING** Data are scarce and precise season unknown. Clutch size two and eggs are probably white (not freckled brown, as reported by Alexander, 1898). **HABITAT** Breeds under roofs and in crevices in cliffs, both coastal and inland, from sea level to the highest peaks. **STATUS** Common endemic resident, most numerous on Fogo and Brava.

PLAIN SWIFT *Apus unicolor* Plate 50 (3)
Sp. Vencejo Unicolor **P.** Andorinhão-da-serra

Monotypic. **L** 14–15 cm (5½–6"); **WS** 38–39 cm (15"). **IDENTIFICATION** Sexes similar. Smaller than Common Swift with a narrower body, proportionately longer tail with a deeper fork, slimmer wings and an even more rapid and frantic flight action. Throat generally darker than Common Swift but when appears pale is never as white or as well defined as in most Common. However, care must be taken as a certain amount of overlap occurs in throat-patch size and coloration. Comparisons of structure and flight action are probably the best method to confirm an Common Swift, but some lone individuals probably cannot be identified safely in the field. **VOICE** Call is a hoarse, screaming *shree* very similar to Common Swift but possibly slightly higher pitched. **RANGE** Breeds in the Madeiran and Canarian archipelagos, and probably in very small numbers in Morocco. Some move to NW Africa in winter but otherwise the main wintering area is unknown. **DISTRIBUTION** In the Canaries breeds on all main islands. Also breeds throughout the Madeiran archipelago and possibly on the Salvages. **HABITAT** Present in all habitats from sea level to *c.* 2,500 m. **BREEDING** Season quite prolonged, starting Apr and finishing Sep. Nest is located in an area of difficult access, such as a crevice or hole in a barranco wall, on a cliff, building or bridge. Usually has basic construction of feathers, and clutch size is 1–2. Pairs can have up to three broods. **STATUS** Recorded from three archipelagos but breeds only on two. **Madeira** Common breeder but some birds depart in winter to an unknown destination, presumably in Africa. **Canaries** A common breeder, recorded year-round, although numbers in winter are much reduced and can be extremely hard to locate Oct–Jan. **Cape Verdes** Four, Rabil lagoon (Boavista), 22 Feb 1999, and one there, 13 Mar 2000.

COMMON SWIFT *Apus apus* Plate 50 (4)
Sp. Vencejo Común **P.** Andorinhão-preto

A. a. apus. **L** 16–17 cm (6–7"); **WS** 42–48 cm (16½–19"). **IDENTIFICATION** Sexes similar. Dark plumage, small but well-defined whitish throat, forked tail, scythe-shaped wings and fast, darting flight are familiar features, but great care should be taken to identify this species, as Plain Swift can appear very similar. Differences between them are dealt with under Plain Swift. Similarities also exist between Common and Pallid Swifts, and their separation is dealt with under the latter. **VOICE** Call is a familiar high-pitched screaming *shreeee*. **RANGE** Breeds in a broad band across most of the Palearctic, except the north and extreme west. Winters almost entirely in sub-Saharan Africa. **DISTRIBUTION** Breeds only in the Canaries. **HABITAT** Can be seen over any habitat on passage but when breeding most likely to be observed around rocky ravines (barrancos) where it nests. **BREEDING** Almost no data from our region. **STATUS** Varies between archipelagos. **Azores** Accidental visitor with *c.* 15 records, the most recent one, Farol des Rosais (São Jorge), 15 Sep 2003. **Madeira** Occasional passage migrant. **Canaries** A rare breeder on Gran Canaria and has also bred on Lanzarote and probably on Tenerife. Otherwise a common passage migrant, sometimes in large numbers, but far more numerous in spring than autumn. **Cape Verdes** Regular, but uncommon, passage migrant and winter visitor.

PALLID SWIFT *Apus pallidus* Plate 50 (5)
Sp. Vencejo Pálido **P.** Andorinhão-pálido

A. p. brehmorum. **L** 16–17 cm (6–7"); **WS** 42–46 cm (16½–18"). **IDENTIFICATION** Sexes similar. Appears more bulky with slightly broader, more blunt-tipped wings and shallower tail-fork than Common Swift. Plumage paler than Common Swift, the throat patch more extensive and the lores and forehead also white, giving the species an owl-like appearance to the face. There is also significant contrast between the inner and outer wing, which is not shown by Common Swift, and the flight action is more laboured than either Common or Plain Swifts. **VOICE** Harsher and lower pitched than Common Swift, also falling in pitch, making it almost disyllabic, unlike Common or Plain Swifts. **RANGE** Breeds mostly within W Palearctic, where it is most common in the Mediterranean Basin. Winters predominantly in Africa just south of the Sahara. **DISTRIBUTION** Summer breeder in the Canaries and Madeira. **HABITAT** Like other swifts could be encountered in almost any habitat but breeds in rocky ravines. In the Canaries recorded up to 1,200 m. **BREEDING** Very little data from our region. **STATUS** Varies on the three archipelagos where it has been recorded. **Madeira** Breeding only proven on Madeira but does occur on the Porto Santo group. Present status on the Salvages unknown but certainly used to breed. **Canaries** Confirmed to breed on Lanzarote, Fuerteventura, Gran Canaria, Tenerife and El Hierro. Formerly more common and now very rare on Tenerife as a breeder, but is also a common passage migrant. **Cape Verdes** Only two confirmed records: two, Cova (Santo Antão), 15 Apr 1999; and two, east of Praia harbour (Santiago), 2 Dec 2001. Three previous records are now considered uncertain.

ALPINE SWIFT *Apus melba* Plate 50 (6)
Sp. Vencejo Real **P.** Andorinhão-real

A. m. melba. **L** 20–22 cm (8–9″); **WS** 54–60 cm (21–23½″). **IDENTIFICATION** Sexes similar. The largest species of swift in the region. Combination of large size and white throat and belly make it totally unmistakable. Upperparts and breast-band drab grey-brown. Sometimes the white throat can be hard to see due to shadow or reduced size. **VOICE** A high-pitched twittering trill that usually rises and falls in pitch, and accelerates before slowing. **RANGE** Breeds in S Europe, the Middle East to C Asia, Arabia, India and Sri Lanka, and in N, C and S Africa. Winters predominantly in sub-Saharan W and E Africa. **STATUS** Varies between the archipelagos. **Madeira** Occasional passage migrant. **Canaries** Regular but uncommon passage migrant recorded from all of main islands. **Cape Verdes** Six records, of which five were in 1997, and perhaps occurs more regularly than records suggest.

LITTLE SWIFT *Apus affinis* Plate 50 (8)
Sp. Vencejo Moro (Vencejo Culiblanco) **P.** Andorinhão-pequeno

A. a. galilejensis. **L** 12 cm (5″); **WS** 34–35 cm (13½–14″). **IDENTIFICATION** Sexes similar. A small swift best identified by its small size, prominent white throat and rump, and almost square-ended tail. Rump patch broad and extends onto flanks, thus visible from below as well as from above. **VOICE** A high-pitched twitter that fluctuates in volume. **RANGE** Predominantly a sub-Saharan and S Asian breeder, but also breeds in NW Africa, SW Arabia and the Middle East. Mostly sedentary but some move south in winter from more northerly regions. **STATUS** Vagrant with only 13 records, 12 from the same archipelago. **Madeira** One record from the Salvage Islands. **Canaries** Accidental with 12 records, from Fuerteventura (5), Gran Canaria (2) and Tenerife (5), the most recent one, Las Parcelas (Fuerteventura), 27 Feb 2000.

WHITE-RUMPED SWIFT *Apus caffer* Plate 50 (7)
Sp. Vencejo Cafre **P.** Andorinhão-cafre

Monotypic. **L** 14 cm (5½″); **WS** 34–36 cm (13½–14″). **IDENTIFICATION** Sexes similar. Intermediate in size between Common Swift and Little Swift. Very dark plumage (virtually black) except white throat patch and narrow white rump-band. Tail deeply forked when held open, but otherwise appears as a long spike, which combined with the relatively long wings and thin body affords the species its very characteristic shape. **VOICE** A rather staccato, guttural chattering. **RANGE** A predominantly African species with a sub-Saharan distribution. Small breeding populations also exist in S Iberia and Morocco, these migrating to the tropics in winter. **STATUS** Vagrant with just seven records, and all but one from the same archipelago. **Madeira** One, Crown Plaza Resort in Funchal (Madeira), 29 Jun 2004. **Canaries*** Four, north of Tindaya (Fuerteventura), 5 Feb 1989; one, Cigüeña (Tenerife), 4 Sep. 1990; one, Guaza (Tenerife), 26 May 1991; one, Garajonay (La Gomera), 5 Jul 1992; one, Garachico (Tenerife), 4 Apr 1995; and one, Mirador del Río (Lanzarote), 16 Oct 1998.

GREY-HEADED KINGFISHER *Halcyon leucocephala* Plate 51 (1)
Sp. Alción Cabeciblanco **P.** Passarinha

H. l. acteon. **L** 21–22 cm (8–9″); **WS** 32–34 cm (12½–13½″). **IDENTIFICATION** Unlikely to be mistaken for any other species. A medium-sized kingfisher with a large red bill, pale grey to white head and breast, chestnut-orange belly and vent, black back and predominantly blue wings and tail. In flight prominent white patches at base of primaries. Sexes similar but ♀ is usually duller. Juvenile similar to adult but has a buff tinge to head and underparts with some dark mottling, and bill is blackish with a reddish base. The local subspecies is whiter headed and paler on underparts than the nominate and there are also slight differences in measurements. **VOICE** A rapid reeling *it-it-it-it-it-it…* and staccato *chi-chi-chi-chi-chi-chi*. **RANGE** Breeds over most of sub-Saharan Africa and SW Arabia.

Highly migratory within its main breeding range, undertaking wet- and dry-season movements. **DISTRIBUTION** Breeds only in the Cape Verdes where resident. **HABITAT** Quite widespread, being found in well-vegetated valleys, cultivated areas, dry country with scattered trees and bushes, and in towns and villages. **BREEDING** Breeds late Jul–Dec and Feb–May, in excavated tunnels in banks and road cuttings, holes in trees and stonewalls. Clutch size normally 3–4, laid in an unlined nest. **STATUS** Only found on one archipelago. **Cape Verdes** Common and widespread on Santiago, Brava and Fogo, the only three islands on which it occurs.

COMMON KINGFISHER *Alcedo atthis* Plate 51 (2)
Sp. Martín Pescador **P.** Guarda-rios (Guarda-rios-comum)

A. a. ispida. **L** 16–17 cm (6–7″); **WS** 24–26 cm (9½–10″). **IDENTIFICATION** Sexes similar. Unmistakable. Combination of small size; metallic greenish-blue crown, nape and wings; brilliant shining blue mantle, rump and tail, with orange-chestnut underparts and ear patch are unique, in our region. Note that colours change significantly with the angle and intensity of light. Breeding ♀ can be separated from ♂ by reddish coloration on lower mandible. Juvenile similar to the adult but generally has duller and greener upperparts. **VOICE** A short, shrill whistle *cheee*, sometimes with a *cheee-ti* repeated when excited. **RANGE** Breeds from Europe and N Africa east, discontinuously, to Japan and the Solomons. Northernmost populations are migratory but most winter in the breeding range. **STATUS** Accidental visitor recorded from two archipelagos. **Madeira** Considered exceptional by Zino *et al.* (1995). **Canaries★** Nowadays accidental with records from Lanzarote, Fuerteventura, Gran Canaria, Tenerife and La Palma, the most recent record one, Maspalomas (Gran Canaria), 7 Oct 1995–15 Jan 1996.

BELTED KINGFISHER *Ceryle alcyon* Plate 51 (3)
Sp. Martín Gigante Norteamericano **P.** Guarda-rios cintada (Guarda-rios-de-crista)

Monotypic. **L** 28–35 cm (11–14″); **WS** 47–52 cm (18½–20½″). **IDENTIFICATION** Sexually dimorphic. Large size and combination of blue-grey upperparts and white underparts should prove unmistakable. ♂ has single grey breast-band, whereas ♀ has double breast-band, the upper one grey like ♂ but the lower is chestnut and this coloration continues onto flanks. Juvenile closely resembles adult of same sex but has rufous feather fringes to upper breast-band. **VOICE** Typical call is a loud, far-carrying, harsh, clattering rattle *kekity-kek-kek-kek*. **RANGE** Breeds across most of N America, northern populations moving south in winter as far as N South America. **STATUS** Extreme vagrant. **Azores** ♀ shot, Flores, Mar 1899; one, Ponta da Caldeira (São Jorge), 21 Oct 1996; and one, Porto Pim (Faial), 2 Oct–9 Dec 2001.

BLUE-CHEEKED BEE-EATER *Merops persicus* Plate 51 (4)
Sp. Abejaruco Papirrojo **P.** Abelharuco-de-faces-azuis (Abelharuco-de-garganta-vermelha)

M. p. chrysocercus. **L** 27–31 cm (11–12″); **WS** 46–49 cm (18–19″). **IDENTIFICATION** Sexes similar. Could only be mistaken for European Bee-eater, but differs in being predominantly green with much longer central tail feathers, and a predominantly red-brown throat with little yellow on chin, no black breast-band and a narrower black face mask bordered above and below with bluish-white. In flight shows a uniform narrow dark trailing edge to underwing. Juvenile is much duller than adult, with pale whitish-green around facial mask and lacks elongated central tail feathers. **VOICE** A rapid, rolling trill *diririp*, similar to European Bee-eater but less liquid and distinctly polysyllabic. **RANGE** Discontinuous throughout Africa and Madagascar, east through Middle East to Kazakhstan and NW India. Migratory over most of range; the NW African population winters in W Africa from Senegambia to Nigeria. **STATUS** Vagrant with just six records. **Canaries★** Four records. On Tenerife, one, Laguna, 25 Apr 1890, with second shot in May, presumably the same year, and six, Ten Bel, 8–12 Sep 1991; and nine, Famara and La Corona (Lanzarote), 22 Oct 1988, then, Timanfaya National Park, 24 Oct 1988. **Cape Verdes** Two records: two, São Jorge dos Orgãos (Santiago), 5 Mar 1985; and three, between Rabil and Curral Velho (Boavista), 28 Feb 1997.

EUROPEAN BEE-EATER *Merops apiaster* Plate 51 (5)
Sp. Abejaruco Común **P.** Abelharuco (Abelharuco-comum)

Monotypic. **L** 27–29 cm (11–11½″); **WS** 44–49 cm (17–19″). **IDENTIFICATION** Sexes similar. Multicoloured plumage is unlike any other bird in the region. Chestnut on crown, nape and inner wing, with yellowish-white shoulder patches. Chin and throat yellow bordered below with a black band and rest of underparts greenish-blue. In flight very sharply pointed wings, almost triangular in shape, and longish tail with two central tail feathers elongated and terminating in a point. Characteristic shape in flight should be sufficient to identify the species from all others except Blue-cheeked Bee-eater. Juvenile duller with a more greenish tone to upperparts. **VOICE** A rather liquid, almost bubbling, mellow *pruik*, repeated frequently. **RANGE** Breeds from Iberia and N Africa east to Pakistan,

Kashmir, south of Lake Balkhash and the Irtysh River. Also breeds uncommonly in S Africa and Namibia. Winters south of the Sahara from Senegal to Ghana, and in E and S Africa. **STATUS** Varies between the archipelagos. **Azores** Two shot, Porto Formoso (São Miguel), August prior to 1903. **Madeira** Scarce passage migrant. **Canaries** Regular passage migrant, more frequent in spring and often in small flocks. Reportedly bred at Arguineguín (Gran Canaria) in May 1856. **Cape Verdes** Three records; 2 ♀♀, 25 Feb and ♂, 2 Mar, obtained, Vila da Ribeira Brava (São Nicolau), in 1924; two, Santa Maria (Sal), 23–25 May 1997; and one, Estância de Baixo (Boavista), mid-May 1999.

EUROPEAN ROLLER *Coracias garrulus* Plate 51 (6)
Sp. Carraca **P.** Rolieiro

C. g. garrulus. **L** 30–32 cm (12–12½″); **WS** 66–73 cm (26–29″). **IDENTIFICATION** Sexes similar. Unmistakable, given predominantly turquoise head, underparts and wing panel, darker blue-black flight feathers and tail, and chestnut-brown mantle. Juvenile is much duller with a brown tinge to head and breast, greyer saddle and more extensive pale forehead and face. **VOICE** Typical call is a rather harsh and somewhat crow-like *rak-rak*. Display call unlikely to be heard in the region. **RANGE** Breeds in Spain and N Africa east to W Pakistan, Kashmir, W China and Kazakhstan, but absent in most of C Europe. Winters predominantly in E Africa but also locally in W Africa. **STATUS** Varies between the archipelagos. **Azores** Two records: immature obtained, Ferraria (São Miguel), 16 Sep 1909; and one obtained, Santa Maria (no other details). **Madeira** Occasional visitor. **Canaries** Rare and irregular passage migrant recorded from all main islands except El Hierro. **Cape Verdes** A wing found, Raso, 14 Mar 1986.

BROAD-BILLED ROLLER *Eurystomus glaucurus* Plate 51 (7)
Sp. Carraca Picogorda **P.** Peito-lilás

E. g. afer. **L** 29–30 cm (11½–12″); **WS** 55–58 cm (22–23″). **IDENTIFICATION** Sexes similar. Unlikely to be mistaken with mantle, crown, nape, face and upperwing-coverts dark rufous-chestnut, flight feathers dark blue, underparts deep lilac, tail, rump and undertail-coverts blue, and bill yellow. Juvenile much duller with a darker bill and more extensive pale turquoise on underparts. **VOICE** In flight gives a harsh cackling *crik-crik-crik-crik* but at rest a rather guttural and nasal series of growls. **RANGE** Most of tropical Africa and Madagascar. Resident only in the central areas. **STATUS** Vagrant with just two old records. **Cape Verdes** ♂ shot, Maio, 22 Nov 1897, and ♀ shot, Praia (Santiago), 28 Apr 1924.

HOOPOE *Upupa epops* Plate 51 (8)
Sp. Abubilla **P.** Poupa

U. e. epops. **L** 26–28 cm (10–11″); **WS** 42–46 cm (16–18″). **IDENTIFICATION** Sexes similar. Unmistakable combination of long, decurved bill, erectile crest and pinkish-brown head, neck, upper mantle and underparts to flanks, with rest of upperparts and wings boldly patterned black and white. Undulating flight, the broad rounded wings and very floppy flight combine to give an appearance similar to a gigantic butterfly. Adult ♀ slightly duller than ♂. Juvenile similar to adult ♀ but initially has a shorter bill, and is even duller and greyer on head and body. **VOICE** A repeated trisyllabic *ooop-ooop-ooop*, which is rather quiet but carries. Also a rather harsh, high-pitched *scheer* when agitated. **RANGE** Breeds over most of S and C Europe, Africa, and C and S Asia, but avoids deserts, montane regions and heavy forests. Migratory over most of its range, wintering in sub-Saharan Africa, the Indian subcontinent and SE Asia. **DISTRIBUTION** Breeds on the Canaries and on Porto Santo in Madeira. **HABITAT** Semi-desert scrub, cultivated areas, golf courses and even in towns and villages. **BREEDING** Nests in cracks in walls, abandoned buildings, sides of barrancos, between rocks or in holes in trees. Clutch size 4–6. Normally nests from Feb but has been recorded in Jan and even Dec, and in Jun it is still possible to find nests with chicks. **STATUS** Varies in the different archipelagos. **Azores** Accidental with 19 records: São Miguel (14), Santa Maria (4) and São Jorge (1), and the most recent in Mar 1984. **Madeira** Common resident on Porto Santo, but rare on Madeira, from where no proof of breeding. Also recorded on the Salvages. **Canaries** Breeding resident on all main islands and also a frequent and often numerous passage migrant. Far more frequent on the eastern islands, particularly Fuerteventura, but on other islands is generally scarce. **Cape Verdes** Five records: one, Mindelo (São Vicente), 22 Jan 1966; one, Riberão Fundo, Topo Verde (Santo Antão), 23 Aug 1988; one, Santa Maria (Sal), 17 Sep 1990; one, near Sal Rei (Boavista), 9 Apr 1999; and one, Ponta do Ervatão (Boavista), 11 Apr 1999.

(EURASIAN) WRYNECK *Jynx torquilla* Plate 51 (9)
Sp. Torcecuello **P.** Torcicolo

J. t. torquilla. **L** 16–17 cm (6–7″); **WS** 25–27 cm (10–11″). **IDENTIFICATION** Sexes similar. Combination of small size, long tail, terrestrial habits and complex plumage pattern should prove unmistakable. Generally greyish, mottled brown and buff, with a conspicuous dark patch on mantle. Throat and upper breast barred, with arrowhead marks on lower breast and flanks, and belly usually unmarked. Juvenile very similar but is slightly duller with black barring on tail more pronounced. **VOICE** Advertising call is a repeated, clear, ringing *quee-quee-quee-que*, each note falling in pitch. Alarm call is a series of hard *teck* notes. **RANGE** Breeds from N Africa and Europe east across temperate and boreal Asia to Japan. Winters in C Africa, the Indian subcontinent and SE Asia. **STATUS** Only recorded from two archipelagos and status varies. **Madeira** Fewer than five records in the last 50 years; also recorded from the Salvages. **Canaries** Scarce and irregular passage migrant with the bulk of records from Fuerteventura, but also Lanzarote, Gran Canaria, Tenerife and La Palma, and most in spring.

GREAT SPOTTED WOODPECKER *Dendrocopos major* Plate 51 (10)
Sp. Pico Picapinos **P.** Pica-pau-malhado (Pica-pau-malhado-grande)

D. m. canariensis (Tenerife) and *D. m. thanneri* (Gran Canaria). **L** 22–23 cm (9″); **WS** 34–39 cm (13½–15″). **IDENTIFICATION** Sexes differ. Unmistakable as it is the only woodpecker, apart from Wryneck, in our region. Prominently pied upperparts and plain off-white underparts, with black patches on the breast-sides and a red ventral area. ♂ has small red patch on nape and juvenile a red crown. Subspecies *canariensis* differs from the nominate *major* in being darker brown or grey-brown on belly, with a more orangey-red vent and reduced white in outer tail feathers; *thanneri* similar to *canariensis* but is noticeably paler on underparts. **VOICE** Call is a short, sharp, explosive *kik* sometimes repeated in a series with *c.* 1 call per second. **RANGE** Most of the Palearctic from the Canaries and N Africa through Europe and C Asia to Kamchatka, Japan and Hainan. **DISTRIBUTION** Confined to Tenerife and Gran Canaria in the Canaries. **HABITAT** Most common in native pine forest where the dominant tree is *Pinus canariensis*. On Gran Canaria also in palm groves and agricultural areas planted with chestnuts. **BREEDING** Normally nests in holes in dead (sometimes live) pines and also recorded breeding in *Eucalyptus* and palms, as well as telephone poles and nest boxes. Clutch 2–5, although 4–5 is more usual. Eggs laid Apr, rarely May, with young leaving nest in Jun–Jul. **STATUS** Uncommon breeder on one archipelagos. **Canaries** Scarce and localised resident in forests of mature *Pinus canariensis*, although on Gran Canaria recorded breeding in almond *Amygdales communis* and even in *Eucalyptus globulus*. Population on Tenerife apparently expanding gradually, but is commoner on Gran Canaria.

BLACK-CROWNED SPARROW-LARK *Eremopterix nigriceps* Plate 52 (1)
Sp. Terrera Negrita **P.** Pastor

E. n. nigriceps. **L** 10–11 cm (4″). **IDENTIFICATION** Sexually dimorphic. Adult ♂ unmistakable given small size, black underparts, sandy upperparts and pied head pattern. ♀ is a small sandy lark with some indistinct streaking on crown, nape and breast, and a dark line on wing formed by dark centres to median coverts. ♀ and juvenile could be confused with other larks, but small size, predominantly black tail and black underwing-coverts (a feature also of ♂) diagnostic. Juvenile resembles ♀ but has more scaly upperparts. **VOICE** Call is a short *chep*. Song is a repeated phrase *wit-teewee*, the last note being slightly lower in pitch. **RANGE** Breeds discontinuously in Sahel zone across Africa and Arabia to the Indian subcontinent. Mostly sedentary but nomadic in parts of range. **DISTRIBUTION** Breeds only on the Cape Verdes. **BREEDING** Season is late Aug–Mar but much is dependent on the irregular rains and there is a definite decrease in activity Dec–Jan. Clutch 2–3 and nests under a large stone or vegetation. **HABITAT** Usually encountered on arid plains, in cultivation, open savanna and along roadsides. **STATUS** Breeds in only one archipelago. **Cape Verdes** Breeds on Santiago, Fogo, Boavista and Maio, where it is locally common. Also recorded on Brava, where breeding suspected but not proven, and from São Nicolau.

BAR-TAILED (DESERT) LARK *Ammomanes cincturus* Plate 52 (2)
Sp. Terrera Colinegra **P.** Calhandra-das-dunas (Calhandra-elegante)

A. c. cincturus (Cape Verdes) and *A. c. arenicolor*. **L** 14–15 cm (5½–6″). **IDENTIFICATION** Sexes similar. A small, sandy and completely unstreaked lark, the most distinctive features being the obvious, clear-cut, blackish terminal bar on the pale red-brown tail and dark tips to primaries, which are often visible in flight and, at rest, contrast with the pale rufous tertials. Juvenile distinguished by narrow pale fringes to feathers of upperparts and indistinct or complete lack of dark tips to the primaries. Nominate *cincturus* is confined to the Cape Verdes and is distinguished by its darker and more sandy-rufous plumage. **VOICE** Call is a short, nasal *chup* or a trilling *cherr*. Song is a couple of quiet *zik* notes, only audible at close range, followed by a prolonged squeaky note, and the complete phrase is repeated a few times. **RANGE** Breeds discontinuously across N Africa, the Middle East and Arabia to the Indian subcontinent. Mainly resident or dispersive. **DISTRIBUTION** Breeds only in the Cape Verdes but is accidental on the Canaries. **BREEDING** Season Sep– Mar, with a decrease in Dec–Jan, and very dependent on timing of the irregular rains. Clutch 2–3 and nest is usually sheltered by a rock or vegetation. **HABITAT** Occurs in bare stony or sandy deserts, with or without sparse vegetation. **STATUS** Varies between the two archipelagos where it has been recorded. **Canaries** Two records: one, Dársena Pesquera, Santa Cruz (Tenerife), 21 & 28 Mar 1990; and one, near Playa de Avalo (La Gomera), 27 Feb 1998. **Cape Verdes** Abundant on Sal, Boavista and Maio, locally common on Santiago and São Nicolau, but rare on Fogo.

(GREATER) HOOPOE LARK *Alaemon alaudipes* Plate 52 (3)
Sp. Alondra Ibis (Sirli Desértico)   **P.** Calhandra-de-bico-curva (Calhandra-ibis)

A. a. boavistae and *A. a. alaudipes*. **L** 18–20 cm (7–8″). **IDENTIFICATION** Sexes similar. Unmistakable given large size, long pale legs, long tail and distinctly long, slightly decurved bill. Also striking black-and-white wing pattern when seen in flight. Two colour morphs, one rufous and the other grey. The rufous form is the most common in the region, and has sandy upperparts with grey restricted to nape. Prominent dark spots on breast, pale supercilium, dark eye-stripe and dark moustachial stripe add to distinctive appearance. Grey morph has grey upperparts and lacks any sandy tones. ♀ is smaller and slightly shorter billed than ♂ with less distinct spotting on breast. Juvenile lacks breast markings, has broad pale scalloping on upperparts and bill is shorter and pinker than adult. Rufous morph of *boavistae* has broader and deeper rufous pink-cinnamon fringes to feathers of upperparts than *alaudipes*. Grey morph is darker and purer grey than nominate. **VOICE** Song is a series of mournful, piping whistles, broken by a squeaky trill, the sequence at first accelerating and then slowing after the trill: *too-too-too-too-too-trrrrreeeeeeee-teew-teew-teew-teew-teew*. Alarm call is a plaintive *too* and the contact a buzzing *zeee*. **RANGE** Similar to Bar-tailed Lark but more extensive, particularly in Arabia and W Asia. Mainly resident species but disperses in response to severe climatic conditions. **DISTRIBUTION** Breeds only in Cape Verde and is an accidental visitor to the Canaries. **BREEDING** Season Sep–Mar, with a decrease in activity Dec–Jan. Clutch 2–3 and the nest is placed atop some low vegetation or on ground where, like other lark species, it is sheltered by a rock or vegetation. Rarely nests in open. **HABITAT** Both sandy and stony deserts, dunes and occasionally sandy beaches. On Maio even seen in the villages. **TAXONOMIC NOTE** Hazevoet (1995) considered *boavistae* to be indistinct from *alaudipes*, and that the Cape Verde population forms the western extreme of a cline running through Africa into Asia. **STATUS** Distinctly differs in the two archipelagos where it has been recorded. **Canaries** Seven records: one, near Tías (Lanzarote), 20 Mar 1967; one, Peñas de Chache (Lanzarote), 14 Dec 1967; one, in northern Lanzarote, 22 Jan 1988; one, Maspalomas (Gran Canaria), 22 Nov 1996; one, south of Corralejo (Fuerteventura), 21 Feb 1997; one, near Castillos de Fustes (Fuerteventura), 1–4 Mar 1998; and one, near La Oliva (Fuerteventura), 9 Mar 1998. **Cape Verdes** Formerly confined as a breeder to Boavista and Maio, where common and widespread. Recently discovered breeding on Sal where probably fewer than ten pairs.

DUPONT'S LARK *Chersophilus duponti* Plate 52 (4)
Sp. Alondra de Dupont   **P.** Calhandra de Dupont

C. d. duponti. **L** 17–18 cm (6½–7″). **IDENTIFICATION** Sexes similar. The distinctive shape of this species should be sufficient to enable easy separation from other larks. Long, slightly decurved bill, crestless head with a faint crown-stripe and rather short tail are all useful features. Heavily streaked upperparts and breast, the rest of underparts being white. Sometimes adopts a very upright stance. Juvenile almost unstreaked above with pale feather fringes and dark feather centres giving rather scaly appearance. Streaking on breast also much reduced. **VOICE** Song is a mixture of a few clear notes followed by a penetrating nasal *wheeeee* that rises in pitch. Calls include a rather human-like disyllabic whistle with the second syllable rising significantly, and a wheezy, nasal *hoo hee*. Alarm call is a quiet, high-pitched *tsii*. **RANGE** Breeds in E and C Spain and locally in N Africa. Resident but some dispersal occurs. **STATUS** Extreme vagrant with only two records, one involving possible breeding. **Canaries*** One, between Puerto del Carmen and Tías (Lanzarote), 20 Mar 1990; and two adults possibly feeding young, Punta de Matorral (Fuerteventura), 17 Mar 1999, and again on 19 and 22 Mar.

CALANDRA LARK *Melanocorypha calandra* Plate 52 (5)
Sp. Calandria Común   **P.** Calhandra-real (Calhandra-comum)

M. c. calandra. **L** 18–19 cm (7–7½″). **IDENTIFICATION** Sexes similar. A large, stocky lark with a heavy yellowish-brown bill and dark breast-side patches most prominent in spring ♂. Upperparts grey-brown with dark streaks, underparts predominantly white with a few fine black streaks on breast. In flight shows rather broad, triangular wings with a prominent white trailing edge and white outer tail feathers. Birds soon obtain adult plumage but juvenile has head and upperparts mottled and breast patches obscured. **VOICE** A characteristic harsh, rolling *schrrrreeep* and nasal *kleeerp*. Other calls resemble those of Skylark. Song, which also resembles Skylark, is unlikely to be heard in our region. **RANGE** Breeds from Iberia and N Africa through the Middle East, and Afghanistan to E Kazakhstan. In the west, this species is resident but not always sedentary. **STATUS** Vagrant with records from just two archipelagos. **Madeira** ♂ and ♀ obtained, Madeira, 20 Nov 1945. **Canaries** Accidental on Lanzarote, Fuerteventura and Tenerife, and possibly Gran Canaria, with most records in Oct–Mar, occasionally in small groups.

(GREATER) SHORT-TOED LARK *Calandrella brachydactyla*   Plate 52 (6)
Sp. Terrera Común   **P.** Calhandrinha (Calhandrinha-comum)

C. b. brachydactyla.   **L** 13–14 cm (5–5½").   **IDENTIFICATION** Sexes similar. Adult is a fairly small, pale, relatively unstreaked and crestless lark with a small, but sometimes very indistinct, dark breast-side patch. Can be very variable in coloration but those in our region are usually more sandy than resident Lesser Short-toed Lark, unstreaked on upperparts and underparts, and have no obvious primary projection. They are also slightly, but noticeably, larger than Lesser Short-toed Lark. Juvenile has upperparts mottled dark and scaled white, and has indistinct neck patches. By first autumn very similar to adult but some breast streaking can remain well into first winter. **VOICE** Call when flushed is usually a dry, rattling *chirrup*. Song has phrases similar to Skylark but more slurred and well spaced, often interspersed with call notes and mimicry. General delivery slower than Lesser Short-toed Lark. **RANGE** Breeds from W Europe and N Africa across S Europe and the Middle East through C Asia to N Mongolia, Tibet and C China. **STATUS** Varies between the three archipelagos where it has been recorded. **Madeira** Three records: two, Caniçal (Madeira), 8 Mar 1987; two, Prainha (Madeira), 3 Oct 1993, with one until 5 Oct; and one, Ponta do Pargo (Madeira), 5 Jul 2003. **Canaries★** Scarce passage migrant that is probably annual and has occasionally been recorded in quite large flocks. Recently only reported from La Graciosa, Fuerteventura and Tenerife. **Cape Verdes** One video-taped, Barril (São Nicolau), 11–13 Apr 2001.

LESSER SHORT-TOED LARK *Calandrella rufescens*   Plate 52 (7)
Sp. Terrera Marismeña   **P.** Calhandrinha-das-marismas

C. r. rufescens and *C. r. polatzeki.*   **L** 13–14 cm (5–5½").   **IDENTIFICATION** Sexes similar. The best distinction between Lesser and Greater Short-toed Larks is the extent to which the primaries project beyond the folded tertials. Lesser Short-toed has a very noticeable primary projection whereas Greater Short-toed has very little or none. Additionally, Lesser Short-toed is usually more streaked on underparts and lacks dark neck patches of Greater Short-toed. Adult brown above with darker streaking, white below with small dark streaks on breast and a rather short, stubby bill. Juvenile similar but has whitish scaling and dark mottling on upperparts. Subspecies *rufescens* has upperparts rufous-brown and chest rufous-cinnamon. Upperparts broadly and heavily streaked, and it also has heavy markings on chest; *polatzeki* is slightly smaller, has narrower streaking on upperparts and ground colour is pink-cinnamon or pale sandy-rufous. **VOICE** The flight call, though similar to Greater Short-toed Lark, is usually distinctly more buzzing, *drrrrd*. Song similar to Greater Short-toed Lark but faster, with shorter spaces between phrases, and is more varied and melodious. Also interspersed with call notes and mimicry. **RANGE** Breeds discontinuously in a band between 25° and 50°N, from Spain and the Canaries across N Africa through the Middle East and C Asia to Manchuria. Those in our region are resident but in other areas they are nomadic or migratory. **DISTRIBUTION** Breeds only on the Canaries. **HABITAT** Dry grassland, cultivated areas and semi-deserts. **BREEDING** Breeds as early as Jan but normally the season is Feb–May, sometimes into Jun. Clutch 2–5 (mean three), with most pairs having two broods. Nest a shallow scrape, usually protected by plants and lined with vegetation. **STATUS** Only recorded from one archipelago. **Canaries** Breeds on Tenerife, where decreasing and very local, and on Gran Canaria, Fuerteventura and Lanzarote. On the latter two islands it is widespread and common. Subspecies *rufescens* is confined to N Tenerife and only an estimated five birds remain at the time of writing! This subspecies possibly breeds on Gran Canaria, but some authorities consider that all those on Gran Canaria are *polatzeki*, which form breeds on Lanzarote, Fuerteventura, and S Tenerife. Recorded on La Gomera, where probably bred in the past.

CRESTED LARK *Galerida cristata*   Plate 52 (8)
Sp. Cogujada Común   **P.** Cotovia-de-poupa

Subspecies unknown.   **L** 17–19 cm (6½–7½").   **IDENTIFICATION** Sexes similar. Similar to Skylark but easily distinguished by its longer crest, visible even when flattened, and buff outer tail feathers. Greyer than Skylark and lacks white trailing edge to secondaries. Also very similar to Thekla Lark but has longer bill, rather less distinct streaking on breast and buffy underwing-coverts. Adult is greyish-brown above with dark-streaked mantle, buffish-white underparts with dark breast and flank streaking. Juvenile has a shorter crest and crown, mantle and scapulars are spotted white. **VOICE** Usual call is a musical, liquid *twee-tee-too* delivered from ground, in flight or when perched. **RANGE** Widespread across Europe, Asia, Africa and Arabia. Most populations sedentary but local movements and some dispersal occur. **STATUS** Accidental visitor recorded from two archipelagos. **Azores★** One, park in Ponta Delgada (São Miguel), 8 Sep 1998. **Canaries★** Vagrant, mainly in winter months and recorded from Lanzarote, Gran Canaria and Tenerife, but details are few. On Lanzarote, classified by Trotter (1970) as common around Tías in Mar & Dec 1967. On Tenerife, near Las Galletas, on 21 Dec 1976, a few small groups of larks were probably this species. In autumn 1995, five were found at Costa Botija (Gran Canaria).

THEKLA LARK *Galerida theklae* Plate 52 (9)
Sp. Cogujada Montesina **P.** Cotovia-escura

Subspecies unknown. **L** 15–17 cm (6–7″). **IDENTIFICATION** Sexes similar. Very similar to Crested Lark but streaking on underparts always distinct, bill shorter and crest less spiky. Adult generally greyer on upperparts than Crested, with whiter underparts, more strongly contrasting tertials, warmer russet-brown uppertail-coverts contrasting with rump, and underwing greyish. Juvenile has white-spotted upperparts and is very similar to juvenile Crested Lark. **VOICE** Typical call is like that of Crested Lark but is lower pitched, the last note the longest and the first rises then falls in pitch. **RANGE** Breeds in Iberia, NW and NE Africa. **STATUS** Only recorded from one archipelago. **Canaries★** Reported from Lanzarote, with Lesser Short-toed Larks, by Trotter (1970), on 2 Nov 1966, 25 Nov 1967, and 20 Feb 1968, but no other details available.

(EURASIAN) SKYLARK *Alauda arvensis* Plate 52 (10)
Sp. Alondra Común **P.** Laverca

A. a. arvensis and *A. a. cantarella*. **L** 17–18 cm (7″). **IDENTIFICATION** Sexes similar. Adult a medium-sized lark with comparatively short bill and longish tail. ♂ has short crest (only obvious when crown feathers raised). Generally brownish above (from warm rufous-brown to cold greyish-brown) and whitish below with heavy streaking on upperparts. Underparts buff on breast and white on belly; the breast and flanks are streaked, the mark on breast ending abruptly. In flight reveals conspicuous white outer tail feathers and whitish trailing edge to secondaries. Juvenile plumage lost quickly and unlikely to be seen in our region. **VOICE** Typical call given when flushed is a liquid, rattling, rippling *chirrrup* and gives a variety of other scratchy, dry notes. Song unlikely to be heard in our region. **RANGE** Breeds from W Europe and NW Africa east to Kamchatka, Sakhalin, Korea and Japan. W European populations are partially migratory. **STATUS** Recorded from three archipelagos but the status varies between them. **Azores** Five records, all of specimens: Flores, May 1904, Faial, 5 Jan 1922 (*cantarella*), Faial, 14 Jan 1928 (*cantarella*), Ponta Garça (São Miguel) (no date), São Miguel, 2 Jan 1951. **Madeira** Regular winter visitor in variable numbers, more common on Porto Santo than Madeira. Also recorded on the Salvages. **Canaries** Regular winter visitor in variable numbers and recorded from all main islands.

RASO LARK *Alauda razae* Plate 52 (11)
Sp. Terrera de Razo **P.** Laverca do Razo

Monotypic. **L** 12–13 cm (4½–5″). **IDENTIFICATION** Sexes differ. Unmistakable being the only resident lark on the small island of Raso. Similar to Skylark but obviously smaller, shorter winged, shorter tailed and longer billed, affording the species a rather front-heavy appearance. ♂ has noticeably larger, stouter and more strongly curved bill than ♀. Juvenile mottled and spotted. **VOICE** Calls resemble Skylark. Song also resembles Skylark but has shorter phrases and is less varied. **RANGE** Confined to Cape Verde. **DISTRIBUTION** Breeds only on Raso (7 km^2), where mostly confined to vegetated areas of the island's flat plain (*c.* 4 km^2). **HABITAT** Restricted to vegetated areas of Raso, which comprise less than 50% of the total surface area of the islet. **BREEDING** Probably rather erratic and dependent on irregular rains. Possibly has two breeding periods, the main one in autumn in Sep–Nov, and then in spring in Feb–Apr. Nest a shallow scrape lined with grass stems and often placed at base of a rock or small shrub. Clutch 1–3. Nest success has been estimated at less than 5% and the species is considered Critically Endangered. It is believed that the near-endemic Giant Gecko is the main nest predator. **STATUS** Present on only one small islet. **Cape Verdes** Endemic to Raso, where the population seems to fluctuate at 20–250 birds. Whether the lower figure represents an incomplete count or the result of a prolonged drought is unclear. A recent survey, in Oct 2001, found 128–138 birds, of which 61–66% were ♂♂.

PURPLE MARTIN *Progne subis* Plate 53 (1)
Sp. Golondrina Purpúrea **P.** Andorinha-roxa

Subspecies unknown but *P. s. subis* would be the most likely. **L** 19–20 cm (7½–8″); **WS** 46 cm (18″). **IDENTIFICATION** Sexually dimorphic. The large size of this transatlantic vagrant should make identification relatively straightforward, but beware confusion with similarly sized Common Starling. ♂ a uniform glossy violet-blue. ♀ plumage is duller with a grey forehead, grey collar and smudgy grey-brown underparts. Juvenile resembles the ♀ but is even duller with paler underparts, the whitish belly contrasting with the greyish throat. **VOICE** Common calls include a descending *cherr* along with various low whistles and chortles. **RANGE** Breeds from Canada to Mexico but is most widespread in E USA. Winters mostly in E South America. **STATUS** Extreme vagrant. **Azores★** Two, Corvo, 28 Sep 1996, was the first record for the W Palearctic, but is still under consideration by the Portuguese rarities committee. More recently, juvenile, Facho (Flores), 6 Sep 2004.

PLAIN MARTIN (BROWN-THROATED SAND MARTIN) *Riparia paludicola*
Sp. Avión Paludícola   P. Andorinha-dos-charcos   Plate 53 (2)

Subspecies unknown but *R. p. mauretanica* likely. **L** 10–11 cm (4″); **WS** 26–27 cm (10–10½″). **IDENTIFICATION** Sexes similar. The smallest hirundine, marginally smaller than Sand Martin. Plumage similar to Sand Martin but Plain Martin lacks distinctive breast-band. Tail is not as forked and wings appear shorter, broader and more rounded, giving this species a shape similar to Crag Martin. Generally greyer on upperparts than Sand Martin with a paler rump. Underparts dusky on chin, throat and breast with belly and undertail-coverts white. This latter feature is a useful distinction from Crag Martin if size cannot be judged, which is often the case with a lone vagrant. Underwing dark grey-brown, like Sand Martin and Crag Martin, but underwing-coverts less contrasting than either of these two species, particularly Crag Martin. Juvenile resembles adult but has buff fringes to upperpart feathers and a buff tinge to underparts. **VOICE** Contact call is a low, rasping *churrr*, similar to Sand Martin, and a harsh *svee-svee*. **RANGE** Breeds in NW Africa and discontinuously over much of sub-Saharan Africa, in the Indian subcontinent and east to Taiwan and the Philippines. NW African population apparently sedentary. **STATUS** Vagrant with records from two archipelagos. **Canaries★** One, near Punta Uña Gata (Fuerteventura), 15 Jul 1994; two, Puerto del Carmen (Lanzarote), 17–18 June 1999; one, Embalse de Los Molinos (Fuerteventura), 20 Apr 2000; one, Antigua (Fuerteventura), 14 Apr 2001; and one, Risco de Paso (Fuerteventura), 16 Apr 2001. **Cape Verdes** Three or four, Terra Boa (Sal), 14 Apr 1996.

SAND MARTIN   *Riparia riparia*   Plate 53 (3)
Sp. Avión Zapador   P. Andorinha-das-barreiras

R. r. riparia. **L** 12–13 cm (5″); **WS** 26–29 cm (10–11½″). **IDENTIFICATION** Sexes similar. A small hirundine, grey-brown above and white below with a distinctive well-marked grey-brown breast-band, which separates white throat from white belly. Underwings dark grey-brown with contrastingly darker underwing-coverts. Most noticeably differs from Plain Martin in the dark breast-band and white throat, and from Crag Martin by smaller size, white vent and lack of white tail spots. Juvenile resembles adult but has buff fringes to upperparts feathers and buff wash to face and throat. **VOICE** Common call is a harsh, rasping *tschrr*. Song a harsh twitter, similar to a series of rapid call notes. **RANGE** Breeds across N America, Europe and Asia east to Kamchatka, Sakhalin, N Japan and E China. European birds winter in sub-Saharan Africa. **STATUS** Recorded from all four archipelagos but status varies. **Azores** Five records including four in 1998: one, Ilhéu do Vila (Santa Maria), 3 Jun 1990; three, Ponta Delgada (Flores), 2 Sep 1998; one, Santa Cruz (Flores), 2–3 Sep 1998; one, Cabo da Praia (Terceira), 3 Sep 1998; and four, Fajã Grande (Flores) 3 Sep 1998, with one remaining next day. **Madeira** An exceptional visitor that has also been recorded on the Salvages. **Canaries** Regular passage migrant but numbers vary and is more frequent in spring, although there are records in all months and from all main islands. **Cape Verdes** Fourteen records, of which was a bird one ringed in Britain, recovered Tarrafal (Santiago), 25 Aug 1969.

(EURASIAN) **CRAG MARTIN**   *Ptyonoprogne rupestris*   Plate 53 (4)
Sp. Avión Roquero   P. Andorinha-das-roches

Monotypic. **L** 14½ cm (6″). **WS** 32–34½ cm (12½–13½″). **IDENTIFICATION** Sexes similar. Large martin with grey upperparts, dusky, faintly streaked throat, pale breast and belly, and dark vent. Underwings dark and underwing coverts even darker. Tail has a shallow fork and white spots in the centre that are only visible when tail spread. Larger than Sand Martin, from which also distinguished by its lack of a breast-band, much greyer plumage, dark vent, white tail spots and more contrastingly dark underwing-coverts, which also separate this species from Plain Martin. Juvenile resembles adult but has rufous fringes to upperparts and a rufous tinge to underparts from chin to belly. **VOICE** Commonest call is a hard *prrrit* or disyllabic *pritit* used in contact, and a variable *zrrr* given in mild alarm. **RANGE** Breeds from Iberia and NW Africa east through S Europe and Turkey to China. Many S European populations winter in NW Africa. **STATUS** Accidental visitor recorded from two archipelagos. **Madeira** One record from the Salvages. **Canaries** Records from Fuerteventura and Tenerife, and possibly Gran Canaria, but only two recently: one, Corralejo (Fuerteventura), 19 Mar 1989, and one, near Las Galletas (Tenerife), 24 Aug 1991.

BARN SWALLOW   *Hirundo rustica*   Plate 53 (5)
Sp. Golondrina Común   P. Andorinha-das-chaminés (Andorinha-chaminés)

H. r. rustica. **L** 17–19 cm (7–7½″); **WS** 32–34½ cm (12½–13½″). **IDENTIFICATION** Sexes similar. A very familiar species with characteristic outline in flight, long, pointed wings and deeply forked tail with narrow, elongated streamers on outer feathers. Adult has dark glossy blue-black upperparts, white or off-white underparts, dark blue breast-band and deep red forehead and throat. When tail is spread it is possible to observe small white 'windows' on outer tail feathers. Juvenile recalls adult but is duller with shorter tail streamers and reddish-buff forehead and throat. In our region, only likely to be confused with Red-rumped Swallow, which has pale rump, rufous nape, black vent and broader, more blunt-tipped tail streamers. **VOICE** Song is a melodious twittering mixed with a harsh rattle. Common call is a sharp *vit* often repeated, sometimes in long series. **RANGE** Breeds over most of N Hemisphere. European breeders winter in sub-Saharan Africa. **STATUS** Recorded from all four archipelagos and a regular visitor to most of region. **Azores** Accidental with records from Santa Maria, São Miguel, Terceira, Pico and Flores. **Madeira** Occasional visitor and also recorded on the Salvages. **Canaries** Common passage migrant and occasional winter visitor. More numerous in spring than in autumn. The first breeding record involved two pairs at El Matorral (Gran Canaria) in 2000, and is thought to have also bred in 2001 and 2002. **Cape Verdes** Regular passage migrant and winter visitor, in small numbers, but uncommon.

RED-RUMPED SWALLOW *Cecropis daurica* Plate 53 (6)
Sp. Golondrina Dáurica   **P.** Andorinha-dáurica

C. d. rufula. **L** 16–17 cm (6–7″); **WS** 32–34 cm (12½–13½″). **IDENTIFICATION** Sexes similar. Slightly smaller than Barn Swallow with less pointed wings and tail, but obvious pale rump. Adult has upperparts blue-black but browner wings and tail, rump pale rust-coloured but at distance can look almost white, and a dark rusty-brown band on nape. Underparts pale buff with narrow dark streaking (only visible at close range) and black undertail-coverts. Soaring birds easily identified from below by combination of pale throat and breast, and square-cut black vent. Flight action somewhat intermediate between Barn Swallow and House Martin. Juvenile resembles adult but is less glossy with buff fringes to upperparts, paler on underparts with less obvious streaking; rump, collar and supercilium also paler rufous-buff and tail streamers shorter. **VOICE** Contact call is a sparrow-like *djuit* but also has a long, nasal *djuui* peculiar to this species. **RANGE** Breeds discontinuously in C and E Africa, NW Africa and S Europe east to China and Japan. European birds are presumed to winter in tropical Africa. **STATUS** Recorded from all four archipelagos but the status varies. **Azores** One in a group of Barn Swallows, south of the airport at Horta (Faial), 9 May 1977. **Madeira** First records on main island in 2002, with three, Ponta do Pargo, 2 Mar, and one, Santa Cruz, 8 Apr; previously recorded only on the Salvages. More recently one, Ponta do Garajau (Madeira), 9 Feb 2004. **Canaries** Scarce but probably annual passage migrant, most frequently noted in spring with records from all main islands and the first for La Palma being three, El Paso, 26 Feb 2000. **Cape Verdes** First recorded 1996 and still only nine records.

(AMERICAN) CLIFF SWALLOW *Petrochelidon pyrrhonota* not illustrated
Sp. Avión Roquero Americano   **P.** Andorinha-de-testa-branca

Subspecies unknown. **L** 13 cm (5″). **WS** 34–35 cm (13½″). **IDENTIFICATION** Sexes similar. Small, rather martin-like swallow with some features resembling Red-rumped Swallow. Adult has an almost square-ended tail, pale whitish forehead, deep rufous ear-coverts and throat, paler rufous collar and rump, black breast patch and pale stripes on mantle. Mantle and crown dark bluish-black with upperwings and tail duller blackish-brown. Underparts predominantly whitish with greyish-brown upper breast and flanks. Undertail coverts dusky with white margins, thus appearing rather spotted. Juvenile duller and less glossy with a less pronounced pale forehead and less rufous face. Most likely to be confused with Red-rumped Swallow but tail shape and pale vent identify it. **VOICE** Contact calls include a low *chur* and rather nasal *nyew*. **RANGE** Breeds across much of N America south to C Mexico. Winters mostly in S America, from Brazil to C Argentina. **STATUS** Extreme vagrant with just two certain records. **Azores** One, Cabo da Praia (Terceira), 28 Sep 2001, joined by a second, 11 Oct 2001. **Madeira** One, Tanque (Porto Santo), 27 Oct 2001. **Canaries** Immature, Roquito del Fraile (Tenerife), 30 Sep 1991, but Cave Swallow *P. fulva* was not eliminated.

(COMMON) HOUSE MARTIN *Delichon urbicum* Plate 53 (7)
Sp. Avión Común   **P.** Andorinha-dos-beirais

D. u. urbicum. **L** 12½ cm (5″); **WS** 26–29 cm (10–11½″). **IDENTIFICATION** Sexes similar. Adult has glossy blue-black upperparts, with short-forked tail, conspicuous white rump and white underparts. Flight not as swift as Barn Swallow, more fluttering with frequent glides on straight wings. Juvenile similar to adult but is duller above with pale tips to tertials and greyish wash on underparts, except belly and vent. **VOICE** Common call is a rather twittering *prrit*. **RANGE** Breeds throughout Europe and in NW Africa east to E Siberia, Sakhalin, Japan, E China and Taiwan. The European and NW African populations winter in sub-Saharan Africa. **STATUS** Recorded from all four archipelagos but status varies. **Azores** Twelve records from Santa Maria (2), São Miguel (6), Terceira (1) and Flores (3). **Madeira** Occasional visitor and also recorded from the Salvages. **Canaries** Common passage migrant, more numerous in spring and records from all main islands. **Cape Verdes** Scarce and irregular passage migrant and winter visitor that is unrecorded from Fogo and Santa Luzia.

RICHARD'S PIPIT *Anthus richardi* Plate 54 (1)
Sp. Bisbita de Richard **P.** Petinha de Richard

Subspecies unknown. **L** 18 cm (7″). **IDENTIFICATION** The largest pipit recorded in the region. Similar to Tawny Pipit but has a very long hindclaw, heavier bill, longer and thicker legs and longer tail, but these are of little use when dealing with a lone individual. Adult Richard's Pipit much more heavily streaked on upperparts and underparts than Tawny and has darker brown upperparts, buff breast and flanks, and often a dark wedge on lower-throat-sides. Juveniles more difficult to separate but lack of a dark loral stripe (hard to see at certain angles) distinguishes Richard's. Juvenile darker on upperparts than adult and further separated by the thin white or pale buff fringes to median coverts, greater coverts and tertials. In adults these fringes are broad and warm buff. First-winter resembles adult but retains some juvenile wing feathers. **VOICE** The call, a loud and explosive *schreep*, is distinctly different from normal call of Tawny. **RANGE** Breeds over most of C and S Asia, sub-Saharan Africa, Indonesia and Australasia. Northern races are long-distance migrants but winter in the range of the southern races, mostly from Pakistan to Indochina. **STATUS** Vagrant with only four records. **Canaries** One, Puerto de la Cruz (Tenerife), late Jan–23 Feb 1983; first-winter, Golf del Sur (Tenerife), 20 Nov 1994; one, Morro Jable (Fuerteventura), 15–16 Nov 2000; and two first-winters, Amarilla Golf (Tenerife), 5–6 Nov 2001, with one until 23 Mar 2002.

TAWNY PIPIT *Anthus campestris* Plate 54 (2)
Sp. Bisbita Campestre **P.** Petinha-dos-campos

A. c. campestris. **L** 17 cm (7″). **IDENTIFICATION** Sexes similar. A large, wagtail-sized pipit, with long legs and tail, and an upright stance. The dark loral stripe and a relatively short hindclaw help distinguish this species, in all plumages, from similar Richard's Pipit. Adult greyish-sandy above with some indistinct streaking and underparts buffish-white with a few streaks on breast-sides. Median coverts contrastingly dark with broad buff-white tips in fresh plumage. Juvenile has streaked upperparts and dark streaks on upper breast and therefore very similar to juvenile Richard's. Apart from the loral stripe, these birds usually have a more heavily streaked rump and cleaner, whiter underparts than similarly aged Richard's. **VOICE** Typical flight calls are a rather House Sparrow-like *tchilp* or shorter *chup*. **RANGE** Breeds from W Europe and NW Africa through C Asia east to the Tien Shan of China. W Palearctic birds winter predominantly in the Sahel of Africa. **STATUS** Varies between the three archipelagos where it has been recorded. **Madeira** Two, Ponta de São Lourenço (Madeira), 23 Jul 1981. **Canaries** Scarce but regular passage migrant, more frequent in spring, and unrecorded on La Palma and El Hierro. **Cape Verdes** One, Boavista, 12 Mar 1986.

BERTHELOT'S PIPIT *Anthus berthelotii* Plate 54 (3)
Sp. Bisbita Caminero **P.** Corre-caminho

A. b. berthelotii (Canaries and Salvages) and *A. b. madeirensis* (Madeira, Porto Santo group and Desertas). **L** 14 cm (5½″). **IDENTIFICATION** Sexes similar. The only breeding pipit in the region. Upperparts dull greyish and almost unstreaked, lacking olive tones of similar-sized *Anthus*. Underparts dirty white with narrow, clear, dark streaking on breast. Also a distinct dirty white supercilium and two pale wingbars, with that on median coverts more pronounced. Juvenile browner above than adult, with upperparts distinctly mottled and supercilium less well defined. Subspecific variation in *berthelotii* and *madeirensis* is very slight, the difference being in the length of the bill and middle toe, which are both significantly longer in *madeirensis*. Those breeding on the Salvages are generally smaller than those on the Canaries. **VOICE** Common call is a *tsliree* given from ground or in flight, and a shorter and softer *chup*. Song, given in flight or from ground, is a *tsliree-tsliree…* repeated with variations. **RANGE** Confined to the Canaries and Madeira. **DISTRIBUTION** Breeds on Madeira, Porto Santo, the Desertas, Salvages and on all of the Canaries and eastern islets, except Roque del Este and Roque del Infierno. **BREEDING** Breeds Dec–Jul. Nests on ground, protected by low vegetation or rocks. Clutch size is 2–6 but more usually 3–4, and often double-brooded. **HABITAT** From coasts to the high alpine zone, found in all habitats except heavily forested areas. **STATUS** Common resident in both archipelagos. **Madeira** On Madeira more local and seems to favour higher altitudes, but on other islands in the group status is similar to that on Canaries. **Canaries** Widespread and common resident.

OLIVE-BACKED PIPIT *Anthus hodgsoni* Plate 54 (4)
Sp. Bisbita de Hodgson **P.** Petinha de Hodgson (Petinha-silvestre)

A. h. yunnanensis. **L** 14.5 cm (6″). **IDENTIFICATION** Sexes similar. Most likely to be confused with Tree Pipit but differs principally in having more olive and almost unstreaked mantle. Head pattern is also quite distinct from tree Pipit: crown less heavily streaked in centre but bordered by a heavy dark stripe; supercilium broad, buffy in front of eye and creamy-white above and behind; on rear ear-coverts there is a prominent pale spot above a distinct dark spot, which are usually indistinct or lacking in Tree Pipit. Streaking on underparts can be more extensive and bolder than on Tree Pipit but this is not constant. Juvenile separated from adult by darker feather centres to upperparts, but resembles adult in first autumn. **VOICE** Flight call is similar to Tree Pipit but alarm call is a shorter, softer and higher pitched *tsiit*. **RANGE** Breeds from approximately the Pechora River east to Kamchatka and south to Mongolia, Manchuria, Ussuriland and Hokkaido. Winters in the Indian subcontinent, SE Asia, the Philippines and S Japan. **STATUS** Extreme vagrant. **Canaries★** One, Ten Bel (Tenerife), 28 Nov 1996–8 Mar 1997.

TREE PIPIT *Anthus trivialis* Plate 54 (5)
Sp. Bisbita Arbóreo **P.** Petinha-das-árvores

A. t. trivialis. **L** 15 cm (6″). **IDENTIFICATION** Sexes similar. Larger than Berthelot's Pipit and has more olive upperparts and buff wash on throat and breast. Supercilium less distinct and streaking on underparts and mantle much bolder. Very similar to Meadow Pipit but has bolder head markings, narrower streaking on flanks, and shorter, more curved hindclaw. Also similar to Red-throated Pipit but latter has variable red throat in breeding plumage and always a streaked rump (unstreaked in Tree Pipit). Juvenile soon indistinguishable from adult but initially has slightly scaly upperparts. **VOICE** The call is a high-pitched and rather drawn-out *speeez* and, in alarm, a short *siit*. **RANGE** Breeds from Britain and N Spain through Europe and N Asia east to the Kolyma River, in Siberia, and south to the S Caspian and N India. Winters almost entirely in the Afrotropics and Indian subcontinent. **STATUS** Recorded from three archipelagos with status varying between them. **Madeira** Fewer than five records, the most recent being one, near Parque Ecologico de Funchal (Madeira), 17 Sep 2003. **Canaries** Regular passage migrant, occurring annually in small numbers, and an irregular winter visitor. Recorded from all main islands. **Cape Verdes** Four records: one, near Curral Velho (Boavista), 20 Sep. 1988; four, near Casas Velhas (Maio), 4 Apr 1991; one, Rabil lagoon (Boavista), 14–15 Mar 1997; and one, Barril (São Nicolau), 13 Apr 2001.

MEADOW PIPIT *Anthus pratensis* Plate 54 (6)
Sp. Bisbita Común **P.** Petinha-dos-prados

A. p. pratensis. **L** 14½ cm (6″). **IDENTIFICATION** Sexes similar. Intermediate in size between Berthelot's Pipit and Tree Pipit. Upperpart coloration olive-green to greyish and underparts off-white to buff. Streaking on flanks more prominent than on Berthelot's or Tree Pipits and is a useful feature if bird does not call. Juvenile inseparable from adult by first autumn but has brighter legs and underparts entirely washed yellowish-buff. **VOICE** Call is a thin, high-pitched *siip siip siip*, distinctly different from its confusion species. **RANGE** Breeds from SE Greenland and Iceland across N and C Europe east to the Ob River and south to C Italy and Georgia. Most winter around the Mediterranean in S Europe and N Africa, but some winter east to Pakistan. **STATUS** Recorded from three archipelagos but status varies significantly. **Azores★** Two recent records: one, Flores, 26–30 Sep 2003; and 3–4, Cabo da Praia (Terceira), 1 Feb 2005. **Madeira** Similar to Tree Pipit. **Canaries** Scarce but regular winter visitor, recorded from all main islands except La Palma and El Hierro.

RED-THROATED PIPIT *Anthus cervinus* Plate 54 (7)
Sp. Bisbita Gorrirrojo **P.** Petinha-de-garganta-ruiva

Monotypic. **L** 15 cm (6″). **IDENTIFICATION** Sexes similar. Unmistakable in breeding plumage when throat, and sometimes breast, distinctly red (lost or less intense in non-breeding plumage). However, extent and shade of coloration very variable. In non-breeding plumage most likely to be confused with Meadow Pipit or Tree Pipit, but one diagnostic feature is the heavy streaking on rump and uppertail-coverts. Red-throated Pipit is also usually darker on upperparts, lacking any strong olive tones, and has more heavily streaked mantle with prominent creamy-white 'tramlines' on sides. Underparts more heavily streaked with a more pronounced malar stripe and wingbars whiter. Generally more contrastingly marked than either confusion species. Juvenile and first-winter not safely distinguished from adults that lack any coloration on throat. **VOICE** Calls are a loud, long and drawn-out, high-pitched *speeeeze* similar to Tree Pipit, but higher, longer and less hoarse; also a short, abrupt *chup*. **RANGE** Breeds from N Scandinavia across N Palearctic to E Siberia and Kamchatka, also in extreme NW Alaska. Most winter in northern tropical Africa and SE Asia. **STATUS** Regular only on Canaries, elsewhere a vagrant. **Madeira** One, Ponta de São Lourenço (Madeira), 2 Jan 1984. **Canaries** Recently discovered to be a regular winter visitor to Golf del Sur/Amarilla Golf, Tenerife, but elsewhere a scarce and irregular passage migrant and winter visitor. Recorded from all main islands except El Hierro. **Cape Verdes** Four records: one, near Santa Maria (Sal), 25 Feb 1997; two, Mindelo sewage farm (São Vicente), 9 Mar 1997, with there, 22–27 Jan 1999, increasing to three, 24 Feb–2 Mar, then four, 13–14 Mar; and one, Barril (São Nicolau), 13 Apr 2001.

ROCK PIPIT   *Anthus petrosus*   Plate 54 (8)
Sp. Bisbita Costero (Bisbita Ribereño Costero)   **P.** Petinha-marítima

A. p. petrosus.   **L** 16½–17 cm (6½–7″).   **IDENTIFICATION** Sexes similar. A large, dark pipit with a long, dark bill and dark legs. Most frequently encountered on coasts but could occur inland. Adult is dark dusky-olive above, appearing almost unstreaked, and underparts are dirty buffish-white with long, broad smudgy streaks extending to rear flanks. Outer tail feathers greyish, not pure white, and wingbars rather indistinct. Supercilium usually short and indistinct, with a narrow white eye-ring. Breeding and non-breeding plumages very similar, but non-breeders lose any olive tones in upperparts and any yellowish tones on underparts. Juvenile similar to adult but streaking on upperparts darker and better defined. **VOICE** Flight call similar to Meadow Pipit, a short, explosive *wiiist* often repeated, but not as rapidly as Meadow Pipit, and usually sounding flatter. **RANGE** Breeds on coasts around Scandinavia, British Isles, Faeroe Is., NW and W France. Winters south to Algeria and Morocco. **STATUS** Vagrant to one archipelago. **Canaries★** Four records, the first three all on Fuerteventura: one, Barranco de Esquinzo, 13–14 Sep 1993; one, Playa del Matorral, 29 Jan 1994; one, Corralejo, 28 Oct 1995; and one, Las Galletas (Tenerife), 17 Nov 2002.

WATER PIPIT   *Anthus spinoletta*   Plate 54 (9)
Sp. Bisbita Ribereño (Bisbita Ribereño Alpino)   **P.** Petinha-ribeirinha

A. s. spinoletta.   **L** 16½–17 cm (6½–7″).   **IDENTIFICATION** Sexes similar. In breeding plumage has almost unstreaked brownish-grey mantle, ash-grey head and nape, white supercilium, whitish underparts with pink wash on throat and breast, with some indistinct streaking on breast-sides and upper flanks, two whitish wingbars and white outer tail feathers. Non-breeder closely resembles Rock Pipit but obviously browner (and almost unstreaked) on upperparts, whiter on underparts with more distinct streaking, whiter wingbars and supercilium, and outer tail feathers white. Juvenile like non-breeding adult but more streaked above. **VOICE** A monosyllabic *weest*, similar to *A. petrosus* but slightly higher pitched. **RANGE** Breeds in alpine areas of C and S Europe, and from Turkey east to Transbaikalia and Nan Shan. European birds winter at lower elevations in Europe and N Africa south to *c.* 30°N. **STATUS** Vagrant to one archipelago. **Canaries★** Three records: one, Arrecife salt works (Lanzarote), 12 May 1967; three, La Santa (Lanzarote), 15 Mar 1970; and one, Santa Cruz (Tenerife), 12 Sep 1988.

YELLOW WAGTAIL   *Motacilla flava*   Plate 55 (1)
Sp. Lavandera Boyera   **P.** Alvéola-amarela

M. f. flava, *M. f. flavissima*, *M. f. iberiae*, *M. f. thunbergi* and *M. f. feldegg*.   **L** 17 cm (7″).   **IDENTIFICATION** Sexually dimorphic. Adult ♂ in breeding plumage can be identified with relative ease to subspecific level. Non-breeding ♂, ♀ and juvenile more difficult to separate and individual variation further complicates the issue. Blue-headed Wagtail *flava* in breeding plumage has blue-grey head with white supercilium, white submoustachial stripe and yellow throat. ♀ much duller with a whiter throat and breast. Non-breeding ♂ resembles ♀ but has more extensively yellow underparts. First-winter resembles ♀ but is browner above with a dusky wash to breast and yellow on underparts restricted to vent. Juvenile darker than ♀ with a dark malar stripe and dark gorget on lower throat. Yellow Wagtail *flavissima* in breeding plumage has crown, nape and ear-coverts yellowish-green with rest of head yellow. ♀ is usually darker than *flava* with yellow supercilium and more extensive yellow underparts. Non-breeding ♂ resembles ♀ and first-winter and juvenile similar to those of *flava*. Spanish Wagtail *iberiae* has forehead, crown and nape dark ashy-grey, separated from darker blackish lores and ear-coverts by a narrow white supercilium. Throat usually white. ♀ resembles *flava* but has darker cheeks and a whiter chin and throat. Non-breeding ♂ resembles ♀ but is duller with a yellowish throat. First-winter and juvenile resemble those of *flava*. Grey-headed Wagtail *thunbergi* has grey crown and nape, blackish ear-coverts and yellow throat. Sometimes shows a white supercilium but only behind eye. ♀ resembles that of *flava* but is darker on crown, ear-coverts and upperparts with more prominent necklace on upper breast and a weaker supercilium. Non-breeding ♂ resembles ♀ and first-winter and juvenile similar to those of *flava*. Black-headed Wagtail *feldegg* is slightly larger than *flava* with a longer bill. Breeding ♂ unmistakable with its shiny black hood, deep yellow underparts, olive mantle and bright wing markings. ♀ and non-breeding ♂ duller than breeding ♂ but retain a dusky hood and exhibit bright wing markings. Juvenile plumage lost early and in first-winter plumage closely resembles adult ♀. **VOICE** Call is a rather thin, monosyllabic *pseet* or more disyllabic, longer *swi-eet*. **RANGE** *flava* breeds over most of continental Europe and winters in Africa; *flavissima* breeds in British Isles and nearby continental coasts, and winters mainly in W Africa; *iberiae* breeds in Iberia, SW France and NW Africa, and winters in W Africa; *thunbergi* breeds from Norway east to the Gydanskiy peninsula in N Siberia, and winters in Africa and India; *feldegg* breeds from Serbia east to Afghanistan, through Greece, Turkey and the Middle East, and winters in E Africa. **STATUS** Recorded from three archipelagos but status varies. **Madeira** An exceptional visitor that has also been recorded on the Salvages. **Canaries** Regular passage migrant, sometimes in flocks of 50–100, often of more than one subspecies; *feldegg* is an extreme vagrant to the region with just one record from the Canaries, an adult ♂, Laguna de Janubio (Lanzarote), 17 Apr 1999. **Cape Verde** Five records: one, near Santa Marta (Sal),

25 Nov 1988; one, Raso, 13–14 Mar 1990; one, Praia (Santiago), 3–14 Mar 1997; one, Santa Maria (Sal), 17–19 Oct 1998; and one, Ponta do Ervatão (Boavista), 11 Apr 1999.

CITRINE WAGTAIL *Motacilla citreola* Plate 55 (2)
Sp. Lavandera Cetrina **P.** Alvéola-citrina

Subspecies unknown. **L** 17 cm (7″). **IDENTIFICATION** Sexually dimorphic. Breeding ♂ unmistakable with bright yellow head and underparts, except whitish undertail-coverts, a grey mantle, two prominent white wingbars and black lower nape extending to neck-sides. ♀ differs in having crown, nape and ear-coverts dirty greyish, the ear-coverts bordered with yellow, and yellow of underparts less intense and less extensive. Non-breeding ♂ resembles ♀. First-winter similar to same-age Yellow Wagtail but upperparts plain grey, wingbars are brighter and broader, ear-coverts have distinct pale surround and underparts whiter with a greyish wash on breast-sides and flanks. Although there is sometimes a yellowish-buff wash on breast, first-winters never have pure yellow on underparts. Juvenile very similar to juvenile Yellow Wagtail. **VOICE** Call resembles Yellow Wagtail but is louder, shorter and more buzzing. **RANGE** Breeds from E Europe to N China and C Siberia and south to Tibet and the Himalayas. Winters in the Indian subcontinent, S China, Myanmar, Thailand and Laos. **STATUS** Extreme vagrant. **Canaries** First-winter, Roquito del Fraile (Tenerife), 10 Nov 2000.

GREY WAGTAIL *Motacilla cinerea* Plate 55 (3)
Sp. Lavandera Cascadeña **P.** Alvéola-cinzenta

M. c. patriciae (Azores), *M. c. schmitzi* (Madeira) and *M. c. canariensis* (Canaries). **L** 18–19 cm (7–7½″). **IDENTIFICATION** Sexes differ. The shortest-legged and longest-tailed wagtail. In all plumages bright yellow vent, grey upperparts, black wings, yellowish-green rump and lack of obvious wingbars are important characters. ♂ in breeding plumage has bright yellow underparts, except black chin and throat, prominent white submoustachial stripe, narrow white supercilium and eye-ring, white outer tail feathers and whitish fringes to tertials. ♀ is paler and less yellow on underparts, the throat is usually buff-white mottled black (some have throat almost as black as ♂) and less distinct submoustachial stripe. Non-breeding ♂ loses black bib, supercilium is buffy and indistinct, and breast is buff-yellow contrasting with pale throat. First-winter resembles ♀ but breast is buffier. Juvenile distinguished by indistinct buffy face markings, buff fringes and tips to wing-coverts and greyish tertial fringes. Subspecies *canariensis* is a deeper colour on underparts than nominate race, and has facial and wing markings more contrasting; *schmitzi* is darker above than *canariensis*, the supercilium is less pronounced and amount of white on outer tail feathers reduced; *patriciae* is similar to *schmitzi* but has an obviously longer bill. **VOICE** Flight call is a repeated short, harsh *tzip* or an anxious, shrill *ziss-ziss-ziss-iss*. Song is a series of high-pitched, trilled, phrases mixed with anxious call notes, and a phrase highly variable in length and pitch rather reminiscent of Atlantic Canary. **RANGE** Breeds from the Atlantic Islands discontinuously east through Europe to Kamchatka, Japan and N China. Western populations are partially migratory with some wintering in Africa south to S Malawi. Insular populations mostly resident but local movements occur. **DISTRIBUTION** Breeds on all main islands and islets in the Azores. In the Madeiran archipelago only breeds on Madeira but has occurred on Porto Santo. In the Canaries breeds on Gran Canaria, Tenerife, La Gomera and La Palma. **BREEDING** Can nest almost anywhere but most frequently in a hole in a wall. Clutch 3–6, but four is most common. Breeds Feb–Jul and is usually double-brooded, occasionally treble. **HABITAT** Strongly associated with areas of fresh water including lakes, streams, reservoirs, water tanks, levadas (man-made canals for transporting water) and even hotel swimming pools. **STATUS** A common breeder and resident in three archipelagos. **Azores** Common and widespread on the main islands. **Madeira** Common and widespread on Madeira but scarce on Porto Santo. Also recorded on the Salvages. **Canaries** Fairly common and widespread on those islands where it breeds, but has decreased due to loss of habitat. On the eastern islands it is a scarce winter visitor and, on El Hierro, accidental. It is unknown if these are representatives of the local subspecies or the nominate form.

WHITE WAGTAIL *Motacilla alba* Plate 55 (4)
Sp. Lavandera Blanca **P.** Alvéola-branca

M. a. alba and *M. a. yarrellii*. **L** 18 cm (7″). **IDENTIFICATION** Sexes similar. ♂ in breeding plumage has black rear crown, nape, chin, throat, upper breast, tail and wings. Grey upperparts and flanks, and white face, forecrown, forehead, lower breast, belly, vent, double wingbar, tertial fringes and outer tail feathers. ♀ in breeding plumage is duller, and the border between the black nape and grey mantle is a diffuse transition rather than sharply defined. ♂ in non-breeding plumage has black on underparts reduced to a breast-band. Non-breeding ♀ also has black on underparts reduced to a band on upper breast, with none on crown and nape, and face greyer than ♂. First-winter resembles ♀ but upperparts tinged olive and has yellowish wash to head. Juvenile similar to first-winter but has a grey-brown breast-band and most of head greyish. **VOICE** Call is a monosyllabic *tsit* or disyllabic *tsli-wee* or *tsli-vitt*, both types often repeated. **RANGE** Breeds throughout Europe to E Siberia, Japan and E China, and south to NW Africa, S Iran and S China. European birds winter in the breeding range and in Africa north of the equator. **STATUS** Recorded from all archipelagos but status varies. **Azores** Four records: one obtained, prior to 1903 (no other details); one obtained, Ribeira Seca (São Miguel), 5 May 1908; one, Graciosa, 2 Aug 1959; and one, at the airport (Santa Maria), 22 Mar 1968. **Madeira** Scarce winter visitor. **Canaries** *alba* occurs in small numbers as a regular winter visitor; one showing characters of *yarrellii* was found at Amarilla Golf (Tenerife), 13 Feb 2004. **Cape Verdes** Eleven records from São Vicente (4), Sal (3), Santiago (2), São Nicolau (1) and Raso (1).

(WHITE-THROATED) DIPPER *Cinclus cinclus* Plate 55 (5)
Sp. Mirlo Acuático **P.** Melro-d'água

Subspecies unknown. **L** 18cm (7"). **IDENTIFICATION** Sexes similar. Rather dumpy, short-tailed bird usually associated with running water. Adult mainly black-brown except gleaming white throat and breast. Juvenile slate-grey above with dark mottling, and white below with grey barring (except on throat). **VOICE** Call is a short, rasping *zrets* and song a scratchy warbling. **RANGE** Breeds discontinuously through Europe, NW Africa, Turkey and C Asia. Mainly resident but some post-breeding dispersal. **STATUS** Extreme vagrant. **Canaries★** One, Lanzarote, 27 Dec 1966, but no details concerning location.

(WINTER) WREN *Troglodytes troglodytes* Plate 56 (1)
Sp. Chochín **P.** Carriça

Subspecies unknown. **L** 9–10 cm (3½–4"). **IDENTIFICATION** Sexes similar. A very small bird with a relatively long, thin bill and short tail, which is often cocked. Dull rufous-brown above, with a pale off-white supercilium, and paler buff-brown below with black vermiculations on back, rump, wings, tail, flanks and vent. Juvenile similar but is warmer and less well marked. **VOICE** Call is a loud *tek* or a harsh, rattling *tzurrrr*. Song is a series of rattling trills and warbles, incredibly loud for such a small bird. **RANGE** Breeds across most of Europe and N North America, also in Asia east to Kamchatka, Japan, E China and Taiwan. Northern populations migratory but mostly resident further south. **STATUS** Vagrant recorded from two archipelagos. **Madeira** Fewer than five records in the last 50 years. **Canaries** One, northeast Lanzarote, 19 Jan 1967.

GREY CATBIRD *Dumetella carolinensis* Plate 55 (6)
Sp. Pájaro Gato **P.** Pássaro-gato

Monotypic. **L** 21 cm (8"). **IDENTIFICATION** Sexes similar. Rather like a long-tailed thrush in shape. Generally slate-grey except black cap and tail, and chestnut undertail-coverts. Irides rich reddish-brown. First-winter has eye colour duller and flight feathers tinged brownish. **VOICE** Call is a nasal, downslurred, cat-like *mew* or a harsh *kek kek kek*. **RANGE** Breeds across most of C North America. Winters in coastal states of USA, West Indies and the Caribbean slope of C America. **STATUS** Extreme vagrant. **Canaries** One, La Mareta (Tenerife), 4 Nov 1999.

DUNNOCK (HEDGE SPARROW) *Prunella modularis* Plate 55 (7)
Sp. Acentor Común **P.** Ferreirinha (Ferreirinha-comum)

Subspecies unknown. **L** 14½ cm (6"). **IDENTIFICATION** Sexes similar. A rather drab, brown and grey species that habitually spends much time on, or near, the ground. Adult has slate-grey head and breast with a brown cast to crown and ear-coverts. Rest of plumage mostly rufous-brown except white belly, the mantle is heavily streaked and flanks more finely streaked. Rump unstreaked and contrasts slightly with dark blackish-brown tail. Two rather indistinct wingbars, one on greater coverts and the other on median coverts. ♀ often indistinguishable from ♂ but some are noticeably duller and less richly coloured. Juvenile has much less grey on head, streaking on underparts is much more extensive and buff tips to greater coverts form distinct wingbar. **VOICE** Common calls include a loud *seep* and a quieter shivering *ti-ti-ti-ti* frequently given in flight. **RANGE** Mostly confined to W Palearctic. N and C European populations are migratory and winter in S Europe and Turkey. **STATUS** Extreme vagrant. **Canaries** One, near Embalse de la Cruz Santa (Tenerife), 26 Jan 2002.

ALPINE ACCENTOR *Prunella collaris* Plate 55 (8)
Sp. Acentor Alpino **P.** Ferreirinha-alpina

Subspecies probably *P. c. collaris*. **L** 18 cm (7"). **IDENTIFICATION** Sexes similar. Larger, dumpier and more colourful than Dunnock. Adult has greyish head, breast and belly with a white throat barred black, heavy brownish-orange streaks on flanks, brownish-grey upperparts with heavy streaking on mantle, two white wingbars formed by white tips to median and greater coverts, and dark panel in centre of wing formed by black greater coverts. Juvenile similar but is duller with a brown hue to greyish areas, dark brown greater coverts and buff throat. **VOICE** Most common call is a rather lark-like *chirr* or *chirrup* often repeated. **RANGE** Breeds in high mountains from Morocco and Spain east through S and C Europe, Turkey and C Asia to Taiwan. Some altitudinal and short-distance movements in winter. **STATUS** Extreme vagrant. **Canaries★** One, El Fraile (Tenerife), 1 Apr 1998.

RUFOUS BUSH ROBIN (RUFOUS-TAILED BUSH CHAT) *Cercotrichas galactotes*
Sp. Alzacola   **P.** Solitário (Rouxinol-do-mato)   Plate 56 (2)

C. g. galactotes. **L** 15 cm (6″). **IDENTIFICATION** Sexes similar. Adult has sandy-brown upperparts, off-white underparts with a buff wash on breast, prominent facial pattern and orange-brown rump and tail, the latter broadly tipped with white spots and a subterminal black bar. At times, raises and fans tail, clearly displaying the pattern. Juvenile similar to adult but is paler above, has indistinct and faint mottling on throat, breast and fore flanks, and tail spots are smaller. **VOICE** Typical calls are a hissing *tseeeeet* (or shorter *tsip*) and hard, harsh *teck*. Song is rich, melodious and warbling with well-spaced phrases. **RANGE** Breeds in S Spain and N Africa, also in isolated areas in C Africa, and in SE Europe, the Middle East, east to C Asia and S Kazakhstan. Winters in the Sahel zone of W Africa and in NE and E Africa. **STATUS** A vagrant to two archipelagos. **Madeira** Fewer than five in the last 50 years, the first being one obtained, Ribeira Brava (Madeira), 28 May 1960. **Canaries★** Six or seven records: one, near Janubio (Lanzarote), 14 Jun 1983; one, Alegranza, autumn 1983; one, Barranco de Esquinzo (Fuerteventura), 12 Jun 1987; one, Los Gigantes (Tenerife), 8 Mar 1989; one, Barranco de Esquinzo (Fuerteventura), 6 May & 25 Jun 1992; and one, Alegranza, Apr 1995.

(EUROPEAN) ROBIN *Erithacus rubecula*   Plate 56 (3)
Sp. Petirrojo   **P.** Pisco-de-peito-ruivo

E. r. rubecula (Azores, Madeira and W Canaries) and *E. r. superbus* (Tenerife and Gran Canaria). **L** 14 cm (5½″).
IDENTIFICATION Adult *rubecula* easily identified from other species by its orange face, throat and breast. Rest of underparts white with buffish flanks. Upperparts olive-brown, separated from orange bib by narrow blue-grey feathering. Wings and tail also olive-brown. First-winter generally indistinguishable but has a more obvious wingbar and is slightly duller. Juvenile lacks orange and grey of adult and has upperparts spotted buff with throat, breast and flanks brownish-buff, scalloped darker and spotted paler. Adult *superbus* resembles *rubecula* but has darker brown upperparts. The bib is more reddish and the rest of underparts whiter; there is an obvious white eye-ring particularly noticeable behind the eye and the blue-grey band separating bib from upperparts is wider. **VOICE** Call of *rubecula* is a short, hard *tic* often repeated and frequently delivered in rapid series. Song starts with a few high-pitched, drawn-out notes, then drops in pitch and increases in speed with a mix of clear warbling phrases and trills, interrupted by abrupt halts, varying in speed and pitch with no two verses the same. Call of *superbus* higher pitched and slower, lacking fast series of *rubecula*. Song also different, being shorter, with more frequent high-pitched notes and much mimicry; by contrast, *rubecula* mimics only rarely. **RANGE** Breeds throughout most of Europe east to the Ob River in W Siberia, and in N Africa. Most populations are at least partially migratory. Wintering areas include N Africa, S Turkey and the Middle East. **DISTRIBUTION** Resident breeder on the Azores, Madeira and the Canaries. **BREEDING** Usual season Apr–Jun, but nests noted Feb–Jul. Clutch 2–5, even seven in La Palma but three is the most common number. Nest usually located below 3 m among branches of bushes and small trees, in banana plantations, holes in walls and, rarely, on ground. **HABITAT** Prefers areas of native woodland and scrub but also occurs in thickets in barrancos or hedgerows in agricultural areas. Scarce in higher dry pine forests on Tenerife. **TAXONOMIC NOTE** Previously, birds on Madeira and the W Canaries were treated subspecifically, under the name *microrhynchus*. It has recently been suggested that the birds breeding on Tenerife and Gran Canaria are sufficiently different to be treated specifically as Canary Islands Robin, *E. superbus*, based on their morphology and vocalisations (see Bergmann and Schottler 2001, for example). **STATUS** Breeds on three archipelagos, where generally common in its preferred habitats, and more widespread on the Azores than in the Canaries or Madeira. **Azores** Breeds on all main islands except Flores and Corvo. **Madeira** Only breeds on Madeira but has been recorded on Porto Santo and the Salvages. **Canaries** *superbus* breeds only on Gran Canaria and Tenerife, and *rubecula* on La Gomera, La Palma and El Hierro. On Fuerteventura and Lanzarote it is a regular but scarce winter visitor.

THRUSH NIGHTINGALE (SPROSSER) *Luscinia luscinia*   Plate 56 (4)
Sp. Ruiseñor Ruso   **P.** Rouxinol-oriental (Rouxinol-russo)

Monotypic. **L** 16½ cm (6½″). **IDENTIFICATION** Sexes similar. Closely resembles Common Nightingale in all plumages. Adult duller than Common Nightingale with greyer brown upperparts and darker brown tail. Underparts diffusely mottled grey-brown on breast and flanks with a dusky malar stripe. Often some indistinct barring on undertail-coverts but this can be very difficult to observe. The first primary is shorter than the primary-coverts and it usually shows eight exposed primary tips on the folded wing, but these last features are of little use in the field. Juvenile also similar to Common Nightingale but is often darker with more contrasting pale tips to tertials and wing-coverts. First-winter resembles adult but retains buff tips to greater coverts. **VOICE** Similar to that of Common Nightingale. Calls include a throaty croak, a piercing whistle, *eeehp* or a low *chuk*. Song is louder but less musical, more monotonous, and more slowly delivered than Common Nightingale. **RANGE** Breeds from Denmark and extreme S Norway east to C Asia and south to Armenia. Winters in E Africa from Ethiopia south to Transvaal. **STATUS** Extreme vagrant. **Canaries★** One, Embalse de Los Molinos (Fuerteventura), 12 Apr 1988.

COMMON NIGHTINGALE *Luscinia megarhynchos* Plate 56 (5)
Sp. Ruiseñor Común **P.** Rouxinol (Rouxinol-comum)

L. m. megarhynchos. **L** 16½ cm (6½"). **IDENTIFICATION** Much warmer brown upperparts than Thrush Nightingale with a much more contrasting orange-brown rump and tail. Underparts whiter and plainer, lacking any greyish vermiculations, with a pale buff wash on breast. First primary longer than the primary-coverts and usually shows only seven exposed primary tips. Juvenile brown above and buff and white below with extensive spotting and scalloping on head, mantle, breast and flanks. First-winter resembles adult but usually has buff tips to greater coverts. **VOICE** Calls very similar to those of Thrush Nightingale. Song also similar but is more melodic and consists of warbling, trilling and gurgling notes with an extended, whistled, crescendo, *pew-pew-pew-pew*. Song loud, clear and variable, and most prolonged at night. **RANGE** Breeds in NW Africa, Spain and S England east to C Asia and Mongolia. Winters in tropical sub-Saharan Africa from Senegal to Kenya. **STATUS** Varies between the archipelagos. **Madeira** Listed as exceptional by Zino *et al.* (1995), and has been recorded on the Salvages. **Canaries** Rare but regular passage migrant, more frequently on the eastern islands and in spring. **Cape Verdes** Two records: one on board ship north of Sal, 7 Sep 1985; and one, Curral Velho oasis (Boavista), 20–21 Sep 1988.

BLUETHROAT *Luscinia svecica* Plate 56 (6)
Sp. Pechiazul **P.** Pisco-de-peito-azul

L. s. svecica and *L. s. cyanecula* **L** 14 cm (5½"). **IDENTIFICATION** Sexually dimorphic. Breeding ♂ unmistakable with its blue bib bordered below with narrow black and white band and a broader rufous one. Throat has either a red spot (*svecica*) or white one (*cyanecula*). Upperparts cold grey-brown with a prominent white supercilium and a rusty tail base that is quite obvious in flight. Rest of underparts off-white with a dusky wash on flanks. Adult ♀ variable as regards amount of colour on breast but many completely lack any blue or the rufous breast-band. Always shows a moustachial stripe, unlike ♂. Non-breeding ♂ can resemble older ♀, as much of the blue throat is replaced and breast-bands are less obvious. First-winter ♂ resembles non-breeding ♂ but outer greater coverts retain buff tips. First-winter ♀ lacks any colour on breast and just has a dark malar stripe with a gorget of dark spots on upper breast and buff-white underparts. Juvenile distinctly spotted and scaled on both upper- and underparts but has distinctive tail pattern of adult. **VOICE** Usual call a harsh *tchack tchack* or higher pitched *tsi* often mixed with first call, also a whistled *hwiit* when agitated. Song, sometimes given in winter, varied and powerful and includes much mimicry. Resembles that of an *Acrocephalus* warbler but is more musical and sustained. Usually commences with a loud, repeated, metallic note that slowly increases in tempo before turning into a cascade of both melodious and squeaky, rattling notes, interspersed with mimicry. **RANGE** Breeds discontinuously in Spain and C Europe, then from E and N Europe across rest of Palearctic. European birds winter across northern tropical Africa, discontinuously in Mediterranean Basin to Arabia and the Indian subcontinent. **STATUS** Accidental. **Canaries** About 12 records in total from Lanzarote, Fuerteventura and Tenerife.

BLACK REDSTART *Phoenicurus ochruros* Plate 56 (7)
Sp. Colirrojo Tizón **P.** Rabirruivo (Rabirruivo-preto)

P. o. gibraltariensis. **L** 14½ cm (6"). **IDENTIFICATION** Sexually dimorphic. A dark, rather upright species with blackish body and dark rusty-red tail with darker central rectrices. Breeding ♂ dark greyish above and below, except greyish-white central belly and orange-buff undertail, with a blacker face, prominent white wing panel and rusty-red tail. ♀ more uniform smoky-grey and lacks darker face and wing panel. First-winter similar to adult ♀ but browner with no grey tone in upperparts. Juvenile only indistinctly mottled and best separated from ♀ by pale gape and browner plumage. **VOICE** Calls a high-pitched *sip* and harsher *tic-tic-tic* often run together, *sip tic-tic*. Song, a medley of whistled and rattled notes with a strange nasal cicada-like sound, is unlikely to be heard in our region. **RANGE** Breeds from Europe east to the Volga River, in NW Africa, Turkey east to S Siberia and China. Main wintering area for European birds is the Mediterranean Basin. **STATUS** Recorded from all archipelagos, but the status varies. **Azores★** One, Mosteiros (São Miguel), 7–8 Dec 2001. **Madeira** Fewer than five records in the last 50 years, the most recent being one, Ponta Delgada (Madeira), 6 Mar 2004, and also recorded from the Salvages. **Canaries** Uncommon but regular passage migrant and winter visitor, recorded from all main islands except La Gomera, and more regular on the eastern islands. **Cape Verdes** ♂, Santa Maria (Sal), 29 Oct 1998.

COMMON REDSTART   *Phoenicurus phoenicurus*   Plate 56 (8)
Sp. Colirrojo Real   **P.** Rabirruivo-de-testa-branca

P. p. phoenicurus.   **L** 14 cm (5½"). **IDENTIFICATION** Sexually dimorphic. Breeding ♂ unmistakable given combination of black face, orange breast, ash-grey crown and back, white forehead, and rufous rump and tail with dark central rectrices. ♀ shares tail pattern but is generally brown above: a narrow white eye-ring is the only facial feature, and buffy-white below with an orange wash to breast. Non-breeding ♂ has bright colours partially obscured by pale fringes to feathers, although unique pattern still discernable. First-winter ♂ looks like non-breeding adult but lacks grey back and crown, and white supercilium, and has closely barred black cheeks and greyish bars on throat. First-winter ♀ resembles adult ♀. Juvenile is typically pale-spotted above and dark-scalloped below. **VOICE** Usual calls are a whistled *hooeet* and short, harsh *tec*, the two often combined. **RANGE** Breeds in Europe and NW Africa east to Lake Baikal and south to the Balkans, S Turkey and Iran. European birds winter in sub-Saharan Africa north of the equator. **STATUS** Recorded from all archipelagos but status varies. **Azores** Two records: one, Lagoa da Paul (Pico), 3–4 Oct 1989; and one, Mosteiros (São Miguel), 5 Dec 2001. **Madeira** Occurs seasonally but only under unusual climatic conditions. Also recorded from the Salvages. **Canaries** Fairly common passage migrant, recorded from all main islands except El Hierro. **Cape Verdes** Two records, both from Sal, with one, near Espargos, 18 Oct 1982, and a ♂, Santa Maria, 16 Oct 1998.

WHINCHAT   *Saxicola rubetra*   Plate 57 (1)
Sp. Tarabilla Norteña   **P.** Cartaxo-d'arribação (Cartaxo-nortenho)

Monotypic. **L** 12½ cm (5"). **IDENTIFICATION** Sexually dimorphic. A rather plump short-tailed species that perches upright on low vegetation. ♂ in breeding plumage has blackish head-sides with a white supercilium and white malar stripe that curves around lower edge of ear coverts. Underparts rich orange-buff on throat and breast with white belly and vent. Upperparts dark brown with blackish feather centres and tail blackish-brown with extensive off-white base to all but central pair of rectrices. Breeding ♀ is duller with browner cheeks, less contrasting head pattern, paler and less orange underparts, and white tail-sides less pronounced. Non-breeding ♂ similar to breeding ♂ but has contrasts in plumage much reduced. First-winter resembles non-breeding adult but has some fine spotting and streaking on breast and lacks white line on wing-coverts. Juvenile duller than ♀ with a reduced supercilium and more spotted plumage. **VOICE** Calls are a short, whistled *fu* and harsh *tek*, often combined as *fu-tek-tek* or *fu-fu-tek-tek*. **RANGE** Breeds from Europe east to W Siberia south to Crete and Turkey. Winters across sub-Saharan Africa and in E Africa south to Malawi. **STATUS** Recorded from all archipelagos but status varies. **Azores** One, Mosteiros (São Miguel), 3 Nov 2001. **Madeira** ♂, Paúl da Serra (Madeira), 26 Sep 2003. Also recorded on the Salvages. **Canaries** Regular passage migrant with records from Lanzarote, Fuerteventura, Tenerife and La Palma. **Cape Verdes** Both records are from Sal, on 30 Sep 1998, with one at Terra Boa and another just to the north.

CANARY ISLANDS STONECHAT   *Saxicola dacotiae*   Plate 57 (2)
Sp. Tarabilla Canaria   **P.** Caldeireta

S. d. dacotiae (Fuerteventura) and formerly *S. d. murielae* (Alegranza, now extinct). **L** 12½ cm (5"). **IDENTIFICATION** Sexually dimorphic. Breeding ♂ has dark brown upperparts with blacker cheeks and streaked mantle, narrow white supercilium not extending to rear ear-coverts, white throat and white half-collar. Underparts have a roughly triangular orange-buff patch on breast, pale buff flanks and white belly and vent. Tail all dark and bill rather long and thin compared to other *Saxicola*. ♀ rather nondescript with only a faint buffish wash on breast, paler upperparts, less distinct supercilium and no white half-collar. First-winter ♂ resembles adult but is browner on head and upperparts and whiter on underparts. Juvenile resembles ♀ but has head and mantle spotted buffish and chest spotted brownish. **VOICE** Calls are a short, harsh *chut* and high-pitched *seit*, often run together. Song when perched is a *bic-bizee-bizeeu* with a very scratchy quality. Sometimes given in flight when it contains a lark-like *liu* and has rather richer and more varied phrasing. **RANGE** Endemic to the Canaries. **DISTRIBUTION** Breeds only on Fuerteventura. **BREEDING** Nests in holes in walls or under rocks and plants. Main breeding season Jan–Apr, although nests with eggs and young chicks found as early as Dec. Clutch size 2–5 but 3–4 most frequent. Young fledge in 16–18 days. **HABITAT** Tends to occupy rocky zones with a bushes and thickets, such as barrancos, hillsides and even in cultivated areas. **STATUS** Only occurs on one archipelago. **Canaries** Population on Fuerteventura estimated at more than 750 pairs spread throughout the island. In recent years reported on various occasions in S Lanzarote.

COMMON STONECHAT *Saxicola torquatus* Plate 57 (3)
Sp. Tarabilla Común **P.** Cartaxo (Cartaxo-comum)

S. t. rubicola **L** 12½ cm (5"). **IDENTIFICATION** Sexually dimorphic. Breeding (worn) ♂ has head blackish with broad white half-collar, extensive orange underparts with only central belly and vent off-white, upperparts brownish-black with small whitish rump and dark-spotted uppertail-coverts, and an all-dark tail. In flight reveals prominent white innerwing-coverts. Breeding ♀ duller with a brownish head and throat, and smaller half-collar. Non-breeding (fresh) ♂ resembles breeding ♂ but has feathers tipped brownish or buffish, obscuring both colour and pattern. Non-breeding ♀ duller than breeding ♀. First-years resemble worn adults. Juvenile spotted on head and upperparts. **VOICE** Calls very similar to those of Canary Islands Stonechat, a harsh *chak* and a shrill *wheet*, but are perhaps not quite as sharp. Song is also similar to Canary Islands Stonechat but is unlikely to be heard in our region. **RANGE** Breeds from NW Africa and W Europe east to E Siberia, Japan, Korea and W China. Some European populations are at least partially migratory, although some are certainly resident in the breeding range. **STATUS** Recorded from two of the archipelagos but status varies. **Madeira** Fewer than five records in the last 50 years. **Canaries** Scarce and irregular winter visitor, more common on eastern islands and not recorded from either La Palma or El Hierro.

ISABELLINE WHEATEAR *Oenanthe isabellina* Plate 57 (4)
Sp. Collalba Isabel **P.** Chasco-isabel

Monotypic. **L** 16½ cm (6½"). **IDENTIFICATION** Unusually amongst *Oenanthe* the sexes are similar. In all plumages resembles ♀ and first-year Northern Wheatear, particularly the larger race *O. o. leucorhoa*. Adult has sandy-brown upperparts, white supercilium and whitish underparts with a sandy wash on breast and flanks. Legs and bill black. ♂ is normally larger with a black loral stripe, the latter feature being weaker or dull brownish in ♀. Juvenile resembles adult ♀ but is duller with weaker face pattern and a few dark brown marks on throat- and breast-sides, and faint mottling on upperparts. Best distinguished from ♀ Northern Wheatear by combination of plumage features and overall jizz. Isabelline is usually larger than Northern Wheatear with longer legs, a shorter tail, thicker bill, a shorter primary projection and appears rather pot-bellied. Plumage differences include a pronounced white or creamy supercilium in front of the eye, fading quickly above ear-coverts. The eye-stripe also fades rapidly behind eye. Less contrast between wings and the upperparts, with the dark brown alula being the darkest and most contrasting feather in the wing. Tail-band usually broader than that of Northern Wheatear and white rump less extensive. Finally, underwing of Isabelline is creamy-white or pale buff, unlike dark grey underwing of Northern Wheatear, but this can be very hard to discern. Also similar to ♀ plumages of other wheatears, such as Desert and Black-eared Wheatears, but tail pattern and extent of white rump are useful identification features. **VOICE** Calls include a nasal *cheet* or whistled *wheet* similar to Northern Wheatear. Song, which is longer and more varied than other wheatears, unlikely to be heard in the region. **RANGE** Breeds in extreme SE Europe, Turkey and the Middle East to Transbaikalia and W Heilongjiang and south to the NW Himalayas. Winters in NW India, Arabia and in sub-Saharan Africa mostly north of the equator. **STATUS** Accidental visitor to two archipelagos. **Madeira** One record on the main island on 7 Dec 1983. **Canaries*** Five dated records: one, La Tejita (Tenerife), 9 Oct 1988; one, near Montaña Amarilla (Tenerife), 21 Mar 1992; one, Amarilla Golf (Tenerife), 25 Sep 1993; one, Alegranza, early April 1995; and one, Tenerife, 11 Oct 1996. Also reported to be an occasional visitor to Timanfaya National Park, on Lanzarote (Concepción 1992, 1993).

NORTHERN WHEATEAR *Oenanthe oenanthe* Plate 57 (5)
Sp. Collalba Gris **P.** Chasco-cinzento

O. o. leucorhoa and *O. o. oenanthe*. **L** 14½–15½ cm (6"). **IDENTIFICATION** Sexually dimorphic. Breeding ♂ of race *oenanthe* has bluish-grey upperparts with a white rump and white base to tail feathers except central rectrices, which like tips of other feathers are black, forming a T-pattern. White supercilium, black mask, black wings, buffish throat and breast, and white belly and vent. ♀ is similar but is generally duller and browner with less distinct facial pattern. Non-breeding ♂ buffy-brown above with a white supercilium. Differs from non-breeding ♀ in blacker lores and ear-coverts. First-winter very similar to non-breeding adult and indistinguishable from non-breeding ♀. Juvenile spotted above with dark scaling on underparts. Greenland race *leucorhoa* larger than nominate with larger bill, longer legs and darker, richer, more uniform buff underparts. **VOICE** Calls are a whistled *wheet* and short, clicking *chack*. **RANGE** Breeds in Greenland, NE Canada, Alaska, W Europe east to NE Siberia, Altai, Transbaikalia and Mongolia. Winters in E and sub-Saharan Africa. **STATUS** Recorded from all four archipelagos and is a regular visitor. **Azores** Rare but probably regular passage migrant, but vagrant in winter. **Madeira** Occasional visitor usually encountered during exceptional climatic conditions. Also recorded on the Salvages. **Canaries** Regular passage migrant with records from all main islands. **Cape Verdes** Accidental visitor, but possibly regular in winter, with 18 records from Sal (6), Boavista (5), Maio (2), São Nicolau (2), Santiago (1), Raso (1) and Branco (1).

BLACK-EARED WHEATEAR *Oenanthe hispanica* Plate 57 (6)
Sp. Collalba Rubia **P.** Chasco-ruivo

O. h. hispanica. **L** 14½ cm (6"). **IDENTIFICATION** Sexually dimorphic and has two forms, dark-throated and pale-throated. Adult ♂ of dark-throated form has black face and throat, black wings, deep buff crown, mantle and breast (which can fade to almost white with wear), whitish belly and vent, white lower back and rump, and a tail pattern that has distinctly more black on outer feathers: thus band is less uniform than in Northern Wheatear. Pale-throated form differs in having throat white and black face reduced to a mask through eye, narrower than in Northern Wheatear. ♀ of dark throated form in worn breeding plumage resembles dull ♂. In fresh non-breeding plumage pale tips to feathers conceal dark face and throat, with an overall gingery tone to breast and upperparts. ♀ of pale-throated form sometimes shows a trace of dark mask, but more often both worn and fresh plumages resemble fresh plumage of dark-throated form. First-years resemble adults but ♂ is slightly darker on upperparts with retained pale tips and fringes to duller black feathers. First-year ♀ probably indistinguishable from adult. Juvenile resembles fresh ♀ but upperparts spotted paler and some indistinct dusky scaling on breast and fore flanks. **VOICE** Typical calls are a buzzing *dsched*, a clicking *chek* and whistled *wheet* often mixed with the clicking call. **RANGE** Breeds in Iberia and NW Africa through S Europe and Turkey to the Caspian and Iran. Winters in sub-Saharan Africa in the semi-desert belt across the northern tropics. **STATUS** Recorded from only one archipelago. **Canaries** Scarce and irregular passage migrant with records from all islands except El Hierro, and no precise details for Gran Canaria.

DESERT WHEATEAR *Oenanthe deserti* Plate 57 (7)
Sp. Collalba Desértica **P.** Chasco-do-deserto

O. d. homochroa. **L** 14–15 cm (5½–6"). **IDENTIFICATION** Sexually dimorphic. Both sexes easily identified as they have an almost all-black tail. ♂ is similar to dark-throated form of Black-eared Wheatear in having sandy upperparts with black wings, face and throat, but distinguished by black tail, whitish scapulars and by black of face and throat meeting that on wings. ♀ resembles ♀ Black-eared as well as ♀ Northern and Isabelline Wheatears, and is best distinguished by lack of white in tail. First-year ♂ distinguished by creamy-white tips obscuring dark face and throat and browner wings with extensive pale fringes. First-year ♀ similar to adult. Juvenile also resembles adult ♀ in autumn but has whiter rump, some faint pale spotting on upperparts and indistinct brown scaling on throat and breast. **VOICE** Calls include a fluty whistled *huiiee*, hard *tchack* and rattling lark-like *trrip*. **RANGE** Breeds from NW Africa discontinuously east through C Asia to Inner Mongolia. Winters in NW India, Arabia and N Africa. **STATUS** Accidental on two archipelagos. **Madeira** Fewer than five records in the last 50 years and recorded from both Madeira and Porto Santo. **Canaries** More than ten records, from La Graciosa, Fuerteventura, Gran Canaria and Tenerife, in Dec–Mar; its appearance usually coincides with dust storms from the Sahara.

MOURNING WHEATEAR *Oenanthe lugens* Plate 57 (8)
Sp. Collalba Núbica **P.** Chasco-fúnebre

O. l. halophila. **L** 14½ cm (5"). **IDENTIFICATION** Sexually dimorphic. Adult ♂ has forehead, crown and nape white with greyish tinge to crown: the mantle and folded wings are black, the lower back and rump white and the tail has a pattern identical to Northern Wheatear. Throat and head-sides black and rest of underparts white except pale buffish undertail-coverts. In flight white inner webs to flight feathers form a grey band across upperwing. Adult ♀ is sandy-grey on crown, upperparts and lesser wing-coverts, contrasting with blackish face and throat, and often has a pale creamy supercilium and eye-ring. Underparts are white. Some pale individuals have brown cheeks and a greyish-white throat. First-year birds resemble adults but retain brown flight feathers and tertials. Juvenile resembles pale-throated ♀ but has some indistinct streaking on upperparts and scaling on breast, and broader pale fringes to wing feathers. Juvenile ♂ has distinct dusky mask. **VOICE** Typical call is a quiet *chack-chack*. Song consists of short, sweet warbled phrases, some ending with a trill mixed with the call note and grating noises. **RANGE** Breeds discontinuously across N Africa, the Middle East, E Africa, Arabia and Iran. N African birds are partially migratory but only involving short-distance movements. **STATUS** Vagrant, recorded from only one island and all records by the same observer. **Canaries*** Three to four records, all from Lanzarote: one, Miradores in Parque Nacional de Timanfaya, 28 Apr 1987; 2 ♂♂, near Teguise 15 Mar & 18 Apr 1989; and one, Femés, 2 Apr 1990.

WHITE-CROWNED (BLACK) WHEATEAR *Oenanthe leucopyga* not illustrated
Sp. Collalba Negra de Brehm **P.** Chasco-de-barrete-blanco

Monotypic. **L** 17–18½ cm (6–7"). **IDENTIFICATION** Sexes similar. Adult is a large *Oenanthe* with all-black plumage except white crown, vent and rump. The predominantly white tail has black central rectrices and small black marks on outer feathers, rarely on all. Immature differs in having a black crown, sometimes with an occasional white feather. **VOICE** Rather variable with mimetic qualities. Calls include a shrill *hiit*, a high cracked *bizz* and throaty *tschreh*.

RANGE Breeds discontinuously across N Africa and Arabia. Mostly sedentary, although some populations make short-distance movements in winter. STATUS Extreme vagrant. Canaries Sole record is an immature photographed at Fuencaliente (La Palma), 11 Jan 2005. Cape Verdes First-winter, Fort Real (Santiago), 16 Jan 2005.

(RUFOUS-TAILED) ROCK THRUSH   *Monticola saxatilis*   Plate 58 (1)
Sp. Roquero Rojo   **P.** Macuco (Melro-das-rochas)

Monotypic. **L** 18½ cm (7"). **IDENTIFICATION** Sexually dimorphic. ♂ in breeding plumage unmistakable with blue-grey head, mantle, scapulars and rump; white band across mid back; and orange-rufous uppertail-coverts and tail (except brown central rectrices). Underparts from breast to vent bright orange-chestnut. Non-breeding ♂ retains same general coloration but is paler and entire plumage, except tail, is intensely mottled, scaled and barred black, brown and buff. ♀ and first-year ♂ similar to non-breeding ♂ but lack blue tones and white on mid back. Juvenile similar to fresh adult ♀ but has greyer, more intensely mottled upperparts, and buff tips to wing-coverts forming indistinct wingbars. **VOICE** Calls include a low *chack* and squeaky *siit*. **RANGE** Breeds from Spain and NW Africa east to N China and Transbaikalia, and winters almost exclusively in the Afrotropics. **STATUS** Vagrant recorded from two archipelagos. **Madeira** One on the Salvages is the only record. **Canaries*** Four records: adult ♂, Juan Adalid (La Palma), 26 Mar 1988; 2♂♂, 1♀, Ye (Lanzarote), 7 Mar 1989; ♀, Guinate (Lanzarote), 10–13 Mar 1989; and an adult and immature, La Graciosa, 1 Oct 1995. Also reported from Timanfaya National Park and the species' remains are said to appear regularly among prey of Eleonora's Falcons on Alegranza.

BLUE ROCK THRUSH   *Monticola solitarius*   Plate 58 (2)
Sp. Roquero Solitario   **P.** Melro-azul

M. s. solitarius. **L** 22 cm (9"). **IDENTIFICATION** Sexually dimorphic. ♂ uniform dark slaty-blue with darker, blacker wings and tail. ♀ grey-brown above and lacks any blue, and underparts paler and distinctly scaled buff or buff-white. First-years resemble adults but ♂ has broader, browner fringes to feathers. Juvenile dark brown with even more pronounced spotting and scaling than ♀. **VOICE** Calls include a whistled *peet* and deep grating *tchack-tchack*. **RANGE** Breeds from Spain and NW Africa east through C Asia to Japan. In the west the species is only a partial migrant. **STATUS** Vagrant only certainly recorded from one archipelago. **Madeira** One, Madeira, 18 Sep 1838, is considered dubious. **Canaries** Four records, all from Fuerteventura: one, near Pico de la Zarza, 22 Mar 1987, then in 1988, ♂, Barranco de Joroz, 3 Apr, one, Montaña Vigán, 18 Jun, and ♂ and ♀/immature, Montaña Cardón, 10 Jul.

WOOD THRUSH   *Hylocichla mustelina*   Plate 58 (3)
Sp. Zorzalito Maculado   **P.** Tordo-dos-bosques

Monotypic. **L** 19 cm (7½"). **IDENTIFICATION** Sexes similar. A small thrush with rusty crown and nape, brown upperparts, pure white underparts with large black spotting on throat-sides, breast, fore belly and flanks, and a narrow white eye-ring. First-winter resembles adult but distinguished by pale wingbar on tips of greater coverts. **VOICE** Calls include an explosive, staccato *pit-pit-pit* and low *tuck-tuck*. **RANGE** Breeds in E USA and extreme SE Canada, and winters in C America. **STATUS** Vagrant recorded on two archipelagos. **Azores** One obtained, Ponta Delgada (São Miguel), prior to 1903 (in the museum in Ponta Delgada). **Madeira** One, Porto Santo, 18 Jan 1986.

GREY-CHEEKED THRUSH   *Catharus minimus*   Plate 58 (4)
Sp. Zorzalito Carigrís   **P.** Tordo-de-cara-cinza

Subspecies unknown. **L** 18.5 cm (7"). **IDENTIFICATION** Sexes similar. A small thrush, not much bigger than Northern Wheatear and recalls a small, dumpy Song Thrush. Upperparts fairly uniform dull grey-brown, underparts greyish-white with little if any obvious buff coloration, and small dark spotting on throat and breast. Face distinctly greyish with a narrow and indistinct pale eye-ring. First-winter separated from adult by pale buff tips to greater coverts and inner tertials, and bright olive-brown fringes to flight feathers. **VOICE** Usually silent but gives a low, downward, slurred *wee-ah* or *pe-uu* and thin *wheesp* in contact on migration. **RANGE** Breeds from NE Siberia through Alaska and N Canada to Newfoundland. Winters from Colombia to Brazil and N Peru. **STATUS** Extreme vagrant. **Azores** One, Lagoa Rasa (Flores), 22 Oct 2002.

RING OUZEL   *Turdus torquatus*   Plate 58 (5)
Sp. Mirlo de Collar (Mirlo Capiblanco)   **P.** Melro-de-peito-branco

Subspecies unknown. **L** 23–24 cm (9–9½"). **IDENTIFICATION** Sexually dimorphic. ♂ predominantly black with a mostly yellow bill, large white crescent on chest and pale fringes to coverts and flight feathers, forming a distinct panel on folded wing. ♀ is browner with more distinct scaling on underparts, a smaller and less distinct chest crescent and duller bill. In adult plumage unlikely to be confused with any other thrush except a partial albino Blackbird with a white chest band. First-years have more extensive white edges to body feathers when fresh, and breast-band almost obscured, particularly in ♀♀. Juvenile browner than ♀, lacks breast-band and has mottled underparts without distinct scaling. **VOICE** Call is a hard *tak-tak-tak* sometimes lengthened into repeated rattling chatter. **RANGE** Breeds discontinuously from the British Isles, Scandinavia, Brittany and N Spain through montane C and S Europe to E Turkey, the Caucasus and east to the Kopet Dag in Turkmenistan. Some move south to winter in Spain, NW Africa and Iran. **STATUS** Recorded from two archipelagos as an accidental visitor. **Madeira** One shot, Paúl da Serra (Madeira), Oct 1993. **Canaries** Formerly considered a vagrant but recent records suggest that it is annual on Fuerteventura in autumn/winter. Also recorded on Alegranza, Fuerteventura, Tenerife, La Gomera, La Palma and El Hierro.

(COMMON) BLACKBIRD *Turdus merula* Plate 58 (6)
Sp. Mirlo Común **P.** Melro (Melro-preto)

T. m. azorensis (Azores) and *T. m. cabrerae* (Madeira and the Canaries). **L** 24–25 cm (9½–10″). **IDENTIFICATION** Sexually dimorphic. Probably one of the most familiar of European birds. ♂ all black with a yellow bill and eye-ring. ♀ all dark brown, slightly paler on underparts, except pale chin and throat, and has a dark bill. First-year ♂ duller than adult with a dusky bill. First-year ♀ distinguished by pale tips to the retained greater coverts. Juvenile browner, more rufous than ♀ with distinct dark-mottled underparts and whitish shaft-streaks on upperparts. Subspecies *azorensis* differs from nominate *merula* in being smaller and more round-winged, ♂ is blacker and ♀ darker blackish-brown, and throat is greyish-white with well-defined dark streaks; *cabrerae* differs similarly, but ♀ is even darker and has a much smaller pale area on throat. **VOICE** Calls include a harsh *chack* in series and accelerated into a maniacal, hysterical chattering when alarmed, and also a thin, high-pitched *seeee*. Song is a melodic series of warbling and fluty notes and phrases, not repeated like those of Song Thrush, and slower. **RANGE** Breeds across most of Europe and N Africa, discontinuously through Middle East to India, Sri Lanka, the Himalayas, C Asia and C China. Introduced in Australia and New Zealand. Most populations resident but some move south in winter. **DISTRIBUTION** Breeds on all main islands in the Azores, on Madeira and Porto Santo, and on the Canaries, except Fuerteventura and Lanzarote, where it is a rare and irregular winter visitor. **BREEDING** In the Azores breeds Apr–Jul but in the Canaries season more extended and lasts Dec–Jul. Clutch size 2–4 but three is most usual. Nests in trees or bushes and even in walls, with *c.* 30 different species of tree or shrub noted. **HABITAT** Found in all natural, as well as man-made habitats from sea level to above 2,000 m. **TAXONOMIC NOTE** Birds from La Palma and El Hierro have been treated as a separate race, *agnetae* on account of ♀ plumage being even darker on those islands. **STATUS** Common resident on three archipelagos. **Azores** Common species throughout, from Santa Maria west to Flores and Corvo. **Madeira** Very common on Madeira but rare on Porto Santo. **Canaries** One of the commonest species in the islands and in some agricultural areas is considered a pest.

DARK-THROATED THRUSH *Turdus ruficollis* Plate 58 (7)
Sp. Zorzal Papinegro **P.** Tordo-de-garganta-preta (Tordo-de-papo-preto)

T. r. atrogularis. **L** 24 cm (9½″). **IDENTIFICATION** Sexually dimorphic. Breeding ♂ *atrogularis* unmarked brownish-grey above; face, throat and breast black and rest of underparts off-white with some slight and indistinct streaking on flanks. Breeding ♀ duller and more olive above, with an indistinct supercilium and whitish chin and throat streaked black. Non-breeding ♂ similar to breeding ♂ but black throat and breast partially obscured by pale tips to feathers. Non-breeding ♀ has black on underparts reduced to streaks on upper breast, extending onto the flanks but less intensely. First-years resemble fresh (non-breeding) adults but distinguished by retained juvenile wing feathers. Juvenile resembles adult ♀ but is streaked and barred above, the greater coverts have pale tips and underparts spotted and barred. **VOICE** Gives a *chak-chak* similar to Ring Ouzel and a chattering alarm like Blackbird. **RANGE** Breeds from E European Russia through C Siberia to Nizhnyaya Tunguska River and south to Tarbagatay, Altai and Tannu Ola. Winters from Iraq and Arabia east through Iran, Afghanistan, Pakistan, N India and the Himalayas to Myanmar. **STATUS** Extreme vagrant. **Madeira** One reported by Schmitz (1903) is included in Zino *et al.* (1995).

FIELDFARE *Turdus pilaris* Plate 58 (8)
Sp. Zorzal Real **P.** Tordo-zornal

Monotypic. **L** 25½ cm (10″). **IDENTIFICATION** Sexes similar. Distinguished from other *Turdus* by combination of grey head and rump, dark chestnut mantle and wing-coverts, and black or brownish-black tail. Underparts warm buff on throat, breast and flanks, boldly marked with black spots on throat and breast, and black chevrons on flanks. In flight reveals white underwing-coverts, a feature shared only with Mistle Thrush. Sexes very similar but ♀ is generally duller with less heavily marked underparts. The retained, pale-tipped greater coverts can sometimes identify first-years. Juvenile much duller than adult, with pale shaft-streaks on upperparts and more rounded spots on underparts. **VOICE** Flight call is a continuous, chattering *tchack-tchack* similar to that given on ground. **RANGE** Breeds in a broad band across the Palearctic from the UK east to Lake Baikal and the Lena River. Winters mainly in Europe (except the north), Turkey and Iran. **STATUS** Accidental visitor to the region. **Azores** Three records: one, between Santa Rita and Ponta de Espatel (Terceira), 18 Jan 1948; ♂ obtained, Candelária (São Miguel), 8 Mar 1963; and adult obtained, Santa Maria, 18 Nov 1972. **Madeira** Occasional, recorded seasonally but only under unusual climatic conditions. Recently a group of at least 20, Paúl de Serra (Madeira), 10 Jan 2004. Also seen on the Salvages. **Canaries** Very scarce or accidental winter visitor, recorded from Lanzarote, Fuerteventura and Tenerife.

SONG THRUSH *Turdus philomelos* Plate 58 (11)
Sp. Zorzal Común **P.** Tordo-pinto (Tordo-comum)

T. p. philomelos. **L** 22 cm (9″). **IDENTIFICATION** Sexes similar. Very familiar to European birdwatchers. Adult has generally grey-brown upperparts lacking any warm tones, and buffy breast and flanks with black-brown spotting. In flight reveals orange-buff wing linings. First-winter similar to adult and not safely separable in the field, unless close enough to see retained juvenile greater coverts with buff shaft-streak. Juvenile has some pale streaking on upperparts. Separation from Redwing and Mistle Thrush is dealt with under those species. **VOICE** Typical calls include a sharp *tsit* and chattering *chock-chock*. Song is a mixture of varied phrases with each note being repeated three times or more before next phrase starts; pauses are short. **RANGE** Breeds from the British Isles and Spain east to Lake Baikal and south to Turkey and Iran. Moves south in winter as far as N Africa, Middle East, Arabia and SE Iran. Introduced in Australia and New Zealand. **STATUS** Varies in the different archipelagos. **Azores** Two records: a specimen in the museum at Ponta Delgada lacks details but was undoubtedly collected in the islands; one, Sete Cidades (São Miguel), 31 Oct 2003. **Madeira** Fewer than five records in the last 50 years and recently one, east of Ponta da Calheta (Porto Santo), 5 Dec 2003. Also recorded from the Salvages. **Canaries** Regular winter visitor in small numbers and recorded on all main islands. **Cape Verdes** One, Santa Maria (Sal), 13–26 Feb 1996.

REDWING *Turdus iliacus* Plate 58 (9)
Sp. Zorzal Alirrojo **P.** Tordo-ruivo (Tordo-ruivo-comum)

T. i. iliacus. **L** 21 cm (9″). **IDENTIFICATION** Sexes similar. Easily distinguished from similar-sized Song Thrush by prominent whitish supercilium and submoustachial stripe, reddish flanks and underwing-coverts, and whiter, less heavily marked underparts. Upperparts are warmer brown than in Song Thrush. Some first-winters have pale spots on tips of tertials and pale tips to greater coverts, both retained from juvenile plumage, but others are inseparable from adults. Juvenile has buff streaking on upperparts and more rounded spots on underparts. **VOICE** In flight a high-pitched *seeeip*, otherwise a scolding rattle *trrrt-trrrt-trrrt* or muffled *chuk-chuk*. **RANGE** Breeds across N Palearctic from Iceland east to the Kolyma River in Siberia. Winters almost exclusively in the W Palearctic, from the British Isles and NW Africa to Turkey, the Middle East and N Iran. **STATUS** Recorded from three archipelagos but status varies. **Azores** First-winter ♀ of nominate race *iliacus* collected, Santa Maria, 20 Nov 1972. Note dates for this species and Fieldfare are only two days apart. **Madeira** Status as that for Fieldfare. Also recorded from the Salvages. **Canaries** Rare and irregular winter visitor recorded on Lanzarote, Fuerteventura and Tenerife.

MISTLE THRUSH *Turdus viscivorus* Plate 58 (10)
Sp. Zorzal Charlo **P.** Tordoveia (Tordeia)

T. v. viscivorus. **L** 27 cm (11″). **IDENTIFICATION** Sexes similar. Much larger than similar Song Thrush and also differs in its white underwing-coverts and pale corners to tail. Upperparts greyer than Song Thrush and underparts more extensively white with larger, more irregularly spaced, black spotting. First-year very similar to adult and often inseparable but retained juvenile feathers on rump, scapulars and forewing often have pale shaft-streaks. Also retained wing-coverts and flight feathers have pale buff fringes. Juvenile well marked with whitish spots and streaks on upperparts. **VOICE** Flight call is a rasping, rattling, chattering *churr* also delivered from the ground, as well as a rapid, staccato *tuck-tuck-tuck*. Song similar to Blackbird, a series of short, clear, loud, fluty phrases, but more monotonous with a faster delivery and shorter pauses. **RANGE** Breeds from the British Isles and NW Africa across Europe east to Lake Baikal and south to Turkey, Iran and W Himalayas. In winter found as far south as N Africa, Israel and C Asia. **STATUS** Possibly breeds on one archipelago but otherwise a vagrant. **Azores** Two very old records: one obtained, prior to 1888 (no other details), and one collected, Ribeira Grande (São Miguel), 1903, presented to the British Museum, 1903. **Madeira** Recorded regularly in Funchal and both adults carrying food and young have been observed. However, as no nest has been found, it is only considered a possible breeder. **Canaries★** Three recent records all from Tenerife: one, El Portillo, 20 Feb 1994; up to three, Las Lagunetas, 3–23 Jan 1996; and one, Macizo de Teno, 20 Feb 1996.

ZITTING CISTICOLA (FAN-TAILED WARBLER)   *Cisticola juncidis*   Plate 59 (1)
Sp. Buitrón   **P.** Fuínha-dos-juncos

C. j. cisticola.   **L** 10 cm (4″).   **IDENTIFICATION** Sexes similar. The smallest warbler likely to be encountered in the region, except for the *Regulus* spp. Superficially resembles a small, bright, short-winged and short-tailed Sedge Warbler. Rather pale buffy-brown or sandy-brown upperparts with heavy dark streaking on crown, nape and mantle. Underparts plain buffish-white, slightly paler on throat and vent. Tail brown with prominent black-and-white tips to underside except central rectrices, whilst upperside has less distinct off-white tips. Rather skulking and often difficult to observe. Juvenile similar but has more prominent eye-ring and streaking is browner and less well defined. Most obvious in song-flight when it delivers its characteristic song; the short wings and tail are also particularly obvious. **VOICE** Song is a repeated, rather monotonous, *zit*. Call is a single *zip* or *chip*, very similar to that used in the song. **RANGE** Breeds in W France, S Europe, NW and most of sub-Saharan Africa, the Middle East and the Indian subcontinent to Taiwan, the Philippines, Indonesia and Australia. Mainly resident but dispersive. **STATUS** Vagrant with only two certain records. **Madeira** An old record based on a clutch of eggs is considered dubious. **Canaries** One, Haría (Lanzarote), 25 Mar 1980. **Cape Verdes** One, on board ship *c.* 52 km northwest of Sal, 7 Sep 1985.

(COMMON) GRASSHOPPER WARBLER   *Locustella naevia*   Plate 59 (2)
Sp. Buscarla Pintoja   **P.** Cigarrinha-malhada (Felosa-malhada)

L. n. naevia.   **L** 12½ cm (5″).   **IDENTIFICATION** Sexes similar. Upperparts usually olive-brown with darker spotting and streaking from crown to back, rump olive-brown with faint streaks fading away on uppertail-coverts, tail warmer brown than rest of upperparts and faintly barred. Underparts off-white to buff with slight streaking on breast-sides and on undertail-coverts. Plumage rather variable and some birds can have buffier or even yellowish underparts with more noticeable or occasionally almost no streaking. Juvenile resembles adult but is often darker with more distinctly streaked upperparts and heavier streaking on breast. **VOICE** Calls include a harsh *tchik* and rather liquid *whit*. Song is a high-pitched, reeling trill sounding rather insect-like and highly variable in length, from *c.* 30 seconds to well over an hour. **RANGE** Breeds from the British Isles and N Spain east to W Siberia. Wintering areas of race *naevia* poorly known but apparently mostly in W Africa south of the Sahara. **STATUS** Recorded from three archipelagos but is only an accidental visitor. **Madeira** One, Selvagem Grande, autumn 1978. **Canaries** Accidental visitor recorded from all islands between Tenerife and Roque del Este. Probably more common than records suggest but skulking habits make observations difficult. **Cape Verdes** One record: ♀ collected, Sal, 1901, but no other details.

RIVER WARBLER   *Locustella fluviatilis*   Plate 59 (3)
Sp. Buscarla Fluvial   **P.** Cigarrinha-ribeirinha (Felosa-fluvial)

Monotypic.   **L** 13 cm (5″).   **IDENTIFICATION** Sexes similar. Slightly larger than Grasshopper Warbler with unmarked dark brown upperparts, a short and indistinct yellowish-buff supercilium, and a slightly paler but still indistinct eye-ring. Underparts off-white with some indistinct greyish mottling on throat and upper breast. Vent and undertail-coverts, which nearly reach tail tip, brownish-buff with broad dull whitish tips often forming distinct crescents. Juvenile similar to adult but is yellower around the face and underparts, and has fresh flight feathers and tail in autumn. **VOICE** A buzzing reel, slower and lower pitched than Grasshopper Warbler but also rather insect-like and can be repeated continuously for over an hour. **RANGE** Breeds from E Europe and S Finland east to the Irtysh River. Winters in E Africa from S Malawi to N South Africa. **STATUS** Extreme vagrant. **Madeira** One, Selvagem Grande, 2 Oct 1980.

SAVI'S WARBLER   *Locustella luscinioides*   Plate 59 (4)
Sp. Buscarla Unicolor   **P.** Cigarrinha-ruiva (Felosa-unicolor)

L. l. luscinioides.   **L** 14 cm (5½″).   **IDENTIFICATION** Sexes similar. Upperparts uniform dull rufous-brown with traces of faint barring on tail. Short and faint buff supercilium and dusky lores only visible at close range. Underparts pale brownish on throat-sides, breast and flanks to undertail-coverts. Chin, throat and belly buffy-white. Undertail-coverts show some very indistinct pale crescents on feather tips. Tail quite long and graduated over half its visible length. Does not show a contrasting paler rump, unlike some similar-plumaged *Acrocephalus* warblers. Juvenile resembles adult but is more rufous with fresh flight feathers and rectrices. **VOICE** Calls include a hard *pwitt*, when alarmed repeated in series, and resembles the chattering rattle of Blackbird. Song starts with a series of short, hard ticking notes that develop into a buzzing trill. Lower pitched than Grasshopper Warbler but with a faster delivery. **RANGE** Breeds discontinuously in NW Africa, Europe and the Middle East to C Asia. Wintering grounds poorly known but apparently in sub-Saharan Africa, from Senegal to Eritrea. **STATUS** Extreme vagrant. **Cape Verdes** One, Santa Maria (Sal), 23 Sep 1997. **Canaries★** One, Fuerteventura, 23 Apr 2002.

MOUSTACHED WARBLER   *Acrocephalus melanopogon*   Plate 59 (5)
Sp. Carricerín Real   **P.** Felosa-real

Subspecies unknown. **L** 12½ cm (5″). **IDENTIFICATION** Sexes similar. Quite similar to Sedge Warbler but has rather rustier brown upperparts, with a darker crown, whiter supercilium (broader behind eye) and shorter primary projection. Juvenile can be aged in late summer by fresher wing and tail feathers. Otherwise resembles adult. **VOICE** Calls include a Blackcap-like *tak* used in contact and churring *trrrp* or *trrrt* in alarm. **RANGE** Breeds from NW Africa and Spain discontinuously east to Iran and Kazakhstan. Western populations mostly sedentary and eastern populations winter from the Middle East to NW India. **STATUS** The only claimed record was one on board a ship. **Cape Verdes** One, at noon position 17°05′N, 23°18′W, 28 nm northwest of Sal, 7 Sep 1985.

AQUATIC WARBLER   *Acrocephalus paludicola*   Plate 59 (6)
Sp. Carricerín Cejudo   **P.** Felosa-aquática

Monotypic. **L** 13 cm (5″). **IDENTIFICATION** Sexes similar. Resembles a pale, sandy-coloured Sedge Warbler but separated by the prominently striped head, with long pale supercilium and central crown-stripe and dark, almost black, lateral crown-stripes. Aquatic also has prominent pale 'tramlines' bordering the mantle, a streaked rump which is not as warm as in Sedge Warbler, wing feathers fringed paler and, in adults, finely streaked flanks. Juvenile recalls adult but shows more contrast and lacks fine streaking on underparts. Easily identified given good views, but with brief views extreme care should be taken with the similar juvenile Sedge Warbler. **VOICE** Typical call is a *tchak*, lower pitched than that of Sedge Warbler. **RANGE** Breeds from E Europe to the central Ob River. Winter range poorly known but occurs in W sub-Saharan Africa. **STATUS** Accidental visitor to one archipelago. **Canaries★** Ten or 11 records from Roque del Este, Alegranza, Montaña Clara, Lanzarote, Fuerteventura, Gran Canaria (doubtful) and Tenerife, with the most recent one, Amarilla Golf (Tenerife), 5 Mar 1997, and one, Teguise golf course (Lanzarote), 19 Mar 1997.

SEDGE WARBLER   *Acrocephalus schoenobaenus*   Plate 59 (7)
Sp. Carricerín Común   **P.** Felosa-dos-juncos

Monotypic. **L** 13 cm (5″). **IDENTIFICATION** Sexes similar. Adult has prominent supercilium, crown darker than streaked mantle and unstreaked deep buff rump. Generally less buff-yellow or striped than Aquatic Warbler and never shows such a prominent central crown-stripe. However, some juveniles do exhibit a much less conspicuous one. Juvenile resembles adult but is brighter, yellower or buffier with some spotting/streaking on breast. At this age confusion with Aquatic is more possible but unstreaked rump and lack of pale 'tramlines' on sides of the mantle are features of Sedge Warbler, which is far more likely in our region than Aquatic. **VOICE** Call is a *tchack* similar to Aquatic but noticeably higher pitched. Song less likely to be heard in the region but is a fast-delivered and varied mix of chattering, trilling and more musical phrases, interspersed with some mimicry and grating sounds. **RANGE** Breeds from the British Isles and France through Europe to the Yenisey River and south to Lake Balkhash and the Tien Shan. Winters in most of sub-Saharan Africa south to N Namibia and E Cape Province. **STATUS** Scarce visitor recorded from two archipelagos. **Madeira** Exceptional visitor that has been recorded on the Salvages. **Canaries** Considered accidental by Martín and Lorenzo (2001); more likely to be a scarce passage migrant having been recorded from the eastern islets, Lanzarote, Fuerteventura, Tenerife and La Gomera.

CAPE VERDE (SWAMP) WARBLER   *Acrocephalus brevipennis*   Plate 59 (8)
Sp. Carricero de Cabo Verde   **P.** Chota-de-cana

Monotypic. **L** 13½ cm (5″). **IDENTIFICATION** Sexes similar. Cape Verde Warbler is larger than Reed Warbler, with a noticeably longer bill, longer tail and very short wings. Grey-brown above and greyish-white below. Otherwise rather featureless, apart from a short, thin and indistinct greyish supercilium, which is only visible at close range. Behaves more like a *Hippolais* than an *Acrocephalus*. Juvenile more rufous above and has a yellowish wash to underparts with a buffish suffusion on breast and flanks. **VOICE** Calls include a disyllabic throaty *pichow*, harsh croaking *churr* and sharp high-pitched *wik wik*. Song is loud and resonant, including a variety of notes with a Nightingale-like quality, and lacks any harsh notes, unlike congeners in the region. One rendering is *put dud-loo* with the last two syllables repeated up to five times at increasing speed. **RANGE** Confined to Cape Verdes. **DISTRIBUTION** Known only from Santiago, São Nicolau and Brava, but only certain extant on the first two, and only recently rediscovered on São Nicolau, in Feb 1998, after an absence of 74 years, apart from a specimen collected there in 1970. **BREEDING** Breeds Aug–Nov, sometimes from early July and continuing into Mar, in response to rainfall. Nests in reeds, sugarcane, and various types of trees and bushes. Clutches of 2–3 have been recorded. Occasionally pairs will raise two broods, depending on extent of summer rains. **HABITAT** Restricted to well-vegetated valleys, cultivated areas such as plantations, and settlements with sufficient vegetation. **STATUS** Endemic to one archipelago. **Cape Verdes** Unrecorded from Brava since one was collected, 21 Oct 1969, and on

São Nicolau population estimated at less than 20 pairs. On Santiago, it is not uncommon but confined to areas with sufficient vegetation. Population is thought not to exceed *c*. 500 pairs.

MARSH WARBLER   *Acrocephalus palustris*   Plate 59 (9)
Sp. Carricero Políglota   **P.** Felosa-palustre

Monotypic.   **L** 13 cm (5″).   **IDENTIFICATION** Similar to Reed Warbler, from which it is best distinguished by a complex combination of features. Marsh Warbler generally lacks rich colour of Reed Warbler and is often more olive above, whilst the underparts are whiter in Marsh Warbler, washed buff or pale yellowish and lacking any rufous tones on flanks. Structurally, Marsh Warbler has longer wings and a heavier bill, and the primary projection is approximately equal to, or longer than, the exposed tertials and considerably longer than the uppertail-coverts. Also, Marsh Warbler shows whitish tips to the primaries and pale borders to the tertials, features that are rarely exhibited by Reed Warbler and never to the same extent. Apart from being heavier, the bill is also rather pale. The legs of Marsh Warbler are pale straw with pale yellowish feet, and if visible the pale claws are diagnostic. Juvenile browner than adult with a more rufous rump and pale yellowish-pink legs. **VOICE** Calls include a sharp *tchack* and softer, lower *tuk*. Song, unlikely to be heard in the region, is the richest and most mimetic of all *Acrocephalus*. **RANGE** Breeds from N France and S England east to Azerbaijan and Kazakhstan. Winters in E and SE Africa. **STATUS** Extreme vagrant. **Madeira** One recorded from Madeira by Schmitz in 1906–08, but no other details.

(EUROPEAN) REED WARBLER   *Acrocephalus scirpaceus*   Plate 59 (10)
Sp. Carricero Común   **P.** Rouxinol-dos-caniços (Rouxinol-pequeño-dos-caniços)

A. s. scirpaceus.   **L** 12½ cm (5″).   **IDENTIFICATION** Sexes similar. Combination of generally richly coloured plumage, earth-brown upperparts with more rufous rump and uppertail-coverts, and buff underparts, plus the shorter primary projection should help to separate this species from Marsh Warbler. The lack of pale outer tail feathers, rounded tail tip and long undertail-coverts distinguish it from another potential confusion species, Western Olivaceous Warbler. Juvenile in first autumn is slightly warmer brown than adult with rich buff flanks; the whole plumage is fresh, whereas adult has worn primaries and tail feathers. **VOICE** Calls include a quiet *churr* often lengthened into a harsh rattle when alarmed. The song is a slow, rather mechanical, series of repeated churring and grating notes, *kerr-kerr-kerr chirruc chirruc chirruc kek-kek-kek chirr-chirr*. **RANGE** Breeds from NW Africa, Spain and Ireland east through Europe and the Middle East to W China. Winters mainly in sub-Saharan Africa. **STATUS** Only recorded from two archipelagos and status varies. **Madeira** Less than five records in the last 50 years, the most recent being one, Deserta Grande (Desertas), 30 Aug 2001, and one, Machico (Madeira), 31 Aug 2002. Also recorded from the Salvages. **Canaries** Scarce, irregular passage migrant, recorded from all main islands bar La Palma.

GREAT REED WARBLER   *Acrocephalus arundinaceus*   Plate 59 (11)
Sp. Carricero Tordal   **P.** Rouxinol-grande-dos-caniços

A. a. arundinaceus.   **L** 19 cm (7½″).   **IDENTIFICATION** Sexes similar. A large unstreaked *Acrocephalus*, like a giant Reed Warbler, and similarly coloured. The only warbler of its size, so unlikely to be confused. Adult brown above with a rufous tinge to rump, whitish below with some indistinct streaking on upper breast. Juvenile more reddish-brown on upperparts, more buff on underparts and lacks any fine streaking on upper breast. **VOICE** Calls include varied harsh notes such as a croaking *churr*, short *chack* and soft *kri-kri*. Song similar to Reed Warbler in composition but is louder and more croaking, rendered as *trr trr garra-garra krie-krie guruk-guruk-guruk krie crrk-crrk…* etc. **RANGE** Breeds over much of Europe, NW Africa, and east to NW Mongolia and Tajikistan. Entire population winters in sub-Saharan Africa. **STATUS** Only recorded from two archipelagos. **Madeira** Fewer than five records in the last 50 years. **Canaries** Accidental visitor recorded from Alegranza, Lanzarote, Fuerteventura, Gran Canaria and Tenerife.

Hippolais warblers

These species are easily mistaken for other warblers in different genera, especially *Acrocephalus* or *Phylloscopus* but can best be recognised as follows. *Hippolais* from *Acrocephalus*: *Hippolais* warblers have relatively short undertail-coverts and square-ended tails. *Hippolais* from *Phylloscopus*: *Hippolais* warblers are larger, with a longer and broader based bill, and are more 'clumsy' when moving through vegetation.

EASTERN OLIVACEOUS WARBLER   *Hippolais pallida*   not illustrated
Sp. Zarcero Pálido Oriental   **P.** Felosa-pálida-oriental

H. p. elaeica.   **L** 12–13 cm (5″).   **IDENTIFICATION** Sexes similar. This and the following species are very similar and extremely difficult to separate in the field. Generally, Eastern Olivaceous is slightly smaller, has a thinner and often shorter bill, weaker legs and proportionately shorter tail. It has greyer upperparts, darker flight feathers and, often, a pale wing panel on the secondaries. The most characteristic feature of Eastern Olivaceous is the continual dipping movements made by the tail. **VOICE** Calls are very similar to Western Olivaceous Warbler but the song differs in being more cyclic in structure, increasing and decreasing in pitch in a repetitive pattern. **RANGE** The race *elaeica* breeds in SE Europe and the Middle East to C Asia. In Africa there are three races, *laeneni* breeds in CN Africa, *reiseri* in Algeria south of the Atlas and possibly in S Tunisia and SE Morocco, and *pallida* along the Nile in Egypt and N Sudan. Most winter across sub-Saharan Africa, mainly north of the equator. **STATUS** Vagrant with only two records from the same archipelago. **Canaries** One collected by von Thanner, at Vilaflor (Tenerife), Sep 1902, and one, Nuevo Horizonte (Fuerteventura), 24 Apr 2002.

WESTERN OLIVACEOUS WARBLER *Hippolais opaca* Plate 61 (11)
Sp. Zarcero Pálido Occidental **P.** Felosa-pálida

Monotypic. **L** 13–13½ cm (5"). **IDENTIFICATION** Sexes similar. Upperparts generally grey-brown but variable. Underparts off-white, often with a buff or greyish wash on flanks and breast. Similar to grey forms of Icterine and Melodious Warblers, and best distinguished by lack of yellow on underparts and lack of olive tones in upperparts. Also similar to Reed Warbler, but structural features such as the shorter undertail-coverts and squarer ended tail, plus whitish edges and corners to the outer tail feathers are useful distinguishing features. Habitually flicks its tail downwards when calling, but this is not diagnostic. Juvenile distinguished from adult by its fresh plumage and more extensive buff wash to underparts. **VOICE** Common call is a short, sharp *tek* but a sparrow-like *churrr* also given. Song similar to Reed Warbler but is faster, more monotonous with less distinct individual notes. **RANGE** Breeds in NW Africa and Iberia. Winters in W Africa from Senegal to N Cameroon. **STATUS** Recorded from three archipelagos but status varies. **Madeira** The only records are from the Salvages. **Canaries** Rare and irregular passage migrant with records from Alegranza, Lanzarote, Fuerteventura, Gran Canaria and Tenerife. **Cape Verdes** One, oasis near Curral Velho (Boavista), 19–20 Sep 1988.

ICTERINE WARBLER *Hippolais icterina* Plate 61 (10)
Sp. Zarcero Icterino **P.** Felosa-amarela (Felosa-icterina)

Monotypic. **L** 13 cm (5"). **IDENTIFICATION** Sexes similar. Upperparts vary from quite bright greenish-olive to distinctly greyish according to season and the bird's age. Also underparts vary from predominantly yellow to almost white. Grey-and-white individuals closely recall Eastern Olivaceous Warbler but a close inspection should reveal some olive tones in the upperparts. Most easily confused with Melodious Warbler but differs in having a usually pronounced pale wing panel formed by the pale fringes to the tertials and inner secondaries. Also, wing structure is noticeably different. On Icterine the wing has a long primary projection, approximately equal to the length of the exposed tertials, with the spacing between the primary tips further apart nearer the wing-point. The wings when folded completely cover the uppertail-coverts. Juvenile often paler below than adult with a browner tone to upperparts, and wing panel less obvious than in fresh adult. Some worn autumn adults can almost completely lack the wing panel. **VOICE** Calls include a short clicking *tek* and *Phylloscopus*-like *hooeet*. Song unlikely to be heard in the region but distinguished from that of Melodious Warbler by presence of distinctive, repeated, shrill, nasal *gie gie...* notes. **RANGE** Breeds from C and N Europe east to C Russia and N Iran. Winters in Africa south of the equator. **STATUS** Vagrant with only three records. **Madeira** On the Salvage Islands, one was ringed, 1–2 Oct 1980, and four birds seen and one of these trapped, Selvagem Grande, 6 Sep 1990. **Canaries*** One, La Lejita (Fuerteventura), 17 Apr 1999.

MELODIOUS WARBLER *Hippolais polyglotta* Plate 61 (9)
Sp. Zarcero Común **P.** Felosa-poliglota

Monotypic. **L** 13 cm (5"). **IDENTIFICATION** Sexes similar. Adult very similar to Icterine Warbler and best distinguished by differences in wing structure, as plumage differences are not diagnostic. Wing shorter than Icterine and when folded does not reach tip of uppertail-coverts. Primary projection shorter, only half length of the exposed tertials or less, and spacing between primary tips more equal. Head of Melodious Warbler usually more rounded, with peak of the crown just behind eye compared to the flatter head of Icterine, where the peak of the crown is well behind the eye. Melodious Warbler occasionally has a pale panel in wing but very rarely as contrasting as in Icterine. Juvenile in autumn is in fresh plumage, whereas adult is worn, but also has a more defined wing panel, is more olive or olive-brown above and has less intensely coloured underparts. **VOICE** Calls include a sparrow-like *churrr* and rather soft metallic *tuk*. Song resembles an *Acrocephalus* warbler but is less harsh and churring. Compared to Icterine Warbler it is less varied but faster, more rattling and more sustained. Lacks characteristic repeated shrill, nasal notes of Icterine. **RANGE** Breeds in C and W Europe and NW Africa. Winters entirely in sub-Saharan W

Africa. **STATUS** Recorded from two archipelagos. **Madeira** Exceptional visitor, the most recent record being one, Deserta Grande, 3 Sep 2001. Also recorded on the Salvages. **Canaries** Scarce and irregular passage migrant with records from islets off Lanzarote, Lanzarote itself, Fuerteventura, Gran Canaria and Tenerife, and is more frequently recorded in autumn.

TRISTRAM'S WARBLER *Sylvia deserticola* Plate 60 (1)
Sp. Curruca de Tristram **P.** Toutinegra do Atlas

Subspecies unknown but *S. d. maroccana* would seem most likely. **L** 12½ cm (5″). **IDENTIFICATION** Sexually dimorphic. ♂ plumage is quite distinctive with bluish-grey upperparts, vinous-brown throat (with faint white spotting), breast and flanks, paler pink to off-white belly and vent, very obvious orange-rufous wing-patch, ill-defined whitish submoustachial stripe and fairly broad white eye-ring. ♀ resembles ♂ but in spring is generally duller and the whitish belly is more extensive. In winter ♀ has sandier upperparts and extensive whitish tips to the underpart feathers, which partially conceal the vinous-brown coloration. In this plumage, confusion with Spectacled Warbler is possible, but the presence of any pinkish tones on the throat is a useful feature, as is the generally more sandy-coloured upperparts. First-winter ♀ very similar to same-age Spectacled but may usually be identified by its sandier upperparts, generally longer tail that is frequently cocked, and more extensively buffish underparts. Juvenile almost inseparable from juvenile Spectacled. **VOICE** Usual call is a soft but harsh *chit* or *chit-it*. **RANGE** Breeds discontinuously across Morocco, Algeria and Tunisia. A partial migrant, only making relatively short-distance or altitudinal movements. **STATUS** Extreme vagrant. **Canaries** Adult ♂, Barranco de los Canarios (Fuerteventura), 30 Oct 1995; and adult ♂, Presa de las Peñitas (Fuerteventura), 29 Oct–2 Nov 1999.

SPECTACLED WARBLER *Sylvia conspicillata* Plate 60 (2)
Sp. Curruca Tomillera **P.** Toutinegra-tomilheira

S. c. orbitalis. **L** 12.5 cm (5″). **IDENTIFICATION** Sexually dimorphic. ♂ in breeding plumage has slate-grey head, black lores, a very prominent white eye-ring, white chin and grey-centred white throat. Mantle rufous-brown and wings also predominantly bright rufous-brown. Tail blackish with prominent white outer tail feathers and underparts pinkish with a paler central belly and vent. ♀ in breeding plumage is occasionally similar to male but loral region is never black, the throat has less grey and underparts are paler. Normally ♀ tends to be less intensely coloured, with much paler lores and olive-brown replacing the slate-grey on upperparts, the eye-ring is less pronounced and underparts noticeably paler. ♀ and first-winter plumages can closely resemble congeners, particularly Subalpine Warbler, Common Whitethroat and Tristram's Warbler. First-winter distinguished from both Subalpine Warbler and Common Whitethroat by careful observation of the middle tertial, which has a well-defined dark arrowhead pattern in the centre, but note Tristram's Warbler can be similar. Wing panel usually more rufous than Subalpine. Primary projection of Spectacled Warbler distinctly shorter than either Subalpine Warbler or Common Whitethroat and, compared to latter, it is more delicate with a shorter, finer bill. Main differences between ♀ Spectacled and Tristram's Warbler are dealt with under that species, but separation can be difficult and juveniles are almost inseparable. Spectacled race *orbitalis* is generally darker and more richly coloured than the nominate and ♂ has a more solidly dusky head. However, there is a cline from north to south, with birds in Madeira being the most obvious, while those on Cape Verde closely resemble the nominate. Birds from the Canaries seem to approach the coloration of the nominate on the underparts and have the upperparts intermediate between Madeiran and Cape Verde populations. **VOICE** Common call is a high-pitched rattling *tchrrrrrr*, but also gives a short *tek-tek*. The song, which can be heard all year, is usually a high, rapid warble, frequently preceded by several clear, whistled notes. Often given in flight. **RANGE** Breeds discontinuously through the Mediterranean Basin and NW Africa. Most of those breeding in Europe migrate to N Africa. **DISTRIBUTION** Breeds on Madeira, the Canaries and Cape Verdes. **BREEDING** Season varies between archipelagos. In Cape Verde it has been recorded in all months except Jun–Jul. In the Canaries the period seems to be Dec–May, whereas in Madeira Apr–Jul is the season. Clutch size is 2–6 but the most frequent number is four. Nest usually located in vegetation between the ground and 1 m up. In the Canaries is known to have at least two broods. **HABITAT** In Madeira commoner at higher altitudes than lower elevations. On Tenerife occurs from sea level to over 2,000 m and is found in agricultural areas as well as in low-elevation *Euphorbia* scrub and the high montane zone dominated by species of broom. In Cape Verde it also occupies a wide range of habitats including towns and villages. **TAXONOMIC NOTE** Birds on Madeira were formerly separated subspecifically as *bella*, but this name is now usually considered a synonym of *orbitalis*. **STATUS** A resident on three archipelagos. **Madeira** Uncommon resident that breeds only on Madeira. **Canaries** Common resident on all main islands, plus the islets of Alegranza and La Graciosa. Most common on the lower, less vegetated, eastern islands of Fuerteventura and Lanzarote. **Cape Verdes** Breeds on all main islands except Santa Luzia and Sal. Generally common and widespread but on São Vicente uncommon and local.

SUBALPINE WARBLER   *Sylvia cantillans*   Plate 60 (3)
Sp. Curruca Carrasqueña   **P.** Toutinegra-de-bigodes (Toutinegra-carrasqueira)

S. c. cantillans, but *S. c. inornata* might also occur in the region. **L** 12 cm (5″). **IDENTIFICATION** Sexually dimorphic. ♂ in breeding plumage unmistakable with blue-grey upperparts, dark grey-brown wings and tail, which has white outer feathers: narrow white moustachial stripe and bright pink-chestnut chin, throat, breast and flanks, with a paler, almost white-centred belly. ♀ similar to others of genus, particularly Spectacled Warbler, but usually identified as it lacks rufous wing panel of latter. Some do show some rufous in the wing but can be identified by the pattern of the tertials, which on ♀ Subalpine have more extensive, but less contrasting, dark centres. Juvenile and first-winter plumages are similar to congenerics, and are best separated from Subalpine by the dark tertials with pale brown fringes, and from Tristram's Warbler by the shorter tail, which is not frequently cocked, and lack of a bright rufous wing panel. **VOICE** Commonest calls likely to be heard are a hard *tek* and rattling *trrrrt*. **RANGE** Breeds in S Europe to W Turkey and in N Africa. Some winter in N Africa but the majority migrate to wintering grounds on the southern edge of the Sahara. **STATUS** Varies between the three archipelagos where it has been recorded. **Madeira** Exceptional visitor with the first record as recent as spring 1983, and also recorded on the Salvages. **Canaries** Scarce but regular passage migrant, most often encountered on the eastern islands in spring, but recorded from all main islands except El Hierro. **Cape Verdes** Two records: ♂, Raso, 14 Mar 1990; and, ♂, Rabil lagoon (Boavista), 19 Mar 1990.

MÉNÉTRIES'S WARBLER   *Sylvia mystacea*   Plate 60 (4)
Sp. Curruca de Menetries   **P.** Toutinegra-rosada (Toutinegra de Menetries)

Subspecies unknown. **L** 13½ cm (5″). **IDENTIFICATION** Sexually dimorphic. ♂ in breeding plumage is very similar to Sardinian Warbler, but if any pinkish coloration is present on underparts then this aids identification. Base of lower mandible pinkish or straw-coloured, not bluish as in Sardinian, and upperparts grey with a slightly bluish tone, quite different to the dusky appearance of Sardinian. Also, dark blackish hood slightly less extensive and certainly less clearly demarcated on Ménétries's Warbler. ♀ much paler and plainer than ♂ and is generally sandy-grey above. It differs from ♀ Sardinian in having the head the same colour as the mantle. Underparts are buffy on breast and flanks but throat almost white. Juvenile resembles ♀ but is even plainer and lacks bright white edges to tail. **VOICE** Calls include a clicking *tsak* and rattling *tcherrr*. **RANGE** Breeds from SE Turkey to Tadzhikhistan and south to N Afghanistan, wintering in Iran, Arabia and NE Africa. **STATUS** Extreme vagrant. **Canaries*** ♀ ringed, Barranco Ajuy (Fuerteventura), 7 Apr 1977.

SARDINIAN WARBLER   *Sylvia melanocephala*   Plate 60 (5)
Sp. Curruca Cabecinegra   **P.** Toutinegra-dos-valados (Toutinegra-de-cabeça-preta)

S. m. leucogastra. **L** 13 cm (5″). **IDENTIFICATION** Sexually dimorphic. ♂ has glossy black head and face with red-brown eye and striking crimson eye-ring. Rest of upperparts dark grey but the white-edged tail is nearly as black as the head. Wings browner than mantle but tertials and inner secondaries distinctly fringed ash-grey. Chin and throat white but rest of underparts basically greyish-white with a faint pinkish wash on breast. ♀ has grey head and face, greyish-olive-brown upperparts, grey-brown breast-sides, flanks and undertail-coverts, white chin and throat, and tail similar to ♂ but not as black. Juvenile ♂ quite variable and can resemble both adult ♂ or adult ♀ but most have a dark hood like ♂; iris dull olive-brown. Juvenile ♀ similar to adult ♀ but is generally browner on head and has a duller iris. Race *leucogastra* is quite variable but generally both sexes are darker on upperparts than nominate. **VOICE** Common call is a series of rattling notes likened to the sound of machine-gun fire, *tutrrr-trr-trr-trr-trr-trr*; also when alarmed a more nasal *djj-djj-djj-djj-djj*.... Song is a musical chattering mixed with clear whistled notes and interspersed with sections of the rattling call. **RANGE** Breeds from NW Spain through S Europe to N Turkey, the Middle Eastern coast of the Mediterranean, and in N Africa discontinuously from Morocco to Libya. Some move south to winter in N Africa south of the breeding range. **DISTRIBUTION** Resident breeder on the Canaries. **BREEDING** Feb–Jul. Clutch size 2–4 but three eggs most frequent and pairs can raise at least two broods. Nests in a great variety of different shrubs, among them species of tamarisk and heather. **HABITAT** Any areas with sufficient vegetation, but not in towns and rarely in villages, and also tends to avoid desert scrub and high mountains. **TAXONOMIC NOTE** *leucogastra* is believed to be the form occurring on all of the Canaries, although some authors suggest that the nominate is present on the eastern islands, a hypothesis supported by the apparent vocal differences of those birds (Martín and Lorenzo, 2001). In contrast, Shirihai *et al.* (2001) do not recognise *leucogastra*, regarding the morphological differences observed in Sardinian Warblers from the Canary Islands as merely representing clinal variation. **STATUS** Varies considerably between the three archipelagos where it has been recorded. **Madeira** One very old record: a pair reported as collected by W. R. Ogilvie-Grant, prior to 1890, but the specimens appear to no longer exist. **Canaries** Breeds on all main islands and is a fairly common resident. **Cape Verdes** ♀, Raso, 11–12 Oct 2001.

AFRICAN DESERT WARBLER *Sylvia deserti* Plate 60 (6)
Sp. Curruca Sahariana **P.** Toutinegra-do-deserto (Toutinegra do Sara)

Monotypic. **L** 11½ cm (4½"). **IDENTIFICATION** Sexes similar. The smallest *Sylvia* warbler recorded in the region. Both sexes have pale sandy upperparts; rufous rump, uppertail-coverts and tail, the latter with white outer feathers. Whitish underparts are often tinged pinkish. Bare-parts coloration diagnostic: pale yellow iris, straw-coloured legs and straw-coloured bill, sometimes with an indistinct dark tip. Juvenile similar to adult, but in autumn the wing and tail feathers are often more abraded. **VOICE** Calls include a weak purring *drrrrrrrrrrrrr* and high-pitched *chee-chee-chee*. **RANGE** Breeds locally in NW Africa where mainly resident although some move in winter. **STATUS** Vagrant recorded from three archipelagos, but with only seven records in total, four from one island. **Madeira** One obtained, Porto Santo, 26 Jan 1901. **Canaries** One, beside the road to Taca (Fuerteventura), 19 Apr 1993; one, Playa de las Americas (Tenerife), 3 Mar 1997; one, San Sebastian (La Gomera), 7 Mar 1997 (possibly the same individual, as both were accompanied by a Sedge Warbler *Acrocephalus schoenobaenus*); one Caleta de Fustes (Fuerteventura), 16 Feb 2001; and one, *c*.5 km west of Corralejo (Fuerteventura), 23 Jul 2004. **Cape Verdes** One shot, 5 km north of Santa Maria (Sal), 9 Mar 1924.

(WESTERN) ORPHEAN WARBLER *Sylvia hortensis* Plate 60 (7)
Sp. Curruca Mirlona **P.** Toutinegra-real

Monotypic. **L** 15 cm (6"). **IDENTIFICATION** Sexes similar. The largest *Sylvia* warbler recorded in the region. Both sexes dark dusky-brown on upperparts, with an obvious pale iris in a dark face and obvious white sides to the dusky-coloured tail. Face of ♂ black but that of ♀ is dusky. Underparts have a slight pinkish tinge on lower throat, breast, flanks, vent and undertail-coverts in ♂, but in ♀ these tracts are browner. First-winter resembles adult ♀ but shows no black on head: ♂ is grey and ♀ brownish-grey, and iris is always dark. **VOICE** Call is a harsh *tak* or *teck*, similar to Blackcap, and also a buzzing *churrrr*. **RANGE** Western race breeds in the W Mediterranean from Spain and NW Africa discontinuously east to Italy in the north and Libya in the south. Winters in the Sahel, from Senegal and S Mauritania east to Niger and Chad. **STATUS** Vagrant with records from two archipelagos. **Madeira** One shot, south coast of Madeira, 25 Mar 1903. Also been recorded on the Salvages, 7 Sep 1990. **Canaries** About eleven records and recorded with certainty from Fuerteventura, Tenerife and La Gomera.

LESSER WHITETHROAT *Sylvia curruca* Plate 60 (8)
Sp. Curruca Zarcerilla **P.** Papa-amoras-cinzento

S. c. curruca. **L** 12.5–13.5 cm (5–5¼"). **IDENTIFICATION** Sexes similar. Small to medium-sized warbler, generally grey-brown above and dull white below, with white sides to dusky tail. Head dusky with an obvious blackish mask on ear-coverts, contrasting with the white throat. ♀ is similar to ♂ but occasionally has a less distinct mask. Juvenile similar to adult but mask definitely less prominent and white edges to tail not as bright. **VOICE** Similar to others in genus; has a sharp, harsh *teck* and also gives a short *churr* in alarm. **RANGE** Breeds from Britain, C France and S Scandinavia east to the Lena River in Siberia and Ningxia in China, and south through Europe and Turkey to Iran, Afghanistan and the NW Himalayas. Western populations winter in NE Africa, mainly in Sudan and Chad, and in smaller numbers in Arabia. **STATUS** Vagrant with records from two archipelagos. **Madeira** Exceptional visitor and also recorded on the Salvages. **Canaries★** One, Morro Jable (Fuerteventura), 30 Oct 1995; one, Erjos (Tenerife), 13 Dec 1997; one, Tenerife, 26 Mar 1998; and one, Morro Jable (Fuerteventura), 10 Nov 2000.

COMMON WHITETHROAT *Sylvia communis* Plate 60 (9)
Sp. Curruca Zarcera **P.** Papa-amoras-comum

S. c. communis. **L** 14 cm (5½"). **IDENTIFICATION** Sexes similar. ♂ is rather rufous brown above with a grey head, white throat and pinkish-buff wash to most underparts. ♀ is similar but has head concolorous with mantle and pinkish tinge to underparts less pronounced. First-winter resembles adult ♀ and many are inseparable using plumage characters, but iris is dark olive-brown not the pale orange-brown or yellowish-brown of the adult. Most likely to be confused with Spectacled Warbler, but differences should be sufficiently obvious in most cases to make identification relatively easy. ♂ is similar in general coloration to the smaller, slimmer Spectacled Warbler but never has black lores, the pattern of the tertials is not as contrasting as in Spectacled Warbler and the dark centres are more extensive and more rounded, particularly on the inner two. The overall size and shape of Common Whitethroat is distinctive and the primary projection is much longer than in Spectacled Warbler. The same structural differences apply to the more similarly plumaged ♀♀ and juveniles. **VOICE** A sharp *tac* softer than the similar call of Lesser Whitethroat, also a harsh *churr* that varies in intensity, and a nasal *wheet* often in series. **RANGE** Breeds from Ireland across Europe east to Lake Baikal and in NW Africa. The entire population winters in Africa south of the Sahara. **STATUS** Varies between the two archipelagos where it has been recorded. **Madeira** Exceptional visitor that has also been recorded from the Salvages. **Canaries** Regular and not uncommon passage migrant recorded on all islands except La Palma.

GARDEN WARBLER *Sylvia borin* Plate 60 (11)
Sp. Curruca Mosquitera **P.** Felosa-das-figueiras

S. b. borin. **L** 14 cm (5½"). **IDENTIFICATION** Sexes similar. The dullest *Sylvia*, both sexes being dull grey-brown above and pale buff below. The virtual lack of any distinct markings is perhaps the best identification feature. The only character of any prominence is the short, pale buff to greyish, fore supercilium and eye-ring, which slightly emphasises the dark eye. Structurally, a medium-sized warbler with a rounded head, plump body and relatively short but strong-looking bill. First-winter almost identical to adult but slightly more olive above and buff below, with the plumage fresher than that of adult at same season. **VOICE** Call is a sharp *check* often repeated in series. **RANGE** Breeds across Europe east to the Caspian and W Siberia. Winters extensively in Africa from *c.* 10°N in the west and 3°N in the east, south to S Africa. **STATUS** Varies between the three archipelagos where it has been recorded. **Madeira** Exceptional visitor that has been recorded from the Salvages. **Canaries** Scarce but regular passage migrant, unrecorded from La Palma and El Hierro. **Cape Verdes** Two records: one, Raso, 1 Oct 2001, joined by a second next day, with one until 4 Oct 2001; and one, Pedra de Lume (Sal), 9 Mar 2004.

BLACKCAP *Sylvia atricapilla* Plate 60 (10)
Sp. Curruca Capirotada **P.** Toutinegra-de-barrete (Toutinegra-de-barrete-preto)

S. a. heineken (Madeira and Canaries), *S. a. gularis* (Azores and Cape Verdes) and *S. a. atricapilla*. **L** 13 cm (5"). **IDENTIFICATION** Sexually dimorphic. Both sexes have dusky-brown upperparts and are greyish-buff below with no white edges to tail. ♂ has a jet-black cap, not extending onto the nape or face. ♀ and juvenile have same-shaped cap but brown. Race *heineken* generally smaller and darker than the nominate, although some birds from the Canaries approach *atricapilla*. Subspecies *gularis* is hardly separable in colour from nominate, but the highly variable wing measurements are generally shorter, and the bill longer, more like *heineken*. A melanistic form occasionally occurs in Azores, Madeira and the Canaries, but has not been recorded from the latter for many years. **VOICE** Common call is a sharp, loud *tack* sometimes repeated in series, and a *churr* in alarm. Song starts with a chattering phrase, changing into clear, rich, melodious, fluty whistles, rising and falling in pitch at end. **RANGE** Breeds from NW Africa and Iberia north to N Norway and east to the southern Caspian and SW Siberia. Sedentary over much of its western range, but other populations winter in NW, CW, NE and E Africa. **DISTRIBUTION** Resident breeder throughout the region. **BREEDING** Season varies: in Cape Verde Aug–Jun, whereas in the Canaries it is shorter and limited to Feb–Jul, and in the Azores shorter still (Apr–Jun). Clutch usually 3–4 but can range from 2–6, and there are often two broods. Nest is sited in a tree or shrub, and is usually placed between 0.5–4.5 m above ground. **HABITAT** Prefers reasonably dense vegetation. **STATUS** Common resident on all archipelagos. **Azores** Common resident on all islands. **Madeira** Very common resident and also recorded on the Salvages. **Canaries** Common resident on the western and central islands, but on Fuerteventura and Lanzarote its status is unclear. It is here, and on Tenerife, that *atricapilla* has been reported, and it is considered a regular passage migrant and winter visitor. **Cape Verdes** Common resident on Santiago, Fogo, Brava, Santo Antão and São Nicolau. Possibly extinct on São Vicente, where not recorded since 1966, and there is one record (possibly a migrant) from Maio.

YELLOW-BROWED WARBLER *Phylloscopus inornatus* Plate 61 (1)
Sp. Mosquitero Bilistado **P.** Felosa-listada (Felosa-bilistada)

Monotypic. **L** 10 cm (4"). **IDENTIFICATION** Sexes similar. A small, but usually very active, warbler often detected by call. Combination of small size (marginally bigger than a *Regulus* spp.), prominent pale supercilium and double wingbar are all important identification features. Generally olive-green above and whitish below, with a long yellowish-white supercilium that extends almost to the nape, a long, dark eye-stripe, two conspicuous off-white wingbars, off-white fringes to tertials and pale tips to primaries. First-winter very similar to adult. **VOICE** Call a loud and frequently given *tsweeet*. **RANGE** Breeds in N Siberia west to the N Urals and winters in S China, SE Asia and the Philippines. **STATUS** Accidental with records from three archipelagos. **Azores★** One, near Sete Cidades (São Miguel), 8–9 Dec 2001. **Madeira** Three, Porto Santo, 21 Nov 1900, with one, 9 & 14 Dec of that year (one collected on Porto Santo in 1900 was probably one of these); one, Porto Santo, 30 Jan 1902; and one, Funchal Botanic Gardens (Madeira), 28 Nov 2003. **Canaries** Accidental visitor in autumn with at least 14 records, mostly from Lanzarote (4) and Fuerteventura (6), but also been recorded on Tenerife (4).

HUME'S (YELLOW-BROWED) WARBLER *Phylloscopus humei* Plate 61 (2)
Sp. Mosquitero de Hume   **P.** Felosa-listada de Hume

P. h. humei.   **L** 10 cm (4″).   **IDENTIFICATION** Sexes similar. Very similar to Yellow-browed Warbler and perhaps best identified by call. However, Hume's is duller, with greyer upperparts than Yellow-browed, the yellow tones on supercilium and underparts being replaced with buffish-white, the upper wingbar less distinct and the bill normally more uniformly dark. First-winter indistinguishable in the field from adult. **VOICE** Usual call a short, loud, descending disyllabic *wesoo* or *wisslo*, also a sparrow-like *ch'wee*, unlike any vocalisation of Yellow-browed Warbler. **RANGE** Breeds in C Asia from the Sayan and Altai ranges south to the NW Himalayas, and winters in Pakistan and the Himalayan foothills east to Bangladesh and south in India to *c.* 16°N. **STATUS** Extreme vagrant. **Canaries★** One, Tachas Blancas (Fuerteventura), 22 Feb 1997, showed all the characters of this species.

WESTERN BONELLI'S WARBLER *Phylloscopus bonelli* Plate 61 (3)
Sp. Mosquitero Papialbo Occidental   **P.** Felosa-de-papo-branco

Monotypic.   **L** 11.5 cm (4½″).   **IDENTIFICATION** Sexes similar. Slightly larger than Common Chiffchaff and has striking white underparts. Head and mantle plain grey-brown with only a very indistinct pale supercilium, the lower back and rump greenish-yellow (sometimes difficult to see), the underparts strikingly silky white and the wings and tail basically dark brown. First-winter has duller rump, with wing-coverts, flight feathers and tail feathers narrowly fringed greenish-yellow, which shows as a distinct panel on folded wing. **VOICE** Call a fairly loud, disyllabic *hoo-eet*. **RANGE** Breeds in NW Africa, W and C Europe east to Austria, NW Balkans and Italy. Winters in a relatively narrow band across Africa at the southern edge of the Sahara. **STATUS** Varies between the various archipelagos where it has been recorded. **Madeira** Vagrant listed as exceptional by Zino *et al.* (1995), and also recorded on the Salvages. **Canaries** Scarce but regular passage migrant on Fuerteventura and Lanzarote, otherwise only recorded from El Hierro. **Cape Verdes** One, Baía da Mordeira (Sal), 20 Sep 1997.

WOOD WARBLER *Phylloscopus sibilatrix* Plate 61 (4)
Sp. Mosquitero Silbador   **P.** Felosa-assobiadeira

Monotypic.   **L** 12–13 cm (5″).   **IDENTIFICATION** Sexes similar. Noticeably larger than Common Chiffchaff or Willow Warbler, with much brighter plumage. Upperparts green with dark brown wings and tail. Tail and wing feathers all fringed bright yellow-green, showing as a panel on folded wing. Supercilium yellow; eye-stripe dark; and cheeks, ear-coverts, chin, throat and upper breast yellow. Rest of underparts almost pure white. First-winter like adult but upperparts slightly darker, yellow of underparts slightly paler and wing and tail feathers fresher and brighter. **VOICE** Call is a plaintive *piu*. **RANGE** Breeds from the British Isles and France east to W Siberia and south to S Italy, Bulgaria and Georgia. Winters in sub-Saharan, C and southern W Africa. **STATUS** Only recorded from two archipelagos. **Madeira** Listed in the checklist as exceptional. **Canaries** Regular passage migrant but not recorded from the western islands of La Gomera and El Hierro.

The chiffchaff complex
Recent taxonomic revisions have brought some major changes in the classification of *Phylloscopus collybita* and its various subspecies. Studies based on DNA sequencing and vocalisations support the promotion to species status of three chiffchaffs, two of which occur in the region covered here, namely Iberian Chiffchaff and Canary Islands Chiffchaff, the latter a Canary Islands endemic. Identification of all forms within the complex is difficult and is based on slight structural and colour differences, as well as the often more obvious vocal distinctions. Anyone attempting to identify one of the migrant species should exercise great care and perhaps leave unidentified any problematic individuals. Whilst the following text is as accurate as possible, it is not feasible to include descriptions of all the variations in colour that can occur.

COMMON CHIFFCHAFF *Phylloscopus collybita* Plate 61 (5)
Sp. Mosquitero Común   **P.** Felosinha (Felosa-comum)

P. c. collybita and *P. c. abietinus*.   **L** 10–11 cm (4″).   **IDENTIFICATION** Sexes similar. Nominate *collybita* has dull greenish upperparts and off-white underparts, with a yellowish wash on breast and flanks. Legs mainly black but can be dark brown or greyish-brown. Wings and tail feathers dark brown, finely fringed olive-green. Supercilium thin, yellowish in colour and extends from the bill base to the ear-coverts. In fresh autumn plumage both adults and first-winters have brighter upperparts than in spring. Subspecies *abietinus* is very similar to the nominate and many are inseparable. However, the most obvious *abietinus* show a distinctive greyish wash to the crown, nape and mantle, and the underparts are generally whiter than in *collybita*. **VOICE** Common call is a monosyllabic *hweet*. Song is a repetitive and monotonous *chiff-chaff-chiff-chaff-chiff-chaff-chiff-chaff* lasting 20 seconds or more, and with one note higher pitched than the other. **RANGE** *collybita* breeds from the British Isles and NE Spain through Europe and S Sweden to Bulgaria, N Greece and NW Turkey. Winters mainly throughout the Mediterranean east to Iraq and in N and W Africa. Race *abietinus* breeds from NE Poland and coastal Norway and Sweden (except the south) east to the Pechora River and to around Odessa on the Black Sea. Winters mainly in the E Mediterranean to Arabia and Iran, in the east, and Kenya and Uganda in the south, but also reaches W Africa. **STATUS** Recorded from all archipelagos but status varies. **Azores★** An influx of c. 20 birds noted on São Miguel, 6–9 Dec 2001, with records from Lagoa das Furnas and Sete Cidades, and subsequent records in autumn 2003 and 2004. **Madeira** Irregular and scarce passage migrant. **Canaries** *collybita* is regular in small numbers on both spring and autumn passage, but is much more uncommon in winter. Can occur in large numbers during abnormal weather conditions; *abietinus* has been recorded, but its status is far from clear and it may just be a vagrant. **Cape Verdes** One bird seen and heard singing, in *Acacia* near Estância de Baixo (Boavista), 10 Dec 1989; and one dead, south shore of Raso, 26 Sep 1997.

IBERIAN CHIFFCHAFF *Phylloscopus ibericus* Plate 61 (7)
Sp. Mosquitero Ibérico **P.** Felosinha-ibérica

Monotypic. **L** 10–11 cm (4"). **IDENTIFICATION** Sexes similar. Very similar to Common Chiffchaff but can appear whiter on belly with a distinctly lemon-yellow vent and undertail-coverts. **VOICE** The most distinctive feature. Song has an entirely different structure to Common Chiffchaff lacking any clear *chiff* or *chaff* notes. Slow delivery, a slow descent in pitch and series of repeated rising or falling notes often delivered 3+3+3, but the number in each phrase may vary; rendered *tit-tit-tit-tit-tswee-tswee-chit-it-it-it-it*. One call is a nasal whistling *piu* which, like the song, is totally unlike any call of Common Chiffchaff. **RANGE** Breeds in extreme SW France, across N and C Spain, south to Murcia, Andalucía and around Gibraltar, over most of Portugal, and possibly in NW Africa. Few data concerning the wintering areas due to the difficulties in identifying silent individuals. **STATUS** Uncertain due to identification problems. **Canaries** Recorded from the archipelago but its status is unclear.

CANARY ISLANDS CHIFFCHAFF *Phylloscopus canariensis* Plate 61 (6)
Sp. Mosquitero Canario **P.** Felosinha das Canárias

P. c. canariensis (W Canaries) and *P. c. exsul* (formerly Lanzarote, but now probably extinct). **L** 10–11 cm (4"). **IDENTIFICATION** Sexes similar. The only species of chiffchaff to breed in the region. Compared to Common Chiffchaff it has a noticeably longer bill, shorter wings and slightly longer tail. The long-tailed appearance may be exaggerated by the short wings. Underparts have a strong brownish-buff tone, particularly on breast and flanks, a feature not shown by Common Chiffchaff. Upperparts generally darker and more brownish than in Common Chiffchaff and the pale supercilium is more distinct and slightly longer. Leg colour of Canary Islands Chiffchaff is variable, ranging from greenish-brown to yellowish-brown, and usually distinctly paler than Common Chiffchaff. Subspecies *exsul* was paler than *canariensis* and had darker legs. **VOICE** Some calls similar to the *hweet* of Common Chiffchaff but also gives a short, metallic *chek* or *chik*. Song distinctly different from Common Chiffchaff being harsher, shorter and more variable, with a more rapid delivery. Sometimes commences singing in a manner likened to the explosive start of Cetti's Warbler. **RANGE** Confined to the Canaries where resident. **DISTRIBUTION** Breeds on El Hierro, La Palma, La Gomera, Tenerife and Gran Canaria. **BREEDING** Season Jan–Jun and most pairs have two broods, with some even a third. Clutch 2–5 but four eggs most frequent. Spherical nest with a side entrance placed at varying heights according to habitat. In coastal areas nests near ground but in laurel forests nests recorded at heights above 8 m. **HABITAT** Occurs in most habitats from coastal gardens to high-altitude vegetation, but is mostly absent from semi-desert scrub in lower zones. **STATUS** Endemic resident in one archipelago. **Canaries** Common and widespread on all of the islands where it is found.

WILLOW WARBLER *Phylloscopus trochilus* Plate 61 (8)
Sp. Mosquitero Musical **P.** Felosa-musical

P. t. trochilus. **L** 10.5–11.5 cm (4–4½"). **IDENTIFICATION** Sexes similar. Very similar to Common Chiffchaff but has a longer primary projection and, usually, paler legs. Also yellow on underparts is more extensive than Common Chiffchaff, particularly on juveniles. First-winters distinguished by their almost completely yellow underparts, which are paler yellow on adults with at least the belly white. Perhaps the best feature for identification is **VOICE**. **VOICE** Song, unlikely to be heard in the region, different from typical monotonous *chiff-chaff* of Common Chiffchaff and call is distinctly disyllabic *hoo-eet*, compared to monosyllabic *hweet* of Common Chiffchaff. **RANGE** Breeds from Ireland across most of C and N Europe east through Russia to the Anadyr River and in NW Spain. Entire population winters in sub-Saharan Africa. **STATUS** Varies in the different archipelagos. **Azores★** Four records: one, Lagoa das Furnas (São Miguel), 29–30 Nov 1998; one, Fajã Grande (Flores), 15 Sep 2001; one, Sete Cidades (São Miguel), 7 Dec 2001, and one also there, 2 Nov 2004. **Madeira** Scarce and irregular passage migrant. **Canaries** Regular in spring and autumn, but numbers vary with weather conditions and is occasionally very common.

Cape Verdes Considered a very rare passage migrant with just six records. However, on Boavista, 19–22 Sep 1988, there were at least ten at three locations, implying that the species might be more common than records indicate.

GOLDCREST *Regulus regulus* Plate 62 (1)
Sp. Reyezuelo Sencillo **P.** Estrelinha-de-poupa

R. r. azoricus, *R. r. inermis* and *R. r. sanctaemariae* (Azores). **L** 9 cm (3½″). **IDENTIFICATION** Sexes differ. Small size and rather dumpy shape combined with pale olive-green upperparts and off-white underparts eliminate most species except Firecrest. ♂ has orange and yellow crown-stripe bordered at sides with black, a greyish face with prominent pale eye-ring, double pale wingbar and pale tips to tertials. ♀ similar but lacks any orange in crown. Juvenile lacks distinctive crown pattern but is inseparable from adult by first autumn. Subspecies *azoricus* has darker upperparts than nominate and darker more olive-buff underparts; *inermis* similar to *azoricus* but underparts darker, more brownish-olive; *sanctaemariae* is palest race, with upperparts paler yellowish-olive and underparts whitish. All three have a longer bill than nominate. **VOICE** Call is a very high-pitched *zree-zree-zree*. Song is a high-pitched *pit-eee-tily* repeated four to six times and terminating in a trilled *zi-zi-zreeu*. **RANGE** Breeds across C and N Europe discontinuously east to Sakhalin and Japan. **DISTRIBUTION** Confined to the Azores. **BREEDING** Season seems fairly short and is concentrated in Apr–Jun, with second broods fairly frequent. Clutch size appears to be smaller than in mainland Europe with a maximum of six. Height of nest above ground level depends on vegetation in which it is placed. **HABITAT** Favours wooded areas and areas of tall scrub with junipers. **STATUS** Resident breeder on one archipelago. **Azores** Fairly common resident breeder on all islands except Graciosa and Corvo.

TENERIFE GOLDCREST (TENERIFE KINGLET) *Regulus teneriffae* Plate 62 (2)
Sp. Reyezuelo Tinerfeño **P.** Bandeirita

Monotypic. **L** 9 cm (3½″). **IDENTIFICATION** Sexes differ. Very similar in all plumages to Goldcrest, the only really obvious difference being that the black crown-stripes of Tenerife Goldcrest meet on the forehead. Also has a relatively long bill, less white on tertials and underparts darker buff. Juvenile differs from adult in its lack of head pattern. **VOICE** Calls and song similar to Goldcrest but harsher and lower pitched. Song is also shorter and somewhat intermediate between Goldcrest and Firecrest. **RANGE** Confined to the Canaries. **DISTRIBUTION** Breeds on Tenerife, La Gomera, La Palma and El Hierro. **BREEDING** Season fairly restricted to Mar–Jun. Clutch size is 3–6, but 4–5 most frequent. Nest placed in various trees, but *Erica arborea* is the most popular, at a height of 2.5–14 m. **HABITAT** Found in both laurel and pine forests, but favours the habitat known locally as fayal-brezal, where the heathers *Erica arborea* and *E. scoparia* are dominant. **STATUS** Only found on one archipelago. **Canaries** Fairly common on Tenerife and the western islands, but has also wandered to Gran Canaria.

MADEIRA FIRECREST *Regulus madeirensis* Plate 62 (3)
Sp. Reyezuelo de Madeira **P.** Estrelinha-de-Madeira

Monotypic. **L** 9 cm (3½″). **IDENTIFICATION** Sexes differ. Similar to Goldcrest but has whiter underparts, a white supercilium, black eye-stripe and bronzy patches on neck-sides. ♂ has much orange on central crown but in ♀ this area is yellow, sometimes partially orange. Black lateral crown-stripes in both sexes meet on forehead. Compared to Firecrest *R. ignicapilla* (which is unrecorded in our region and from which Madeira Firecrest has recently been split), Madeira Firecrest has a narrower supercilium that terminates just behind the eye. **VOICE** Call similar to that of Goldcrest but is slower, less penetrating and lower pitched. Song also simpler, lower and less varied in pitch, shorter than that of Goldcrest and lacks its terminal flourish; transcribed *tsi-tsi-tsi-tsitsitsit*. **RANGE** Breeds from W Europe and N Africa east to Turkey. Northern and eastern populations migrate south and west, wintering mainly in the Mediterranean region. **DISTRIBUTION** Confined to the Madeiran archipelago. **BREEDING** Season May–Jul. Clutch size is 5–7, but six eggs most frequent. Nest is most often placed in *Erica* bushes and trees, but other species such as laurel are also used. **HABITAT** All zones with sufficient trees, but most common in tree heath at higher altitudes. **STATUS** Resident on only one archipelago. **Madeira** Confined to the island of Madeira where fairly common in heath forest and other wooded areas.

SPOTTED FLYCATCHER *Muscicapa striata* Plate 62 (4)
Sp. Papamoscas Gris **P.** Papa-moscas-cinzento

M. s. striata. **L** 14½ cm (6″). **IDENTIFICATION** Sexes similar. A rather nondescript species with greyish-brown upperparts and off-white underparts, and indistinct streaking on breast, throat, forehead and forecrown. First-winter distinguished from adult by presence of a buffish wingbar on tips of greater coverts. **VOICE** Common calls are a squeaky, shrill *zee* and sharp, clicking *chick*. **RANGE** Breeds from NW Africa, Iberia and Ireland east to NW

China and south to the Himalayas. Winters in sub-Saharan Africa mostly south of the equator. **STATUS** Recorded from three archipelagos but status varies. **Madeira** Listed by Zino *et al.* (1995) as being exceptional; recently one, Porto Santo, 24 Sep 2003. Also recorded on the Salvages. **Canaries** Regular passage migrant, particularly in autumn. More numerous on the eastern islands but has been recorded from all main islands except of El Hierro. **Cape Verdes** Four records: one, near Porto Ferreira (Boavista), 21 Sep 1988; one, Terra Boa (Sal), 27 Apr 1991, another there, 30 Sep 1998; and one, Palmeira (Sal), 17 Oct 1998.

RED-BREASTED FLYCATCHER   *Ficedula parva*   Plate 62 (5)
Sp. Papamoscas Papirrojo   **P.** Papa-moscas-pequeno

F. p. parva. **L** 11½ cm (4½″). **IDENTIFICATION** Sexually dimorphic. ♂ is unmistakable with reddish-orange throat and upper breast, and greyish head. Adult ♀ and first-winter have dull brown upperparts and off-white underparts with a buff wash to breast, narrow buff eye-ring and first-winters also have a thin buff wingbar on tips of greater coverts. Most conspicuous feature of both ♀ and first-winter plumages is tail pattern, which is like that of Northern Wheatear, dark central rectrices but outer feathers have extensive white bases and dark tips, forming a T when tail opened. Can be rather difficult to see on occasion. **VOICE** Call is a rather soft Wren-like rattle of short *chick* notes slurred together. Also a plaintive *weeit*. **RANGE** Race *parva* breeds from C and E Europe to the S Urals, Balkans and S Caspian, and winters in Pakistan and India. **STATUS** Recorded from three archipelagos. **Azores** Juvenile, on board ship in Azorean waters, 7 Nov 1995. **Madeira** One, at sea 10 miles north of Madeira, 19 Oct 1973; and one, Caminho Velho da Ajuda, Funchal (Madeira), 24 Apr 1974. **Canaries** Eleven records from Lanzarote, Fuerteventura, Tenerife and La Gomera, the most recent a first-winter, Ten Bel (Tenerife), 26 Jan–23 Mar 2002.

COLLARED FLYCATCHER   *Ficedula albicollis*   not illustrated
Sp. Papamoscas Collarino   **P.** Papa-moscas-de-colar

Monotypic. **L** 13 cm (5″). **IDENTIFICATION** Sexually dimorphic. Breeding ♂ resembles Pied Flycatcher but broad white collar on nape is diagnostic. Collared also has a white rump, relatively large white spot on forehead, large white patch at base of primaries and an all-black tail, but all these can be shown by Atlas Flycatcher *Ficedula speculigera* of NW Africa (which was formerly considered a race of Pied Flycatcher). ♀ very similar to that of Pied Flycatcher but usually paler and greyer above with a large white patch at base of primaries. Non-breeding ♂ also very similar to that of Pied Flycatcher but wings blacker and white patch at base of primaries is larger. First-winter Collared Flycatcher can show a more distinct second wingbar and large white patch at base of primaries but is otherwise indistinguishable from first-winter Pied. **VOICE** Shares quiet *tec* call given by Pied Flycatcher but also has an inhaled whistle, *seeb*. **RANGE** Breeds in C Europe east to W Russia. Winters in E Africa mostly between the equator and 20°S. **STATUS** Extreme vagrant. **Canaries** ♂, Barranco de la Torre (Fuerteventura), 14 May 2000.

PIED FLYCATCHER   *Ficedula hypoleuca*   Plate 62 (6)
Sp. Papamoscas Cerrojillo   **P.** Papas-moscas preto

F. h. hypoleuca. **L** 13 cm (5″). **IDENTIFICATION** Sexually dimorphic. ♂ in breeding plumage black above and white below with small white patch on forehead, white sides to base of tail, white patch in wing on tertials and inner greater coverts, and an indistinct white patch at base of primaries, which feature is often absent. ♀ has brown upperparts, lacks white spot on forehead and amount of white in wing is much reduced. Non-breeding ♂ resembles ♀. First-winter also resembles ♀ but has broader white outer edge to tertials and buffy-white tips to some median coverts. **VOICE** Call is a short metallic *pik* frequently repeated in series, or a soft clicking *tec*. **RANGE** Breeds from N Europe, Iberia and N Africa east to the Yenisei River in Siberia. Winters in W Africa south of the Sahara and mostly north of the Gulf of Guinea. **STATUS** Records from three archipelagos. **Madeira** Listed in the checklist as exceptional but, more recently, eight in Sep 2003, including four in Funchal on 17th, suggests that, under favourable conditions, the species is more common than records belie. Also recorded from the Salvages. **Canaries** Regular passage migrant sometimes in large numbers, more frequent in autumn and recorded from all main islands except El Hierro. **Cape Verdes** Five records: ♀ collected, Sal, 1901; two, on board a ship with noon position 17°05′N 23°18′W, 28 nm northwest of Sal, 7 Sep 1985; then, in 1988, on Boavista individuals, Fonte Vicente, 19 Sep, Curral Velho, 20 Sep, and Porto Ferreira, 22 Sep.

AFRICAN BLUE TIT   *Cyanistes ultramarinus*   Plate 63 (1)
Sp. Herrerillo Africano   **P.** Chapim-azul

C. u. degener, C. u. ombriosus, C. u. palmensis and *C. u. teneriffae* (all confined to the Canaries). **L** 11.5 cm (4½″). **IDENTIFICATION** Sexes similar. This species is markedly different from the familiar European Blue Tit. In general it is much more blackish on crown, with darker, more slate-blue upperparts, and a longer bill. Four subspecies occur in the region and all can be identified on plumage, as well as range. The most widespread is *teneriffae*, which has no wingbar and very indistinct pale fringes to tertials. Upperparts dark slaty-blue and underparts deep yellow with a much-reduced dark ventral line. Most similar to *teneriffae* is *ombriosus*, which has a greenish tone to upperparts and indistinct greyish-white wingbar; *palmensis* can also show a greenish tone to upperparts but wingbar and tertial fringes are more pronounced than *ombriosus*. Underparts of *palmensis* are paler yellow and belly extensively white. Finally, *degener* of the eastern islands has paler and greyer upperparts with a broad white wingbar and obvious white fringes to tertials. Underparts predominantly yellow, apart from a small whitish area on belly. **VOICE** Songs differ from those of European Blue Tit, principally in that they comprise of repetitions of the same units or groups of units without tremolo, and with a metallic timbre; *teneriffae* has a principal alarm call that is a loud, vibrant, scolding *cher-er-er-er-er-g*, at times introduced with a *chi* or *chi-chi-chivichi*. Like other subspecies it gives various types of song emphasising a fine trill and hurried *iss-sirichichiu* with some variation; a powerful, cheerful and musical *chi-chi-chirichi-chichiri* which is repeated continuously and *chi-iss-chi-chi* or *iss-chichichiuu*, more deliberate and less musical than the previous with apparent variation between islands (in La Gomera seems slightly different in tone and rhythm). Other common calls are a lively *chivichi-chivichi* and resonant, musical *chillu-chillu chi*. Subspecies *palmensis* has a totally distinct repertoire to that of other Canarian Blue Tits, indeed the voice has been likened to that of Great Tit. Alarm call is a harsh, broken *tcher-tcher-cher-cher-cher*, unlike the typical purring of other subspecies. A song or call exclusive to this subspecies is an unmistakable *tsa-za-chúz*, sometimes with the final note repeated and the entire phrase preceded with an introductory *er-er-r*. Other calls include a nasal and scraping *chié-chié-chié-chié-chié* and *chiff-chichichi* with a blowing effect on the first syllable. Subspecies *ombriosus* has an alarm call similar to *teneriffae*. The song consists of an accelerated musical trill similar to that of *teneriffae* but the twittering is quicker and more continuous. Subspecies *degener* has a call like that of *teneriffae* and *ombriosus* but louder and more scolding. Also gives a *chíi-ererer chíi-ererer chíi-ererer*, which is appreciably different to other subspecies. **RANGE** Resident of the Canaries and NW Africa. **DISTRIBUTION** Resident breeder on all main islands in the Canarian archipelago; *degener* breeds on the eastern islands of Lanzarote and Fuerteventura; *teneriffae* on the central islands of Gran Canaria, Tenerife and La Gomera, whilst *palmensis* and *ombriosus* breed on the western islands of La Palma and El Hierro respectively. **BREEDING** Usually commences Feb and continues until Jul, but there are records of nests with chicks in Oct–Dec. Some pairs have even been recorded nesting in Jan. Nest is normally sited below 3 m, in holes and cracks in trees, walls, barrancos, beside tracks and even on the ground. Also uses next boxes where provided. Clutch size 3–5 but six eggs also recorded. **HABITAT** Occurs in coniferous and laurel woodland, fayal-brezal, parks and gardens, from sea level to the tree line. **TAXONOMIC NOTE** African Blue Tit is a recent split from European Blue Tit *C. caeruleus*, following genetic research by Salzburger *et al*. (2002). **STATUS** Common breeder on the W and C Canaries but rarer and more locally distributed on the less well-vegetated eastern islands of Fuerteventura and Lanzarote.

GREAT TIT *Parus major*   not illustrated
Sp. Carbonero Común **P.** Chapin-real

Subspecies unknown but *P. m. excelsus* would seem the most likely. **L** 14 cm (5½″). **IDENTIFICATION** Sexes similar. Much larger and longer tailed than African Blue Tit, the only species likely to cause any identification problems, especially as African Blue Tit has a much darker crown than the more familiar European Blue Tit, hence superficially resembling Great Tit. Adult has a glossy black cap extending to bill, broad central black ventral band on bright yellow underparts, greenish back and white sides to outer tail, all of which separate this species from African Blue Tit. Juvenile duller than adult with a sooty-brown cap, yellowish rather than white cheeks, a smaller, duller sooty-brown bib and ventral line indistinct or even lacking. **VOICE** Has a very complex and varied vocabulary, but common calls include a variety of churrs mixed with other monosyllabic unmusical notes, including a collection of *tsee*, *pee* and *tink* calls. **DISTRIBUTION** Breeds across the Palearctic, except the most northern regions to Japan, and south to N Africa, Sri Lanka and Indonesia; mainly resident but occasionally irruptive. **STATUS** Vagrant with only three records. **Canaries** Adult, near Poris de Abona (Tenerife), 16 Mar 1997; adult, Presa de las Peñitas (Fuerteventura), 23–25 Apr 1997; and adult, Barranco de Ruiz (Tenerife), 15 Apr 2000.

EURASIAN NUTHATCH *Sitta europaea*   Plate 62 (7)
Sp. Trepador Azul **P.** Trepadeira-azul

Subspecies unknown. **L** 14 cm (5½″). **IDENTIFICATION** Sexes similar. In all plumages unmistakable due to unique shape, short tail and rather long, chisel-like bill. Adult blue-grey above with long black eye-stripe which curves down neck-sides; white cheeks and throat, and rest of underparts orangey-buff with darker, more chestnut, rear flanks and undertail-coverts, showing some white spotting. ♀ resembles ♂ but paler on rear flanks and undertail-coverts, and has slightly less distinct eye-stripe. Juvenile resembles adult of same sex but is duller overall with a less distinct eye-stripe. **VOICE** Contact call a high-pitched *tsit* or *seet* but common call is an excited, loud *chwett* or *twett*. **RANGE** Breeds from W Europe and NW Africa east to Kamchatka, Japan, China and Taiwan. Mostly resident but dispersal, by juveniles, can form irruptions, although these usually avoid even short water crossings. **STATUS** Extreme vagrant. **Canaries★** One, in NE Lanzarote, 20 Feb 1968. Mentioned historically by Ledrú for Tenerife, in 1796, and also by Bolle for the same island, prior to 1857.

PENDULINE TIT *Remiz pendulinus* Plate 62 (8)
Sp. Pájaro Moscón **P.** Chapim-de-mascarilha (Chapim-de-faces-pretas)

R. p. pendulinus. **L** 11 cm (4"). **IDENTIFICATION** Sexes similar. Adult ♂ unmistakable with its grey head, black forehead and mask, chestnut mantle and buffish underparts, with some chestnut markings on breast. ♀ duller with a less extensive mask and very little, if any, chestnut marking on breast. Juvenile distinctly different from adult: lacks black mask and has rather plain grey-brown head, paler brown mantle and off-white underparts. **VOICE** Common call a high-pitched, thin, whistle *tsiiiuu*. **RANGE** Breeds from Iberia through C and S Europe and the Middle East, east to the Ob River and Kazakhstan. Most of the European population winters in the N Mediterranean. **STATUS** Extreme vagrant. **Canaries★** One, Tejina Ponds (Tenerife), 8 Apr 1998.

GOLDEN ORIOLE *Oriolus oriolus* Plate 63 (2)
Sp. Oropéndola **P.** Papa-figos

O. o. oriolus. **L** 24 cm (9½"). **IDENTIFICATION** Sexually dimorphic. ♂ unmistakable with a dark pink bill, bright yellow head and body, and black wings and tail. ♀ and immature less striking but equally distinctive. ♀ has bill darker than that of ♂ but still pinkish, the upperparts are green with a brighter, yellower, rump and uppertail-coverts. Underparts greyish-white with pale yellow flanks and vent, and overall dull brown streaking. Juvenile resembles adult ♀ but has yellow wash to underparts and a dark brown bill. **VOICE** Song is a very distinctive fluty whistle, *wheela-weeoo*. Calls include a harsh, drawn-out cat-like note and a falcon-like *gigigigigi*. **RANGE** Breeds throughout most of Europe east to the Yenisei River and the Tien Shan. Winters in S Africa and in the Indian subcontinent. **STATUS** Recorded from three archipelagos but only regular on one. **Azores** Five records, three of which are pre-1903, and the most recent a ♀ obtained, Fajã de Baixo (São Miguel), 24 May 1963. **Madeira** Exceptional vagrant. **Canaries** Scarce but regular passage migrant, more common in spring and on the eastern islands.

ISABELLINE SHRIKE *Lanius isabellinus* Plate 63 (3)
Sp. Alcaudón Isabel **P.** Picanço-isabel

Subspecies unknown. **L** 17½ cm (7"). **IDENTIFICATION** In the nominate form sexes are similar, but in subspecies *phoenicuroides* they differ. Race *isabellinus* has pale sandy-brown upperparts and buffy-white underparts. Rump and uppertail distinctly rusty, with a black mask and small white patch at base of primaries. ♀ has weaker mask and sometimes shows some slight vermiculations on breast and flanks. First-winter resembles ♀ but has more markings on underparts and usually some distinct barring on upperparts. In *phoenicuroides* ♂ has crown and nape rufous-brown, the facial pattern well defined with a blacker mask and more obvious white supercilium, darker wings and upperparts, and whiter underparts. ♀ browner and has a smaller mask, more indistinct supercilium, less distinct primary patch and usually some indistinct barring on breast and flanks. First-winter resembles ♀ but is more extensively barred on underparts. **VOICE** Rather silent away from breeding grounds but call similar to Red-backed Shrike. **RANGE** Breeds in Iran and Kazakhstan east to N China and Mongolia. Winters in NW India and Pakistan through S Arabia to E Africa, thence across the Sahel to Senegambia. **STATUS** Extreme vagrant. **Canaries★** One, near El Matorral (Fuerteventura), 12 Jul 1989.

RED-BACKED SHRIKE *Lanius collurio* Plate 63 (4)
Sp. Alcaudón Dorsirrojo **P.** Picanço-de-dorso-ruivo

L. c. collurio. **L** 17cm (7"). **IDENTIFICATION** Sexually dimorphic. ♂ unmistakable with grey crown and nape, reddish-brown mantle and wing-coverts, grey rump, broad black mask, pinkish tone to white underparts and black tail with white sides at base. ♀ has coarsely vermiculated off-white underparts, dark brown mantle, grey-brown crown and small brown mask. Immature similar to ♀ but usually has more heavily vermiculated upperparts. **VOICE** Most common call a harsh *tshak* or *shak-shak*, but song unlikely to be heard in our region. **RANGE** Breeds over most of Europe east to W Siberia and winters in E and S Africa. **STATUS** Vagrant with records from two archipelagos. **Madeira** Sole record is from the Ilhas Selvagens. **Canaries★** Seven records: one obtained by Thanner, at Vilaflor (Tenerife), 16 Oct 1907; one, Lanzarote, 4 Apr 1967; then four records in 1978, all by same observer; one, Boca de Tauce (Tenerife), 29 Mar; two, between El Poris and El Médano (Tenerife), 2 Apr; juvenile, Tecina (La Gomera), 7 Oct; juvenile, Playa Vallehermoso (La Gomera), 17 Oct; and one, Alegranza, Apr 1995.

SOUTHERN GREY SHRIKE   *Lanius meridionalis*   Plate 63 (5)
Sp. Alcaudón Real   **P.** Picanço-real (Picanço-real-comum)

L. m. koenigi and *L. m. elegans.*   **L** 24–25 cm (9½–10″).   **IDENTIFICATION** Sexes similar. Unlikely to be confused with any other resident species in the region. Combination of grey, black and white plumage, plus habit of perching conspicuously atop rocks and vegetation, unique to this species. Resident race *koenigi* has medium to dark grey upperparts, narrow white supercilium (often lacking), and pale greyish breast and flanks that can contrast quite distinctly with white chin and throat. Unlike other races of Southern Grey Shrike, immature of *koenigi* shows quite distinctly vermiculated underparts. Race *elegans* is larger than *koenigi*, with paler grey upperparts, whiter underparts, more extensive white in wing and tail, and more extensive white on scapulars. **VOICE** Song is a mixture of squeaks, whistles, trills, clicks and chattering notes. Calls include a harsh *schreee* and rather liquid *cli-cli*. **RANGE** Breeds in N Africa, Iberia, the Middle East, C Asia and N Indian subcontinent. Mostly resident but eastern populations migratory. **DISTRIBUTION** Confined to the Canaries where it breeds on Tenerife, Gran Canaria, Fuerteventura, Lanzarote and the islet of Graciosa. **BREEDING** Season starts Jan on the eastern islands and continues until May in the highest zones of Tenerife. Clutch size 3–6 but 4–5 are the most frequently recorded. Nest is placed in various types of trees and shrubs at heights lower than 3 m. Occasionally noted nesting in stone walls and even on ground. **HABITAT** Dry areas such as *Euphorbia* scrub, abandoned agricultural land and high-altitude vegetation of Las Cañadas National Park on Tenerife. Avoids forested areas and cultivated land. **STATUS** Recorded from two archipelagos but status varies greatly between them. **Canaries** Local and uncommon on Tenerife and Gran Canaria, but noticeably more numerous on the eastern islands of Fuerteventura and Lanzarote. **Cape Verdes** One, Santa Maria (Sal), 24 Jan 1997, which was accepted as being of race *elegans*.

WOODCHAT SHRIKE   *Lanius senator*   Plate 63 (6)
Sp. Alcaudón Común   **P.** Picanço-barreteiro

L. s. badius and *L. s. senator.*   **L** 18 cm (7″).   **IDENTIFICATION** Sexes similar. ♂ has red-brown crown and nape, broad black mask, black forehead and back, and predominantly black wings and tail, large white patch on scapulars and generally white underparts. ♀ is similar but duller, occasionally with some vermiculations on underparts. Immature resembles immature Red-backed Shrike but has generally greyer upperparts with a pale rump and a suggestion of the pale scapular patch of adults. Race *badius*, which breeds on islands in W Mediterranean, distinguished from nominate by lack of white patch at base of primaries. **VOICE** Call is a series of short, harsh notes, *gek gek gek*, or a hoarse trill *trrrr*. **RANGE** Breeds around Mediterranean and east to Iran. Winters across tropical Africa between S Sahara and the equator. **STATUS** Recorded from three archipelagos but status differs between them. **Madeira** Very rare vagrant with fewer than five records in the last 50 years. The first record, documented by Sarmento, was assigned to *badius*. **Canaries** Fairly regular spring passage migrant and sometimes very common, but scarce in autumn. Both subspecies have been recorded. **Cape Verdes★** One, Pedra de Lume (Sal), 9 Mar 2004.

(SPOTTED) NUTCRACKER   *Nucifraga caryocatactes*   Plate 64 (1)
Sp. Cascanueces   **P.** Quebra-nozes

Subspecies unknown.   **L** 32 cm (12½″).   **IDENTIFICATION** Sexes similar. Unmistakable: adult has predominantly dark chocolate-brown body, black wings, a white-tipped black tail and white vent. Entire body, except crown, covered in white spots, particularly dense on head-sides. Juvenile browner than adult with less white spotting. Like adult by first autumn but separated by white fringes to inner primaries and white spots on median coverts. **VOICE** Common call is a drawn-out, harsh, rolling *kraaaaak*, often repeated in rapid succession. **RANGE** Breeds from S Scandinavia, C and SE Europe east through the taiga zone to E Siberia, Kamchatka, China, Japan and Taiwan. Irruptive, with vagrants appearing far from breeding range. **STATUS** Extreme vagrant, although the one record was perhaps not of wild origin. **Canaries★** One, near the Faculty of Biology at the University of La Laguna (Tenerife), 30 Aug 1989.

RED-BILLED CHOUGH   *Pyrrhocorax pyrrhocorax*   Plate 64 (2)
Sp. Chova Piquirroja   **P.** Gralha-de-bico-vermelho

P. p. barbarus.   **L** 39–40 cm (15–16″).   **IDENTIFICATION** Sexes similar. Adult completely glossy black except long, downcurved red bill and pinkish-red legs and feet. Wings uniformly broad throughout their length, blunt-tipped and deeply fingered. From below, underwing-coverts visibly darker than rest of underwing. Immature similar to adult but plumage less glossy and bill lacks bright red colour, having a yellowish hue. The race in the Canaries and NW Africa, *barbarus*, differs from the nominate in being larger with a thicker, longer bill and a green gloss to wing-coverts, mantle and scapulars. **VOICE** Common call is a harsh, rather high-pitched *chee-aw* given in flight and on ground. Can be repeated with varying tones and pitches, thus producing a series of closely related calls. **RANGE**

Breeds from W Europe and NW Africa discontinuously east to C Asia and China, with an isolated population in Ethiopia. Mostly resident but some local movements occur. **DISTRIBUTION** Restricted to La Palma in the Canaries where fairly widespread and relatively common. **BREEDING** Normally Mar–Apr. Clutch 4–5. Sometimes breeds in isolated pairs but more often in small colonies, and nests on sea cliffs or inland in dry ravines. **HABITAT** Can be encountered almost anywhere on the island, from sea cliffs to mountain tops, as well as in pine forests and agricultural areas. **STATUS** Only recorded from one archipelago. **Canaries** Resident breeder on La Palma and has also been recorded on Tenerife and La Gomera, where probably only a vagrant.

WESTERN JACKDAW   *Corvus monedula*   Plate 64 (3)
Sp. Grajilla   **P.** Gralha-de-nuca-cinzenta

C. m. soemmerringii. **L** 33 cm (13"). **IDENTIFICATION** Sexes similar. Slightly smaller than Red-billed Chough. Adult is dark grey, not black, with paler nape and neck-sides. Forehead and crown noticeably darker than nape and eye greyish-white. Bill short and relatively narrow compared to other corvids. In flight shows uniform grey underwing. Juvenile has dark iris and nape and neck-sides less pale. E European *soemmerringii* has neck-sides paler with whitish patches at base. **VOICE** Commonest call is a short, hard *chack*. **RANGE** Breeds from W Europe and NW Africa east to Lake Baikal and nw Mongolia. Sedentary over most of its European range but large eruptions occur. **STATUS** Vagrant with records from three archipelagos. **Azores** ♀ collected, São Miguel, 13 Nov 1947. **Madeira** One obtained, Funchal, 9 May (prior to 1948), listed by Sarmento in *Vertebrados da Madeira*. The specimen is in the municipal museum. **Canaries** Three records: one obtained, Laguna (Tenerife), Feb 1830; one, Playa de Las Arenas (Tenerife), 10–12 Jan 1998; and one showing characters of *soemmerringii*, Alcalá (Tenerife), 7 Nov 2001 until early 2002 at least.

ROOK   *Corvus frugilegus*   Plate 64 (4)
Sp. Graja   **P.** Gralha-calva

C. f. frugilegus. **L** 46–47 cm (18–18½"). **IDENTIFICATION** Sexes similar. Adult is all black and has distinct metallic gloss to plumage varying from blue to green and purple depending on light. Also obvious whitish bill base and face are diagnostic. Immature, before developing the bare bill base, very similar to Carrion Crow but distinguished by bill and head shape. Rook has much straighter culmen and more pointed tip to bill, and head is less rounded than Carrion Crow. In flight, the almost wedge-shaped tail tip and narrow base to wings are useful identification pointers. **VOICE** Calls include a harsh, nasal *kaah* and high-pitched *kraa-ah* lacking rolling quality of Carrion Crow. **RANGE** Nominate race breeds from W Europe east to the Altai region of C Asia. Some are resident but others winter a little to south of the main breeding range. Another race breeds in E Asia. **STATUS** Vagrant with records from three archipelagos. **Azores** Eight records, of which seven are from São Miguel and one on Santa Maria. Most recent records are three, Arrifes (São Miguel), 24–30 Jan 1984; possibly one of those birds at Lagoa de Congro (São Miguel), 11 Mar 1984; and a specimen, São Miguel, 1998. **Madeira** One recorded by Padre Schmitz published in his list of birds seen since 1896. **Canaries*** All records relate to reports by Trotter on Lanzarote: one, southeast of the island, 27 Nov 1966, one, 6 Dec 1966, and one, Arrecife, 5 Dec 1967.

CARRION CROW   *Corvus corone*   Plate 64 (5)
Sp. Corneja Negra   **P.** Gralha-preta

C. c. corone. **L** 47 cm (18½"). **IDENTIFICATION** Sexes similar. Another all-black species but metallic gloss is less prominent than in some corvids. The most likely confusion species is Common Raven but latter is much larger and has an obvious wedge-shaped tail. Confusion also possible with immature Rook (which see). Juvenile has dull black plumage, not glossy like adult. **VOICE** Common call is a rather harsh, rolling croak *krraa-krraa-kraaa*. **RANGE** Breeds in W Europe and E Asia. Populations mostly sedentary but some dispersal in winter. **STATUS** Vagrant with records from two or three of the archipelagos. **Azores** Three records: one obtained, Arrifes (São Miguel), prior to 1903 (specimen now lost); one, Angra (Terceira), 1927; and flock of 13, Lajes (Terceira), 25 Oct 2003. **Madeira** Retained in the 1995 checklist, but now considered dubious. The record relates to birds seen by Forster, in 1772, which according to Bannerman (1965) were actually in an aviary. **Canaries*** One, Guinate Forest Park (Lanzarote), 28 Jul 1999.

BROWN-NECKED RAVEN   *Corvus ruficollis*   Plate 64 (7)
Sp. Cuervo Desertícola   **P.** Corvo-do-deserto

C. r. ruficollis. **L** 50–52 cm (19¾–20½"). **IDENTIFICATION** Sexes similar. Much like Common Raven; the two species are difficult to separate. Brown-necked Raven is smaller than Common Raven, with the tail tip only extending to the end of the primaries at rest. Bill comparatively thin, the throat hackles less obvious and a bronzy tone to the nape visible at close **RANGE**. In flight, generally slimmer with a narrower outer wing than Common Raven. Perhaps the best feature is the calls, which are distinctly more like Carrion Crow than Common Raven, although it should be noted that the subspecies of Common Raven in our region also has some calls similar to Carrion Crow. Juvenile resembles adult but when both are in fresh plumage juvenile lacks glossy appearance. **VOICE** Calls include a hoarse, high-pitched, crow-like *karr-karr-karr*. **RANGE** Breeds across desert plains of N Africa, the Middle East and C Asia. **DISTRIBUTION** Found only on Cape Verde where it breeds on all main islands and larger islets. **BREEDING** Season Nov–Apr with a peak in late Jan–early Mar. Nests on cliffs, rocks, buildings and trees, although electricity pylons and other industrial installations are also used. Clutch normally 4–5 but six eggs also recorded. **HABITAT** Present around towns, in cultivated areas, deserts, mountains or shores; rather catholic. **STATUS** Resident breeder in Cape Verde but a vagrant elsewhere. **Madeira** Type specimen of *Corvus leptonyx* Peale 1838 (= *C. ruficollis*) was said to have been collected "within a short distance of Funchal", whilst a description given to Schmitz (1899) must be regarded as rather dubious. It is almost certain that the specimen was actually collected in Cape Verdes. **Canaries★** Three records but it has been suggested that these may relate to Common Raven in atypical plumage. A breeding pair, near the Presa de Ayagaures (Gran Canaria), 21 Apr 1976; two, Antigua (Fuerteventura), 27 Oct–10 Nov 1988; and one, Pajara (Fuerteventura), 22 Mar 1999. **Cape Verdes** Common resident on all islands, except Sal where it is scarce.

COMMON RAVEN *Corvus corax* Plate 64 (6)
Sp. Cuervo **P.** Corvo

C. c. tingitanus. **L** 64 cm (25″). **IDENTIFICATION** The largest of all northern hemisphere corvids. Plumage is all glossy black but bleaches to dark brown on head and body. Bill heavy, tail long and wedge-shaped and feathers on throat elongated and pointed, affording a bearded appearance. Race *tingitanus*, which occurs in N Africa and on the Canaries, is the smallest race with the shortest throat hackles. Thus, in bleached plumage it very closely resembles Brown-necked Raven and great care must be taken with identification. **VOICE** Common call a rather hollow, croaking honk *pruk-pruk-pruk*, but has a complex and varied vocabulary Subspecies *tingitanus* has calls not normally heard from the nominate, such as a duck-like quacking and disyllabic *teer-do* dropping in pitch on the second syllable. Generally, higher pitched and not as rough as those of nominate. **RANGE** Breeds across much of Holarctic. The most widespread corvid. **DISTRIBUTION** Breeds on all main islands in the Canaries and islets north of Lanzarote. **BREEDING** Usually Mar–Apr but can start Feb and rarely even May–Jun. Normally nests on cliffs, crags and sides of ravines, making use of holes and crevices. Has been noted nesting in *Pinus canariensis* but this is exceptional, and on Fuerteventura even uses electric pylons. Clutch 3–7. **HABITAT** Found from sea level to high montane areas but most often encountered around cliffs and mountainous areas. **STATUS** Recorded from two archipelagos, breeding on one and is a vagrant to the other. **Madeira** Fewer than five records in the last 50 years. **Canaries** On Tenerife is in rapid decline but on other islands the population seems relatively stable, being particularly common on Fuerteventura.

COMMON STARLING *Sturnus vulgaris* Plate 65 (1)
Sp. Estornino Pinto **P.** Estorninho-malhado

S. v. granti (Azores) and *S. v. vulgaris* (Canaries). **L** 21 cm (8″). **IDENTIFICATION** Sexes similar. A familiar species, the adult in breeding plumage is basically black with a pointed yellow bill and pinkish legs. An obvious green and purple sheen to the plumage with some yellowish spotting, particularly on the upperparts and vent. In non-breeding plumage the bill is dark and the plumage is covered in white and buff spots. Juvenile rather dirty brown with a paler throat, indistinct streaking on underparts and dark bill. **VOICE** Song is a complex mixture of whistles, rattles and squawks, with a varied repertoire of mimicry and even man-made noises interwoven. Calls include a harsh *tcheeerr* and sharp *kyett*. **RANGE** Breeds from W Europe (except most of Iberia) east to Lake Baikal and south to Sind, Pakistan. Northern populations move south in winter and occupy areas south and west of the breeding **RANGE**, from N Africa to N India. **DISTRIBUTION** Resident breeder in the Azores, and on Tenerife and Gran Canaria in the Canaries. **BREEDING** Few data from the Canaries but egg laying starts late Mar and the clutch size is usually 4–5. In the Azores clutch 2–7 and nests late Apr–Jul, with most pairs having two broods. **HABITAT** Towns, parks and gardens, golf courses, agricultural areas and coastal cliffs. **TAXONOMIC NOTE** *granti* on the Azores is considered by some authors insufficiently distinct to warrant naming; they include it within the nominate subspecies *vulgaris*. **STATUS** Recorded from all four archipelagos but status varies considerably. **Azores** Common resident on all islands. **Madeira** Occasional visitor in winter and also recorded from the Salvages. **Canaries** Scarce and local breeder on Tenerife and Gran Canaria. A few pairs breed on Tenerife, around La Laguna, and on Gran Canaria at Maspalomas. In winter more common and widespread, being recorded from all main islands except La Palma. **Cape Verdes** One collected, São Nicolau, 30 Oct 1970, and one, Mindelo sewage farm (São Vicente), 8–9 Mar 1997.

SPOTLESS STARLING *Sturnus unicolor* Plate 65 (2)
Sp. Estornino Negro   **P.** Estorninho-preto

Monotypic. **L** 21 cm (8"). **IDENTIFICATION** Sexes similar. In all plumages very similar to Common Starling. Adult in breeding plumage is blacker than Common Starling and lacks pale spotting. The blacker appearance is caused by the less iridescent feathering and the sheen is more uniformly purple. In winter plumage also blacker than Common Starling with spotting greatly reduced, often almost absent, on upperparts and crown, and much less distinct on underparts. Also wings lack buff fringes to feathers of Common Starling. Immature is slightly darker than Common Starling with, usually, no streaking on underparts. **VOICE** Similar to Common Starling but whistles in the song clearer and louder. Like Common Starling sings year-round. **RANGE** Resident in Iberia, NW North Africa, Corsica, Sardinia and Sicily. **STATUS** Vagrant with records from two archipelagos. **Madeira** One record, 26, Salvages, winter 1980/81. **Canaries*** Three records: one, Laguna (Tenerife), prior to 1893; one, San Isidro (Tenerife), 26 Nov 1995; and one, La Lajita (Fuerteventura), 29 Mar 1998.

ROSE-COLOURED STARLING (ROSY STARLING) *Sturnus roseus* Plate 65 (3)
Sp. Estornino Rosado   **P.** Estorninho-rosado

Monotypic. **L** 21 cm (8¼"). **IDENTIFICATION** Sexes similar. Adult unmistakable with pink body and black head, wings and tail. The head has a purple gloss and the wings a greenish gloss, and a long crest on nape. ♀ duller than ♂ with a shorter crest and brown-toned upperparts. First winter has a short crest, paler greyish-pink body and lacks gloss on head and wings. Juvenile resembles juvenile Common Starling but is a paler milky-tea colour, with a whiter rump and shorter, yellowish bill. **VOICE** Calls include a Common Starling-like *tcheerrr* and harsh, rattling *chik-ik-ik-ik…* **RANGE** Breeds from Turkey east through C Asia to W China. Winters entirely in the Indian subcontinent. **STATUS** Vagrant with records from two archipelagos. **Madeira** Immature ♂ obtained, Horgulho (Madeira), 8 Dec 1906. **Canaries** Adult, Puerto de la Cruz (Tenerife), 9 Jan–18 Feb 1981.

COMMON MYNA *Acridotheres tristis* Plate 65 (4)
Sp. Miná Común   **P.** Mainato

Subspecies unknown. **L** 24 cm (9½"). **IDENTIFICATION** Sexes similar. Adult predominantly brown with a glossy black head, black wings with a white patch in primaries, white tip to black tail, white undertail and central belly, and patch of bare yellow skin behind and below eye. Juvenile similar but head is browner and generally duller overall. **VOICE** Noisy and varied, with loud whistles and harsh calls mixed with other noises and some mimicry. **RANGE** In native range it breeds in India and in C and S Asia. **DISTRIBUTION** Breeds only on Tenerife. **BREEDING** Only one nest has been found in our region. **HABITAT** Confined to parks and gardens. **STATUS** Introduced species present on one archipelago. **Canaries** Feral population on Tenerife numbers no more than a couple of pairs.

HOUSE SPARROW *Passer domesticus* Plate 65 (5)
Sp. Gorrión Doméstico (Gorrión Común)   **P.** Pardal (Pardal-comum)

P. d. domesticus. **L** 14–15 cm (5½–6"). **IDENTIFICATION** Sexually dimorphic. Breeding ♂ has underparts dull greyish-white with a black bib extending onto upper breast. Crown and forehead grey with a chestnut border and nape also chestnut, the head-sides dull greyish, the rump grey and the upperparts brown with dark streaking. Bill black. Non-breeding ♂ has black bib restricted to chin and upper throat, a yellowish bill and less pronounced head pattern. ♀ has dull greyish-brown underparts, brown upperparts streaked black and a dirty buffish supercilium. Juvenile resembles ♀ but usually has broader buff-brown fringes to upperparts and generally appears rather scruffy. **VOICE** A wide variety of chirps and chattering notes including a monotonous *chirrup* and soft *swee-swee*. When excited or alarmed gives a rattling *chur-r-r-it-it-it*. Song is a repeated series of call notes mixed with other similar notes. **RANGE** Original range was from W Europe and NW Africa east to Arabia, the Indian subcontinent and Yakutsk and Khabarovsk in Siberia. Now a common and widespread introduction throughout the Americas, Australia, New Zealand, S Africa and many other areas. **DISTRIBUTION** Azores, Canaries and Cape Verdes. **BREEDING** Nests in palms and holes in buildings. Clutch usually 4–5. Season is poorly studied, although on Gran Canaria nests in March. **HABITAT** Common around habitation and in agricultural areas, but avoids desert and heavily vegetated zones. **STATUS** Introduced on the Azores. Those in Cape Verdes are almost certainly ship-assisted colonists, as are those on the Canaries. **Azores** Breeds on all islands in the archipelago with Corvo and Santa Maria both being colonised since 1986. Originally introduced to Terceira in 1960 or 1961. **Madeira** Listed by Zino *et al* (1995) as being of exceptional occurrence. **Canaries** A small colony in the docks at Las Palmas (Gran Canaria) was first noted Mar 1998 but had been present some time, as hybrids with Spanish Sparrow were seen. This colony still exists but there is no sign of expansion. **Cape Verdes** Found only on São Vicente, where it presumably arrived by ship. Common around Mindelo, Ribeira da Vinha and oases in the interior. Hybrids with Spanish Sparrow occur.

SPANISH SPARROW *Passer hispaniolensis* Plate 65 (6)
Sp. Gorrión Moruno **P.** Pardal-espanhol

P. h. hispaniolensis (Madeira, Canaries and Cape Verdes). **L** 15 cm (6"). **IDENTIFICATION** Sexually dimorphic. Breeding ♂ has chestnut crown and white stripe through upper part of eye separated from white cheeks by a chestnut line through the lower part. Chin, throat and breast black, rest of underparts white with heavily and boldly streaked flanks and lower breast. Mantle, back and scapulars also heavily and boldly streaked black and there is a white wingbar on median coverts. Non-breeding ♂ has intensity and density of black streaking much reduced and a duller crown and nape. ♀ very similar to ♀ House Sparrow and is almost impossible to separate in the field, although ♀ Spanish is somewhat paler below and has more obvious streaking on underparts. Juvenile very similar to ♀ and very similar to corresponding plumage of House Sparrow. **VOICE** Song similar to House Sparrow but slightly higher pitched and more metallic. Calls also similar with *chirp* flight call somewhat harsher. **RANGE** Breeds in Spain and NW Africa east to C Asia and NW China. Resident in many areas but also winters south of breeding range. **DISTRIBUTION** Breeds in Madeira, the Canaries and in Cape Verde. **BREEDING** A colonial nester. On the Canaries season protracted, starting as early as Jan and continuing until Aug, with pairs having up to three broods. Clutch 3–7 but 4–5 eggs are the most frequent. Nests are in holes in walls of buildings, under bridges and in trees such as palms, laurels and *Eucalyptus*. **HABITAT** Occurs in a wide range of habitats but usually absent from more heavily vegetated areas. Common around towns and villages, as well as in abandoned agricultural areas. **STATUS** Recorded from three archipelagos, where it is a resident breeder. **Madeira** A scarce breeder on Madeira but abundant on Porto Santo. **Canaries** Common breeder on all islands. **Cape Verdes** Recorded from all islands but regular only on Santiago, Fogo, São Vicente, Maio and Boavista, and is rare and local on Sal and São Nicolau.

IAGO SPARROW (CAPE VERDE SPARROW) *Passer iagoensis* Plate 65 (7)
Sp. Gorrión de Cabo Verde **P.** Pardal de Cabo Verde

Monotypic. **L** 13 cm (5"). **IDENTIFICATION** Sexually dimorphic. Adult ♂ has blackish forehead and crown, becoming grey on nape and upper mantle, a whitish face, black eye-stripe, narrow white stripe above lores, and a rich chestnut line from eye around rear of ear-coverts. Underparts whitish except for neat black bib. Upperparts brown streaked black with chestnut scapulars and broad white tips to median coverts. Adult ♀ similar to ♀ House Sparrow but is smaller. **VOICE** Calls include a *chirp* similar to House Sparrow but lower pitched, and a slightly nasal *cheesp*. Song is a varied series of combined call notes. **RANGE** Confined to Cape Verde. **DISTRIBUTION** Breeds on all main islands except Fogo. **BREEDING** Aug–Apr with a peak in Sep–Nov and is related to the onset of seasonal rains. Nests in holes in cliffs and rocks, under stones, in trees and in artificial holes in buildings and in streetlights. Clutch 3–5. Sometimes forms small colonies. **HABITAT** Occurs on lava plains, in gorges, on cliffs, at edges of cultivated areas and around man-made constructions. **STATUS** Endemic to one archipelago. **Cape Verdes** Common resident on all main islands except Sal and Santa Luzia where it is scarce.

(EURASIAN) TREE SPARROW *Passer montanus* Plate 65 (8)
Sp. Gorrión Molinero **P.** Pardal-montês (Pardal-montez)

P. m. montanus. **L** 14 cm (5½"). **IDENTIFICATION** Sexes similar. Adults have a chestnut cap, white head-sides, black spot on ear-coverts, near-complete whitish collar, a small neat black bib, brown upperparts streaked black, double white wingbar, and dull greyish or dirty white underparts. Bill black in breeding season but with a paler yellowish base in non-breeding season. Juvenile similar but duller with greyer head-sides. **VOICE** Calls and song similar to House Sparrow but higher pitched. In flight gives a distinctive hard *tet*. **RANGE** Breeds from W Europe east to Sakhalin, Japan, E China and south to Turkey, NW India and Indonesia. Mostly sedentary but some local movements. **DISTRIBUTION** In our region confined to Gran Canaria. **BREEDING** Little known on Gran Canaria but season starts Feb and continues until Jun, during which period perhaps more than one brood is raised. The only nests found on Gran Canaria have been in tiled roofs, buildings, under bridges and on electric pylons. **HABITAT** Parks and gardens, and agricultural areas such as orchards and tomato plantations. **STATUS** Recorded from only one archipelago. **Canaries** Small breeding populations at Maspalomas, Castillo del Romeral, Sardina, Lomo de Las Crucitas, El Tablero, El Troncón, Albercón de la Virgen, Puerto de Mogán, El Cercado and Puerto de las Nieves, on Gran Canaria, with one record of three birds, Santa Cruz (Tenerife), 11 Jul 1992.

ROCK SPARROW *Petronia petronia* Plate 65 (9)
Sp. Gorrión Chillón **P.** Pardal-francês

P. p. madeirensis. **L** 14 cm (5½"). **IDENTIFICATION** Sexes similar. Adult is rather like a plump, short-tailed ♀ House Sparrow with a distinctive head pattern. Crown dark blackish-brown with a pale, broad central stripe and bordered by a prominent creamy supercilium and dark eye-stripe, producing a very stripy head pattern. Rather grey-brown upperparts streaked darker with buff braces on sides of mantle. Underparts are whitish streaked brown with a yellow spot on upper breast only present in fresh plumage, and even then can be exceedingly difficult to see. Tail greybrown with pale subterminal spots on all rectrices, especially visible in flight. Juvenile similar but has less streaking on underparts with a greyish wash to breast, more prominent double wingbar and browner upperparts. Subspecies *madeirensis* has upperparts darker than any other subspecies. **VOICE** Most characteristic call is a wheezy, nasal *peeeooeee*, at times with 2–3 syllables and at others very monosyllabic. Also gives calls similar to those of *Passer* sparrows. Song is a series of call notes run together. **RANGE** Breeds discontinuously from Madeira and the Canaries east through S Europe and NW Africa to W China. Sedentary in our region. **DISTRIBUTION** Only breeds in Madeira and the Canaries. **BREEDING** Courtship observed from late Jan through Feb, with egg laying Mar–Jun. Clutch 4–6. Nest is normally located in holes or cracks in houses, stone walls and sides of ravines. **HABITAT** Still found around habitation, although more common nowadays in more remote areas, particularly areas where Spanish Sparrow has not yet colonised. Also in areas of fields and rocky hillsides, and in more mountainous regions. Often forms flocks in non-breeding season, sometimes numbering hundreds. **TAXONOMIC NOTE** Birds in our region are afforded subspecific status as *madeirensis*, but some authors consider the differences too slight and group them with nominate *petronia* (Cramp and Perrins 1994a). **STATUS** Breeds on two archipelagos and is a vagrant to a third. **Azores** Adult ♀ collected, Lagoa (São Miguel), prior to 1903. **Madeira** Common breeder on Madeira, Porto Santo and the Desertas. **Canaries** Generally scarce and local on the central and western islands having has declined since colonisation by Spanish Sparrow. Vagrant on Lanzarote and Fuerteventura.

(WHITE-WINGED) SNOWFINCH *Montifringilla nivalis* Plate 67 (8)
Sp. Gorrión Alpino **P.** Pardal-alpino

M. n. nivalis. **L** 17 cm (7"). **IDENTIFICATION** Sexes similar. Breeding ♂ has grey head with darker lores, black bib, brown mantle and scapulars, and extensive white in wings and tail. Breeding ♀ resembles ♂ but bib less well defined and bill yellowish-brown with a dark tip. Non-breeding ♂ and ♀ are alike; the throat is mostly whitish, the head paler and the bill yellow with a black tip. Juvenile resembles ♀ but head is browner and, when very young, has white in tail more buff. **VOICE** Call is a sharp, nasal *pschieu*, also a *tsee* and softer *pruuk*, or when alarmed a purring *pchrrrt*. **RANGE** Breeds discontinuously in mountains from N Spain to C Asia and W China. Mostly sedentary but some altitudinal movements in winter. **STATUS** Extreme vagrant. **Canaries*** One obtained, La Orotava (Tenerife), by a Mr A. Diston, prior to 1842.

VILLAGE WEAVER *Ploceus cucullatus* not illustrated
Sp. Tejedor Cogullado **P.** Cacho-caldeirão

P. c. cucullatus. **L** 17 cm (7"). **IDENTIFICATION** Sexually dimorphic. ♂ in breeding plumage is a large sparrowlike bird with a yellow nape and underparts, black hood that comes to a point on breast, red eye, stout black bill and yellow fringes to black wing feathers. Breeding ♀ has yellow underparts, greenish-olive upperparts streaked darker and yellowish supercilium. Non-breeding ♀ and ♂ resemble breeding ♀ but have less yellow on underparts. Juvenile similar to non-breeding adult but browner above and has brown eye. **VOICE** A range of chattering noises. **RANGE** Resident over most of sub-Saharan Africa except southwest. **STATUS** Recorded on two archipelagos and may have bred. **Canaries** Individuals on Tenerife and Gran Canaria presumably relate to escapes. **Cape Verdes** 5♂♂, 2♀♀ collected, Praia (Santiago), 1 May 1924; *c.* 10 displaying and nest building, Mindelo (São Vicente), Jun–Jul 1993, with further records from the same locality in 1995–99. Thus, a small population may have become established at Mindelo and, as this is a major port, ship-assisted origin must be a real possibility.

RED-BILLED QUELEA *Quelea quelea* Plate 66 (1)
Sp. Tejedor de Pico Rojo (Quelea Común) **P.** Bico-carmim

Subspecies unknown. **L** 13 cm (5"). **IDENTIFICATION** Sexually dimorphic. Breeding ♂ has bright red bill with a black face and throat; tawny crown, nape and breast, with buff and black-streaked upperparts. ♀ lacks black face and throat, and has a browner crown and nape, pale supercilium and less tawny underparts. Non-breeding ♂ resembles breeding ♂ but has face patch streaked buff, and crown and nape streaked dark. Juvenile resembles ♀ but bill is brownish-pink. **VOICE** Call is a harsh *chack-chack* and flocks produce a noisy chattering. **RANGE** Breeds from Senegal to Ethiopia and south to S Africa. Highly migratory depending on weather conditions and food supply. **STATUS** Extreme vagrant with only two records that might relate to wild birds. **Canaries** A group of a few hundred, near some tomato plantations at Maspalomas (Gran Canaria), 23 Nov 1965 with a few dozen still present on 29 Nov. Other records in the archipelago all relate to individuals that probably escaped from captivity. **Cape Verdes** Pair, Mindelo sewage farm (São Vicente), 2 Mar 2004. Probably escapes from captivity but the proximity to a large port would imply that ship-assisted passage is a distinct possibility.

RED-CHEEKED CORDON-BLEU *Uraeginthus bengalus* Plate 66 (2)
Sp. Coliazul Bengali **P.** Peito-celeste

U. b. bengalus. **L** 13 cm (5″). **IDENTIFICATION** Sexes differ. ♂ has brown upperparts except blue rump and tail, predominantly blue underparts except buffy belly and vent, a diagnostic kidney-shaped red patch on ear-coverts and a pinkish bill. ♀ is similar but lacks red on ear-coverts and has less extensive blue on underparts. Juvenile resembles ♀ but has even less extensive and paler blue on underparts, and a blackish bill. **VOICE** Call is a weak, high-pitched *tsee-tsee* and the song a high-pitched *seeseedelee-seedelee-see-see* or variations. **RANGE** Resident across most of C and E Africa. **STATUS** Perhaps introduced in the past but currently only escapes from captivity are recorded. **Canaries** Escapes seen on Tenerife and Fuerteventura. Recorded breeding in the wild on the latter, but does not appear to have become established. **Cape Verdes** ♂ and ♀ collected, Mindelo (São Vicente), 20 Jan & 13 Oct 1924. It is unknown whether these were the last of an introduction or merely escapes, but it has not been recorded since.

ORANGE-CHEEKED WAXBILL *Estrilda melpoda* not illustrated
Sp. Estrilda de Carita Naranja **P.** Face-laranja

Subspecies possibly *E. m. melpoda*. **L** 10 cm (4″). **IDENTIFICATION** Sexes similar. Adult has brown upperparts with a red rump and black tail, grey head and underparts with a bright orange face and red bill. Juvenile similar but has a black bill, paler orange face, duller rump and more buff on underparts. **VOICE** Call is a high-pitched *sree-sree-sree* and the song consists of short notes, a repeated *tsee-ree-ree*. **RANGE** Resident from Senegal to N Congo and south to N Angola and N Zambia. **STATUS** Sightings in the 1980s suggest that it may have bred in one of the archipelagos. **Canaries** Records from Tenerife include a flock of *c*. 25, Puerto de la Cruz, 1980, comprising both adults and first-years, suggesting local breeding. Birds were also seen there in 1981 and 1985 but not since.

COMMON WAXBILL *Estrilda astrild* Plate 66 (3)
Sp. Pico de Coral **P.** Bico-de-lacre

E. a. jagoensis. **L** 11 cm (4″). **IDENTIFICATION** Sexes similar. The adult is a very small finch, with grey-brown upperparts and more buff-brown underparts, finely vermiculated dark throughout, a pinkish flush to belly, dark red stripe through eye and bright red bill. ♀ sometimes distinguished by its paler coloration but individual variation is considerable. Juvenile resembles adult but is paler with a orange stripe through eye and blackish bill. **VOICE** Commonest call is a sharp *pit* or a twittering given constantly by small flocks. **RANGE** Resident throughout most of sub-Saharan Africa. **DISTRIBUTION** Breeds in the Canaries and Cape Verdes. **BREEDING** In Cape Verde the main breeding season is Aug–Nov but there is often a second brood following the rains. Clutch 4–6 and nests in low bushes. In the Canaries recorded nesting Jun, Aug and Dec. Clutch size identical but nests have been discovered up to 5 m above ground in Canary Island Palm *Phoenix canariensis*. **HABITAT** In Cape Verde occurs in cultivated and irrigated valleys and fields. In the Canaries also found in man-made habitats such as parks, gardens and golf courses. **STATUS** Recorded from all four archipelagos although all records refer to introductions or escapes. **Azores** Recent records from Ponta Delgada (São Miguel) and Cabo da Praia (Terceira) suggest that the species is becoming established. **Madeira** Some have recently become established on Madeira where a self-sustaining population now exists. **Canaries** Introduced on Gran Canaria where very common at various sites. However, on Tenerife it is very scarce and is certainly not well established. **Cape Verdes** Previously reported on Fogo, Brava, Santo Antão and São Nicolau. Currently extant only on Santiago, where it is widespread and locally abundant, and on São Vicente where occurs in the vicinity of the sewage ponds.

COMMON CHAFFINCH *Fringilla coelebs* Plate 67 (1)
Sp. Pinzón Vulgar **P.** Tentilhão (Tentilhão-comum)

F. c. canariensis, *F. c. ombriosa* and *F. c. palmae* (Canaries), *F. c. maderensis* (Madeira), *F. c. moreletti* (Azores) and *F. c. coelebs*. **L** 14½ cm (6″). **IDENTIFICATION** Sexually dimorphic. ♂♂ of the insular forms are very different compared to ♂ of the nominate subspecies. ♂ *coelebs* has vinaceous-red underparts and face, a white vent, the crown and nape slaty blue-grey, mantle and back chestnut-brown, green lower back and rump, double white wingbars and a dark tail with white outer tail feathers. ♀ *coelebs* is dull grey-brown above and greyish-white below with narrower wingbars than ♂. ♂ *canariensis* has upperparts deep slate-blue with a black forehead, dull yellowish-green rump, black wings with prominent white bars on median coverts and tips of greater coverts, the underparts and face are peachy-buff with a whitish centre to belly and undertail-coverts. ♀ has dull grey-brown upperparts with an olive tone to rump and bluish-grey uppertail-coverts. Throat dull peachy-buff with throat- and breast-sides washed brownish, and central belly and undertail whitish. Wingbars as in ♂. ♂ *palmae* is similar to ♂ *canariensis* but has rump concolorous with back, upperparts slightly paler and less blackish on head. Underparts are more pinkish on breast and rest of underparts from lower breast are whitish. ♀ is similar to ♀ *canariensis*. ♂ *ombriosa* is very similar to ♂ *palmae* but has a greenish rump and less white on underparts.

♀ is similar to ♀ *palmae*. ♂ *maderensis* is similar to ♂ *canariensis* but lower mantle, scapulars, back and rump are bright green. ♀ resembles ♀ *canariensis* but has white on only the outermost tail feather. ♂ *moreletti* resembles ♂ *maderensis* but has more extensive green upperparts, browner breast, less vinous belly and less white in outer tail feathers. ♀ resembles ♀ *maderensis* but has white on two outermost tail feathers. **VOICE** *canariensis* has a song consisting of a rapid stream of notes in an accelerating cascade *chiu-chiu-chiu-chiu*… that terminates with a series of variable notes …*chichichiaar* or …*chiyaa-chiyaa*. The call of *canariensis* on Tenerife is a *chiu-chiu*, whereas those on La Gomera usually give a clear *chivi-chivi*. Race *ombriosa* has a similar song to *canariensis* but some different notes appear in the middle, *chie-chie-chie-chie-chia-chia-chie-chia-chiiar* and the terminal flourish is less clear. Also, uniquely in this subspecies, a song has been noted with distinct sections and a vibrant *titiirrr* in the middle of the phrase, similar to that of *coelebs*. The call of *ombriosa* is like that of *canariensis* on Tenerife. Race *palmae* has a song similar to *canariensis* but can be distinguished in that it often ends in a clear and musical …*chio-chu-gui*, which is unmistakable and found only in *palmae*. Call is similar to that of *canariensis* on La Gomera but is more of a *chiri-chiri*. Race *moreletti* has a similar song to *canariensis* but the trill is occasionally split into sections and sometimes there is a short terminal flourish. Both these are characters of the song of nominate European birds. The call of *moreletti* is also quite distinct and is sometimes transcribed *gai*. Race *maderensis* has a call similar to *moreletti* but is quieter and more musical. **RANGE** Breeds from W Europe, the Atlantic Islands and NW Africa east to the region of Lake Baikal and south to the Middle East and the southern Caspian. Resident and migratory but mainly winters in the breeding range. **DISTRIBUTION** *canariensis* breeds on Gran Canaria, Tenerife and La Gomera, *palmae* on La Palma and *ombriosa* on El Hierro; *moreletti* breeds on all of the main Azorean islands; *maderensis* breeds only on the island of Madeira. **BREEDING** In the Canaries laying is from late Mar until May or even Jun at highest altitudes. Clutch 2–4 but 2–3 eggs most common. Nest is usually placed in a tree, 2.5–9 m above ground. In Madeira the season is later as ♀♀ have been seen carrying nesting material in mid May. In the Azores egg laying occurs May–late Jun. Clutch size is 1–4 with 3–4 being the most frequent, and there is sometimes a second brood. **HABITAT** In the Canaries and Madeira common in laurel forest and denser vegetated areas. In the Azores also abundant in agricultural areas. **STATUS** Common resident on three archipelagos. **Azores** Common breeding resident on all islands. **Madeira** Common breeding resident but restricted to the main island. **Canaries** Common resident on the western and central islands, and occasional vagrants from the continent reach the eastern islands.

BLUE CHAFFINCH *Fringilla teydea* Plate 67 (2)
Sp. Pinzón Azul **P.** Tentilhão-azul

F. t. polatzeki (Gran Canaria) and *F. t. teydea* (Tenerife). **L** 16 cm (6″). **IDENTIFICATION** Sexually dimorphic. Blue Chaffinch is larger and longer legged than Common Chaffinch, with a noticeably larger bill. ♂ *teydea* is slaty-blue overall except paler whitish belly and striking white vent. The narrow but prominent white eye-ring is broken in front of the eye. Wings dark with two pale bluish-grey wingbars, one on tips of greater coverts and the other on median coverts. Bill rather metallic steel-blue. ♀ has dull olive-brown upperparts and slightly paler grey-brown underparts. Wingbars are whiter than in ♂, but are buffish or whitish, and never pure white. Juvenile very similar to ♀ but is darker above with paler wingbars. ♂ *polatzeki* smaller than nominate, with greyer plumage, a blacker band over base of bill and whiter wingbars. ♀ very similar to nominate ♀ but has paler underparts, especially belly. **VOICE** Common call of *teydea* is a disyllabic *tchap-chiie*. Song resembles Common Chaffinch, an accelerating trill followed by a terminal flourish, transcribed as *chip-chip-chip-chip-chip-chip-chip-chip-chiu-chuioo* and described as sounding like a drunken Common Chaffinch. Also a high-pitched trill, *sschi-errrr*, a clear *djio* and rarely a fine *chip-chip*. Song of *polatzeki* distinctly higher with the trill notes almost disyllabic and much sweeter. Call also different, being a weaker, rather monosyllabic *ui* or *uit* with an almost *Phylloscopus*-like quality. Also a *ui-chuui* similar to that of *teydea*. **RANGE** Confined to the Canaries. **DISTRIBUTION** Resident only on Tenerife and Gran Canaria. **BREEDING** Normally early Jun but nests have been found Apr–Aug, and are normally in native pines but have been noted in laurel and heather. Clutch normally two but one and three (rarely) have been recorded. A second brood has been recorded recently on Gran Canaria. **HABITAT** Restricted to forests of the endemic pine *Pinus canariensis*. **STATUS** Only found on one archipelago. **Canaries** *teydea* is fairly common within its restricted range on Tenerife, but *polatzeki* on Gran Canaria is now very rare and close to extinction.

BRAMBLING *Fringilla montifringilla* Plate 67 (3)
Sp. Pinzón Real **P.** Tentilhão-montês (Tentilhão-montez)

Monotypic. **L** 14 cm (5½″). **IDENTIFICATION** Sexually dimorphic. Combination of black head, orange underparts, orange scapulars and lesser coverts, and large white rump make breeding ♂ unmistakable. Non-breeding ♂ has head, neck, mantle and back with paler greyish-brown fringes, nape pale grey with a pale grey area on head-sides that extends around rear of ear-coverts. Underparts duller, as are the lesser coverts, and the median coverts are less extensively white. Bill pale yellowish with a darker tip. Breeding ♀ resembles non-breeding ♂ but is duller still with a browner head and mantle, grey head-sides extending from the eye around the rear of ear-coverts, dull orange scapulars and lesser coverts with a brownish tone. Non-breeding ♀ similar to non-breeding ♂ but with a paler head, duller orange underparts and more diffuse spotting on flanks. Juvenile is like ♀ but browner with a less pure white rump and belly. **VOICE** Common call is a rasping, nasal *tsweek* or *tswee-ik*, in flight a repeated *chuck* or *chup*. **RANGE** Breeds across the N Palearctic from Scandinavia to E Siberia, Kamchatka and Sakhalin. In west of range winters south to extreme N Africa and the Middle East. **STATUS** Accidental visitor with records from three archipelagos. **Azores★** ♀, Serra da Tronqueira (São Miguel), 12 Dec 2003. **Madeira** Exceptional with the most recent record a ♂, Bica de Cana (Madeira), 16 Apr 2004. **Canaries** Seven records, six of them from Fuerteventura, including a group of *c.*20, Barranco de Esquinzo, 22–24 Nov 1993. The only record from Tenerife concerns a non-breeding ♂, El Bailadero, 3–5 Dec 1995.

EUROPEAN SERIN *Serinus serinus* Plate 66 (4)
Sp. Verdecillo   **P.** Milheirinha (Chamariz)

Monotypic. **L** 11½ cm (4½″). **IDENTIFICATION** Sexually dimorphic. The smallest finch of the region and superficially similar to Canary. Adult ♂ has bright yellow head and breast, with a yellowish-green crown and nape, and yellowish-green ear-coverts with a small yellow crescent below eye; the belly, flanks and undertail-coverts are white with prominent black streaking on breast-sides and flanks. Mantle greenish with bold dark streaking, rump bright yellow and wings and tail dark blackish-brown with two narrow yellowish wingbars. Adult ♀ similar to ♂ but duller, less yellow, with more extensive streaking, which is blackish, and less striking on breast-sides and flanks. Rump yellow or greenish-yellow but much duller than ♂. Juvenile browner and even duller than ♀ with all yellowish areas buffy-white. **VOICE** Call is a rapid, high-pitched trill *tirrillilit*. Song resembles that of Corn Bunting in being a jangling mixture of notes, but is much higher pitched. **RANGE** Breeds over most of S and C Europe, NW Africa, Turkey and Israel. Northern populations winter mostly around Mediterranean. **DISTRIBUTION** Breeds only on Gran Canaria and Tenerife. **BREEDING** Season imperfectly known but is probably late Feb–May. Clutch 3–4. Uses both native and introduced trees and shrubs, with the nest often being placed *c.* 2 m above ground. **HABITAT** Usually found around habitation, in parks, gardens, golf courses and agricultural areas. **STATUS** Recorded from two archipelagos. **Madeira** The only record is from the Salvages where five were recorded in winter 1980/81. **Canaries** Scarce and local breeder on Gran Canaria and Tenerife, and also recorded on Alegranza, Lanzarote, Fuerteventura, La Gomera and La Palma as an accidental migrant.

(ATLANTIC) CANARY *Serinus canaria* Plate 66 (5)
Sp. Canario   **P.** Canario-da-terra

Monotypic. **L** 13 cm (5″). **IDENTIFICATION** Sexes differ. ♂ has yellow forehead and supercilium, yellow face with dusky ear-coverts, yellow on most of underparts with a greyish tinge to breast-sides and flanks on some, and some blackish streaking on rear flanks; lower belly and vent white. Crown and nape greenish-yellow, mantle, back and scapulars greyish-green with dark streaks and rump dull yellow. Legs and bill pale brownish-pink. ♀ similar to ♂ but is generally duller, greyer on head and breast, the rump is duller and greener, and the underparts less yellow. Juvenile resembles ♀ but is browner overall with streaked breast and lacks any pure yellow on face and underparts or any green tones to upperparts; rump pale brown with some darker streaking. **VOICE** Most common is the contact call, a high-pitched *sooeeet* or *sweeet*. Song identical to that of the well-known cage variety, a mixture of melodious, fluty trills and whistles, sometimes interspersed with twitters or churrs. **RANGE** Confined to Macaronesia. **DISTRIBUTION** Breeds in the Azores, Madeira and Canaries. **BREEDING** Commences Jan and continues until Jul, during which period several broods may be raised. Nests in a wide variety of trees including native pine, various fruit trees, cypress, broom, heather and even banana, and is usually placed 3–4 m above ground (although heights of 1–6 m have been recorded). Clutch 3–5 but four eggs commonest. **HABITAT** Occurs in all habitats from sea level to high mountains including parks, gardens, orchards, pine forests and laurel forests. **STATUS** Fairly common resident on three archipelagos. **Azores** Common breeder on all islands. **Madeira** Breeds on Madeira, Porto Santo and the Desertas where it is common, and has been recorded on the Salvages. **Canaries** Common breeder on the central and western islands, but very scarce and local on the eastern islands, and on Fuerteventura 12–15 birds were deliberately released around Betancuria in *c.* 1980.

(EUROPEAN) GREENFINCH *Carduelis chloris* Plate 66 (6)
Sp. Verderón (Verderón Común)   **P.** Verdilhão (Verdilhão-comum)

C. c. aurantiiventris. **L** 15 cm (6″). **IDENTIFICATION** Sexes similar. Adult ♂ has yellowish-green underparts paler on vent, greyish-green upperparts, greyish head-sides, prominent yellow base to the sides of the tail, bright yellow outer webs on the primaries, a grey wing panel formed by the greater coverts and fringes to the secondaries and tertials, and a fairly large pale pinkish bill. ♀ duller than ♂ with a browner tone to upperparts, yellow wing flash less prominent and less yellow on the tail-sides. Juvenile even duller than ♀ with browner more mottled upperparts and paler underparts with distinct streaking. **VOICE** Common call a twittering *djururut* but also a Canary-like *tsooeet*. Song is of two distinct types: an unmusical, wheezy *dzweeeee* repeated with long pauses, or a far more pleasant mixture of trills, whistles and twitters, sometimes with the wheezy song interwoven. **RANGE** Breeds across Europe and in NW Africa east to C Asia and S Kazakhstan. Local birds are mostly resident but some arrive in winter due to dispersal by continental populations. **DISTRIBUTION** Breeds on the Azores, where introduced *c*. 1890, Madeira and Canaries. **BREEDING** In the Canaries breeds Mar–Jun, during which periods 2–3 broods may be raised. Clutch 3–5. Nest normally constructed in an introduced tree species, at a height of 2–6 m. **HABITAT** Inhabits anywhere with trees, such as parks, gardens and wooded areas as well as cultivation. **STATUS** Recorded from three archipelagos. **Azores** Introduced on São Miguel, Terceira and Faial, but is a rather uncommon resident. **Madeira** Very rare breeder on main island of Madeira where breeding first confirmed in 1968. Also recorded on the Salvages. **Canaries** Breeds on Fuerteventura, Gran Canaria, Tenerife, La Gomera and El Hierro, and recorded from Lanzarote and La Palma but without evidence of breeding. Migrants boost the local population in winter.

(EUROPEAN) GOLDFINCH *Carduelis carduelis* Plate 66 (7)
Sp. Jilguero **P.** Pintassilgo

C. c. parva. **L** 12½ cm (5″). **IDENTIFICATION** Sexes similar. Adult unmistakable with its red face, white throat and head-sides, black crown and nape, sandy-brown mantle and scapulars, white rump, black tail with white spots, and black wings with a broad yellow bar across central upperwing. Juvenile lacks distinctive head pattern, is rather drab grey-brown on head and upperparts with fine streaks on crown and nape broadening on the mantle. Underparts are yellowish-brown with streaks on breast and flanks. **VOICE** Usual call is a twittering *sti-ki-lit* with a rather tinkling quality. Song consists of rapid trills and tinkling phrases mixed with a few buzzing notes, all interwoven with the call note. **RANGE** Breeds from Europe and NW Africa east to W China and south to Egypt, S Iran and Kashmir. **DISTRIBUTION** Currently breeds on Azores, Madeira and the Canaries. Formerly bred on Cape Verde but no recent records and is presumably extirpated. **BREEDING** Little known in our region. In the Canaries commences Feb and continues until Jun, whereas on the Azores only commences mid Apr. Clutch 4–5 and pairs frequently have two broods. Nest is located in a tree (usually an introduced species) at a height of 2–8 m. **HABITAT** Normally found in parks, gardens and cultivated areas. **STATUS** Recorded from all archipelagos but only extant on three. **Azores** Presumed to have been introduced to São Miguel around 1860 and a few years later on Terceira, and now breeds on all main islands except Corvo and is a common resident. **Madeira** Breeds on both Madeira and Porto Santo but is much commoner on the main island. **Canaries** Breeds on all main islands except Lanzarote and El Hierro where breeding unproven. Rather uncommon and local on Tenerife, scarce on La Palma and La Gomera, and now rare on El Hierro. On Gran Canaria it is far more numerous and on Fuerteventura it is locally common. **Cape Verdes** Breeding recorded in 1963 and 1965, on Santiago, but no subsequent records.

(EUROPEAN) SISKIN *Carduelis spinus* Plate 66 (8)
Sp. Lúgano **P.** Lugre

Monotypic. **L** 12 cm (5″). **IDENTIFICATION** Sexually dimorphic. Adult ♂ has green mantle with black streaks, black cap and chin, yellow rump, yellow base to sides of black tail, black wings with two prominent yellow wingbars, yellow underparts except lower belly and vent, which are white, and dark-streaked flanks. ♀ is similar but duller, lacks black crown and is more extensively streaked on upper- and underparts. Juvenile resembles ♀ but is even duller and browner with heavy streaking both above and below, yellow flashes at base of tail and prominent pale wingbars. **VOICE** Common call is a clear, thin *dluee*, *diu-li* or *dli-u*. **RANGE** Breeds from Spain east to Sakhalin and E China. In west of range winters south to N Morocco, N Algeria and N Egypt. **STATUS** Recorded from three archipelagos and an extreme vagrant on two of these. **Azores** A group of ♂♂, ♀♀ and juveniles, Ponta Delgada (São Miguel), 5–21 Apr 1982; one, Lagoa das Furnas (São Miguel), 10 Dec 2003, with 2–3 there, 15 Dec 2003; and three, Sete Cidades (São Miguel), 11 Dec 2003. **Madeira** One on the Salvages, winter 1980/81 and seven, including at least four ♂♂, Mont do Pereiro (Madeira), 14 Oct 2002. **Canaries** Scarce and irregular winter visitor recorded from all main islands except La Gomera and El Hierro.

(COMMON) LINNET *Carduelis cannabina* Plate 66 (9)
Sp. Pardillo Común **P.** Pintarroxo (Pintarrôxo-comum)

C. c. guentheri (Madeira), *C. c. harterti* (E Canaries) and *C. c. meadewaldoi* (C and W Canaries). **L** 13–14 cm (5–5½″). **IDENTIFICATION** Sexually dimorphic. ♂ breeding unmistakable with chestnut-brown mantle, back and wing-coverts, grey head and small crimson forehead patch, and a more extensive one on breast. ♀ is duller, lacks crimson patches and is streaked on mantle and underparts. Non-breeding ♂ resembles ♀ but upperparts richer brown and lacks streaking on underparts. Juvenile resembles ♀ but has warmer brown upperparts, particularly the head, and darker streaking on both surfaces. Subspecies *meadewaldoi* differs from nominate mainly in its more richly coloured mantle and brighter red forecrown and chest; *harterti* is paler above than *meadewaldoi*, the sides and flanks lack rufous-cinnamon and are less prominently streaked pale brown; *guentheri* is the darkest of the three races. **VOICE** Common call is a soft and slightly nasal *tett-tett-ett*, frequently given in flight, also a short *tett* and a soft *tsooeet* in alarm. Song is a series of rattles and trills, interspersed with musical whistles, usually given from an exposed position atop a bush. **RANGE** Breeds from W Europe and N Africa east to C Asia. Some populations migratory but in our region is resident. **DISTRIBUTION** Madeira and the Canaries. **BREEDING** Can commence Jan and continue until July but main season Mar–May. Nest usually located below 2 m, in a bush or small tree. Clutch 3–5 but four eggs is the most common. **HABITAT** On the Canaries breeds in all habitats from coasts to high mountains, except heavily forested areas. On Madeira occupies a similar range of altitudes but it is perhaps most common in agricultural zones. **STATUS** Breeds on two archipelagos. **Madeira** *guentheri* is a resident breeder on Madeira, Porto Santo and the Desertas, but is far less numerous than formerly. Recorded on the Salvages but the subspecies was not determined. **Canaries** *harterti* is a fairly common resident on Fuerteventura, Lanzarote, Lobos and Graciosa, whilst *meadewaldoi* is a common resident on Gran Canaria, Tenerife, La Gomera, La Palma and El Hierro.

COMMON CROSSBILL *Loxia curvirostra* Plate 67 (4)
Sp. Piquituerto Común **P.** Cruza-bico (Cruza-bico-comum)

Subspecies unknown. **L** 16½ cm (6½″). **IDENTIFICATION** Sexually dimorphic. The crossed mandibles are an unmistakable feature of this species. ♂ dull red with darker upperparts and a brighter rump. ♀ basically green and, like ♂, darker on upperparts with a brighter, more yellowish, rump. First-winter resembles adult but is duller and the coloration less extensive. Juvenile is like ♀ but browner with streaks on both under- and upperparts. **VOICE** Call is a loud, hard, metallic *chip* sometimes repeated in series, *chip-chip-chip…* Song is a mixture of call notes, trills and warbling phrases. **RANGE** Breeds across much of the Holarctic. Mostly resident but irrupts on a regular basis. **STATUS** Vagrant. **Madeira** Recorded from Madeira, 12 Dec 1930, and Porto Santo, 26 Jul 1943. **Canaries** Immature ♂, singing near Arure (La Gomera), 27 Jan 1991; and one captured, Los Naranjeros (Tenerife), late Jan 1997, when apparently there were other birds present in the same area.

TRUMPETER FINCH *Bucanetes githagineus* Plate 67 (7)
Sp. Camachuelo Trompetero **P.** Trombeteiro (Pintarrôxo-trombeteiro)

B. g. amantum (Canaries) and *B. g. zedlitzi*. **L** 12.5–15 cm (5–6″). **IDENTIFICATION** Breeding ♂ unmistakable with its grey head, thick-based red bill and red forehead. Body plumage plain, unstreaked sandy grey-brown with a dark pink tinge to rump, wings and underparts. Breeding ♀ is plain, unstreaked sandy-grey with faint pinkish tips to wing-coverts and on sides of base of tail. The bill is a straw colour. Non-breeding ♂ similar to breeding ♀ but has slightly more extensive pink on wings, tail and underparts. Juvenile also similar to ♀ but has a darker bill and buffish-brown wing panel. **VOICE** Call is a short *chee* or *chit* and the song a simple drawn-out nasal, monotonous buzzing *cheeeee*, often followed by short toy-trumpetlike phrases, or is mixed with metallic clicks and whistles. **RANGE** Breeds from the Canaries across much of N Africa, the Middle East and C Asia. Mostly resident but some short-distance movements outside breeding season. **DISTRIBUTION** Breeds only on central and eastern Canaries. **HABITAT** Prefers arid areas such as rocky plains, dry ravines, cliffs, rocky hillsides and abandoned cultivation. **BREEDING** Season Jan–May. Nest is sited in holes in walls, under rocks or on ground well concealed by vegetation. Clutch usually 4–5 but extremes of three and six noted. Pairs normally have at least two broods. **STATUS** Recorded from two archipelagos but status differs dramatically. **Canaries** Scarce and local on Tenerife and Gran Canaria, but far more common and widespread on Fuerteventura and Lanzarote. On La Gomera now very scarce and on El Hierro only a vagrant. **Cape Verdes** One, probably an immature, north of Santa Maria (Sal), 25 Nov 1989.

AZORES BULLFINCH *Pyrrhula murina* Plate 67 (5)
Sp. Camachuelo de las Azores **P.** Priolo (Priôlo)

Monotypic. **L** 16 cm (6″). **IDENTIFICATION** Sexes similar. Both sexes resemble ♀ of Eurasian Bullfinch *P. pyrrhula*, with grey-brown upperparts and paler, buffier rump, and underparts and cheeks paler still. Cap, throat, wings and tail black with a purple-blue gloss. Wings also have very broad greyish tips to greater coverts and outer web of shortest tertial is usually orange-pink. Structurally, larger than Eurasian Bullfinch with a longer and deeper bill but shorter, more rounded wings. ♂ distinctly larger than ♀. Occasionally, ♂ can show a very faint reddish-tawny suffusion to flanks and vent. First-years are difficult to age as adults also have buffish fringes to greater coverts. **VOICE** Call is a soft, piping *phew*. Song is a discontinuous mixture of the call note with various squeaky and scraping notes. **RANGE** Confined to the Azores. **DISTRIBUTION** Breeds only on São Miguel. **BREEDING** Still very little known. ♀♀ with brood patches have been found mid Jun to late Aug, suggesting a season later and shorter than in Eurasian Bullfinch. Two young appears to be the norm. **HABITAT** Confined to areas of natural vegetation, thus now mostly found in steep-sided and remote valleys in E São Miguel. Formerly encountered often in orchards but this habit may have been the cause of its decline to near extinction, as it was declared a pest and extensively hunted. **STATUS** Rare and known only from a small area at the east end of São Miguel, in the region of Pico da Vara. Population estimated to fluctuate at 60–200 pairs depending on breeding success in previous years.

HAWFINCH *Coccothraustes coccothraustes* Plate 67 (6)
Sp. Picogordo **P.** Bico-grossudo

C. c. coccothraustes. **L** 18 cm (7"). **IDENTIFICATION** Sexes similar. In all plumages massive conical bill unmistakable. Breeding ♂ has black throat and lores, rich brown head and underparts, grey nape, darker brown mantle and black tail with a broad white tip. Wings black with a blue gloss on secondaries and on curled extensions to inner primaries. Median and greater coverts form a large white patch, with a cinnamon tone to inner greater coverts. Bill steel greyblack. In flight reveals white band across base of primaries. ♀ duller than ♂ and has pale blue-grey secondaries. Non-breeding ♂ resembles breeding ♂ but has duller mantle and dull yellowish bill. Juvenile is a duller version of adult with dark spots and bars on underparts, and can be sexed using same features as adults. **VOICE** Call is a hard, explosive *tzik*. **RANGE** Breeds from Spain and NW Africa east to Kamchatka, Sakhalin and Japan. Northern populations are migratory but European birds usually winter within breeding range of southerly populations. **STATUS** Vagrant. **Madeira** Listed as exceptional by Zino *et al.* (1995), the first was one, Porto Santo, 11 Nov 1906; has also been recorded on the Salvages.

YELLOW WARBLER *Dendroica petechia* Plate 68 (3)
Sp. Reinita Amarilla **P.** Mariquita-dos-mangais (Toutinegra-amarela)

D. p. aestiva. **L** 12.5 cm (5"). **IDENTIFICATION** Sexes similar. ♂ in breeding plumage unmistakable with bright yellow head and rufous streaking on bright yellow underparts. ♀ duller with less prominent streaking and a greenish crown. In first-year plumage ♂ is similar to adult ♀ but ♀ is even duller than adult ♀ with buffy-white rather than yellow underparts. In all plumages has distinctive yellow 'spots' in tail. **VOICE** Call is a soft, clear *chip* and a buzzy *zzee* in flight. **RANGE** Breeds over most of N and C America, the West Indies and coastal NW South America. Northern populations winter in S Mexico and N South America. **STATUS** Vagrant with only three records. **Azores** One, Ponta Delgada (Flores), 20 Aug 1995; and one, near Sete Cidades (São Miguel), 5 Dec 2001. **Madeira** An adult and immature, Selvagem Grande, 10–12 Sep 1993.

MAGNOLIA WARBLER *Dendroica magnolia* Plate 68 (2)
Sp. Reinita Cejiblanca **P.** Mariquita-de-faces-pretas

Monotypic. **L** 12 cm (5"). **IDENTIFICATION** Sexes similar. In all plumages has characteristic band on tail formed by white on inner webs of all rectrices except central pair, which is clearly visible when tail is fanned. ♂ breeding plumage unmistakable: yellow underparts with black breast-band and black streaks on flanks, black ear-coverts, white supercilium, blue-grey crown and nape, blackish upperparts, yellow rump and white patch on wing. Breeding ♀ duller with the wing patch reduced to a double wingbar, olive upperparts and streaking on underparts is less bold. Non-breeding ♂ resembles breeding ♀ but supercilium much less pronounced. Non-breeding ♀ also resembles breeding ♀ but a greyish breast-band replaces streaks on breast. First-year similar to non-breeding ♀ but even duller with duller wingbars and streaks on underparts less distinct. **VOICE** Call is a high, hard *dzip*. **RANGE** Breeds from W Canada to Newfoundland and south to W Virginia. Winters in C America and the West Indies. **STATUS** Extreme vagrant. **Azores★** One, Ponta Delgada (Flores), 21–22 Sep 1999.

YELLOW-RUMPED WARBLER (MYRTLE WARBLER) *Dendroica coronata* Plate 68 (4)
Sp. Reinita Coronada **P.** Mariquita-coroada (Toutinegra-das-murtas)

D. c. coronata. **L** 13.5 cm (5"). **IDENTIFICATION** Sexes similar. ♂ in breeding plumage unmistakable with blue-grey upperparts broadly streaked black, black ear-coverts, narrow white supercilium, white underparts with heavy dark streaking on breast and flanks, yellow rump and yellow patches on head and breast-sides. Breeding ♀ is duller with less heavy streaking, a brownish tone to upperparts and grey-brown ear-coverts. Non-breeding ♂ much duller and browner than breeding ♂, and resembles other non-breeding plumages but yellow breast and crown patches are distinct and streaking is bolder on upperparts. Non-breeding ♀ resembles breeding ♀ but is duller and browner, and usually duller than corresponding plumage of ♂. First-year dullest of all with very indistinct yellow patches on breast-sides and crown, and a duller yellow rump. **VOICE** Call is a loud, hard *chek* or soft *tsee* in flight. **RANGE** Breeds across most of N America except far north and southeast. Winters in S North America, C America and the West Indies. **STATUS** Extreme vagrant. **Azores★** First winter ♀, Fajã Grande (Flores), 15–16 Oct 2000. **Canaries** First-year ♂, Maspalomas (Gran Canaria), 25 Feb–2 Mar 1984.

BLACKPOLL WARBLER *Dendroica striata* Plate 68 (1)
Sp. Reinita Listada **P.** Mariquita-de-perna-clara (Toutinegra-raiada)

Monotypic. **L** 13.5 cm (5"). **IDENTIFICATION** Sexually dimorphic. Breeding ♂ unmistakable with its black cap, white cheeks, grey upperparts and white underparts. Both upper- and underparts have bold, dark, streaking and two prominent white wingbars. Breeding ♀ has a grey crown and nape streaked darker, greyish face and off-white underparts with less bold streaking. Non-breeding ♂ has olive-green head and upperparts with a yellow supercilium, throat and breast, and rest of underparts white. Streaked both on upperparts and underparts but less bold than breeding ♂. Non-breeding ♀ and first-year resemble non-breeding ♂ but streaking is less bold, particularly on breast-sides, and the yellow supercilium, throat and breast are less intense. Individual variation can make ageing and sexing in this plumage rather difficult. In all plumages legs are usually pale. **VOICE** Common calls include a loud *chip* and high-pitched *seet*. **RANGE** Breeds from Alaska across Canada to Newfoundland. Winters in C and N South America. **STATUS** Extreme vagrant. **Canaries★** One, Stella Canaris Hotel at Morro Jable (Fuerteventura), 21 Oct 2000.

AMERICAN REDSTART *Setophaga ruticilla* Plate 68 (5)
Sp. Colirrojo Americano **P.** Mariquita-de-rabo-vermelho (Toutinegra-de-cauda-ruiva)

Monotypic. **L** 13 cm (5″). **IDENTIFICATION** Sexually dimorphic. Adult ♂ unmistakable given combination of black upperparts, head and breast, white belly and vent, and orange patches in wings and tail and on breast-sides. ♀ has greyish head, olive upperparts, white underparts and yellow patches in wings and tail and on breast-sides. First-winter resembles ♀ but yellow wing patch reduced or lacking. **VOICE** Call is a sharp *chip*. **RANGE** Breeds across C and SE North America. Winters West Indies, C America and N South America south to Peru and Surinam. **STATUS** Vagrant with only three records. **Azores** Two captured on board ship near the Azores, Oct 1967, a ♀ on 5th (at 40°30'N 25°48'W), and ♂ (39°58'N 28°46'W) on 14th. **Madeira** One, Selvagem Grande, 2 Oct 1981.

LOUISIANA WATERTHRUSH *Seiurus motacilla* Plate 68 (6)
Sp. Reinita de Louisiana **P.** Mariquita da Louisiana

Monotypic. **L** 15 cm (6″). **IDENTIFICATION** Sexes similar. Adult similar to Northern Waterthrush but separated by supercilium that is whiter and broadens behind eye; bill longer and stouter; flanks and undertail distinctly buff; throat usually unstreaked; and legs bright bubblegum pink. First-winter can have narrow rusty fringes to tertials. **VOICE** Call a loud, sharp *chink* very similar to Northern Waterthrush but louder, slightly lower and less metallic. **RANGE** Breeds E USA and extreme S Canada. Winters C America and the Caribbean. **STATUS** Extreme vagrant. **Canaries** One, Playa Nueva (La Palma), 10–26 Nov 1991.

NORTHERN WATERTHRUSH *Seiurus noveboracensis* Plate 68 (7)
Sp. Reinita Charquera **P.** Mariquita-boreal (Tordo-d'água)

Monotypic. **L** 15 cm (6″). **IDENTIFICATION** Sexes similar. Adult has dark brown upperparts, whitish underparts and supercilium, with dark streaks on throat, breast and flanks. Supercilium is of uniform width and colour throughout, but tapers at rear. Legs usually dark flesh. First-year in autumn resembles adult but can show rufous tips to tertials and buffy tips to greater-coverts. **VOICE** Call is a loud, sharp, metallic *chink*. **RANGE** Breeds from W Alaska across Canada to Newfoundland and extreme NE USA. Winters in Mexico to N South America and in the Caribbean. **STATUS** Extreme vagrant. **Azores** One, on board a ship offshore from Santa Maria, 4 Nov 1996.

SAVANNAH SPARROW *Passerculus sandwichensis* Plate 68 (8)
Sp. Sabanero Zanjero **P.** Tico-tico-dos-prados

Subspecies unknown. **L** 14 cm (5½″). **IDENTIFICATION** Sexes similar. Rather nondescript and rather reminiscent of a small, short-tailed female bunting, as it is fairly featureless and heavily streaked. Differs noticeably from female/immature *Emberiza* in its lack of contrasting white outer-tail feathers. Overall coloration varies from pale sandy-brown to rather dark brown, and intensity of streaking is also variable, being finest on paler birds. Perhaps the most obvious and useful identification characters are the yellow loral stripe and fore supercilium, although this is also variable being least obvious in fresh plumage in autumn, as well as the bright pinkish legs and whitish median crown-stripe. Behaviour more closely resembles a small pipit, but bill shape and lack of white outer tail feathers eliminates that possibility. **VOICE** Calls include a thin *tsi* and very high, sharp *stip*. **RANGE** Breeds in Alaska and Canada, west to NE USA, Mexico and Guatemala. Winters in S USA and C America. **STATUS** Extreme vagrant. **Azores** One, Fajã Grande (Flores), 31 Oct 2002.

LAPLAND BUNTING *Calcarius lapponicus* not illustrated
Sp. Escribano Lapón **P.** Escrevedeira da Lapónia

Subspecies unknown. **L** 14–15.5 cm (5½–6″). **IDENTIFICATION** Sexually dimorphic. Breeding ♂ unmistakable but is unlikely to be seen in our region in this plumage, thus only non-breeding adults and juvenile plumages are described. Possibly the most characteristic feature is the reddish-brown wing panel, formed by the fringes of the greater coverts and tertials, bordered by two narrow but distinct white wingbars on tips of median and greater coverts. Non-breeding adult has rufous nape, pale median crown-stripe, a conspicuous dark border to ear-coverts, pale lores and supercilium, off-white underparts with blackish streaking on breast and flanks and dark upperparts, and pale sandy-brown stripes on mantle. Juvenile very similar but is duller and darker, and slightly more heavily streaked. Lark-like habits and preference for rather open habitats help distinguish this species from similar Reed Bunting, and the reddish-brown wing-panel is a feature not shown by any larks in the region. **VOICE** Common flight call is a short rattling *prrrt*, frequently followed by a short, clear, whistled *chu*. **RANGE** Breeds on tundra throughout the arctic and winters on steppes, prairies and coasts south of the breeding range. **STATUS** Extreme vagrant. **Canaries** One photographed, Amarilla Golf Course (Tenerife), 15–30 Nov 2003.

SNOW BUNTING *Plectrophenax nivalis* Plate 69 (1)
Sp. Escribano Nival **P.** Escrevedeira-das-neves

P. n. nivalis. **L** 16–17 cm (6–7″). **IDENTIFICATION** Sexually dimorphic. ♂ in breeding plumage unmistakable given combination of all-white head and underparts, black back, white rump, and black-and-white wings. Breeding ♀ has browner mantle than ♂, fine dark streaking on crown, nape and ear-coverts, predominantly dark primary-coverts and dark bases to greater coverts. Non-breeding ♂ is a complex mixture of orange-brown, white and black. Non-breeding ♀ is similar to non-breeding ♂ but duller, with a less distinct white supercilium and less white in wing. First-winter ♂ similar to non-breeding ♂ but is less white on face and throat. First-winter ♀ indistinguishable from non-breeding ♀. Juvenile is very grey on head and breast but this plumage is soon lost and is unlikely to be seen in our region. **VOICE** Flight call is a soft ripple, *tirrrrit*, followed by a clear, ringing *pyu*, both of which can be given separately. **RANGE** Breeding range circumpolar and mostly north of 68°N, in Atlantic and Pacific coastal areas occurs south to 51–60°N. Winters mainly from North Sea coasts west to Sakhalin, N Japan and NW China, and across N USA and S Canada. **STATUS** Accidental with records from three archipelagos. **Azores** Irregular winter visitor recorded from all islands except São Jorge and Santa Maria, and is occasionally reported in flocks, but more frequently in smaller groups. **Madeira** Listed as exceptional, the most recent record was a flock of *c*. 15 birds, Pico de Arieiro (Madeira), 15 Mar 2004. **Canaries** About 11 records, all in Oct–Feb and all of lone birds; recorded from Alegranza, Lanzarote, Tenerife and La Gomera.

CIRL BUNTING *Emberiza cirlus* Plate 69 (2)
Sp. Escribano Soteño **P.** Escrevedeira-de-garganta-preta

Monotypic. **L** 16 cm (6″). **IDENTIFICATION** Sexually dimorphic. ♂ in breeding plumage has unmistakable head pattern of yellow; and black-striped face, black throat and dark greyish-green crown streaked darker. On upper breast is a band of yellow, below this a band of greyish-green and chestnut patches on breast-sides. ♀ in breeding plumage is rather dull and streaky, with a greyish rump, faint yellowish wash on underparts and rufous tinge to breast-sides. Non-breeding ♂ resembles breeding ♂ but is duller with a less distinct face pattern. Non-breeding ♀ similar to breeding ♀ but is even duller. First-winter resembles non-breeding adult, as does juvenile. **VOICE** Normal call is a soft, high-pitched *tsi* or *tsiip* repeated in series when agitated. **RANGE** Breeds in Spain, France and extreme S England, around the N Mediterranean, N Turkey and NW Africa. Mostly resident but some winter dispersal. **STATUS** Extreme vagrant. **Canaries★** ♂, Ayagaures Reservoir (Gran Canaria), 21 Apr 1976.

ROCK BUNTING *Emberiza cia* Plate 69 (3)
Sp. Escribano Montesino **P.** Cia

E. c. cia. **L** 16 cm (6″). **IDENTIFICATION** Sexes similar. ♂ in breeding plumage is unlikely to be mistaken for any other bunting. Pale grey head and upper breast with black border to ear-coverts, black lateral crown-stripes and orangey lower breast, flanks and belly are characteristic. Breeding ♀ is similar but has a duller head pattern with often some faint streaking on breast-sides and flanks. Non-breeding ♂ has head pattern slightly obscured by buffy fringes. Non-breeding ♀ also has head pattern obscured by buffy fringes. First-winter resembles respective non-breeding adult and is almost indistinguishable in the field. Juvenile is very streaky, has rufous tinge on rump and belly, a narrow dark border to ear-coverts, and white bar on median coverts. **VOICE** Call is a high-pitched *tsii*, in flight a short *tewp* or rattling *si-tititi*. **RANGE** Breeds from Spain and NW Africa through S Europe to Iran, C Asia and the W Himalayas. Northern populations partially migratory and southern populations undergo some altitudinal and short-distance wanderings. **STATUS** Extreme vagrant. **Canaries★** A group, near Las Palmas (Gran Canaria), 10 May 1910, and mentioned by Trotter as numerous in the mountainous northeast of Lanzarote, 11 Mar 1967.

HOUSE BUNTING *Emberiza striolata* Plate 69 (4)
Sp. Escribano Sahariano **P.** Escrevedeira-domestica (Escrevedeira-africana)

E. s. sahari. **L** 13½ cm (5"). **IDENTIFICATION** Sexes similar. ♂ has a greyish head, neck and upper breast with a pale supercilium, dark eye-stripe, white submoustachial stripe and black moustachial stripe. Throat and breast streaked blackish but rest of underparts unmarked rusty-brown. Upperparts unmarked rufous-brown with a darker tail but no pure white on outer tail feathers. Bill has yellowish lower mandible. ♀ similar but duller overall with a browner head and breast. Juvenile resembles ♀ but is even duller, lacks any yellow on lower mandible and has very little streaking on head and upper breast. **VOICE** Call is a harsh, nasal *dzwee*. **RANGE** *sahari* breeds in Morocco, Algeria, Tunisia and NW Libya. It is mainly resident but local movements occur in winter. **STATUS** Extreme vagrant. **Canaries** one obtained by Cabrera at Punta del Hidalgo (Tenerife) prior to 1893.

ORTOLAN BUNTING *Emberiza hortulana* Plate 69 (5)
Sp. Escribano Hortelano **P.** Sombria

Monotypic. **L** 16½ cm (6½"). **IDENTIFICATION** Sexes similar. Both sexes share distinctive head pattern but that of ♂ is more pronounced. Head predominantly greenish-grey with characteristic yellow submoustachial stripe and throat, and an obvious pale yellowish-white eye-ring. There is a very sharp cut-off between greenish-grey breast and rufous-buff belly and flanks. ♀ is similar to ♂ but has streaking on crown and breast, and the colours are not as strong. Juvenile predominantly brown and buff streaked black, but has a distinctive black malar stripe and pale submoustachial stripe. **VOICE** Call is a metallic, disyllabic *sli-uu* or short *chu*, often given alternatively with a pause of *c.* 2 seconds between them. **RANGE** Breeds from W Europe, through C Asia to the Altai Mountains. Winters in Africa, locally in W Africa and, more commonly, in E Sudan and Ethiopia. **STATUS** Recorded from two archipelagos. **Madeira** One record from the Ilhas Selvagens. **Canaries** Eight records: one from Alegranza and the others from Fuerteventura, and all in spring except one in Sep.

CRETZSCHMAR'S BUNTING *Emberiza caesia* Plate 69 (6)
Sp. Escribano Ceniciento **P.** Sombria-de-capuz (Escrevedeira-cinzenta)

Monotypic. **L** 16 cm (6"). **IDENTIFICATION** Sexes similar. Both sexes have distinctive head pattern similar to Ortolan Bunting. Adult ♂ has head and breast blue-grey with orange-chestnut lores, throat and submoustachial stripe, and obvious pale eye-ring. Rest of underparts rufous-chestnut. ♀ resembles ♂ but has paler crown and nape, browner face, paler lores, throat and submoustachial stripe, with streaking on crown, nape and breast. Underparts also paler than ♂. First-winter very similar to Ortolan and difficult to identify, although Cretzschmar's has whiter eye-ring and throat, and warm rufous tone to rump. Juvenile probably indistinguishable from Ortolan. **VOICE** Call differs from Ortolan in being a harder and sharper but less metallic *tchipp*. **RANGE** Breeds E Mediterranean from NE Greece through Aegean Is., W and S Turkey to N Israel and W Jordan. Winters almost entirely in Sudan and Eritrea. **STATUS** Extreme vagrant. **Canaries★** Adult ♂, Rosa de los Negrines (Fuerteventura), 1 Apr 1980.

LITTLE BUNTING *Emberiza pusilla* Plate 69 (7)
Sp. Escribano Pigmeo **P.** Escrevedeira-pequena (Escrevedeira-pigmeia)

Monotypic. **L** 13–14 cm (5–5½"). **IDENTIFICATION** Sexes similar. A small bunting with chestnut ear-coverts, narrow (but obvious) pale eye-ring, prominent lateral crown-stripes and dark border to ear-coverts, which does not reach bill. Also often has a pale cream spot on rear ear-coverts, which, if present, is another useful identification feature. Only real confusion is with a small ♀ Reed Bunting but identified by its bolder eye-ring, call, colour and pattern of ear-coverts, more prominent wingbars and almost straight, rather than convex, culmen. Juvenile and first-winter similar to adult but crown-stripes more streaked and, on average, head pattern duller. **VOICE** Call is a short, sharp, clicking *tsik*. **RANGE** Breeds from N Scandinavia east to E Siberia, generally above 60°N. Winters from NE India and the Himalayas to S China. **STATUS** Vagrant with four records from one archipelago. **Canaries★** One, El Matorral (Fuerteventura), 25 Feb 1981; one, Alegranza, 5 May 1993; one, Vallebrón (Fuerteventura), 18 Jan 1995; and one, Barranco de la Torre (Fuerteventura), 16 Feb 2004.

(COMMON) REED BUNTING *Emberiza schoeniclus* Plate 69 (8)
Sp. Escribano Palustre **P.** Escrevedeira-dos-caniços

Subspecies unknown. **L** 15–16 cm (6"). **IDENTIFICATION** Sexually dimorphic. ♂ breeding unmistakable with black head, white submoustachial stripe and nuchal collar. Rest of underparts white with some black streaks on flanks. Wings predominantly rufous, rump greyish and tail black with gleaming white outer feathers. ♀ in breeding plumage has brown crown, pale supercilium, brown ear-coverts with a darker border, white moustachial stripe and throat, and a dark malar stripe. Underparts mostly white with prominent streaking on breast and flanks. Wings less rufous than in ♂. ♂ in non-breeding plumage has dark head partially obscured by pale fringes to feathers. First-winter resembles respective adult in fresh autumn plumage but has worn rectrices and flight feathers. Juvenile resembles ♀ but has more yellowish-buff underparts, duller upperparts and better-defined streaking. **VOICE** Common call a *seeoo* falling smoothly in pitch, and a hoarse *chew* often given in flight. **RANGE** Breeds from W Europe east to Kamchatka, Sakhalin and N Japan. European birds winter south to Morocco. **STATUS** Vagrant with just five records from two archipelagos. **Madeira** First-winter ♀, Tanque (Porto Santo), 1 Nov 2001. **Canaries★** ♀, between Puerto del Carmen and Arrecife (Lanzarote), 22 Dec 1992; one, Playa San Juan (Tenerife), 2 Dec 2001; one, Las Galletas (Tenerife), 13 Nov 2003; and one, Barranco de la Torre (Fuerteventura), 16–17 Feb 2004.

CORN BUNTING *Emberiza calandra* Plate 69 (9)
Sp. Triguero **P.** Trigueirão

M. c. calandra. **L** 18 cm (7"). **IDENTIFICATION** Sexes similar but ♂ larger. A rather nondescript bunting, lacking any obvious distinctive features, e.g. facial pattern, prominent wingbars or even white outer tail feathers. Bulky with a relatively large head and bill, and shortish tail. Generally grey-brown above and whitish below, streaked darker throughout, but underparts distinctly buffish on breast and flanks. Face pattern plain apart from conspicuous blackish malar stripe and large pale bill. Juvenile is paler and brighter, and more scalloped and spotted, rather than streaked. Also lores pale and dark surround to rear ear-coverts more pronounced than in adult. **VOICE** Calls include a quiet *pit*, often repeated in short series when taking flight. Song commences with a repeated series of chipping notes, accelerating into a squeaky jangling rattle, and is often likened to a bunch of keys being shaken. **RANGE** Breeds from the Canaries and British Isles east, discontinuously, across C and S Europe to S Iran and extreme W China. Western populations mainly sedentary but, in the east, some winter in the Gulf States, Saudi Arabia and Egypt. **DISTRIBUTION** Resident breeder on all main islands in the Canaries. **BREEDING** Starts late Mar/early Apr but somewhat later at higher altitudes. Clutch size usually 4–5 but a nest with six chicks observed. Nests on ground, well hidden between plants. **HABITAT** Cereal crops and open country from sea level to 1,500 m. **STATUS** Known from only one archipelago. **Canaries** Locally common; noticeably more so on the central and western islands than on Fuerteventura and Lanzarote.

BOBOLINK *Dolichonyx oryzivorus* Plate 68 (9)
Sp. Charlatán **P.** Triste-pia (Tagarela-americano)

Monotypic. **L** 18 cm (7"). **IDENTIFICATION** Sexually dimorphic. Breeding ♂ unmistakable, mainly black with a buff nape, striking white scapulars and large white rump. Breeding ♀ somewhat bunting-like but has rather unusual spike-tipped tail feathers and very conical bill. Generally buffy-brown with heavy dark streaking on head and upperparts, whitish 'braces' on mantle-sides, finer streaks on flanks and breast-sides, a pale central crown-stripe and whitish supercilium. Non-breeding ♂, non-breeding ♀ and first-winter all resemble breeding ♀ but more yellowish-buff on underparts. First-winter sometimes aged by presence of unmoulted worn juvenile tertials. **VOICE** Usual call a rather liquid *blink* often given in flight. **RANGE** Breeds across most of C North America, and winters in S America from Peru to S Brazil and Argentina. **STATUS** Extreme vagrant. **Azores** All records on Flores: one, Fajã Grande, 3 Sep 1998; one, Ponta Delgada, 21–22 Sep 1999; and another there, 5 Oct 2002.

APPENDIX A: UPDATE FOR 2005 AND 2006

THE CUT-OFF DATE FOR RECORDS in the main text was the end of 2004. This section updates the situation for 2005, a year that proved to be one of the most amazing in the history of Macaronesian ornithology. The year started well, with the discovery of a singing Northern Mockingbird *Mimus polyglottos* at Arguineguin on Gran Canaria in the Canary Islands, although, according to locals, it had been present in the area since November 2004. This was not only a first for the Canary Islands, but also a first for Spain and for Macaronesia. Also early in the year, an *Accipiter* hawk seen briefly on the Azores was either the first Eurasian Sparrowhawk for the Azores or, perhaps more intriguingly, a Sharp-shinned Hawk *Accipiter striatus*, which would be the first for the Western Palearctic. The mockingbird would have been the bird of the year in any normal year, but no-one with an interest in the Atlantic Islands could have predicted what the autumn was to bring to the region.

The autumn started its surprises on 18 September when the first Common Crossbill for the Azores, an adult male, was discovered near Pico de Vara on São Miguel. A surprising addition to the regional list was a Common Rosefinch *Carpodacus erythrinus* found on 1 October on Alegranza to the north of Lanzarote. The next Azorean 'first' came on 3 October when a first-year Pallas's Gull *Larus ichthyaetus* was reported from the harbour at Ponta Delgada on São Miguel. But this was about to be overshadowed by an incredible haul of transatlantic vagrants seen on the Azores during October and November. It was not until Peter Alfrey arrived on the small western island of Corvo that things really started to hot up. The first Arctic Redpoll *Carduelis hornemanni* for the region was discovered in fields near the island capital of Vila Nova on 20 October. Then on 22 October he found the first Western Palearctic record of White-eyed Vireo *Vireo griseus* and the region's first Indigo Bunting *Passerina cyanea*. From here on things got hectic with the region's first Rose-breasted Grosbeak *Pheucticus ludovicianus*, a first-winter female, on 23 October, the first Black-throated Blue Warbler *Dendroica caerulescens* on 24 October, the first White-crowned Sparrow *Zonotrichia leucophrys* on 25 October, and then, on an amazing 26 October, the first Red-eyed Vireo *Vireo olivaceus*, Philadelphia Vireo *Vireo philadelphicus* and Hooded Warbler *Wilsonia citrina*, the latter being a stunning male. Meanwhile, on the same day on São Miguel the first Tree Swallow *Tachycineta bicolor* for the region was seen at Sete Cidades. On 28 October the first Lapland Bunting for the Azores was discovered, while 29 October brought the region's first Scarlet Tanager *Piranga olivacea* and Buff-bellied Pipit *Anthus rubescens*. On 30 October the first record of the North American race of Barn Swallow *Hirundo rustica erythrogaster* for the Western Palearctic was found at Ponta Delgada on Flores, and the first Caspian Terns for the Azores were seen at Praia da Vitoria on Terceira. During this exceptional period on the Azores there was a spate of surprising records in the Canaries, including at least 30 White-rumped Sandpipers, 7 Chimney Swifts, and the first Hawfinch for the islands on Fuerteventura on 24 October.

November also started with a bang, with the discovery of the region's first Ovenbird *Seiurus aurocapilla* on Corvo on 1 November, and the first Snowy Egret for the Cape Verdes at Mindelo Sewage Ponds. This was followed on 2 November with the identification of a Mourning Dove *Zenaida macroura* that had been present on Corvo for a while but remained elusive. It was presumed that the best would have passed by this time of year, but Baltimore Oriole *Icterus galbula* on 6 November, Tennessee Warbler *Vermivora peregrina* on 21 November and Common Yellowthroat *Geothlypis trichas* on 30 November (all regional firsts) continued Corvo's remarkable autumn; a staggering 16 species new to Macaronesia were found on this tiny island in 2005. The Azores saw a total of 21 new species for the Azores and 3 for the Canary Islands; 19 of these were new for Macaronesia. Perhaps the most surprising additions to the list were White-eyed Vireo, Black-throated Blue Warbler and Hooded Warbler, but it was the sheer numbers of certain species (a total of 10 Indigo Buntings, for example, and four Tree Swallows) plus the incredible diversity of species that were unprecedented in the Western Palearctic.

2006

This year was not as outstanding as 2005 but there were still some interesting records. In March a European Golden Plover on the Cape Verde Islands was a first and a Spur-winged Plover a second, while on the Canary Islands a Red-backed Shrike on Tenerife was the first since 1995. In May the second Trindade Petrel for the Azores (and for the Western Palearctic) was seen off Faial on the 17th, and a Bridled Tern was found incubating on Praia Islet off Graciosa. In September an immature Black-browed Albatross was seen off Faial in the Azores on the 13th and the first Long-billed Dowitcher for the Azores was discovered at Cabo da Praia on Terceira on the 16th. The first Hudsonian Whimbrel for the Canary Islands was reported from Puerto de la Cruz on the 12th and the third Willet for the Azores was on São Jorge on the 2nd. October produced the first Macaronesian record of Summer Tanager *Piranga rubra* on Corvo on the 26th, and the third Western Palearctic record of Trindade Petrel from Corvo on the 29th. November saw the second and third Western Palearctic records of African Crake, both from Tenerife, on the 15th and 16th respectively, while the night of 12 December saw the return of the Bermuda Petrel to Vila Islet, off Santa Maria in the Azores.

Appendix B: Distribution of Species on the Atlantic Islands

i. The Canary Islands

E = Eastern Islets, L = Lanzarote F = Fuerteventura C = Gran Canaria
T = Tenerife G = La Gomera P = La Palma H = El Hierro

Note that for the purposes of this table, the Eastern Islets include Roque del Este, Alegranza, Montaña Clara, Roque del Oeste and La Graciosa; Fuerteventura includes the islet of Lobos; Tenerife includes the Roques de Anaga and Roque de Garachico; El Hierro includes the Roques de Salmor.

B = current breeding species   **b** = species that has bred   ● = has occurred
? = possibly breeds   **E** = extinct as a breeding species

English name	Scientific name	Spanish name	Portuguese name	E	L	F	C	T	G	P	H
Red-throated Diver	Gavia stellata	Colimbo Chico	Mobelha-pequena					●			
Black-throated Diver	Gavia arctica	Colimbo Ártico	Mobelha-de-garganta-preta				●	●	●		
Great Northern Diver	Gavia immer	Colimbo Grande	Mobelha-grande	●				●		●	
Pied-billed Grebe	Podilymbus podiceps	Zampullín Picogrueso	Mergulhão-caçador					●			
Little Grebe	Tachybaptus ruficollis	Zampullín Común	Mergulhão-pequeno		●	●		●			
Great Crested Grebe	Podiceps cristatus	Somormujo Lavanco	Mergulhão-de-poupa				●				
Black-necked Grebe	Podiceps nigricollis	Zampullín Cuellinegro	Cagarraz		●	●	●	●		●	●
Northern Fulmar	Fulmarus glacialis	Fulmar Boreal	Pombalete	●				●			
Fea's Petrel	Pterodroma feae	Petrel Gon-gon	Gon-Gon				●	●	●	●	
Bulwer's Petrel	Bulweria bulwerii	Petrel de Bulwer	Alma-negra	B	B	?	?	B	B	B	B
Cory's Shearwater	Calonectris diomedea	Pardela Cenicienta	Cagarra	B	B	B	B	B	B	B	B
Cape Verde Shearwater	Calonectris edwardsii	Pardela Cenicienta de Edwards	Cagarra de Cabo Verde					●		●	
Great Shearwater	Puffinus gravis	Pardela Capirotada	Pardela-de-barrete			●		●		●	
Sooty Shearwater	Puffinus griseus	Pardela Sombría	Pardela-preta	●	●	●		●		●	
Balearic Shearwater	Puffinus mauretanicus	Pardela Balear	Fura-bucho					●		●	
Manx Shearwater	Puffinus puffinus	Pardela Pichoneta	Fura-bucho do Atlântico		●	●	●	B	?	B	?
Macaronesian Shearwater	Puffinus baroli	Pardela Chica	Pintainho	B	B	B	?	B	B	?	?
Wilson's Storm-petrel	Oceanites oceanicus	Paíño de Wilson	Casquilho	●	●			●		●	
White-faced Storm-petrel	Pelagodroma marina	Paíño Pechialbo	Calca-mar	B	●			●		●	
European Storm-petrel	Hydrobates pelagicus	Paíño Común	Alma-de-mestre	B	?	B	?	B	B	?	B
Leach's Storm-petrel	Oceanodroma leucorhoa	Paíño de Leach	Paínho-de-cauda-forcada	●	●	●	●		●		
Madeiran Storm-petrel	Oceanodroma castro	Paíño de Madeira	Roquinho	B	B	B		B	?	?	?
Red-billed Tropicbird	Phaethon aethereus	Rabijunco Etéreo	Rabijunco		●	●	●	●			
Brown Booby	Sula leucogaster	Piquero Pardo	Alcatraz-pardo					●			
Northern Gannet	Morus bassanus	Alcatraz Atlántico	Alcatraz	●	●	●	●	●	●	●	●
Cormorant	Phalacrocorax carbo	Cormorán Grande	Corvo-marinho	●	●	●	●	●			●
Shag	Phalacrocorax aristotelis	Cormorán Moñudo	Galheta					●			
Pink-backed Pelican	Pelecanus rufescens	Pelícano Roseo-gris	Pelicano-cinzento					●			
Magnificent Frigatebird	Fregata magnificens	Rabihorcado Magnífico	Rabiforcado					●			
Eurasian Bittern	Botaurus stellaris	Avetoro Común	Abetouro			●	●	●			
American Bittern	Botaurus lentiginosus	Avetoro Lentiginoso	Abetouro-americano					●			
Little Bittern	Ixobrychus minutus	Avetorillo Común	Garçote			●	●	●			
Dwarf Bittern	Ardeirallus sturmii	Avetorillo Plomizo	Garçote-preto					●			
Black-crowned Night Heron	Nycticorax nycticorax	Martinete Común	Goraz	●	●	●	●	●	●	●	●
Squacco Heron	Ardeola ralloides	Garcilla Cangrejera	Papa-ratos			●	●	●			
Cattle Egret	Bubulcus ibis	Garcilla Bueyera	Carraceiro			B	●	●			
Western Reef Egret	Egretta gularis	Garceta Sombría	Garça-dos-recifes			●		●			
Little Egret	Egretta garzetta	Garceta Común	Garça-branca	●	B	●	●	●		●	
Great White Egret	Egretta alba	Garceta Grande	Garça-branca-grande			●	●	●		●	
Grey Heron	Ardea cinerea	Garza Real	Garça-real	●	?	●	●	?	●	●	●
Great Blue Heron	Ardea herodias	Garza Azulada	Garça-real-americana				●				

APPENDIX B: DISTRIBUTION OF SPECIES ON THE CANARY ISLANDS

English name	Scientific name	Spanish name	Portuguese name	E	L	F	C	T	G	P	H
Purple Heron	*Ardea purpurea*	Garza Imperial	Garça-ruiva	●	●	●	●	●	●	●	
Black Stork	*Ciconia nigra*	Cigüeña Negra	Cegonha-preta				●	●	●	●	
White Stork	*Ciconia ciconia*	Cigüeña Común	Cegonha-branca	●	●	●	●	●		●	●
Glossy Ibis	*Plegadis falcinellus*	Morito	Ibis-preto	●	●	●	●	●			
Sacred Ibis	*Threskiornis aethiopicus*	Ibis Sagrado	Ibis-sagrado			B	B	●	●		
Eurasian Spoonbill	*Platalea leucorodia*	Espátula	Colhereiro	●	●	●	●	●		●	
Greater Flamingo	*Phoenicopterus roseus*	Flamenco	Flamingo	●	●	●	●	●			
Lesser Flamingo	*Phoenicopterus minor*	Flamenco Enano	Flamingo-pequeno			●	●				
White-faced Whistling Duck	*Dendrocygna viduata*	Suirirí Cariblanco	Ireré					●			
Mute Swan	*Cygnus olor*	Cisne Vulgar	Cisne-mudo					●		●	
Pink-footed Goose	*Anser brachyrhynchus*	Ánsar Piquicorto	Ganso-de-bico-curto	●							
Greater White-fronted Goose	*Anser albifrons*	Ánsar Careto Grande	Ganso-de-testa-branca					●			
Greylag Goose	*Anser anser*	Ánsar Común	Ganso-bravo				●	●		●	
Barnacle Goose	*Branta leucopsis*	Barnacla Cariblanca	Ganso-marisco						●		
Brent Goose	*Branta bernicla*	Barnacla Carinegra	Ganso-de-faces-pretas					●			
Ruddy Shelduck	*Tadorna ferruginea*	Tarro Canelo	Pato-casarca				B	●			
Common Shelduck	*Tadorna tadorna*	Tarro Blanco	Tadorna					●		●	
Wood Duck	*Aix sponsa*	Pato Joyuyo	Pato-carolino					●		●	
Eurasian Wigeon	*Anas penelope*	Silbón Europeo	Piadeira	●	●	●	●	●	●	●	●
American Wigeon	*Anas americana*	Silbón Americano	Piadeira-americana					●		●	
Falcated Duck	*Anas falcata*	Cerceta de Alfanjes	Pato-falcado					●			
Gadwall	*Anas strepera*	Ánade Friso	Frisada			●	●	●		●	
Eurasian Teal	*Anas crecca*	Cerceta Común	Marrequinha			●	●	●		●	
Green-winged Teal	*Anas carolinensis*	Cerceta Americana	Marrequinha-americano			●	●				
Mallard	*Anas platyrhynchos*	Ánade Azulón	Pato-real			●	●	●		●	
American Black Duck	*Anas rubripes*	Ánade Sombrío	Pato-escuro-americano					●			
Pintail	*Anas acuta*	Ánade Rabudo	Arrábio			●	●	●		●	●
Garganey	*Anas querquedula*	Cerceta Carretona	Marreco			●	●	●		●	
Blue-winged Teal	*Anas discors*	Cerceta Aliazul	Marreca-d'asa-azul			●	●	●			
Shoveler	*Anas clypeata*	Pato Cuchara	Pato-colhereiro	●		●	●	●		●	
Marbled Duck	*Marmaronetta angustirostris*	Cerceta Pardilla	Pardilheira			b	b	●			
Red-crested Pochard	*Netta rufina*	Pato Colorado	Pato-de-bico-vermelho					●			
Common Pochard	*Aythya ferina*	Porrón Común	Zarro				●	●	●	●	
Ring-necked Duck	*Aythya collaris*	Porrón Acollarado	Caturro				●	●	●	●	●
Ferruginous Duck	*Aythya nyroca*	Porrón Pardo	Pêrra			●	●	●			
Tufted Duck	*Aythya fuligula*	Porrón Moñudo	Negrinha				●	●		●	●
Greater Scaup	*Aythya marila*	Porrón Bastardo	Negrelho				●	●		●	
Lesser Scaup	*Aythya affinis*	Porrón Bola	Negrelho-americano					●		●	
Long-tailed Duck	*Clangula hyemalis*	Havelda	Pato-rabilongo		●						
Common Scoter	*Melanitta nigra*	Negrón Común	Negrola		●						
Bufflehead	*Bucephala albeola*	Porron Albeola	Olho-dourado-de-touca		●						
Hooded Merganser	*Mergus cucullatus*	Serreta Capuchona	Merganso-capuchinho							●	
Smew	*Mergus albellus*	Serreta Chica	Merganso-pequeno		●						
Red-breasted Merganser	*Mergus serrator*	Serreta Mediana	Merganso-de-poupa					●		●	
Honey-buzzard	*Pernis apivorus*	Abejero Europeo	Bútio-vespeiro			●	●	●	●	●	
Swallow-tailed Kite	*Elanoides forficatus*	Elanio Tijereta	Gavião-tesoura					●			
Black-shouldered Kite	*Elanus caeruleus*	Elanio Azul	Peneireiro-cinzento					●	●		
Black Kite	*Milvus migrans*	Milano Negro	Milhafre-preto	●	●	●	●	●	●	●	●
Red Kite	*Milvus milvus*	Milano Real	Milhafre-real	●	●	●	E	E	E		E
White-tailed Eagle	*Haliaeetus albicilla*	Pigargo Europeo	Pigargo					●			
Egyptian Vulture	*Neophron percnopterus*	Alimoche Común	Britango	B	B	B	E	E	E		E
Short-toed Eagle	*Circaetus gallicus*	Culebrera Europea	Águia-cobreira				●	●			
Marsh Harrier	*Circus aeruginosus*	Aguilucho Lagunero	Águia-sapeira			●	●	●		●	
Hen Harrier	*Circus cyaneus*	Aguilucho Pálido	Tartaranhão-cinzento			●	●	●	●	●	●
Pallid Harrier	*Circus macrourus*	Aguilucho Papialbo	Tartaranão-pálido					●			
Montagu's Harrier	*Circus pygargus*	Aguilucho Cenizo	Águia-caçadeira			●	●	●		●	
Northern Goshawk	*Accipiter gentilis*	Azor	Açor					●			

English name	Scientific name	Spanish name	Portuguese name	Islands							
				E	L	F	C	T	G	P	H
Eurasian Sparrowhawk	*Accipiter nisus*	Gavilán Común	Gavião	●	●	●	B	B	B	B	B
Common Buzzard	*Buteo buteo*	Busardo Ratonero	Águia-d'asa-redonda	●	E	B	B	B	B	B	B
Long-legged Buzzard	*Buteo rufinus*	Busardo Moro	Búteo-mourisco	●		●	●				●
Golden Eagle	*Aquila chrysaetos*	Águila Real	Águia-real						●	●	
Booted Eagle	*Aquila pennata*	Aguila Calzada	Águia-calçada	●	●	●	●		●	●	●
Bonelli's Eagle	*Aquila fasciata*	Aguila Perdicera	Águia-perdigueira	●			●				●
Osprey	*Pandion haliaetus*	Aguila Pescadora	Águia-pesqueira	B	B	E	E	B	B	E	B
Lesser Kestrel	*Falco naumanni*	Cernícalo Primilla	Francelho	●	●	●	●	●	●		
Common Kestrel	*Falco tinnunculus canariensis*	Cernícalo Vulgar	Peneireiro				B	B	B	B	B
	Falco tinnunculus dacotiae			B	B	B					
Red-footed Falcon	*Falco vespertinus*	Cernícalo Patirrojo	Falcão-vespertino	●	●	●	●	●			
Merlin	*Falco columbarius*	Esmerejón	Esmerilhão				●	●			●
Hobby	*Falco subbuteo*	Alcotán Europeo	Ógea				●	●	●		●
Eleonora's Falcon	*Falco eleonorae*	Halcón de Eleonora	Falcão-da-rainha	B	?	●	●	●	●	●	●
Lanner Falcon	*Falco biarmicus*	Halcón Borní	Alfaneque				●	●			
Peregrine Falcon	*Falco peregrinus*	Halcón Peregrino	Falcão-peregrino				●	●	●	●	●
Barbary Falcon	*Falco pelegrinoides*	Halcón de Berbería	Falcão-tagarote	B	B	B	B	B	B	B	B
Red-legged Partridge	*Alectoris rufa*	Perdiz Común	Perdiz						B		
Barbary Partridge	*Alectoris barbara*	Perdiz Moruna	Perdiz-mourisca	B	B	B	b		B	B	B
Common Quail	*Coturnix coturnix*	Codorniz	Codorniz	b	B	B	B	B	B	B	B
Helmeted Guineafowl	*Numida meleagris*	Pintada Común	Pintada			?	?	b			?
Water Rail	*Rallus aquaticus*	Rascón	Frango-d'água				●				
Spotted Crake	*Porzana porzana*	Polluela Pintoja	Franga-d'água-malhada		●	●	●	●		●	●
Little Crake	*Porzana parva*	Polluela Bastarda	Franga-d'água-bastarda	●			●	●		●	
Baillon's Crake	*Porzana pusilla*	Polluela Chica	Franga-d'água-pequena				●	●		●	
African Crake	*Crex egregia*	Guión Africano	Codornizão-africano					●			
Corn Crake	*Crex crex*	Guión de Cordonices	Codornizão				●				●
Moorhen	*Gallinula chloropus*	Polla de Agua	Galinha-d'água	●	●	B	B	B	B	?	●
Lesser Moorhen	*Gallinula angulata*	Gallineta Chica	Galinha-d'água-pequena						●		
Allen's Gallinule	*Porphyrula alleni*	Calamón de Allen	Camão-pequeno				●	●	●		●
American Purple Gallinule	*Porphyrula martinica*	Calamoncillo Americano	Camão-americano					●			
Eurasian Coot	*Fulica atra*	Focha Común	Galeirão			●	B	B	B	●	●
Red-knobbed Coot	*Fulica cristata*	Focha Cornuda	Galeirão-de-crista				●				
Common Crane	*Grus grus*	Grulla Común	Grou	●		●					
Little Bustard	*Tetrax tetrax*	Sisón	Sisão					●			
Houbara Bustard	*Chlamydotis undulata*	Hubara	Hubara	B	B	B	●				
Eurasian Oystercatcher	*Haematopus ostralegus*	Ostrero	Ostraceiro	●	●	●	●	●			●
Canary Islands Oystercatcher	*Haematopus meadewaldoi*	Ostrero Canario	Ostraceiro das Canárias	E	E	E					
Black-winged Stilt	*Himantopus himantopus*	Cigüeñuela	Pernilongo		B	b	b	●		●	
Avocet	*Recurvirostra avosetta*	Avoceta	Alfaiate		●	●	●	●			●
Stone-curlew	*Burhinus oedicnemus distinctus*	Alcaraván	Alcaravão				B	B	B	B	B
	Burhinus oedicnemus insularum			B	B	B					
Egyptian Plover	*Pluvianus aegyptius*	Pluvial Egipcio	Ave-do-crocodilo						●		
Cream-coloured Courser	*Cursorius cursor*	Corredor	Corredeira	?	B	B	b	b			
Collared Pratincole	*Glareola pratincola*	Canastera	Perdiz-do-mar				●	●	●	●	●
Little Ringed Plover	*Charadrius dubius*	Chorlitejo Chico	Borrelho-pequeno-de-coleira		●	B	B	B		●	●
Ringed Plover	*Charadrius hiaticula*	Chorlitejo Grande	Borrelho-grande-de-coleira	●	●	●	●	●	●	●	●
Killdeer	*Charadrius vociferus*	Chorlitejo Culirrojo	Borrelho-de-coleira-dupla					●			
Kentish Plover	*Charadrius alexandrinus*	Chorlitejo Patinegro	Borrelho-de-coleira-interrompida	B	B	B	B	B	●	●	
Lesser Sand Plover	*Charadrius mongolus*	Chorlitejo Mongol Chico	Borrelho-pequeño-de-colar-ruivo					●			
Dotterel	*Charadrius morinellus*	Chorlito Carambolo	Borrelho-ruivo	●	●	●	●	●			

APPENDIX B: DISTRIBUTION OF SPECIES ON THE CANARY ISLANDS

English name	Scientific name	Spanish name	Portuguese name	E	L	F	C	T	G	P	H
American Golden Plover	*Pluvialis dominica*	Chorlito Dorado Chico	Batuiruçu		●	●	●				●
Pacific Golden Plover	*Pluvialis fulva*	Chorlito Dorado Siberiano	Tarambola-dourada-siberiana				●				
European Golden Plover	*Pluvialis apricaria*	Chorlito Dorado Común	Tarambola-dourada	●	●	●	●	●		●	
Grey Plover	*Pluvialis squatarola*	Chorlito Gris	Tarambola-cinzenta	●	●	●	●	●	●	●	●
Sociable Lapwing	*Vanellus gregarius*	Avefría Sociable	Abibe-sociável								●
White-tailed Lapwing	*Vanellus leucurus*	Avefría Coliblanca	Abibe-de-cauda-branca				●				
Northern Lapwing	*Vanellus vanellus*	Avefría Europea	Abibe	●	●	●	●	●		●	●
Knot	*Calidris canutus*	Correlimos Gordo	Seixoeira		●	●	●	●			●
Sanderling	*Calidris alba*	Correlimos Tridáctilo	Pilrito-das-praias	●	●	●	●	●		●	●
Semipalmated Sandpiper	*Calidris pusilla*	Correlimos Semipalmeado	Pilrito-rasteirinho				●				
Western Sandpiper	*Calidris mauri*	Correlimos de Alaska	Pilrito-miúdo				●				
Little Stint	*Calidris minuta*	Correlimos Menudo	Pilrito-pequeño	●	●	●	●	●		●	●
Temminck's Stint	*Calidris temminckii*	Correlimos de Temminck	Pilrito de Temmink		●	●	●	●			
Least Sandpiper	*Calidris minutilla*	Correlimos Menudillo	Pilrito-anão		●	●					
White-rumped Sandpiper	*Calidris fuscicollis*	Correlimos Culiblanco	Pilrito-de-sobre-branco					●		●	●
Baird's Sandpiper	*Calidris bairdii*	Correlimos de Baird	Pilrito-de-bico-fino				●	●			
Pectoral Sandpiper	*Calidris melanotos*	Correlimos Pectoral	Pilrito-de-colete		●	●	●	●		●	●
Sharp-tailed Sandpiper	*Calidris acuminata*	Correlimos Acuminado	Pilrito-acuminado				●				
Curlew Sandpiper	*Calidris ferruginea*	Correlimos Zarapitín	Pilrito-de-bico-comprido	●	●	●	●	●		●	●
Purple Sandpiper	*Calidris maritima*	Correlimos Oscuro	Pilrito-escuro		●						
Dunlin	*Calidris alpina*	Correlimos Común	Pilrito-de-peito-preta	●	●	●	●	●		●	●
Buff-breasted Sandpiper	*Tryngites subruficollis*	Correlimos Canelo	Pilrito-acanelado				●				
Ruff	*Philomachus pugnax*	Combatiente	Combatente		●	●	●	●		●	●
Jack Snipe	*Lymnocryptes minimus*	Agachadiza Chica	Narceja-galega		●	●	●	●			●
Common Snipe	*Gallinago gallinago*	Agachadiza Común	Narceja		●	●	●	●		●	●
Wilson's Snipe	*Gallinago delicata*	Agachadiza Americana	Narceja-americana					●			
Great Snipe	*Gallinago media*	Agachadiza Real	Narceja-real					●			
Long-billed Dowitcher	*Limnodromus scolopaceus*	Agujeta Escolopácea	Maçarico-de-bico-comprido					●			
Woodcock	*Scolopax rusticola*	Chocha Perdiz	Galinhola	●		●	B	B	B	B	B
Black-tailed Godwit	*Limosa limosa*	Aguja Colinegra	Milherango		●	●	●	●		●	●
Bar-tailed Godwit	*Limosa lapponica*	Aguja Colipinta	Fuselo	●	●	●	●	●		●	●
Whimbrel	*Numenius phaeopus*	Zarapito Trinador	Maçarico-galego	●	●	●	●	●		●	●
Slender-billed Curlew	*Numenius tenuirostris*	Zarapito Fino	Maçarico-de-bico-fino		●						
Eurasian Curlew	*Numenius arquata*	Zarapito Real	Maçarico-real	●	●	●	●	●		●	●
Upland Sandpiper	*Bartramia longicauda*	Correlimos Batitú	Maçarico-do-campo					●			
Spotted Redshank	*Tringa erythropus*	Archibebe Oscuro	Perna-vermelha-bastardo		●	●	●	●		●	●
Common Redshank	*Tringa totanus*	Archibebe Común	Perna-vermelha	●	●	●	●	●		●	●
Marsh Sandpiper	*Tringa stagnatilis*	Archibebe Fino	Perna-verde-fino				●	●			
Greenshank	*Tringa nebularia*	Archibebe Claro	Perna-verde	●	●	●	●	●		●	●
Greater Yellowlegs	*Tringa melanoleuca*	Archibebe Patigualdo Grande	Perna-amarela-grande				●				
Lesser Yellowlegs	*Tringa flavipes*	Archibebe Patigualdo Chico	Perna-amarela-pequeno		●	●	●	●			
Solitary Sandpiper	*Tringa solitaria*	Andarríos Solitario	Maçarico-solitário					●			
Green Sandpiper	*Tringa ochropus*	Andarríos Grande	Maçarico-bique-bique		●	●	●	●		●	●
Wood Sandpiper	*Tringa glareola*	Andarríos Bastardo	Maçarico-de-dorso-malhado		●	●	●	●		●	●
Terek Sandpiper	*Xenus cinereus*	Andarríos del Terek	Maçarico-sovela				●				
Common Sandpiper	*Actitis hypoleucos*	Andarríos Chico	Maçarico-das-rochas		●	●	●	●		●	●
Spotted Sandpiper	*Actitis macularius*	Andarríos Maculado	Maçarico-pintado				●	●			
Turnstone	*Arenaria interpres*	Vuelvepiedras	Rola-do-mar	●	●	●	●	●		●	●
Wilson's Phalarope	*Phalaropus tricolor*	Falaropo de Wilson	Pisa-n'água					●			
Red-necked Phalarope	*Phalaropus lobatus*	Falaropo Picofino	Falaropo-de-bico-fino		●						
Grey Phalarope	*Phalaropus fulicarius*	Falaropo Picogrueso	Falaropo-de-bico-grosso					●		●	●
Pomarine Skua	*Stercorarius pomarinus*	Págalo Pomarino	Moleiro do Árctico			●	●	●	●	●	●
Arctic Skua	*Stercorarius parasiticus*	Págalo Parásito	Moleiro-pequeno		●	●	●	●	●	●	●

English name	Scientific name	Spanish name	Portuguese name	Islands							
				E	L	F	C	T	G	P	H
Long-tailed Skua	*Stercorarius longicaudus*	Págalo Rabero	Moleiro-rabilongo				●		●	●	
Great Skua	*Stercorarius skua*	Págalo Grande	Alcaide		●	●	●	●	●	●	●
South Polar Skua	*Stercorarius maccormicki*	Págalo Polar	Alcaide do Antárctico					At	Sea		
Pallas's Gull	*Larus ichthyaetus*	Gavión Cabecinegro	Gaivotão-de-cabeça-preta					●			
Mediterranean Gull	*Larus melanocephalus*	Gaviota Cabecinegra	Gaivota-de-cabeça-preta		●	●	●	●	●		●
Laughing Gull	*Larus atricilla*	Gaviota Guanaguanare	Gaivota-alegre						●		
Franklin's Gull	*Larus pipixcan*	Gaviota de Franklin	Gaivota-das-pradarias		●						
Little Gull	*Larus minutus*	Gaviota Enana	Gaivota-pequena		●						
Sabine's Gull	*Larus sabini*	Gaviota de Sabine	Gaivota de Sabine		●		●		●	●	
Bonaparte's Gull	*Larus philadelphia*	Gaviota de Bonaparte	Garrincho-americano		●					●	
Black-headed Gull	*Larus ridibundus*	Gaviota Reidora	Garrincho	●	●	●	●	●	●	●	●
Slender-billed Gull	*Larus genei*	Gaviota Picofina	Gaivota-de-bico-fino			b		●			
Audouin's Gull	*Larus audouinii*	Gaviota de Audouin	Gaivota de Audouin		●	●	●	●	●		
Ring-billed Gull	*Larus delawarensis*	Gaviota de Delaware	Gaivota-de-bico-riscado		●					●	
Common Gull	*Larus canus*	Gaviota Cana	Famego		●						
Lesser Black-backed Gull	*Larus fuscus*	Gaviota Sombría	Gaivota-d'asa-escura	●	●	●	●	●	●	●	●
Herring Gull	*Larus argentatus*	Gaviota Argéntea	Gaivota-prateada						●	●	
Yellow-legged Gull	*Larus michahellis*	Gaviota Patiamarilla	Gaivota-de-patas-amarelas	B	B	B	B	B	B	B	B
Iceland Gull	*Larus glaucoides*	Gaviota Polar	Gaivota-branca						●		
Glaucous Gull	*Larus hyperboreus*	Gaviota Hiperbórea	Gaivotão-branco		●		●		●		
Glaucous-winged Gull	*Larus glaucescens*	Gaviota Glauca	Gaivota do Mar de Bering								●
Great Black-backed Gull	*Larus marinus*	Gavión	Gaivotão-real		●		●		●	●	
Kittiwake	*Rissa tridactyla*	Gaviota Tridáctila	Gaivota-tridáctila		●	●	●		●	●	
Gull-billed Tern	*Gelochelidon nilotica*	Pagaza Piconegra	Tagáz		●		●	●	●		
Caspian Tern	*Hydroprogne caspia*	Pagaza Piquirroja	Garajau-grande					●			
Royal Tern	*Sterna maxima*	Charrán Real	Garajau-real					●	●	●	
Lesser Crested Tern	*Sterna bengalensis*	Charrán Bengalí	Garajau-pequeño						●		
Sandwich Tern	*Sterna sandvicensis*	Charrán Patinegro	Garajau	●	●	●	●	●	●	●	●
Roseate Tern	*Sterna dougallii*	Charrán Rosado	Gaivina-rosada		●		●		●	●	b
Common Tern	*Sterna hirundo*	Charrán Común	Gaivina		E	B	B	B	B	B	B
Arctic Tern	*Sterna paradisaea*	Charrán Ártico	Gaivina-ártica	●	●						
Forster's Tern	*Sterna forsteri*	Charrán de Forster	Gaivina de Forster						●		
White-cheeked Tern	*Sterna repressa*	Charrán Cariblanco	Gaivina-arábica						●		
Sooty Tern	*Onychoprion fuscata*	Charrán Sombrío	Gaivina-de-dorso-preto				●		●	●	
Little Tern	*Sternula albifrons*	Charráncito	Chilreta					●			
Whiskered Tern	*Chlidonias hybrida*	Fumarel Cariblanco	Gaivina-dos-pauis				●				
Black Tern	*Chlidonias niger*	Fumarel Común	Gaivina-preta						●		
White-winged Tern	*Chlidonias leucopterus*	Fumarel Aliblanco	Gaivina-d'asa-branca								●
Common Guillemot	*Uria aalge*	Arao Común	Airo						●		
Razorbill	*Alca torda*	Alca Común	Torda-mergulheira		●		●		●	●	
Little Auk	*Alle alle*	Mérgulo Marino	Torda-miúda		●						
Puffin	*Fratercula arctica*	Frailecillo Común	Papagaio-do-mar	●	●		●		●	●	
Black-bellied Sandgrouse	*Pterocles orientalis*	Ganga Ortega	Cortiçol-de-barriga-preta	?	?	B	●				
Pin-tailed Sandgrouse	*Pterocles alchata*	Ganga Ibérica	Ganga					●			
Rock Dove	*Columba livia*	Paloma Bravía	Pombo-das-rochas	B	B	B	B	B	B	B	B
Woodpigeon	*Columba palumbus*	Paloma Torcaz	Pombo-torcaz		●		●	●	●		●
Bolle's Pigeon	*Columba bollii*	Paloma Turqué	Pombo-turqueza				E	B	B	B	B
Laurel Pigeon	*Columba junoniae*	Paloma Rabiche	Pombo-rabil						B	B	?
African Collared Dove	*Streptopelia roseogrisea*	Tórtola de Cabeza Rosa	Rola-rosada		B	B	B	B	B		
Eurasian Collared Dove	*Streptopelia decaocto*	Tórtola Turca	Rola-turca	●	B	B	B	B	B	B	?
Turtle Dove	*Streptopelia turtur*	Tórtola Común	Rola-brava	?	B	B	B	B	B	B	B
Laughing Dove	*Streptopelia senegalensis*	Tórtola Senegalesa	Rola-dos-palmares			B		●	●		
Namaqua Dove	*Oena capensis*	Tortolita Rabilarga	Rolinha-rabilonga					●			
Budgerigar	*Melopsittacus undulatus*	Periquito Común	Periquito						?		
Senegal Parrot	*Poicephalus senegalus*	Lorito Senegalés	Periquito-massarongo						?	●	
Rose-ringed Parakeet	*Psittacula krameri*	Cotorra de Kramer	Periquito-rabijunco		B	?	B	B			
Monk Parakeet	*Myiopsitta monachus*	Cotorra Argentina	Caturrita		●	B	B	B			B
Great Spotted Cuckoo	*Clamator glandarius*	Críalo	Cuco-rabilongo	●	●	●	●	●	●		●
Common Cuckoo	*Cuculus canorus*	Cuco	Cuco	●	●	●	●	●	●	●	●

APPENDIX B: DISTRIBUTION OF SPECIES ON THE CANARY ISLANDS

English name	Scientific name	Spanish name	Portuguese name	E	L	F	C	T	G	P	H
Barn Owl	Tyto alba alba	Lechuza Común	Coruja-das-torres				B	B	B	B	B
	Tyto alba gracilirostris			B	B	B					
European Scops Owl	Otus scops	Autillo	Mocho-d'orelhas	●		●	●	●		●	
Eagle Owl	Bubo bubo	Búho Real	Bufo-real			●	●	●			
Hawk Owl	Surnia ulula	Lechuza Gavilana	Mocho-rabilongo				no details				
Tawny Owl	Strix aluco	Cárabo	Coruja-do-mato						●	●	
Long-eared Owl	Asio otus canariensis	Búho Chico	Bufo-pequeno		?	B	B	B	B	B	B
	Asio otus otus			●							
Short-eared Owl	Asio flammeus	Lechuza Campestre	Coruja-galhofa	●	●	●	●	●	●	●	●
Marsh Owl	Asio capensis	Lechuza Mora	Coruja-galhofa-mourisca					●			
European Nightjar	Caprimulgus europaeus	Chotacabras Gris	Noitibó-cinzento	●	●	●	●	●		●	
Red-necked Nightjar	Caprimulgus ruficollis	Chotacabras Pardo	Noitibó-de-nuca-vermelha				●				
Common Nighthawk	Chordeiles minor	Añapero Yanqui	Bacurau-norte-americano					●			
Chimney Swift	Chaetura pelagica	Vencejo de Chimenea	Rabo-espinhoso			●					
Plain Swift	Apus unicolor	Vencejo Unicolor	Andorinhão-da-serra	?	B	B	B	B	B	B	B
Common Swift	Apus apus	Vencejo Común	Andorinhão-preto	●	●	B	B	?	●	●	●
Pallid Swift	Apus pallidus	Vencejo Pálido	Andorinhão-pálido	?	B	B	B	B	?	?	●
Alpine Swift	Apus melba	Vencejo Real	Andorinhão-real			●	●	●	●	●	●
White-rumped Swift	Apus caffer	Vencejo Cafre	Andorinhão-cafre				●	●			
Little Swift	Apus affinis	Vencejo Culiblanco	Andorinhão-pequeno					●	●		
Common Kingfisher	Alcedo atthis	Martín Pescador	Guarda-rios			●	●	●	●		
Blue-cheeked Bee-eater	Merops persicus	Abejaruco Papirrojo	Abelharuco-de-faces-azuis			●		●			
European Bee-eater	Merops apiaster	Abejaruco Común	Abelharuco	●	●	●	●	●	●	●	●
European Roller	Coracias garrulus	Carraca	Rolieiro	●	●	●	●	●		●	
Hoopoe	Upupa epops	Abubilla	Poupa	B	B	B	B	B	B	B	B
Wryneck	Jynx torquilla	Torcecuello	Torcícolo	●	●	●	●	●			
Great Spotted Woodpecker	Dendrocopos major canariensis	Pico Picapinos	Pica-pau-malhado							B	
	Dendrocopos major thanneri							B			
Bar-tailed Lark	Ammomanes cincturus	Terrera Colinegra	Calhandra-das-dunas						●	●	
Greater Hoopoe Lark	Alaemon alaudipes	Sirli Desértico	Calhandra-de-bico-curva						●	●	
Dupont's Lark	Chersophilus duponti	Alondra de Dupont	Calhandra de Dupont						●	●	
Calandra Lark	Melanocorypha calandra	Calandria Común	Calhandra-real					●	●	●	●
Short-toed Lark	Calandrella brachydactyla	Terrera Común	Calhandrinha	●			●		●		
Lesser Short-toed Lark	Calandrella rufescens rufescens	Terrera Marismeña	Calhandrinha-das-marismas						?	B	
	Calandrella rufescens polatzeki			B	B	B	B	?			
Crested Lark	Galerida cristata	Cogujada Común	Cotovia-de-poupa						●		
Thekla Lark	Galerida theklae	Cogujada Montesina	Cotovia-escura						●		
Skylark	Alauda arvensis	Alondra Común	Alberca	●	●	●	●	●	●	●	●
Plain Martin	Riparia paludicola	Avión Paludícola	Andorinha-dos-charcos	●	●						
Sand Martin	Riparia riparia	Avión Zapador	Andorinha-das-barreiras	●	●	●	●	●		●	
Crag Martin	Ptyonoprogne rupestris	Avión Roquero	Andorinha-das-roches						●	●	
Barn Swallow	Hirundo rustica	Golondrina Común	Andorinha-das-chaminés	●	●	●	●	b	●	●	●
Red-rumped Swallow	Cecropis daurica	Golondrina Dáurica	Andorinha-dáurica	●	●	●	●	●	●	●	●
American Cliff Swallow	Petrochelidon pyrrhonota	Avión Roquero Americano	Andorinha-de-testa-branca						●		
House Martin	Delichon urbicum	Avión Común	Andorinha-dos-beirais	●	●	●	●	●		●	
Richard's Pipit	Anthus richardi	Bisbita de Richard	Petinha de Richard						●		
Tawny Pipit	Anthus campestris	Bisbita Campestre	Petinha-dos-campos	●	●		●	●		●	
Berthelot's Pipit	Anthus berthelotii berthelotii	Bisbita Caminero	Corre-caminho	B	B	B	B	B	B	B	B
Olive-backed Pipit	Anthus hodgsoni	Bisbita de Hodgson	Petinha de Hodgson						●		
Tree Pipit	Anthus trivialis	Bisbita Arbóreo	Petinha-das-árvores	●	●	●	●	●	●	●	●
Meadow Pipit	Anthus pratensis	Bisbita Común	Petinha-dos-prados		●	●	●	●	●	●	●
Red-throated Pipit	Anthus cervinus	Bisbita Gorgirrojo	Petinha-de-garganta-ruiva	●	●		●	●		●	
Rock Pipit	Anthus petrosus	Bisbita Costero	Petinha-marítima					●			
Water Pipit	Anthus spinoletta	Bisbita Ribereño	Petinha-ribeirinha		●			●			

English name	Scientific name	Spanish name	Portuguese name	E	L	F	C	T	G	P	H
Yellow Wagtail	*Motacilla flava*	Lavandera Boyera	Alvéola-amarela	●	●	●	●	●	●	●	●
	Motacilla flava feldegg					●					
Citrine Wagtail	*Motacilla citreola*	Lavandera Cetrina	Alvéola-citrina					●			
Grey Wagtail	*Motacilla cinerea canariensis*	Lavandera Cascadeña	Alvéola-cinzenta				B	B	B	B	●
	Motacilla cinerea cinerea			●	●	●					
White Wagtail	*Motacilla alba*	Lavandera Blanca	Alvéola-branca	●	●	●	●	●	●	●	●
Dipper	*Cinclus cinclus*	Mirlo Acuático	Melro-d'água			●					
Wren	*Troglodytes troglodytes*	Chochín	Cariça			●					
Grey Catbird	*Dumetella carolinensis*	Pájaro Gato	Pássaro-gato						●		
Dunnock	*Prunella modularis*	Acentor Común	Ferreirinha						●		
Alpine Accentor	*Prunella collaris*	Acentor Alpino	Ferreirinha-alpina						●		
Rufous Bush Robin	*Cercotrichas galactotes*	Alzacola	Solitário	●	●	●	●	●			
European Robin	*Erithacus rubecula rubecula*	Petirrojo	Pisco-de-peito-ruivo	●	●	●	●	●	B	B	B
	Erithacus (rubecula) superbus						B	B			
Thrush Nightingale	*Luscinia luscinia*	Ruiseñor Ruso	Rouxinol-oriental					●			
Common Nightingale	*Luscinia megarhynchos*	Ruiseñor Común	Rouxinol	●	●	●	●	●			
Bluethroat	*Luscinia svecica*	Pechiazul	Pisco-de-peito-azul								
Black Redstart	*Phoenicurus ochruros*	Colirrojo Tizón	Rabirruivo	●	●	●	●			●	●
Common Redstart	*Phoenicurus phoenicurus*	Colirrojo Real	Rabirruivo-de-testa-branca	●	●	●	●	●	●	●	
Whinchat	*Saxicola rubetra*	Tarabilla Norteña	Cartaxo-d'arribação	●	●	●		●		●	
Canary Islands Stonechat	*Saxicola dacotiae*	Tarabilla Canaria	Caldeireta	E	●	B					
Common Stonechat	*Saxicola torquatus*	Tarabilla Común	Cartaxo								
Isabelline Wheatear	*Oenanthe isabellina*	Collalba Isabel	Chasco-isabel	●	●		●				
Northern Wheatear	*Oenanthe oenanthe*	Collalba Gris	Chasco-cinzento	●	●	●	●	●		●	●
Black-eared Wheatear	*Oenanthe hispanica*	Collalba Rubia	Chasco-ruivo	●	●	●	●	●			
Desert Wheatear	*Oenanthe deserti*	Collalba Desértica	Chasco-do-deserto	●			●	●			
Mourning Wheatear	*Oenanthe lugens*	Collalba Núbica	Chasco-fúnebre					●			
White-crowned Wheatear	*Oenanthe leucopyga*	Collalba Negra de Brehm	Chasco-de-barrete-branco							●	
Rufous-tailed Rock Thrush	*Monticola saxatilis*	Roquero Rojo	Macuco	●	●					●	
Blue Rock Thrush	*Monticola solitarius*	Roquero Solitario	Melro-azul					●			
Ring Ouzel	*Turdus torquatus*	Mirlo de Collar	Melro-de-peito-branco	●			●		●	●	●
Blackbird	*Turdus merula cabrerae*	Mirlo Común	Melro				B	B	B	B	B
	Turdus merula ssp.						●	●			
Fieldfare	*Turdus pilaris*	Zorzal Real	Tordo-zornal			●	●	●			
Song Thrush	*Turdus philomelos*	Zorzal Común	Tordo-pinto	●	●	●	●	●	●	●	●
Redwing	*Turdus iliacus*	Zorzal Alirrojo	Tordo-ruivo			●		●			
Mistle Thrush	*Turdus viscivorus*	Zorzal Charlo	Tordoveia					●			
Zitting Cisticola	*Cisticola juncidis*	Buitrón	Fuínha-dos-juncos		●						
Grasshopper Warbler	*Locustella naevia*	Buscarla Pintoja	Cigarrinha-malhada			●					
Savi's Warbler	*Locustella luscinioides*	Buscarla Unicolor	Cigarrinha-ruiva					●			
Aquatic Warbler	*Acrocephalus paludicola*	Carricerín Cejudo	Felosa-aquática	●	●	●	●	●			
Sedge Warbler	*Acrocephalus schoenobaenus*	Carricerín Común	Felosa-dos-juncos	●			●	●		●	●
Reed Warbler	*Acrocephalus scirpaceus*	Carricero Común	Rouxinol-dos-caniços				●	●	●	●	
Great Reed Warbler	*Acrocephalus arundinaceus*	Carricero Tordal	Rouxinol-grande-dos-caniços	●	●		●	●			
Eastern Olivaceous Warbler	*Hippolais pallida*	Zarcero Pálido Oriental	Felosa-pálida-oriental					●			
Western Olivaceous Warbler	*Hippolais opaca*	Zarcero Pálido Occidental	Felosa-pálida	●	●	●	●	●			
Icterine Warbler	*Hippolais icterina*	Zarcero Icterino	Felosa-amarela					●			
Melodious Warbler	*Hippolais polyglotta*	Zarcero Común	Felosa-poliglota	●	●	●	●	●			
Tristram's Warbler	*Sylvia deserticola*	Curruca de Tristram	Toutinegra do Atlas				●				

APPENDIX B: DISTRIBUTION OF SPECIES ON THE CANARY ISLANDS

English name	Scientific name	Spanish name	Portuguese name	E	L	F	C	T	G	P	H
Spectacled Warbler	*Sylvia conspicillata*	Curruca Tomillera	Toutinegra-tomilheira	B	B	B	B	B	B	B	B
Subalpine Warbler	*Sylvia cantillans*	Curruca Carrasqueña	Toutinegra-de-bigodes	●	●	●	●	●	●	●	●
Ménétries's Warbler	*Sylvia mystacea*	Curruca de Menetries	Toutinegra-rosada			●					
Sardinian Warbler	*Sylvia melanocephala leucogastra*	Curruca Cabecinegra	Toutinegra-dos-valados	●		B	B	B	B	B	B
African Desert Warbler	*Sylvia deserti*	Curruca Sahariana	Toutinegra-do-deserto			●		●	●	●	
Orphean Warbler	*Sylvia hortensis*	Curruca Mirlona	Toutinegra-real			●	●	●	●	●	
Lesser Whitethroat	*Sylvia curruca*	Curruca Zarcerilla	Papa-amoras-cinzento			●	●	●	●		
Common Whitethroat	*Sylvia communis*	Curruca Zarcera	Papa-amoras-comum	●	●	●	●	●	●		●
Garden Warbler	*Sylvia borin*	Curruca Mosquitera	Felosa-das-figueiras	●		?	?	●	●	●	
Blackcap	*Sylvia atricapilla heineken*	Curruca Capirotada	Toutinegra-de-barrete	●	?	?	B	B	B	B	B
Yellow-browed Warbler	*Phylloscopus inornatus*	Mosquitero Bilistado	Felosa-listada			●	●	●	●		
Hume's Warbler	*Phylloscopus humei*	Mosquitero de Hume	Felosa-listada de Hume					●			
Western Bonelli's Warbler	*Phylloscopus bonelli*	Mosquitero Papialbo Occidental	Felosa-de-papo-branco	●		●	●	●	●		●
Wood Warbler	*Phylloscopus sibilatrix*	Mosquitero Silbador	Felosa-assobiadeira	●		●	●	●	●	●	
Common Chiffchaff	*Phylloscopus collybita collybita*	Mosquitero Común	Felosinha	●		●	●	●	●		
	P. c. abietinus			●							
Iberian Chiffchaff	*Phylloscopus ibericus*	Mosquitero Ibérico	Felosinha-ibérica	●				●			
Canary Island Chiffchaff	*Phylloscopus canariensis*	Mosquitero Canario	Felosinha das Canarias			B	B	B	B	B	B
Willow Warbler	*Phylloscopus trochilus*	Mosquitero Musical	Felosa-musical	●		●	●	●	●		●
Tenerife Kinglet	*Regulus teneriffae*	Reyezuelo Tinerfeño	Bandeirita					B	B	B	B
Spotted Flycatcher	*Muscicapa striata*	Papamoscas Gris	Papa-moscas-cinzento					●			
Red-breasted Flycatcher	*Ficedula parva*	Papamoscas Papirrojo	Papa-moscas-pequeño			●	●	●	●		
Collared Flycatcher	*Ficedula albicollis*	Papamoscas Collarino	Papa-moscas-de-colar					●			
Pied Flycatcher	*Ficedula hypoleuca*	Papamoscas Cerrojillo	Papas-moscas preto	●		●	●	●	●	●	
African Blue Tit	*Cyanistes ultramarinus teneriffae*	Herrerillo Africano	Chapim-azul					B	B	B	
	C. u. degener				B	B					
	C. u. palmensis									B	
	C. u. ombriosus										B
Great Tit	*Parus major*	Carbonero Común	Chapin-real					●			
Eurasian Nuthatch	*Sitta europaea*	Trepador Azul	Trepadeira-azul			●					
Penduline Tit	*Remiz pendulinus*	Pájaro Moscón	Chapim-de-mascarilha					●			
Golden Oriole	*Oriolus oriolus*	Oropéndola	Papa-figos	●		●	●	●	●	●	
Isabelline Shrike	*Lanius isabellinus*	Alcaudón Isabel	Picanço-isabel					●			
Red-backed Shrike	*Lanius collurio*	Alcaudón Dorsirrojo	Picanço-de-dorso-ruivo	●	●			●	●		
Southern Grey Shrike	*Lanius meridionalis koenigi*	Alcaudón Real	Picanço-real	B	B	B	B	B		E	
Woodchat Shrike	*Lanius senator*	Alcaudón Común	Picanço-barreteiro	●		●	●	●	●	●	●
Spotted Nutcracker	*Nucifraga caryocatactes*	Cascanueces	Quebra-nozes						●		
Red-billed Chough	*Pyrrhocorax pyrrhocorax*	Chova Piquirroja	Gralha-de-bico-vermelho						●	●	B
Western Jackdaw	*Corvus monedula*	Grajilla	Gralha-de-nuca-cinzenta						●		
Rook	*Corvus frugilegus*	Graja	Gralha-calva			●					
Carrion Crow	*Corvus corone*	Corneja Negra	Gralha-preta			●					
Brown-necked Raven	*Corvus ruficollis*	Cuervo Desertícola	Corvo-do-deserto			●	●				
Common Raven	*Corvus corax*	Cuervo	Corvo	B	B	B	B	B	B	B	B
Common Starling	*Sturnus vulgaris*	Estornino Pinto	Estorninho-malhado	●	●	●	B	B	●	●	●
Spotless Starling	*Sturnus unicolor*	Estornino Negro	Estorninho-preto					●	●		
Rosy Starling	*Sturnus roseus*	Estornino Rosado	Estorninho-rosado						●		
Common Myna	*Acridotheres tristis*	Miná Común	Mainato						B		
House Sparrow	*Passer domesticus*	Gorrión Doméstico	Pardal					B			
Spanish Sparrow	*Passer hispaniolensis*	Gorrión Moruno	Pardal-espanhol	B	B	B	B	B	B	B	B
Tree Sparrow	*Passer montanus*	Gorrión Molinero	Pardal-montês					B	●		
Rock Sparrow	*Petronia petronia*	Gorrión Chillón	Pardal-francês					B	B	B	B
Snowfinch	*Montifringilla nivalis*	Gorrión Alpino	Pardal-alpino						●		
Red-billed Quelea	*Quelea quelea*	Quelea Común	Bico-carmim					●			
Orange-cheeked Waxbill	*Estrilda melpoda*	Estrilda de Carita Naranja	Face-laranja						b		

BIRDS OF THE ATLANTIC ISLANDS

English name	Scientific name	Spanish name	Portuguese name	Islands								
				E	L	F	C	T	G	P	H
Common Waxbill	*Estrilda astrild*	Pico de Coral	Bico-de-lacre				B	b			
Common Chaffinch	*Fringilla coelebs canariensis*	Pinzón Vulgar	Tentilhão				B	B	B		
	Fringilla coelebs palmae									B	
	Fringilla coelebs ombrosus										B
	Fringilla coelebs coelebs			●	●	●	●	●			
Blue Chaffinch	*Fringilla teydea polatzeki*	Pinzón Azul	Tentilhão-azul				B				
	Fringilla teydea teydea							B			
Brambling	*Fringilla montifringilla*	Pinzón Real	Tentilhão-montês					●		●	
European Serin	*Serinus serinus*	Verdecillo	Milheirinha		●	●	B	B	●	●	
Canary	*Serinus canaria*	Canario	Canario-da-terra		B	B	B	B	B	B	B
Greenfinch	*Carduelis chloris*	Verderón	Verdilhão	●	●	B	B	B	B	?	B
Goldfinch	*Carduelis carduelis*	Jilguero	Pintassilgo		?	B	B	B	B	B	?
Siskin	*Carduelis spinus*	Lúgano	Lugre		●	●	●	●		●	
Linnet	*Carduelis cannabina harterti*	Pardillo Común	Pintarroxo	B	B	B					
	Carduelis cannabina meadewaldoi						B	B	B	B	B
Common Crossbill	*Loxia curvirostra*	Piquituerto Común	Cruza-bico						●	●	
Trumpeter Finch	*Bucanetes githagineus*	Camachuelo Trompetero	Trombeteiro	B	B	B	B	B	B		●
Yellow-rumped Warbler	*Dendroica coronata*	Reinita Coronada	Mariquita-coroada					●			
Blackpoll Warbler	*Dendroica striata*	Reinita Listada	Mariquita-de-perna-clara					●			
Louisiana Waterthrush	*Seiurus motacilla*	Reinita de Louisiana	Mariquita da Louisiana							●	
Lapland Bunting	*Calcarius lapponicus*	Escribano Lapón	Escrevedeira da Lapónia						●		
Snow Bunting	*Plectrophenax nivalis*	Escribano Nival	Escrevedeira-das-neves	●	●				●	●	
Cirl Bunting	*Emberiza cirlus*	Escribano Soteño	Escrevedeira						●		
Rock Bunting	*Emberiza cia*	Escribano Montesino	Cia					●		●	
House Bunting	*Emberiza striolata*	Escribano Sahariano	Escrevedeira-domestica						●		
Ortolan Bunting	*Emberiza hortulana*	Escribano Hortelano	Sombria	●		●					
Cretzschmar's Bunting	*Emberiza caesia*	Escribano Ceniciento	Sombria-de-capuz							●	
Little Bunting	*Emberiza pusilla*	Escribano Pigmeo	Escrevedeira-pequena	●		●					
Reed Bunting	*Emberiza schoeniculus*	Escribano Palustre	Escrevedeira-dos-caniços		●					●	
Corn Bunting	*Emberiza calandra*	Triguero	Trigueirão		B	B	B	B	B	B	B

ii. Madeira and the Salvage Islands

M = Madeira
S = Salvage Islands

B = current breeding species
b = species that has bred
? = possibly breeds
E = now extinct as a breeding species
● = has occurred

English name	Scientific name	Spanish name	Portuguese name	Islands	
				M	S
Great Northern Diver	*Gavia immer*	Colimbo Grande	Mobelha-grande	●	
Little Grebe	*Tachybaptus ruficollis*	Zampullín Común	Mergulhão-pequeno	●	
Slavonian Grebe	*Podiceps auritus*	Zampullín Cuellirrojo	Mergulhão-de-penachos	●	
Black-necked Grebe	*Podiceps nigricollis*	Zampullín Cuellinegro	Cagarraz	●	
Fea's Petrel	*Pterodroma feae*	Petrel Gon-gon	Gon-Gon	B	●
Zino's Petrel	*Pterodroma madeira*	Petrel Freira	Freira	B	
Soft-plumaged Petrel	*Pterodroma mollis*	Petrel Suave	Freira-de-penas-lisas	●	
Bulwer's Petrel	*Bulweria bulwerii*	Petrel de Bulwer	Alma-negra	B	B
Cory's Shearwater	*Calonectris diomedea*	Pardela Cenicienta	Cagarra	B	B
Great Shearwater	*Puffinus gravis*	Pardela Capirotada	Pardela-de-barrete	●	
Sooty Shearwater	*Puffinus griseus*	Pardela Sombría	Pardela-preta	●	
Balearic Shearwater	*Puffinus mauretanicus*	Pardela Balear	Fura-bucho	●	
Manx Shearwater	*Puffinus puffinus*	Pardela Pichoneta	Fura-bucho do Atlântico	B	
Macaronesian Sh'water	*Puffinus baroli baroli*	Pardela Chica	Pintainho	B	B

APPENDIX B: DISTRIBUTION OF SPECIES ON MADEIRA

English name	Scientific name	Spanish name	Portuguese name	Islands M	S
Wilson's Storm-petrel	*Oceanites oceanicus*	Paíño de Wilson	Casquilho	●	
White-faced Storm-petrel	*Pelagodroma marina*	Paíño Pechialbo	Calca-mar	●	B
European Storm-petrel	*Hydrobates pelagicus*	Paíño Común	Alma-de-mestre	●	●
Leach's Storm-petrel	*Oceanodroma leucorhoa*	Paíño de Leach	Paínho-de-cauda-forcada	●	
Swinhoe's Storm-petrel	*Oceanodroma monorhis*	Paíño de Swinhoe	Painho de Swinhoe		●
Madeiran Storm-petrel	*Oceanodroma castro*	Paíño de Madeira	Roquinho	B	B
Red-billed Tropicbird	*Phaethon aethereus*	Rabijunco Etéreo	Rabijunco	●	
Northern Gannet	*Morus bassanus*	Alcatraz Atlantico	Alcatraz	●	●
Cormorant	*Phalacrocorax carbo*	Cormorán Grande	Corvo-marinho	●	
Shag	*Phalacrocorax aristotelis*	Cormorán Moñudo	Galheta	●	
Eurasian Bittern	*Botaurus stellaris*	Avetoro Común	Abetouro	●	
Little Bittern	*Ixobrychus minutus*	Avetorillo Común	Garçote	●	●
Night Heron	*Nycticorax nycticorax*	Martinete Común	Goraz	●	
Squacco Heron	*Ardeola ralloides*	Garcilla Cangrejera	Papa-ratos	●	
Cattle Egret	*Bubulcus ibis*	Garcilla Bueyera	Carraceiro	●	●
Western Reef Egret	*Egretta gularis*	Garceta Sombría	Garça-dos-recifes	●	
Little Egret	*Egretta garzetta*	Garceta Común	Garça-branca	●	●
Grey Heron	*Ardea cinerea*	Garza Real	Garça-real	●	●
Purple Heron	*Ardea purpurea*	Garza Imperial	Garça-ruiva	●	●
Black Stork	*Ciconia nigra*	Cigüeña Negra	Cegonha-preta	●	
White Stork	*Ciconia ciconia*	Cigüeña Común	Cegonha-branca	●	
Glossy Ibis	*Plegadis falcinellus*	Morito	Ibis-preto	●	
Eurasian Spoonbill	*Platalea leucorodia*	Espátula	Colheireo	●	
Bean Goose	*Anser fabalis*	Ánsar Campestre	Ganso-campestre	●	
Pink-footed Goose	*Anser brachyrhynchus*	Ánsar Piquicorto	Ganso-de-bico-curto	●	
Greater White-fronted Goose	*Anser albifrons*	Ánsar Careto Grande	Ganso-de-testa-branca	●	
Greylag Goose	*Anser anser*	Ánsar Común	Ganso-bravo	●	
Ruddy Shelduck	*Tadorna ferruginea*	Tarro Canelo	Pato-casarca	●	
Common Shelduck	*Tadorna tadorna*	Tarro Blanco	Tadorna	●	
Eurasian Wigeon	*Anas penelope*	Silbón Europeo	Piadeira	●	
Eurasian Teal	*Anas crecca*	Cerceta Común	Marrequinha	●	
Mallard	*Anas platyrhynchos*	Ánade Azulón	Pato-real	●	
Pintail	*Anas acuta*	Ánade Rabudo	Arrábio	●	
Garganey	*Anas querquedula*	Cerceta Carretona	Marreco	●	
Shoveler	*Anas clypeata*	Pato Cuchara	Pato-colhereiro	●	
Marbled Duck	*Marmaronetta angustirostris*	Cerceta Pardilla	Pardilheira	●	
Common Pochard	*Aythya ferina*	Porrón Común	Zarro	●	
Ring-necked Duck	*Aythya collaris*	Porrón Acollarado	Caturro	●	
Tufted Duck	*Aythya fuligula*	Porrón Moñudo	Negrinha	●	
Greater Scaup	*Aythya marila*	Porrón Bastardo	Negrelho	●	
Long-tailed Duck	*Clangula hyemalis*	Havelda	Pato-rabilongo	●	
Common Scoter	*Melanitta nigra*	Negrón Común	Negrola	●	
Surf Scoter	*Melanitta perspicillata*	Negrón Careto	Negrola-de-lunetas	●	
Common Goldeneye	*Bucephala clangula*	Porrón Osculado	Olho-dourado	●	
Red-breasted Merganser	*Mergus serrator*	Serreta Mediana	Merganso-de-poupa	●	
Honey-buzzard	*Pernis apivorus*	Abejero Europeo	Bútio-vespeiro	●	
Black Kite	*Milvus migrans*	Milano Negro	Milhafre-preto	●	●
Red Kite	*Milvus milvus*	Milano Real	Milhafre-real	●	
Egyptian Vulture	*Neophron percnopterus*	Alimoche Común	Britango	●	
Marsh Harrier	*Circus aeruginosus*	Aguilucho Lagunero Occidental	Águia-sapeira	●	●
Hen Harrier	*Circus cyaneus*	Aguilucho Pálido	Tartaranhão-cinzento	●	
Montagu's Harrier	*Circus pygargus*	Aguilucho Cenizo	Águia-caçadeira	●	●
Eurasian Sparrowhawk	*Accipiter nisus*	Gavilán Común	Gavião	●	
Common Buzzard	*Buteo buteo*	Busardo Ratonero	Águia-d'asa-redonda	B	●
Long-legged Buzzard	*Buteo rufinus*	Busardo Moro	Búteo-mourisco	●	●
Booted Eagle	*Aquila pennata*	Aguila Calzada	Águia-calçada	●	
Osprey	*Pandion haliaetus*	Aguila Pescadora	Águia-pesqueira	●	
Lesser Kestrel	*Falco naumanni*	Cernícalo Primilla	Francelho	●	

English name	Scientific name	Spanish name	Portuguese name	Islands	
				M	S
Common Kestrel	*Falco tinnunculus tinnunculus*	Cernícalo Vulgar	Peneireiro	●	
	Falco tinnunculus canariensis			B	b
Red-footed Falcon	*Falco vespertinus*	Cernícalo Patirrojo	Falcão-pés-vermelhos	●	
Merlin	*Falco columbarius*	Esmerejón	Esmerilhão		●
Hobby	*Falco subbuteo*	Alcotán Europeo	Ógea	●	●
Eleonora's Falcon	*Falco eleonorae*	Halcón de Eleonora	Falcão-da-rainha	●	●
Peregrine Falcon	*Falco peregrinus*	Halcón Peregrino	Falcão-peregrino	●	●
Barbary Falcon	*Falco pelegrinoides*	Halcón de Berbería	Falcão-tagarote	●	●
Red-legged Partridge	*Alectoris rufa*	Perdiz Común	Perdiz	●	
Barbary Partridge	*Alectoris barbara*	Perdiz Moruna	Perdiz-mourisca	E	
Common Quail	*Coturnix coturnix*	Codorniz	Codorniz	B	●
Water Rail	*Rallus aquaticus*	Rascón	Frango-d'água	●	
Spotted Crake	*Porzana porzana*	Polluela Pintoja	Franga-d'água-malhada	●	●
Little Crake	*Porzana parva*	Polluela Bastarda	Franga-d'água-bastarda	●	
Baillon's Crake	*Porzana pusilla*	Polluela Chica	Franga-d'água-pequena	●	
Corn Crake	*Crex crex*	Guión de Cordonices	Codornizão	●	
Black Crake	*Limnocorax flavirostra*	Polluela Negra Africana	Franga-d'água-preta	●	
Moorhen	*Gallinula chloropus*	Gallineta Común	Galinha-d'água	●	●
Allen's Gallinule	*Porphyrula alleni*	Calamón de Allen	Camão-pequeno	●	
American Purple Gallinule	*Porphyrula martinica*	Calamoncillo Americano	Camão-americano	●	
Common Coot	*Fulica atra*	Focha Común	Galeirão	●	
Common Crane	*Grus grus*	Grulla Común	Grou	●	
Little Bustard	*Tetrax tetrax*	Sisón	Sisão	●	
Eurasian Oystercatcher	*Haematopus ostralegus*	Ostrero	Ostraceiro	●	
Canary Islands Oystercatcher	*Haematopus meadewaldoi*	Ostrero Canario	Ostraceiro das Canárias	●	
Black-winged Stilt	*Himantopus himantopus*	Cigüeñuela	Pernilongo	●	
Avocet	*Recurvirostra avosetta*	Avoceta	Alfaiate	●	
Stone-curlew	*Burhinus oedicnemus*	Alcaraván	Alcaravão	●	
Cream-coloured Courser	*Cursorius cursor*	Corredor	Corredeira	●	
Collared Pratincole	*Glareola pratincola*	Canastera	Perdiz-do-mar	●	●
Little Ringed Plover	*Charadrius dubius*	Chorlitejo Chico	Borrelho-pequeno-de-coleira	B	●
Ringed Plover	*Charadrius hiaticula*	Chorlitejo Grande	Borrelho-grande-de-coleira	●	●
Semipalmated Plover	*Charadrius semipalmatus*	Chorlitejo Semipalmeado	Batuíra-de-bando	●	
Killdeer	*Charadrius vociferus*	Chorlitejo Culirrojo	Borrelho-de-coleira-dupla	●	
Kentish Plover	*Charadrius alexandrinus*	Chorlitejo Patinegro	Borrelho-de-coleira-interrompida	B	●
Eurasian Dotterel	*Charadrius morinellus*	Chorlito Carambolo	Borrelho-ruivo	●	●
European Golden Plover	*Pluvialis apricaria*	Chorlito Dorado Común	Tarambola-dourada	●	●
Grey Plover	*Pluvialis squatarola*	Chorlito Gris	Tarambola-cinzenta	●	
Northern Lapwing	*Vanellus vanellus*	Avefría Europea	Abibe	●	●
Knot	*Calidris canutus*	Correlimos Gordo	Seixoeira	●	
Sanderling	*Calidris alba*	Correlimos Tridáctilo	Pilrito-das-praias	●	●
Semipalmated Sandpiper	*Calidris pusilla*	Correlimos Semipalmeado	Pilrito-rasteirinho	●	
Western Sandpiper	*Calidris mauri*	Correlimos de Alaska	Pilrito-miúdo	●	
Little Stint	*Calidris minuta*	Correlimos Menudo	Pilrito-pequeño	●	
Temminck's Stint	*Calidris temminckii*	Correlimos de Temminck	Pilrito de Temminck	●	
White-rumped Sandpiper	*Calidris fuscicollis*	Correlimos Culiblanco	Pilrito-de-sobre-branco	●	
Pectoral Sandpiper	*Calidris melanotus*	Correlimos Pectoral	Pilrito-de-colete	●	
Curlew Sandpiper	*Calidris ferruginea*	Correlimos Zarapitín	Pilrito-de-bico-comprido	●	
Purple Sandpiper	*Calidris maritima*	Correlimos Oscuro	Pilrito-escuro	●	
Dunlin	*Calidris alpina*	Correlimos Común	Pilrito-de-peito-preta	●	●
Ruff	*Philomachus pugnax*	Combatiente	Combatente	●	

APPENDIX B: DISTRIBUTION OF SPECIES ON MADEIRA

English name	Scientific name	Spanish name	Portuguese name	Islands M	S
Jack Snipe	Lymnocryptes minimus	Agachadiza Chica	Narceja-galega	●	
Common Snipe	Gallinago gallinago	Agachadiza Común	Narceja	●	
Great Snipe	Gallinago media	Agachadiza Real	Narceja-real	●	
Dowitcher sp.	Limnodromus sp.	Agujeta sp.	Maçarico sp.		●
Woodcock	Scolopax rusticola	Chocha Perdiz	Galinhola	●	
Black-tailed Godwit	Limosa limosa	Aguja Colinegra	Milherango	●	●
Bar-tailed Godwit	Limosa lapponica	Aguja Colipinta	Fuselo	●	
Whimbrel	Numenius phaeopus	Zarapito Trinador	Maçarico-galego	●	
Eurasian Curlew	Numenius arquata	Zarapito Real	Maçarico-real	●	
Upland Sandpiper	Bartramia longicauda	Correlimos Batitú	Maçarico-do-campo	●	
Spotted Redshank	Tringa erythropus	Archibebe Oscuro	Perna-vermelha-bastardo	●	
Common Redshank	Tringa totanus	Archibebe Común	Perna-vermelha	●	●
Greenshank	Tringa nebularia	Archibebe Claro	Perna-verde	●	
Lesser Yellowlegs	Tringa flavipes	Archibebe Patigualdo Chico	Perna-amarela-pequeno	●	
Green Sandpiper	Tringa ochropus	Andarríos Grande	Maçarico-bique-bique	●	●
Wood Sandpiper	Tringa glareola	Andarríos Bastardo	Maçarico-de-dorso-malhado	●	
Common Sandpiper	Actitis hypoleucos	Andarríos Chico	Maçarico-das-rochas	●	●
Spotted Sandpiper	Actitis macularius	Andarríos Maculado	Maçarico-pintado	●	
Turnstone	Arenaria interpres	Vuelvepiedras	Rola-do-mar	●	
Wilson's Phalarope	Phalaropus tricolor	Falaropo de Wilson	Pisa-n'água	●	
Red-necked Phalarope	Phalaropus lobatus	Falaropo Picofino	Falaropo-de-bico-fino	●	
Grey Phalarope	Phalaropus fulicarius	Falaropo Picogrueso	Falaropo-de-bico-grosso	●	
Pomarine Skua	Stercorarius pomarinus	Págalo Pomarino	Moleiro do Árctico	●	●
Arctic Skua	Stercorarius parasiticus	Págalo Parásito	Moleiro-pequeno	●	●
Long-tailed Skua	Stercorarius longicaudus	Págalo Rabero	Moleiro-rabilongo	●	
Great Skua	Stercorarius skua	Págalo Grande	Alcaide	●	
Pallas's Gull	Larus ichthyaetus	Gavión Cabecinegro	Gaivotão-de-cabeça-preta	●	
Mediterranean Gull	Larus melanocephalus	Gaviota Cabecinegra	Gaivota-de-cabeça-preta	●	
Franklin's Gull	Larus pipixcan	Gaviota de Franklin	Gaivota-das-pradarias	●	
Little Gull	Larus minutus	Gaviota Enana	Gaivota-pequena	●	
Sabine's Gull	Larus sabini	Gaviota de Sabine	Gaivota de Sabine	●	
Black-headed Gull	Larus ridibundus	Gaviota Reidora	Garrincho	●	●
Ring-billed Gull	Larus delawarensis	Gaviota de Delaware	Gaivota-de-bico-riscado	●	
Common Gull	Larus canus	Gaviota Cana	Farnego	●	
Lesser Black-backed Gull	Larus fuscus	Gaviota Sombría	Gaivota-d'asa-escura	●	●
American Herring Gull	Larus smithsonianus	Gaviota Argéntea Americana	Gaivota-prateada-americana	●	
Herring Gull	Larus argentatus	Gaviota Argéntea	Gaivota-prateada	●	
Yellow-legged Gull	Larus michahellis	Gaviota Patiamarilla	Gaivota-de-patas-amarelas	B	B
Iceland Gull	Larus glaucoides	Gaviota Polar	Gaivota-branca	●	●
Glaucous Gull	Larus hyperboreus	Gaviota Hiperbórea	Gaivotão-branco	●	
Great Black-backed Gull	Larus marinus	Gavión	Gaivotão-real	●	
Kittiwake	Rissa tridactyla	Gaviota Tridáctila	Gaivota-tridáctila	●	
Gull-billed Tern	Geochelidon nilotica	Pagaza Piconegra	Tagáz	●	
Caspian Tern	Hydroprogne caspia	Pagaza Piquirroja	Garajau-grande	●	
Lesser Crested Tern	Sterna bengalensis	Charrán Bengalí	Garajau-pequeño		●
Sandwich Tern	Sterna sandvicensis	Charrán Patinegro	Garajau	●	●
Roseate Tern	Sterna dougallii	Charrán Rosado	Gaivina-rosada	B	B
Common Tern	Sterna hirundo	Charrán Común	Gaivina	B	B
Arctic Tern	Sterna paradisaea	Charrán Ártico	Gaivina-ártica	●	
Sooty Tern	Onychoprion fuscata	Charrán Sombrío	Gaivina-de-dorso-preto		●
Little Tern	Sternula albifrons	Charráncito	Chilreta	●	
Whiskered Tern	Chlidonias hybrida	Fumarel Cariblanco	Gaivina-dos-pauis	●	
Black Tern	Chlidonias niger	Fumarel Común	Gaivina-preta	●	
White-winged Tern	Chlidonias leucopterus	Fumarel Aliblanco	Gaivina-d'asa-branca	●	
Little Auk	Alle alle	Mérgulo Marino	Torda-miúda	●	
Puffin	Fratercula arctica	Frailecillo Común	Papagaio-do-mar	●	●
Rock Dove	Columba livia	Paloma Bravía	Pombo-das-rochas	B	B
Woodpigeon	Columba palumbus	Paloma Torcaz	Pombo-torcaz	E	●
Trocaz Pigeon	Columba trocaz	Paloma Torqueza	Pombo de Madeira	B	

English name	Scientific name	Spanish name	Portuguese name	Islands	
				M	S
Eurasian Collared Dove	*Streptopelia decaocto*	Tórtola Turca	Rola-turca	●	●
Turtle Dove	*Streptopelia turtur*	Tórtola Común	Rola-brava	B	●
Rose-ringed Parakeet	*Psittacula krameri*	Cotorra de Kramer	Periquito-rabijunco	?	
Great Spotted Cuckoo	*Clamator glandarius*	Crialo	Cuco-rabilongo	●	
Common Cuckoo	*Cuculus canorus*	Cuco	Cuco	●	●
Barn Owl	*Tyto alba*	Lechuza Común	Coruja-das-torres	B	●
Scops Owl	*Otus scops*	Autillo	Mocho-d'orelhas	●	
Short-eared Owl	*Asio flammeus*	Lechuza Campestre	Coruja-galhofa	●	●
European Nightjar	*Caprimulgus europaeus*	Chotacabras Gris	Noitibó-cinzento	●	●
Red-necked Nightjar	*Caprimulgus ruficollis*	Chotacabras Pardo	Noitibó-de-nuca-vermelha	●	●
Plain Swift	*Apus unicolor*	Vencejo Unicolor	Andorinhão-da-serra	B	●
Common Swift	*Apus apus*	Vencejo Común	Andorinhão-preto	●	?
Pallid Swift	*Apus pallidus*	Vencejo Pálido	Andorinhão-pálido	B	?
Alpine Swift	*Apus melba*	Vencejo Real	Andorinhão-real	●	●
Little Swift	*Apus affinis*	Vencejo Culiblanco	Andorinhão-pequeno		●
White-rumped Swift	*Apus caffer*	Vencejo Cafre	Andorinhão-cafre	●	
Common Kingfisher	*Alcedo atthis*	Martín Pescador	Guarda-rios	●	
European Bee-eater	*Merops apiaster*	Abejaruco Común	Abelharuco	●	●
European Roller	*Coracias garrulus*	Carraca	Rolieiro	●	
Hoopoe	*Upupa epops*	Abubilla	Poupa	B	●
Wryneck	*Jynx torquilla*	Torcecuello	Torcicolo	●	●
Calandra Lark	*Melanocorypha calandra*	Calandria Común	Calhandra-real	●	
Short-toed Lark	*Calandrella brachydactyla*	Terrera Común	Calhandrinha	●	●
Skylark	*Alauda arvensis*	Alondra Común	Alberca	●	
Sand Martin	*Riparia riparia*	Avión Zapador	Andorinha-das-barreiras	●	●
Crag Martin	*Ptyonoprogne rupestris*	Avión Roquero	Andorinha-das-roches		●
Barn Swallow	*Hirundo rustica*	Golondrina Común	Andorinha-das-chaminés	●	●
Red-rumped Swallow	*Cecropis daurica*	Golondrina Dáurica	Andorinha-dáurica	●	●
American Cliff Swallow	*Petrochelidon pyrrhonota*	Avión Roquero Americano	Andorinha-de-testa-branca	●	
House Martin	*Delichon urbicum*	Avión Común	Andorinha-dos-beirais	●	●
Tawny Pipit	*Anthus campestris*	Bisbita Campestre	Petinha-dos-campos	●	●
Berthelot's Pipit	*Anthus berthelotii*	Bisbita Caminero	Corre-caminho	B	B
Tree Pipit	*Anthus trivialis*	Bisbita Arbóreo	Petinha-das-árvores	●	●
Meadow Pipit	*Anthus pratensis*	Bisbita Común	Petinha-dos-prados	●	
Red-throated Pipit	*Anthus cervinus*	Bisbita Gorgirrojo	Petinha-de-garganta-ruiva	●	
Yellow Wagtail	*Motacilla flava*	Lavandera Boyera	Alvéola-amarela	●	●
Grey Wagtail	*Motacilla cinerea*	Lavandera Cascadeña	Alvéola-cinzenta	B	●
White Wagtail	*Motacilla alba*	Lavandera Blanca	Alvéola-branca	●	
Wren	*Troglodytes troglodytes*	Chochín	Carriça	●	
Rufous Bush Robin	*Cercotrichas galactotes*	Alzacola	Solitário	●	
European Robin	*Erithacus rubecula*	Petirrojo	Pisco-de-peito-ruivo	B	●
Common Nightingale	*Luscinia megarhynchos*	Ruiseñor Común	Rouxinol	●	
Black Redstart	*Phoenicurus ochruros*	Colirrojo Tizón	Rabirruivo	●	●
Common Redstart	*Phoenicurus phoenicurus*	Colirrojo Real	Rabirruivo-de-testa-branca	●	●
Whinchat	*Saxicola rubetra*	Tarabilla Norteña	Cartaxo-d'arribação		●
Common Stonechat	*Saxicola torquatus*	Tarabilla Común	Cartaxo	●	
Isabelline Wheatear	*Oenanthe isabellina*	Collalba Isabel	Chasco-isabel	●	
Northern Wheatear	*Oenanthe oenanthe*	Collalba Gris	Chasco-cinzento	●	●
Desert Wheatear	*Oenanthe deserti*	Collalba Desértica	Chasco-do-deserto	●	
Rock Thrush	*Monticola saxatilis*	Roquero Rojo	Macuco		●
Wood Thrush	*Hylocichla mustelina*	Zorzalito Maculado	Tordo-dos-bosques	●	
Ring Ouzel	*Turdus torquatus*	Mirlo de Collar	Melro-de-peito-branco	●	●
Blackbird	*Turdus merula*	Mirlo Común	Melro	B	
Dark-throated Thrush	*Turdus ruficollis*	Zorzal Papinegro	Tordo-de-garganta-preta	●	
Fieldfare	*Turdus pilaris*	Zorzal Real	Tordo-zornal	●	●
Song Thrush	*Turdus philomelos*	Zorzal Común	Tordo-pinto	●	●
Redwing	*Turdus iliacus*	Zorzal Alirrojo	Tordo-ruivo	●	●

APPENDIX B: DISTRIBUTION OF SPECIES ON MADEIRA

English name	Scientific name	Spanish name	Portuguese name	Islands M	Islands S
Mistle Thrush	*Turdus viscivorus*	Zorzal Charlo	Tordoveia	●	
Grasshopper Warbler	*Locustella naevia*	Buscarla Pintoja	Cigarrinha-malhada		●
River Warbler	*Locustella fluviatilis*	Buscarla Fluvial	Cigarrinha-ribeirinha		
Sedge Warbler	*Acrocephalus schoenobaenus*	Carricerín Común	Felosa-dos-juncos	●	●
Marsh Warbler	*Acrocephalus palustris*	Carricero Políglota	Felosa-palustre	●	
Reed Warbler	*Acrocephalus scirpaceus*	Carricero Común	Rouxinol-dos-caniços	●	●
Great Reed Warbler	*Acrocephalus arundinaceus*	Carricero Tordal	Rouxinol-grande-dos-caniços	●	
Western Olivaceous Warbler	*Hippolais opaca*	Zarcero Pálido Occidental	Felosa-pálida		●
Icterine Warbler	*Hippolais icterina*	Zarcero Icterino	Felosa-amarela		●
Melodious Warbler	*Hippolais polyglotta*	Zarcero Común	Felosa-poliglota	●	●
Spectacled Warbler	*Sylvia conspicillata*	Curruca Tomillera	Toutinegra-tomilheira	B	
Subalpine Warbler	*Sylvia cantillans*	Curruca Carrasqueña	Toutinegra-de-bigodes		●
Sardinian Warbler	*Sylvia melanocephala*	Curruca Cabecinegra	Toutinegra-dos-valados	●	
African Desert Warbler	*Sylvia deserti*	Curruca Sahariana	Toutinegra-do-deserto	●	
Orphean Warbler	*Sylvia hortensis*	Curruca Mirlona Occidental	Toutinegra-real	●	
Lesser Whitethroat	*Sylvia curruca*	Curruca Zarcerilla	Papa-amoras-cinzento	●	●
Common Whitethroat	*Sylvia communis*	Curruca Zarcera	Papa-amoras-comum	●	●
Garden Warbler	*Sylvia borin*	Curruca Mosquitera	Felosa-das-figueiras	●	●
Blackcap	*Sylvia atricapilla*	Curruca Capirotada	Toutinegra-de-barrete	B	
Yellow-browed Warbler	*Phylloscopus inornatus*	Mosquitero Bilistado	Felosa-listada		●
Western Bonelli's Warbler	*Phylloscopus bonelli*	Mosquitero Papialbo Occidental	Felosa-de-papo-branco	●	●
Wood Warbler	*Phylloscopus sibilatrix*	Mosquitero Silbador	Felosa-assobiadeira	●	
Common Chiffchaff	*Phylloscopus collybita*	Mosquitero Común	Felosinha	●	●
Willow Warbler	*Phylloscopus trochilus*	Mosquitero Musical	Felosa-musical	●	●
Madeira Firecrest	*Regulus madeirensis*	Reyezuelo de Madeira	Esterlinha-de-Madeira	B	
Spotted Flycatcher	*Muscicapa striata*	Papamoscas Gris	Papa-moscas-cinzento	●	●
Red-breasted Flycatcher	*Ficedula parva*	Papamoscas Papirrojo	Papa-moscas-pequeño	●	●
Pied Flycatcher	*Ficedula hypoleuca*	Papamoscas Cerrojillo	Papas-moscas preto	●	●
Golden Oriole	*Oriolus oriolus*	Oropéndola	Papa-figos	●	●
Red-backed Shrike	*Lanius collurio*	Alcaudón Dorsirrojo	Picanço-de-dorso-ruivo	●	●
Woodchat Shrike	*Lanius senator*	Alcaudón Común	Picanço-barreteiro	●	●
Western Jackdaw	*Corvus monedula*	Grajilla	Gralha-de-nuca-cinzenta	●	
Rook	*Corvus frugilegus*	Graja	Gralha-calva	●	
Common Raven	*Corvus corax*	Cuervo	Corvo	●	
Common Starling	*Sturnus vulgaris*	Estornino Pinto	Estorninho-malhado	●	●
Spotless Starling	*Sturnus unicolor*	Estornino Negro	Estorninho-preto		●
Rosy Starling	*Sturnus roseus*	Estornino Rosado	Estorninho-rosado	●	
House Sparrow	*Passer domesticus*	Gorrión Doméstico	Pardal	●	
Spanish Sparrow	*Passer hispaniolensis*	Gorrión Moruno	Pardal-espanhol	B	
Rock Sparrow	*Petronia petronia*	Gorrión Chillón	Pardal-francês	B	●
Common Chaffinch	*Fringilla coelebs*	Pinzón Vulgar	Tentilhão	B	
Brambling	*Fringilla montifringilla*	Pinzón Real	Tentilhão-montês	●	
European Serin	*Serinus serinus*	Verdecillo	Milheirinha		●
Canary	*Serinus canaria*	Canario	Canario-da-terra	B	●
Greenfinch	*Carduelis chloris*	Verderón	Verdilhão	B	●
Goldfinch	*Carduelis carduelis*	Jilguero	Pintassilgo	B	●
Siskin	*Carduelis spinus*	Lúgano	Lugre		●
Linnet	*Carduelis cannabina*	Pardillo Común	Pintarroxo	B	●
Common Crossbill	*Loxia curvirostra*	Piquituerto Común	Cruza-bico	●	●
Hawfinch	*Coccothraustes coccothraustes*	Picogordo	Bico-grossudo	●	●
Yellow Warbler	*Dendroica petechia*	Reinita Amarilla	Mariquita-dos mangais		●
American Redstart	*Setophaga ruticilla*	Colirrojo Americano	Mariquita-de-rabo-vermelho		●
Snow Bunting	*Plectrophenax nivalis*	Escribano Nival	Escrevedeira-das-neves	●	
Ortolan Bunting	*Emberiza hortulana*	Escribano Hortelano	Sombria		●
Reed Bunting	*Emberiza schoeniculus*	Escribano Palustre	Escrevedeira-dos-caniços	●	

345

iii. The Azores

S = Santa Maria M = São Miguel T = Terceira J = São Jorge G = Graciosa
P = Pico F = Faial Fl = Flores C = Corvo

B = current breeding species b = species that has bred
? = possibly breeds E = now extinct as a breeding species
● = has occurred

English name	Scientific name	Spanish name	Portuguese name	S	M	T	J	G	P	F	Fl	C
Red-throated Diver	Gavia stellata	Colimbo Chico	Mobelha-pequena		●							
Great Northern Diver	Gavia immer	Colimbo Grande	Mobelha-grande		●	●			●	●	●	
Pied-billed Grebe	Podilymbus podiceps	Zampullín Picogrueso	Mergulhão-caçador		●						●	
Great Crested Grebe	Podiceps cristatus	Somormujo Lavanco	Mergulhão-de-poupa		●							
Slavonian Grebe	Podiceps auritus	Zampullín Cuellirrojo	Mergulhão-de-penachos		●	●						
Black-necked Grebe	Podiceps nigricollis	Zampullín Cuellinegro	Cagarraz		●						●	
Northern Fulmar	Fulmarus glacialis	Fulmar Boreal	Pombalete		●							
Trindade Petrel	Pterodroma arminjoniana	Petrel de la Trindade	Freira-da-Trindade						●			
Bermuda Petrel	Pterodroma cahow	Petrel Cahow	Petrel das Bermudas	●								
Fea's Petrel	Pterodroma feae	Petrel Gon-gon	Gon-Gon					●				
Bulwer's Petrel	Bulweria bulwerii	Petrel de Bulwer	Alma-negra	B	●							
Cory's Shearwater	Calonectris diomedea	Pardela Cenicienta	Cagarra	B	B	B	B	B	B	B	B	B
Great Shearwater	Puffinus gravis	Pardela Capirotada	Pardela-de-barrete	●	●	●			●	●	●	●
Sooty Shearwater	Puffinus griseus	Pardela Sombría	Pardela-preta	●	●	●			●	●	●	
Balearic Shearwater	Puffinus mauretanicus	Pardela Balear	Fura-bucho		●						●	
Manx Shearwater	Puffinus puffinus	Pardela Pichoneta	Fura-bucho do Atlántico	●	●						B	B
Macaronesian Sh'water	Puffinus baroli	Pardela Chica	Pintainho	B	?		?	B	B	?	?	?
Wilson's Storm-petrel	Oceanites oceanicus	Paíño de Wilson	Casquilho						●			
White-faced Storm-petrel	Pelagodroma marina	Paíño Pechialbo	Calca-mar	no location details
Leach's Storm-petrel	Oceanodroma leucorhoa	Paíño de Leach	Paínho-de-cauda-forcada	●	●							
Madeiran Storm-petrel	Oceanodroma castro	Paíño de Madeira	Roquinho	B					B			
Red-billed Tropicbird	Phaethon aethereus	Rabijunco Etéreo	Rabijunco						b			
Brown Booby	Sula leucogaster	Piquero Pardo	Alcatraz-pardo		●							
Northern Gannet	Morus bassanus	Alcatraz Atlántico	Alcatraz		●	●			●		●	●
Cormorant	Phalacrocorax carbo	Cormorán Grande	Corvo-marinho		●				●		●	●
Double-crested Cormorant	Phalacrocorax auritus	Cormorán Orejudo	Corvo-marinho-d'orelhas		●				●		●	●
Magnificent Frigatebird	Fregata magnificens	Rabihorcado Magnífico	Rabiforcado		●							
Eurasian Bittern	Botaurus stellaris	Avetoro Común	Abetouro		●	●						
American Bittern	Botaurus lentiginosus	Avetoro Lentiginoso	Abetouro-americano		●					●		
Least Bittern	Ixobrychus exilis	Avetorillo Panamericano	Socoí-vermelho	●	●							
Little Bittern	Ixobrychus minutus	Avetorillo Común	Garçote		●	●	●	●	●		●	
Night Heron	Nycticorax nycticorax	Martinete Común	Goraz	●	●			●			●	
Green-backed Heron	Butorides virescens	Garcita Verdosa	Socó-mirim		●					●		
Squacco Heron	Ardeola ralloides	Garcilla Cangrejera	Papa-ratos		●							
Cattle Egret	Bubulcus ibis	Garcilla Bueyera	Carraceiro	●	●							
Western Reef Egret	Egretta gularis	Garceta Sombría	Garça-dos-recifes		●							
Little Blue Heron	Hydranassa caerulea	Garceta Azul	Garça-morena						●		●	
Tricoloured Heron	Hydranassa tricolor	Garceta Tricolor	Garça-tricolor						●			
Snowy Egret	Egretta thula	Garceta Nívea	Garça-branca-americana		●	●					●	●
Little Egret	Egretta garzetta	Garceta Común	Garça-branca	●	●							
Great White Egret	Egretta alba	Garceta Grande	Garça-branca-grande		●							
Grey Heron	Ardea cinerea	Garza Real	Garça-real	●	●	●		●	●	●	●	●
Great Blue Heron	Ardea herodias	Garza Azulada	Garça-real-americana					●		●	●	●
Purple Heron	Ardea purpurea	Garza Imperial	Garça-ruiva	●	●	●						
Black Stork	Ciconia nigra	Cigüeña Negra	Cegonha-preta		●							
Glossy Ibis	Plegadis falcinellus	Morito	Ibis-preto	●	●							
Bald Ibis	Geronticus eremita	Ibis Eremita	Ibis-pelado		●							

APPENDIX B: DISTRIBUTION OF SPECIES ON THE AZORES

English name	Scientific name	Spanish name	Portuguese name	S	M	T	J	G	P	F	Fl	C
Eurasian Spoonbill	*Platalea leucorodia*	Espátula	Colhereiro	●	●						●	●
Fulvous Whistling Duck	*Dendrocygna bicolor*	Suirirí Bicolor	Marreca-caneleira		●							
Mute Swan	*Cygnus olor*	Cisne Vulgar	Cisne-mudo			●		●	●	●		
Bean Goose	*Anser fabalis*	Ánsar Campestre	Ganso-campestre		●							●
Pink-footed Goose	*Anser brachyrhynchus*	Ánsar Piquicorto	Ganso-de-bico-curto		●	●						
White-fronted Goose	*Anser albifrons*	Ánsar Careto Grande	Ganso-de-testa-branca		●							
Greylag Goose	*Anser anser*	Ánsar Común	Ganso-bravo	●	●							
Snow Goose	*Anser caerulescens*	Ánsar Nival	Ganso-das-neves		●							
Canada Goose	*Branta canadensis*	Barnacla Canadiense	Ganso do Canadá		●							
Barnacle Goose	*Branta leucopsis*	Barnacla Cariblanca	Ganso-marisco		●							
Brent Goose	*Branta bernicla*	Barnacla Carinegra	Ganso-de-faces-pretas		●	●						
Ruddy Shelduck	*Tadorna ferruginea*	Tarro Canelo	Pato-casarca		●							
Wood Duck	*Aix sponsa*	Pato Joyuyo	Pato-carolino		●				●		●	●
Eurasian Wigeon	*Anas penelope*	Silbón Europea	Piadeira		●					●	●	
American Wigeon	*Anas americana*	Silbón Americano	Piadeira-americana		●		●			●	●	
Gadwall	*Anas strepera*	Ánade Friso	Frisada		●							
Eurasian Teal	*Anas crecca*	Cerceta Común	Marrequinha		●					●	b	
Green-winged Teal	*Anas carolinensis*	Cerceta Americana	Marrequinha-americano		●					●	●	
Mallard	*Anas platyrhynchos*	Ánade Azulón	Pato-real		●						b	●
Black Duck	*Anas rubripes*	Ánade Sombrío	Pato-escuro-americano		●						b	●
Pintail	*Anas acuta*	Ánade Rabudo	Arrábio		●						●	
Garganey	*Anas querquedula*	Cerceta Carretona	Marreco		●				●			
Blue-winged Teal	*Anas discors*	Cerceta Aliazul	Marreca-d'asa-azul		●	●	●					
Shoveler	*Anas clypeata*	Pato Cuchara	Pato-colhereiro		●						b	
Common Pochard	*Aythya ferina*	Porrón Común	Zarro		●							
Ring-necked Duck	*Aythya collaris*	Porrón Acollarado	Caturro		●			●		●	●	●
Ferruginous Duck	*Aythya nyroca*	Porrón Pardo	Pêrra		●							
Tufted Duck	*Aythya fuligula*	Porrón Moñudo	Negrinha		●							
Greater Scaup	*Aythya marila*	Porrón Bastardo	Negrelho		●						●	
Lesser Scaup	*Aythya affinis*	Porrón Bola	Negrelho-americano		●							
Common Eider	*Somateria mollissima*	Eider	Eider		●							
King Eider	*Somateria spectabilis*	Eider Real	Eider-real		●							
Long-tailed Duck	*Clangula hyemalis*	Havelda	Pato-rabilongo		●	●						
Common Scoter	*Melanitta nigra*	Negrón Común	Negrola		●							
Surf Scoter	*Melanitta perspicillata*	Negrón Careto	Negrola-de-lunetas						●		●	●
Bufflehead	*Bucephala albeola*	Porron Albeola	Olho-dourado-de-touca					●	●		●	
Common Goldeneye	*Bucephala clangula*	Porrón Osculado	Olho-dourado		●				●			
Hooded Merganser	*Mergus cucullatus*	Serreta Capuchona	Merganso-capuchinho								●	●
Red-breasted Merganser	*Mergus serrator*	Serreta Mediana	Merganso-de-poupa	●	●	●				●		
Egyptian Vulture	*Neophron percnopterus*	Alimoche Común	Britango		●							
Marsh Harrier	*Circus aeruginosus*	Aguilucho Lagunero Occidental	Tartaranhão-ruivo-dos-pauis		●							
Hen Harrier	*Circus cyaneus*	Aguilucho Pálido	Tartaranhão-azulado								●	
Common Buzzard	*Buteo buteo*	Busardo Ratonero	Águia-d'asa-redonda	B	B	B	B	B	B	B		
Rough-legged Buzzard	*Buteo lagopus*	Busardo Calzado	Bútio-calçado			●				●		●
Osprey	*Pandion haliaetus*	Aguila Pescadora	Águia-pesqueira		●	●			●			
Lesser Kestrel	*Falco naumanni*	Cernícalo Primilla	Francelho		●							
Common Kestrel	*Falco tinnunculus*	Cernícalo Vulgar	Peneireiro	●	●							
American Kestrel	*Falco sparverius*	Cernícalo Americano	Peneireiro-americano		●							
Red-footed Falcon	*Falco vespertinus*	Cernícalo Patirrojo	Falcão-vespertino		●							
Merlin	*Falco columbarius*	Esmerejón	Esmerilhão		●	●					●	
Peregrine Falcon	*Falco peregrinus*	Halcón Común	Falcão-peregrino						●			
Red-legged Partridge	*Alectoris rufa*	Perdiz Común	Perdiz	b					b			
Common Quail	*Coturnix coturnix*	Codorniz	Codorniz	B	B	B	B	B	B	B	B	B
Water Rail	*Rallus aquaticus*	Rascón	Frango-d'água		●	●						
Spotted Crake	*Porzana porzana*	Polluela Pintoja	Franga-d'água-grande	●	●							
Sora	*Porzana carolina*	Polluela de Carolina	Franga-d'água-americana					●				
Little Crake	*Porzana parva*	Polluela Bastarda	Franga-d'água-bastarda		●					●		
Baillon's Crake	*Porzana pusilla*	Polluela Chica	Franga-d'água-pequena		●							

English name	Scientific name	Spanish name	Portuguese name	Islands								
				S	M	T	J	G	P	F	Fl	C
Corn Crake	*Crex crex*	Guión de Cordonices	Codornizão	●	●	●				●		
Moorhen	*Gallinula chloropus*	Polla de Agua	Galinha-d'água		●	b	●	●	●	●	●	
Allen's Gallinule	*Porphyrula alleni*	Calamón de Allen	Camão-pequeno	●	●							
American Purple Gallinule	*Porphyrula martinica*	Calamoncillo Americano	Camão-americano			●					●	
Common Coot	*Fulica atra*	Focha Común	Galeirão		●	E	●	●	●	●	●	●
American Coot	*Fulica americana*	Focha Americana	Galeirão-americano		●	●			●			
Common Crane	*Grus grus*	Grulla Común	Grou			●						
Sandhill Crane	*Grus canadensis*	Grulla Canadiense	Grou-americano							●		
Eurasian Oystercatcher	*Haematopus ostralegus*	Ostrero	Ostraceiro	●	●	●	●					
Black-winged Stilt	*Himantopus himantopus*	Cigüeñuela	Pernilongo	●	●	●				●		
Avocet	*Recurvirostra avosetta*	Avoceta	Alfaiate		●	●						
Stone-curlew	*Burhinus oedicnemus*	Alcaraván	Alcaravão	●								
Little Ringed Plover	*Charadrius dubius*	Chorlitejo Chico	Borrelho-pequeno-de-coleira			●						●
Ringed Plover	*Charadrius hiaticula*	Chorlitejo Grande	Borrelho-grande-de-coleira		●	●				●		
Semipalmated Plover	*Charadrius semipalmatus*	Chorlitejo Semipalmeado	Batuíra-de-bando	●						●	●	●
Killdeer	*Charadrius vociferus*	Chorlitejo Culirrojo	Borrelho-de-coleira-dupla		●	●						
Kentish Plover	*Charadrius alexandrinus*	Chorlitejo Patinegro	Borrelho-de-coleira-interrompida	B	B	B	B	B	●	●	●	●
Dotterel	*Charadrius morinellus*	Chorlito Carambolo	Borrelho-ruivo							●		
American Golden Plover	*Pluvialis dominica*	Chorlito Dorado Chico	Batuiruçu		●	●						
European Golden Plover	*Pluvialis apricaria*	Chorlito Dorado Común	Tarambola-dourada		●	●						
Grey Plover	*Pluvialis squatarola*	Chorlito Gris	Tarambola-cinzenta		●	●				●		
Northern Lapwing	*Vanellus vanellus*	Avefría Europea	Abibe	●	●	●						
Knot	*Calidris canutus*	Correlimos Gordo	Seixoeira		●	●						
Sanderling	*Calidris alba*	Correlimos Tridáctilo	Pilrito-das-praias	●	●	●				●	●	●
Semipalmated Sandpiper	*Calidris pusilla*	Correlimos Semipalmeado	Pilrito-rasteirinho		●	●				●		
Western Sandpiper	*Calidris mauri*	Correlimos de Alaska	Pilrito-miúdo			●						
Little Stint	*Calidris minuta*	Correlimos Menudo	Pilrito-pequeño		●	●				●		
Temminck's Stint	*Calidris temminckii*	Correlimos de Temminck	Pilrito de Temminck			●						
Least Sandpiper	*Calidris minutilla*	Correlimos Menudillo	Pilrito-anão		●	●				●	●	
White-rumped Sandpiper	*Calidris fuscicollis*	Correlimos Culiblanco	Pilrito-de-sobre-branco		●	●				●	●	
Baird's Sandpiper	*Calidris bairdii*	Correlimos de Baird	Pilrito-de-bico-fino		●	●						
Pectoral Sandpiper	*Calidris melanotos*	Correlimos Pectoral	Pilrito-de-colete	●	●	●						
Sharp-tailed Sandpiper	*Calidris acuminata*	Correlimos Acuminado	Pilrito-acuminado			●						
Curlew Sandpiper	*Calidris ferruginea*	Correlimos Zarapitín	Pilrito-de-bico-comprido	●		●				●		
Purple Sandpiper	*Calidris maritima*	Correlimos Oscuro	Pilrito-escuro		●	●				●	●	
Dunlin	*Calidris alpina*	Correlimos Común	Pilrito-de-peito-preta		●	●			●	●		
Stilt Sandpiper	*Calidris himantopus*	Correlimos Zancolín	Pilrito-pernilongo		●	●						
Buff-breasted Sandpiper	*Tryngites subruficollis*	Correlimos Canelo	Pilrito-acanelado		●	●						
Ruff	*Philomachus pugnax*	Combatiente	Combatente		●	●				●	●	
Jack Snipe	*Lymnocryptes minimus*	Agachadiza Chica	Narceja-galega		●	●						
Common Snipe	*Gallinago gallinago*	Agachadiza Común	Narceja	●	B	B	B	●	B	B	B	●
Wilson's Snipe	*Gallinago delicata*	Agachadiza Americana	Narceja-americana			●					●	●
Short-billed Dowitcher	*Limnodromus griseus*	Agujeta Gris	Maçarico-de-bico-curto		●	●						
Woodcock	*Scolopax rusticola*	Chocha Perdiz	Galinhola	?	B	B	B	?	B	B	B	B
Black-tailed Godwit	*Limosa limosa*	Aguja Colinegra	Milherango		●	●						
Bar-tailed Godwit	*Limosa lapponica*	Aguja Colipinta	Fuselo		●	●						
Whimbrel	*Numenius phaeopus*	Zarapito Trinador	Maçarico-galego	●	●	●			●	●	●	●
Slender-billed Curlew	*Numenius tenuirostris*	Zarapito Fino	Maçarico-de-bico-fino		●							
Eurasian Curlew	*Numenius arquata*	Zarapito Real	Maçarico-real		●	●			●	●		
Upland Sandpiper	*Bartramia longicauda*	Correlimos Batitú	Maçarico-do-campo		●							
Spotted Redshank	*Tringa erythropus*	Archibebe Oscuro	Perna-vermelha-bastardo		●	●						
Common Redshank	*Tringa totanus*	Archibebe Común	Perna-vermelha	●	●	●				●	●	

APPENDIX B: DISTRIBUTION OF SPECIES ON THE AZORES

English name	Scientific name	Spanish name	Portuguese name	S	M	T	J	G	P	F	Fl	C
Marsh Sandpiper	*Tringa stagnatilis*	Archibebe Fino	Perna-verde-fino	●	●	●			●			
Greenshank	*Tringa nebularia*	Archibebe Claro	Perna-verde		●	●	●		●	●		●
Greater Yellowlegs	*Tringa melanoleuca*	Archibebe Patigualdo Grande	Perna-amarela-grande						●		●	
Lesser Yellowlegs	*Tringa flavipes*	Archibebe Patigualdo Chico	Perna-amarela-pequeno		●	●	●		●	●	●	
Solitary Sandpiper	*Tringa solitaria*	Andarríos Solitario	Maçarico-solitário		●	●					●	
Green Sandpiper	*Tringa ochropus*	Andarríos Grande	Maçarico-bique-bique							●		
Wood Sandpiper	*Tringa glareola*	Andarríos Bastardo	Maçarico-de-dorso-malhado		●	●						
Common Sandpiper	*Actitis hypoleucos*	Andarríos Chico	Maçarico-das-rochas	●	●	●	●	●	●	●	●	●
Spotted Sandpiper	*Actitis macularius*	Andarríos Maculado	Maçarico-pintado		●	●			●	●	●	
Willet	*Catoptrophorus semipalmatus*	Playero Aliblanco	Maçarico-d'asa-branca		●				●			
Turnstone	*Arenaria interpres*	Vuelvepiedras	Rola-do-mar	●	●	●	●		●	●	●	●
Wilson's Phalarope	*Phalaropus tricolor*	Falaropo de Wilson	Pisa-n'água						●		●	
Red-necked Phalarope	*Phalaropus lobatus*	Falaropo Picofino	Falaropo-de-bico-fino		●						●	
Grey Phalarope	*Phalaropus fulicarius*	Falaropo Picogrueso	Falaropo-de-bico-grosso							●		
Pomarine Skua	*Stercorarius pomarinus*	Págalo Pomarino	Moleiro do Ártico	●		●	●		●	●	●	●
Arctic Skua	*Stercorarius parasiticus*	Págalo Parásito	Moleiro-pequeno	●	●	●	●	●	●	●	●	●
Long-tailed Skua	*Stercorarius longicaudus*	Págalo Rabero	Moleiro-rabilongo						●			
Great Skua	*Stercorarius skua*	Págalo Grande	Alcaide	●	●	●	●		●	●	●	●
South Polar Skua	*Stercorarius maccormicki*	Págalo Polar	Alcaide do Antártico						●			
Mediterranean Gull	*Larus melanocephalus*	Gaviota Cabecinegra	Gaivota-de-cabeça-preta		●	●						
Laughing Gull	*Larus atricilla*	Gaviota Guanaguanare	Gaivota-alegre								●	●
Franklin's Gull	*Larus pipixcan*	Gaviota de Franklin	Gaivota-das-pradarias								●	
Little Gull	*Larus minutus*	Gaviota Enana	Gaivota-pequena	●	●					●		
Sabine's Gull	*Larus sabini*	Gaviota de Sabine	Gaivota de Sabine							●		
Bonaparte's Gull	*Larus philadelphia*	Gaviota de Bonaparte	Garrincho-americano						●	●		
Black-headed Gull	*Larus ridibundus*	Gaviota Reidora	Garrincho		●	●			●	●	●	
Ring-billed Gull	*Larus delawarensis*	Gaviota de Delaware	Gaivota-de-bico-riscado						●	●	●	
Common Gull	*Larus canus*	Gaviota Cana	Famego							●		
Lesser Black-backed Gull	*Larus fuscus*	Gaviota Sombría	Gaivota-d'asa-escura							●		
American Herring Gull	*Larus smithsonianus*	Gaviota Argéntea Americana	Gaivota-prateada-americana		●	●				●	●	
Yellow-legged Gull	*Larus michahellis*	Gaviota Patiamarilla	Gaivota-de-patas-amarelas	B	B	B	B	B	B	B	B	B
Iceland Gull	*Larus glaucoides*	Gaviota Polar	Gaivota-branca		●	●			●	●	●	
Glaucous Gull	*Larus hyperboreus*	Gaviota Hiperbórea	Gaivotão-branco							●		
Great Black-backed Gull	*Larus marinus*	Gavión	Gaivotão-real		●					●		
Kittiwake	*Rissa tridactyla*	Gaviota Tridáctila	Gaivota-tridáctila	●	●	●	●		●	●	●	●
Gull-billed Tern	*Gelochelidon nilotica*	Pagaza Piconegra	Tagáz		●							
Royal Tern	*Sterna maxima*	Charrán Real	Garajau-real				●					
Sandwich Tern	*Sterna sandvicensis*	Charrán Patinegro	Garajau		●							
Roseate Tern	*Sterna dougallii*	Charrán Rosado	Gaivina-rosada	B	B	B	B	B	B	B	B	B
Common Tern	*Sterna hirundo*	Charrán Común	Gaivina	B	B	B	B		B	B	B	B
Arctic Tern	*Sterna paradisaea*	Charrán Ártico	Gaivina-ártica		●	●				●		
Forster's Tern	*Sterna forsteri*	Charrán de Forster	Gaivina de Forster						●			
Bridled Tern	*Onychoprion anaethetus*	Charrán Embridado	Gaivina-de-dorso-castanho	●	b	●	b	●				
Sooty Tern	*Onychoprion fuscata*	Charrán Sombrío	Gaivina-de-dorso-preto	B					●	b		●
Little Tern	*Sternula albifrons*	Charráncito	Chilreta		●						●	
Whiskered Tern	*Chlidonias hybrida*	Fumarel Cariblanco	Gaivina-dos-pauis		●	●						
Black Tern	*Chlidonias niger*	Fumarel Común	Gaivina-preta							●		●
Brünnich's Guillemot	*Uria lomvia*	Arao de Brünnich	Airo-de-freio						●			
Razorbill	*Alca torda*	Alca Común	Torda-mergulheira							●	●	
Little Auk	*Alle alle*	Mérgulo Marino	Torda-miúda						●			
Puffin	*Fratercula arctica*	Frailecillo Común	Papagaio-do-mar	●	●				●		●	
Rock Dove	*Columba livia*	Paloma Bravía	Pombo-das-rochas	B	B	B	B	B	B	B	B	B
Woodpigeon	*Columba palumbus*	Paloma Torcaz	Pombo-torcaz	B	B	B	B	B	B			
Turtle Dove	*Streptopelia turtur*	Tórtola Común	Rola-brava		●	●						
Rose-ringed Parakeet	*Psittacula krameri*	Cotorra de Kramer	Periquito-rabijunco		●							
Common Cuckoo	*Cuculus canorus*	Cuco	Cuco	●	●	●		●				

English name	Scientific name	Spanish name	Portuguese name	S	M	T	J	G	P	F	Fl	C
Black-billed Cuckoo	Coccyzus erythrophthalmus	Cuco Piquinegro	Papa-lagarta-de-bico-preto				●				●	
Yellow-billed Cuckoo	Coccyzus americanus	Cuco Piquigualdo	Papa-lagarta-norte-americano	●	●	●	●			●	●	●
Barn Owl	Tyto alba	Lechuza Común	Coruja-das-torres		●							
Snowy Owl	Bubo scandiacus	Búho Nival	Coruja-das-neves						●			
Long-eared Owl	Asio otus	Búho Chico	Bufo-pequeno		B	B	B	B	B	B		
Short-eared Owl	Asio flammeus	Lechuza Campestre	Coruja-galhofa		●	●		●				
European Nightjar	Caprimulgus europaeus	Chotacabras Gris	Noitibó-cinzento		●							
Common Nighthawk	Chordeiles minor	Añapero Yanqui	Bacurau-norte-americano						●		●	
Chimney Swift	Chaetura pelagica	Vencejo de Chimenea	Rabo-espinhoso		●							
Common Swift	Apus apus	Vencejo Común	Andorinhão-preto	●	●	●	●					
Belted Kingfisher	Ceryle alcyon	Martín Gigante Americano	Guarda-rios cintada						●		●	
European Bee-eater	Merops apiaster	Abejaruco Común	Abelharuco									
European Roller	Coracias garrulus	Carraca	Rolieiro	●	●							
Hoopoe	Upupa epops	Abubilla	Poupa	●	●		●					
Crested Lark	Galerida cristata	Cogujada Común	Cotovia-de-poupa		●							
Skylark	Alauda arvensis	Alondra Común	Alberca								●	
Purple Martin	Progne subis	Golondrina Purpúrea	Andorinha-roxa								●	●
Sand Martin	Riparia riparia	Avión Zapador	Andorinha-das-barreiras	●		●					●	
Barn Swallow	Hirundo rustica	Golondrina Común	Andorinha-das-chaminés	●	●	●			●			
Red-rumped Swallow	Cecropis daurica	Golondrina Dáurica	Andorinha-dáurica							●		
American Cliff Swallow	Hirundo pyrrhonota	Avión Roquero Americano	Andorinha-de-testa-branca						●			
House Martin	Delichon urbicum	Avión Común	Andorinha-dos-beirais	●	●	●					●	
Meadow Pipit	Anthus pratensis	Bisbita Común	Petinha-dos-prados					●				
Grey Wagtail	Motacilla cinerea	Lavandera Cascadeña	Alvéola-cinzenta	B	B	B	B	B	B	B	B	B
White Wagtail	Motacilla alba	Lavandera Blanca	Alvéola-branca	●	●			●				
European Robin	Erithacus rubecula	Petirrojo	Pisco-de-peito-ruivo	B	B	B	B	B	B			
Black Redstart	Phoenicurus ochruros	Colirrojo Tizón	Rabirruivo		●							
Common Redstart	Phoenicurus phoenicurus	Colirrojo Real	Rabirruivo-de-testa-branca		●					●		
Whinchat	Saxicola rubetra	Tarabilla Norteña	Cartaxo-d'arribação		●							
Northern Wheatear	Oenanthe oenanthe	Collalba Gris	Chasco-cinzento		●	●				●	●	
Wood Thrush	Hylocichla mustelina	Zorzalito Maculado	Tordo-dos-bosques		●							
Grey-cheeked Thrush	Catharus minimus	Zorzalito Carigrís	Tordo-de-cara-cinza								●	
Blackbird	Turdus merula	Mirlo Común	Melro	B	B	B	B	B	B	B	B	B
Fieldfare	Turdus pilaris	Zorzal Real	Tordo-zornal	●	●	●						
Song Thrush	Turdus philomelos	Zorzal Común	Tordo-pinto		●							
Redwing	Turdus iliacus	Zorzal Alirrojo	Tordo-ruivo	●								
Mistle Thrush	Turdus viscivorus	Zorzal Charlo	Tordoveia		●							
Blackcap	Sylvia atricapilla	Curruca Capirotada	Toutinegra-de-barrete	B	B	B	B	B	B		B	B
Yellow-browed Warbler	Phylloscopus inornatus	Mosquitero Bilistado	Felosa-listada		●							
Common Chiffchaff	Phylloscopus collybita	Mosquitero Común	Felosinha		●	●						
Willow Warbler	Phylloscopus trochilus	Mosquitero Musical	Felosa-musical		●						●	
Goldcrest	Regulus regulus	Reyezuelo Sencillo	Estrelinha-de-poupa	B	B	B	B			B	B	?
Red-breasted Flycatcher	Ficedula parva	Papamoscas Papirrojo	Papa-moscas-pequeño		●							
Golden Oriole	Oriolus oriolus	Oropéndola	Papa-figos		●						●	
Western Jackdaw	Corvus monedula	Grajilla	Gralha-de-nuca-cinzenta		●							
Rook	Corvus frugilegus	Graja	Gralha-calva	●	●							
Carrion Crow	Corvus corone	Corneja Negra	Gralha-preta		●	●						
Common Starling	Sturnus vulgaris	Estornino Pinto	Estorninho-malhado	B	B	B	B	B	B	B	B	B
House Sparrow	Passer domesticus	Gorrión Doméstico	Pardal			B	B	B	B	B	B	B
Rock Sparrow	Petronia petronia	Gorrión Chillón	Pardal-francês			B	B	B	B	B	B	B
Common Waxbill	Estrilda astrild	Pico de Coral	Bico-de-lacre			B	B					
Common Chaffinch	Fringilla coelebs	Pinzón Vulgar	Tentilhão	B	B	B	B	B	B	B	B	B
Brambling	Fringilla montifringilla	Pinzón Real	Tentilhão-montês		●							
Canary	Serinus canaria	Canario	Canario-da-terra	B	B	B	B	B	B	B	B	B
Greenfinch	Carduelis chloris	Verderón	Verdilhão		B	B					B	

APPENDIX B: DISTRIBUTION OF SPECIES ON THE CAPE VERDE ISLANDS

English name	Scientific name	Spanish name	Portuguese name	S	M	T	J	G	P	F	Fl	C
Goldfinch	*Carduelis carduelis*	Jilguero	Pintassilgo	B	B	B	B	B	B	B	B	
Siskin	*Carduelis spinus*	Lúgano	Lugre		●							
Azores Bullfinch	*Pyrrhula murina*	Camachuelo de las Azores	Priolo		B							
Yellow Warbler	*Dendroica petechia*	Reinita Amarilla	Mariquita-dos mangais		●						●	
Magnolia Warbler	*Dendroica magnolia*	Reinita Cejiblanca	Mariquita –de-faces-preta								●	
Yellow-rumped Warbler	*Dendroica coronata*	Reinita Coronada	Mariquita-coroada								●	
American Redstart	*Setophaga ruticilla*	Colirrojo Americano	Mariquita-de-rabo-vermelho					Captured at sea				
Northern Waterthrush	*Seiurus noveboracensis*	Reinita Charquera	Mariquita-boreal	●								
Snow Bunting	*Plectrophenax nivalis*	Escribano Nival	Escrevedeira-das-neves	●	●			●	●	●	●	●
Bobolink	*Dolichonyx oryzivorus*	Charlatán	Triste-pia								●	

iv. THE CAPE VERDE ISLANDS

S   = Sal   B = Boavista   M = Maio   St = Santiago
F   = Fogo   Br = Brava   N = São Nicolau   R = Raso
L   = Santa Luzia   V = São Vicente   A = Santo Antão

The column for Raso includes the neighbouring island of Branco, and the Ilhéus do Rombo are included in the column for Brava.

B = current breeding species   **b** = species that has bred   **?** = possibly breeds
★ = former status unclear   **E** = now extinct as a breeding species **●** = has occurred

English name	Scientific name	Spanish name	Portuguese name	S	B	M	St	F	Br	N	R	L	V	A
Fea's Petrel	*Pterodroma feae*	Petrel Gon-gon	Gon-Gon		●	●	B	B	●	B	●	●	●	B
Bulwer's Petrel	*Bulweria bulwerii*	Petrel de Bulwer	Alma-negra					B		B				
Cory's Shearwater	*Calonectris diomedea*	Pardela Cenicienta	Cagarra			Recorded from Cape Verde seas								
Cape Verde Shearwater	*Calonectris edwardsii*	Pardela Cenicienta de Edwards	Cagarra de Cabo Verde	?	?	●	B	●	B	B	B	●	●	B
Great Shearwater	*Puffinus gravis*	Pardela Capirotada	Pardela-de-barrete			Recorded from Cape Verde seas								
Sooty Shearwater	*Puffinus griseus*	Pardela Sombría	Pardela-preta					●			●			
Manx Shearwater	*Puffinus puffinus*	Pardela Pichoneta	Fura-bucho Atlântico			Recorded from Cape Verde seas								
Macaronesian Shearwater	*Puffinus baroli boydi*	Pardela Chica	Pintainho	E	b		B	?	B	?	B		?	?
Wilson's Storm-petrel	*Oceanites oceanicus*	Paíño de Wilson	Casquilho			Recorded from Cape Verde seas								
White-faced Storm-petrel	*Pelagodroma marina*	Paíño Pechialbo	Calca-mar		B	B			B		B		●	
Leach's Storm-petrel	*Oceanodroma leucorhoa*	Paíño de Leach	Paínho-de-cauda-forcada			Recorded from Cape Verde seas								
Madeiran Storm-petrel	*Oceanodroma castro*	Paíño de Madeira	Roquinho		B				B		B			
Red-billed Tropicbird	*Phaethon aethereus*	Rabijunco Etéreo	Rabijunco	B	B		B		B		B			B
White-tailed Tropicbird	*Phaethon lepturus*	Rabijunco Menor	Rabijunco-pequeno		●									
Red-footed Booby	*Sula sula*	Piquero Patirrojo	Atobá-de-pés-vermelhos								●			
Brown Booby	*Sula leucogaster*	Piquero Pardo	Alcatraz-pardo	E	B		B		B		B		?	
Northern Gannet	*Morus bassanus*	Alcatraz Atlantico	Alcatraz			Recorded from Cape Verde seas								
Cormorant	*Phalacrocorax carbo*	Cormorán Grande	Corvo-marinho		●					●	●		●	
White Pelican	*Pelecanus onocrotalus*	Pelícano Común	Pelicano-branco		●									
Magnificent Frigatebird	*Fregata magnificens*	Rabihorcado Magnífico	Rabiforcado	E	B	●	●				●		E	●

BIRDS OF THE ATLANTIC ISLANDS

English name	Scientific name	Spanish name	Portuguese name	S	B	M	St	F	Br	N	R	L	V	A
Little Bittern	*Ixobrychus minutus*	Avetorillo Común	Garçote						●			●		
Night Heron	*Nycticorax nycticorax*	Martinete Común	Goraz		●		●				●		●	
Squacco Heron	*Ardeola ralloides*	Garcilla Cangrejera	Papa-ratos	●	●	●								
Cattle Egret	*Bubulcus ibis*	Garcilla Bueyera	Carraceiro	●	●	●	b	●	●	●			●	●
Western Reef Egret	*Egretta gularis*	Garceta Sombría	Garça-dos-recifes		●					●			●	
Black Heron	*Hydranassa ardesiaca*	Garceta Azabache	Garça-preta		●									
Little Egret	*Egretta garzetta*	Garceta Común	Garça-branca	●	B	●	B	●	●	B	B	B	●	●
Intermediate Egret	*Egretta intermedia*	Garceta Intermedia	Garça-branca-intermédia	●	●		●							●
Great White Egret	*Egretta alba*	Garceta Grande	Garça-branca-grande		●									
Grey Heron	*Ardea cinerea*	Garza Real	Garça-real	●	●	●		●		●	●	●	●	B
Great Blue Heron	*Ardea herodias*	Garza Azulada	Garça-real-americana		●									
Purple Heron	*Ardea purpurea*	Garza Imperial	Garça-ruiva		●		●						●	
Cape Verde Purple Heron	*Ardea (purpurea) bournei*						B							
Glossy Ibis	*Plegadis falcinellus*	Morito	Ibis-preto		●	●	●							
Bald Ibis	*Geronticus eremita*	Ibis Eremita	Ibis-pelado		●									
Eurasian Spoonbill	*Platalea leucorodia*	Espátula	Colhereiro		●		●							
Greater Flamingo	*Phoenicopterus roseus*	Flamenco	Flamingo	E	E	★								
Eurasian Wigeon	*Anas penelope*	Silbón Europea	Piadeira							●				
American Wigeon	*Anas americana*	Silbón Americano	Piadeira-americano							●				
Eurasian Teal	*Anas crecca*	Cerceta Común	Marrequinha		●								●	
Green-winged Teal	*Anas carolinensis*	Cerceta Americana	Marrequinha-americano							●				
Pintail	*Anas acuta*	Ánade Rabudo	Arrábio	●		●								
Garganey	*Anas querquedula*	Cerceta Carretona	Marreco											
Blue-winged Teal	*Anas discors*	Cerceta Aliazul	Marreca-d'asa-azul										●	
Marbled Duck	*Marmaronetta angustirostris*	Cerceta Pardilla	Pardilheira		●									
Common Pochard	*Aythya ferina*	Porrón Común	Zarro	●			●			●				
Ring-necked Duck	*Aythya collaris*	Porrón Acollorado	Caturro	●										
Ferruginous Duck	*Aythya nyroca*	Porrón Pardo	Pêrra		●									
Tufted Duck	*Aythya fuligula*	Porrón Moñudo	Negrinha	●			●							
Lesser Scaup	*Aythya affinis*	Porrón Bola	Negrelho-americano										●	
Black Kite	*Milvus migrans*	Milano Negro	Milhafre-preto	●	?	E	E	●	E	E	●	●	E	E
Cape Verde Kite	*Milvus fasciicauda*	Milano de Cabo Verde	Milhafre-vermelho		?	●	E		E	E			E	?
Egyptian Vulture	*Neophron percnopterus*	Alimoche Común	Britango	b	B	B	B	B		B	B	b	●	B
Marsh Harrier	*Circus aeruginosus*	Aguilucho Lagunero Occidental	Águia-sapeira	●	●					●	●			●
Montagu's Harrier	*Circus pygargus*	Aguilucho Cenizo	Águia-caçadeira		●	●								
Common Buzzard	*Buteo buteo*	Busardo Ratonero	Águia-d'asa-redonda		E		B	●	●	?			●	B
Osprey	*Pandion haliaetus*	Águila Pescadora	Águia-pesqueira	B	B	B	B	B	B	B	B	B	B	B
Common Kestrel	*Falco tinnunculus*	Cernícalo Vulgar	Peneireiro	●										
Alexander's Kestrel	*Falco tinnunculus alexandri*			B	B	B	B	B	B					
Neglected Kestrel	*Falco tinnunculus neglectus*									B	B	B	B	B
Eleonora's Falcon	*Falco eleonorae*	Halcón de Eleonora	Falcão-da-rainha										●	
Cape Verde Peregrine	*Falco peregrinus madens*	Halcón Peregrino	Falcão-peregrino	●	●		B	?	B	?	?		?	B
Common Quail	*Coturnix coturnix*	Codorniz	Codorniz	B	B	B	B	B	B	B			B	B
Helmeted Guineafowl	*Numida meleagris*	Pintada Común	Pintada					B	B		B			
Spotted Crake	*Porzana porzana*	Polluela Pintoja	Franga-d'água-grande										●	
Moorhen	*Gallinula chloropus*	Gallineta Común	Galinha-d'água		E		E						●	
Eurasian Oystercatcher	*Haematopus ostralegus*	Ostrero	Ostraceiro		●		●			●		●	●	
Black-winged Stilt	*Himantopus himantopus*	Cigüeñuela	Pernilongo	B	●	●	●						●	
Avocet	*Recurvirostra avosetta*	Avoceta	Alfaiate	●	●	●								
Cream-coloured Courser	*Cursorius cursor*	Corredor	Corredeira	B	B	B	B			●	●	●	B	●

APPENDIX B: DISTRIBUTION OF SPECIES ON THE CAPE VERDE ISLANDS

English name	Scientific name	Spanish name	Portuguese name	S	B	M	St	F	Br	N	R	L	V	A
Collared Pratincole	Glareola pratincola	Canastera	Perdiz-do-mar	●		●	●						●	
Little Ringed Plover	Charadrius dubius	Chorlitejo Chico de-coleira	Borrelho-pequeno-		●		●							
Ringed Plover	Charadrius hiaticula	Chorlitejo Grande	Borrelho-grande-de-coleira	●	●	●	●	●		●	●		●	
Semipalmated Plover	Charadrius semipalmatus	Chorlitejo Semipalmeado	Batuíra-de-bando	●									●	
Kentish Plover	Charadrius alexandrinus	Chorlitejo Patinegro	Borrelho-de-coleira-interrompida	B	B	B	B		●	●	●		B	●
American Golden Plover	Pluvialis dominica	Chorlito Dorado Chico	Batuiruçu						●				●	●
Grey Plover	Pluvialis squatarola	Chorlito Gris	Tarambola-cinzenta	●	●	●	●		●	●			●	●
Spur-winged Lapwing	Vanellus spinosus	Avefría Espolada	Tui-tui-ferrão							●				
Northern Lapwing	Vanellus vanellus	Avefría Europea	Abibe	●	●									
Knot	Calidris canutus	Correlimos Gordo	Seixoeira		●		●						●	
Sanderling	Calidris alba	Correlimos Tridáctilo	Pilrito-das-praias	●	●	●	●		●				●	●
Semipalmated Sandpiper	Calidris pusilla	Correlimos Semipalmeado	Pilrito-rasteirinho		●									
Little Stint	Calidris minuta	Correlimos Menudo	Pilrito-pequeño	●	●		●				●		●	●
Temminck's Stint	Calidris temminckii	Correlimos de Temminck	Pilrito de Temmink	●	●									
Least Sandpiper	Calidris minutilla	Correlimos Menudillo	Pilrito-anão										●	
Pectoral Sandpiper	Calidris melanotos	Correlimos Pectoral	Pilrito-de-colete						●					
Curlew Sandpiper	Calidris ferruginea	Correlimos Zarapitín	Pilrito-de-bico-comprido	●	●	●	●			●			●	●
Dunlin	Calidris alpina	Correlimos Común	Pilrito-de-peito-preta	●	●	●	●			●			●	●
Ruff	Philomachus pugnax	Combatiente	Combatente	●										
Jack Snipe	Lymnocryptes minimus	Agachadiza Chica	Narceja-galega						●					
Common Snipe	Gallinago gallinago	Agachadiza Común	Narceja	●	●		●						●	
Black-tailed Godwit	Limosa limosa	Aguja Colinegra	Milherango	●	●		●						●	
Bar-tailed Godwit	Limosa lapponica	Aguja Colipinta	Fuselo		●		●	●	●				●	
Whimbrel	Numenius phaeopus	Zarapito Trinador	Maçarico-galego	●	●	●	●	●	●	●	●	●	●	●
Eurasian Curlew	Numenius arquata	Zarapito Real	Maçarico-real	●	●						●			●
Spotted Redshank	Tringa erythropus	Archibebe Oscuro	Perna-vermelha-bastardo	●	●									
Common Redshank	Tringa totanus	Archibebe Común	Perna-vermelha	●	●		●							
Marsh Sandpiper	Tringa stagnatilis	Archibebe Fino	Perna-verde-fino	●										
Greenshank	Tringa nebularia	Archibebe Claro	Perna-verde	●	●						●		●	●
Greater Yellowlegs	Tringa melanoleuca	Archibebe Patigualdo Grande	Perna-amarela-grande						●					
Lesser Yellowlegs	Tringa flavipes	Archibebe Patigualdo Chico	Perna-amarela-pequeno	●	●									
Solitary Sandpiper	Tringa solitaria	Andarríos Solitario	Maçarico-solitário		●								●	
Green Sandpiper	Tringa ochropus	Andarríos Grande	Maçarico-bique-bique		●								●	
Wood Sandpiper	Tringa glareola	Andarríos Bastardo	Maçarico-de-dorso-malhado											
Common Sandpiper	Actitis hypoleucos	Andarríos Chico	Maçarico-das-rochas	●	●	●	●	●	●	●	●	●	●	●
Spotted Sandpiper	Actitis macularius	Andarríos Maculado	Maçarico-pintado										●	
Turnstone	Arenaria interpres	Vuelvepiedras	Rola-do-mar	●	●	●	●	●	●	●	●	●	●	●
Red-necked Phalarope	Phalaropus lobatus	Falaropo Picofino	Falaropo-de-bico-fino	●										
Grey Phalarope	Phalaropus fulicarius	Falaropo Picogrueso	Falaropo-de-bico-grosso	Recorded from Cape Verde seas									
Pomarine Skua	Stercorarius pomarinus	Págalo Pomarino	Moleiro do Árctico	●		●	●			●	●			
Arctic Skua	Stercorarius parasiticus	Págalo Parásito	Moleiro-pequeno		●					●		●		
Long-tailed Skua	Stercorarius longicaudus	Págalo Rabero	Moleiro-rabilongo		●									
Great Skua	Stercorarius skua	Págalo Grande	Alcaide	●							●		●	●
Sabine's Gull	Larus sabini	Gaviota de Sabine	Gaivota de Sabine		●									●

English name	Scientific name	Spanish name	Portuguese name	S	B	M	St	F	Br	N	R	L	V	A
Black-headed Gull	*Larus ridibundus*	Gaviota Reidora	Garrincho	●	●	●	●	●		●			●	●
Slender-billed Gull	*Larus genei*	Gaviota Picofina	Gaivota-de-bico-fino	●										
Ring-billed Gull	*Larus delawarensis*	Gaviota de Delaware	Gaivota-de-bico-riscado		●									
Lesser Black-backed Gull	*Larus fuscus*	Gaviota Sombría	Gaivota-d'asa-escura	●	●					●				●
Yellow-legged Gull	*Larus michahellis*	Gaviota Patiamarilla	Gaivota-de-patas-amarelas	●	●			●		●	●	●	●	
Kittiwake	*Rissa tridactyla*	Gaviota Tridáctila	Gaivota-tridáctila		●	●	●	●	●					
Gull-billed Tern	*Gelochelidon nilotica*	Pagaza Piconegra	Tagáz		●									
Caspian Tern	*Hydroprogne caspia*	Pagaza Piquirroja	Garajau-grande		●								●	
Royal Tern	*Sterna maxima*	Charrán Real	Garajau-real	●									●	
Sandwich Tern	*Sterna sandvicensis*	Charrán Patinegro	Garajau	●	●			●		●				
Roseate Tern	*Sterna dougallii*	Charrán Rosado	Gaivina-rosada	●	●									
Common Tern	*Sterna hirundo*	Charrán Común	Gaivina	●	●	●								
Arctic Tern	*Sterna paradisaea*	Charrán Ártico	Gaivina-ártica	●								●	●	
Little Tern	*Sternula albifrons*	Charráncito	Chilreta		●	●								
Rock Dove	*Columba livia*	Paloma Bravía	Pombo-das-rochas				B	B	B	B			B	B
Turtle Dove	*Streptopelia turtur*	Tórtola Común	Rola-brava	●	●	●		●						
Namaqua Dove	*Oena capensis*	Tórtolita Rabilarga	Rolinha-rabilonga						●					
Rose-ringed Parakeet	*Psittacula krameri*	Cotorra de Kramer	Periquito-rabijunco				?						●	
Great Spotted Cuckoo	*Clamator glandarius*	Críalo	Cuco-rabilongo	●										
Common Cuckoo	*Cuculus canorus*	Cuco	Cuco			●	●	●						
Barn Owl	*Tyto alba detorta*	Lechuza de Cabo Verde	Coruja de Cabo Verde		B	B	B	?	B	B	B		●	B
Scops Owl	*Otus scops*	Autillo	Mocho-d'orelhas	●										
Short-eared Owl	*Asio flammeus*	Lechuza Campestre	Coruja-galhofa	●				●	●					
Cape Verde Swift	*Apus alexandri*	Vencejo de Cabo Verde	Andorinhão de Cabo Verde	●	●	●	B	B	B	B	●		●	B
Plain Swift	*Apus unicolor*	Vencejo Unicolor	Andorinhão-da-serra		●									
Common Swift	*Apus apus*	Vencejo Común	Andorinhão-preto	●			●	●	●	●	●		●	●
Pallid Swift	*Apus pallidus*	Vencejo Pálido	Andorinhão-pálido			●								●
Alpine Swift	*Apus melba*	Vencejo Real	Andorinhão-real	●			●	●						
Grey-headed Kingfisher	*Halcyon leucocephala*	Alción Cabeciblanco	Passarinha				B	B	B					
Blue-cheeked Bee-eater	*Merops persicus*	Abejaruco Papirrojo	Abelharuco-de-faces-azuis						●					
European Bee-eater	*Merops apiaster*	Abejaruco Común	Abelharuco	●	●					●				
European Roller	*Coracias garrulus*	Carraca	Rolieiro									●		
Broad-billed Roller	*Eurystomus glaucurus*	Carraca Picogorda	Peito-lilás					●	●					
Hoopoe	*Upupa epops*	Abubilla	Poupa	●	●								●	●
Black-crowned Sparrow-lark	*Eremopterix nigriceps*	Terrera Negrita	Pastor	?	B	B	B	B	●	●				
Bar-tailed Lark	*Ammomanes cinctorus*	Terrera Colinegra	Calhandra-das-dunas	B	B	B	B	?	●	B		●		
Greater Hoopoe Lark	*Alaemon alaudipes*	Sirli Desértico	Calhandra-de-bico-curva	?	B	B								
Short-toed Lark	*Calandrella brachydactyla*	Terrera Común	Calhandrinha								●			
Raso Lark	*Alauda razae*	Terrera de Razo	Alberca do Razo									B		
Plain Martin	*Riparia paludicola*	Avión Paludícola	Andorinha-dos-charcos	●										
Sand Martin	*Riparia riparia*	Avión Zapador	Andorinha-das-barreiras	●			●			●			●	
Barn Swallow	*Hirundo rustica*	Golondrina Común	Andorinha-das-chaminés	●	b		●	●	●	●	●		●	●
Red-rumped Swallow	*Cecropis daurica*	Golondrina Dáurica	Andorinha-dáurica	●			●				●		●	

APPENDIX B: DISTRIBUTION OF SPECIES ON THE CAPE VERDE ISLANDS

English name	Scientific name	Spanish name	Portuguese name	S	B	M	St	F	Br	N	R	L	V	A
House Martin	*Delichon urbicum*	Avión Común	Andorinha-dos-beirais	●	b	●	●		●	●	●		●	●
Tawny Pipit	*Anthus campestris*	Bisbita Campestre	Petinha-dos-campos		●									
Tree Pipit	*Anthus trivialis*	Bisbita Arbóreo	Petinha-das-árvores		●	●				●				
Red-throated Pipit	*Anthus cervinus*	Bisbita Gorgirrojo	Petinha-de-garganta-ruiva	●									●	
Yellow Wagtail	*Motacilla flava*	Lavandera Boyera	Alvéola-amarela	●			●				●			
White Wagtail	*Motacilla alba*	Lavandera Blanca	Alvéola-branca	●			●				●		●	●
Common Nightingale	*Luscinia megarhynchos*	Ruiseñor Común	Rouxinol	●	●									
Black Redstart	*Phoenicurus ochruros*	Colirrojo Tizón	Rabirruivo	●										
Common Redstart	*Phoenicurus phoenicurus*	Colirrojo Real	Rabirruivo-de-testa-branca	●										
Whinchat	*Saxicola rubetra*	Tarabilla Norteña	Cartaxo-d'arribação	●										
Northern Wheatear	*Oenanthe oenanthe*	Collalba Gris	Chasco-cinzento	●		●	●				●		●	●
White-crowned Wheatear	*Oenanthe leucopyga*	Collalba Negra de Brehm	Chasco-de-barrete-branco						●					
Song Thrush	*Turdus philomelos*	Zorzal Común	Tordo-pinto	●										
Zitting Cisticola	*Cisticola juncidis*	Buitrón	Fuínha-dos-juncos	●										
Grasshopper Warbler	*Locustella naevia*	Buscarla Pintoja	Cigarrinha-malhada	●										
Savi's Warbler	*Locustella luscinoides*	Buscarla Unicolor	Cigarrinha-ruiva	●										
Moustached Warbler	*Acrocephalus melanopogon*	Carricerín Real	Felosa-real	●										
Cape Verde Warbler	*Acrocephalus brevipennis*	Carricero de Cabo Verde	Chota-de-cana						B	?	B			
Western Olivaceous Warbler	*Hippolais opaca*	Zarcero Pálido Occidental	Felosa-pálida	●										
Spectacled Warbler	*Sylvia conspicillata*	Curruca Tomillera	Toutinegra-tomilheira		B	B	B	B	B	B			B	B
Subalpine Warbler	*Sylvia cantillans*	Curruca Carrasqueña	Toutinegra-de-bigodes		●						●			
Sardinian Warbler	*Sylvia melanocephala*	Curruca Cabecinegra	Toutinegra-dos-valados								●			
African Desert Warbler	*Sylvia deserti*	Curruca Sahariana	Toutinegra-do-deserto	●										
Garden Warbler	*Sylvia borin*	Curruca Mosquitera	Felosa-das-figueiras								●			
Blackcap	*Sylvia atricapilla*	Curruca Capirotada	Toutinegra-de-barrete	B	●	●	B	B	B	B			E	B
Western Bonelli's Warbler	*Phylloscopus bonelli*	Mosquitero Papialbo Occidental	Felosa-de-papo-branco	●										
Common Chiffchaff	*Phylloscopus collybita*	Mosquitero Común	Felosinha		●						●			
Willow Warbler	*Phylloscopus trochilus*	Mosquitero Musical	Felosa-musical		●						●	●		
Spotted Flycatcher	*Muscicapa striata*	Papamoscas Gris	Papa-moscas-cinzento	●	●									
Pied Flycatcher	*Ficedula hypoleuca*	Papamoscas Cerrojillo	Papas-moscas preto	●	●									
Southern Grey Shrike	*Lanius meridionalis*	Alcaudón Real	Picanço-real	●										
Woodchat Shrike	*Lanius senator*	Alcaudón Común	Picanço-barreteiro	●										
Brown-necked Raven	*Corvus ruficollis*	Cuervo Desertícola	Corvo-do-deserto	B	B	B	B	B	B	B	B	B	B	B
Common Starling	*Sturnus vulgaris*	Estornino Pinto	Estorninho-malhado							●			●	
House Sparrow	*Passer domesticus*	Gorrión Doméstico	Pardal										B	
Spanish Sparrow	*Passer hispaniolensis*	Gorrión Moruno	Pardal-espanhol	?	B	B	B	B	?	B	●		?	?
Iago Sparrow	*Passer iagoensis*	Gorrión de Cabo Verde	Pardal de Cabo Verde	B	B	B	B	?	B	B	B	B	B	B
Village Weaver	*Ploceus cucullatus*	Tejedor Cogullado	Cacho-caldeirão							●			●	
Red-billed Quelea	*Quelea quelea*	Tejedor de Pico Rojo	Bico-carmim										●	
Red-cheeked Cordon-bleu	*Uraeginthus bengalus*	Coliazul Bengalí	Peito-celeste										●	
Common Waxbill	*Estrilda astrild*	Pico de Coral	Bico-de-lacre						B	E	E	E		●	●
Goldfinch	*Carduelis carduelis*	Jilguero	Pintassilgo						E					
Trumpeter Finch	*Bucanetes githagineus*	Camachuelo Trompetero	Trombeteiro	●										

Appendix C: Distribution of endemic taxa on the Atlantic Islands

B = breeding species or subspecies **?** = breeding status uncertain **●** = recorded but no proof of breeding
Bold type = endemic species **E** = extinct **b** = has bred **I** = introduced

i. The Canary Islands

L = Lanzarote   F = Fuerteventura   C = Gran Canaria   T = Tenerife
G = La Gomera   P = La Palma   H = El Hierro

English name	Scientific name	L	F	C	T	G	P	H
Macaronesian Shearwater	*Puffinus baroli baroli*	B	B	B	B	B	B	B
Egyptian Vulture	*Neophron percnopterus junoniae*	B	B	b	b	b		b
Eurasian Sparrowhawk	*Accipiter nisus granti*			B	B	B	B	B
Common Buzzard	*Buteo buteo insularum*	B	B	B	B	B	B	B
Common Kestrel	*Falco tinnunculus canariensis*			B	B	B	B	B
	Falco tinnunculus dacotiae	B	B					
Houbara Bustard	*Chlamydotis undulata fuerteventurae*	B	B	b	●			
Canary Islands Oystercatcher	*Haematopus meadewaldoi*	E	E					
Stone-curlew	*Burhinus oedicnemus insularum*	B	B					
	Burhinus oedicnemus distinctus			B	B	B	B	B
Yellow-legged Gull	*Larus michahellis atlantis*	B	B	B	B	B	B	B
Bolle's Pigeon	*Columba bollii*			b	B	B	B	B
Laurel Pigeon	*Columba junoniae*				B	B	B	B
Barn Owl	*Tyto alba gracilirostris*	B	B					
Long-eared Owl	*Asio otus canariensis*		B	B	B	B	B	B
Plain Swift	*Apus unicolor*	B	B	B	B	B	B	B
Great Spotted Woodpecker	*Dendrocopus major canariensis*				B			
	Dendrocopus major thanneri			B				
Lesser Short-toed Lark	*Calandrella rufescens rufescens*			?	B			
	Calandrella rufescens polatzeki	B	B	B	B			
Berthelot's Pipit	*Anthus berthelotii berthelotii*	B	B	B	B	B	B	B
Grey Wagtail	*Motacilla cinerea canariensis*			B	B	B	B	B
Robin	*Erithacus rubecula superbus*			B	B			
Canary Islands Stonechat	*Saxicola dacotiae*		B					
Blackbird	*Turdus merula cabrerae*			B	B	B	B	B
Spectacled Warbler	*Sylvia conspicillata orbitalis*	B	B	B	B	B	B	B
Sardinian Warbler	*Sylvia melanocephala leucogastra*	?	?	B	B	B	B	B
Common Chiffchaff	*Phylloscopus collybita exsul*	E?						
Canary Islands Chiffchaff	*Phylloscopus canariensis*			B	B	B	B	B
Tenerife Goldcrest	*Regulus teneriffae*			B	B	B		B
African Blue Tit	*Cyanistes ultramarinus teneriffae*			B	B	B		
	Cyanistes ultramarinus palmensis						B	
	Cyanistes ultramarinus ombriosus							B
	Cyanistes ultramarinus degener	B	B					
Southern Grey Shrike	*Lanius meridionalis koenigi*	B	B	B	B		B	
Rock Sparrow	*Petronia petronia madeirensis*			B	B	B	B	B
Common Chaffinch	*Fringilla coelebs canariensis*			B	B	B		
	Fringilla coelebs palmae						B	
	Fringilla coelebs ombriosa							B
Blue Chaffinch	*Fringilla teydea teydea*				B			
	Fringilla teydea polatzeki			B				
Canary	*Serinus canaria*	B	I?	B	B	B	B	B
Linnet	*Carduelis cannabina meadewaldoi*			B	B	B	B	B
	Carduelis cannabina harterti	B	B					
Trumpeter Finch	*Bucanetes githagineus amantum*	B	B	B	B	B		●

ii. Madeira

M = Madeira   P = Porto Santo   D = Desertas   S = Salvage Islands

English name	Scientific name	M	S	D	S
Fea's Petrel	*Pterodroma feae*			B	
Zino's Petrel	*Pterodroma madeira*	B			
Macaronesian Shearwater	*Puffinus baroli baroli*		B	B	B
Common Kestrel	*Falco tinnunculus canariensis*	B	B	B	B
Eurasian Sparrowhawk	*Accipiter nisus granti*	B			
Yellow-legged Gull	*Larus michahellis atlantis*	B	B	B	B
Woodpigeon	*Columba palumbus maderensis*	E			
Trocaz Pigeon	*Columba trocaz*	B			
Barn Owl	*Tyto alba schmitzi*	B	B		
Plain Swift	*Apus unicolor*	B	B	●	●
Berthelot's Pipit	*Anthus berthelotii madeirensis*	B	B	B	
	Anthus berthelotii berthelotii				B
Grey Wagtail	*Motacilla cinerea schmitzi*	B	B		
Blackbird	*Turdus merula cabrerae*	B	B		
Spectacled Warbler	*Sylvia conspicillata orbitalis*	B	B		
Blackcap	*Sylvia atricapilla heineken*	B	B		
Madeira Firecrest	*Regulus madeirensis*	B			
Rock Sparrow	*Petronia petronia madeirensis*	B	B		
Common Chaffinch	*Fringilla coelebs maderensis*	B			
Canary	*Serinus canaria*	B	B		
Linnet	*Carduelis cannabina guentheri*	B	B		

iii. The Azores

S = Santa Maria   M = São Miguel   T = Terceira   J = São Jorge   G = Graciosa,
P = Pico   F = Faial   Fl = Flores   C = Corvo

English name	Scientific name	S	M	T	F	G	J	P	Fl	C
Macaronesian Shearwater	*Puffinus baroli baroli*	B	b			B	b			b
Common Buzzard	*Buteo buteo rothschildi*	B	B	B	B	B	B	B		
Common Quail	*Coturnix coturnix conturbans*	B	B	B	B	B	B	B	B	B
Yellow-legged Gull	*Larus michahellis atlantis*	B	B	B	B	B	B	B	B	B
Woodpigeon	*Columba palumbus azorica*	B	B	B	B	B	B	B		
Blackbird	*Turdus merula azorensis*	B	B	B	B	B	B	B	B	B
Blackcap	*Sylvia atricapilla gularis*	B	B	B	B	B	B	B	B	B
Goldcrest	*Regulus regulus sanctaemariae*	B								
	Regulus regulus azoricus		B							
	Regulus regulus inermis			B	B		B	B	B	?
Grey Wagtail	*Motacilla cinerea patriciae*	B	B	B	B	B	B	B	B	B
Common Starling	*Sturnus vulgaris granti*	B	B	B	B	B	B	B	B	B
Common Chaffinch	*Fringilla coelebs moreletti*	B	B	B	B	B	B	B	B	B
Azores Bullfinch	*Pyrrhula murina*		B							
Canary	*Serinus canarius*	B	B	B	B	B	B	B	B	B

iv. The Cape Verde Islands

S = Sal	B = Boavista	M = Maio	St = Santiago
F = Fogo	Br = Brava	N = São Nicolau	R = Raso
Bo = Branco	L = Santa Luzia	V = São Vicente	A = Santo Antão

English name	Scientific name	S	B	M	St	F	Br	N	R	Bo	L	V	A
Fea's Petrel	*Pterodroma feae*				B	B		B					B
Cape Verde Shearwater	*Calonectris edwardsii*	b	?		B		B	B	B	B			B
Macaronesian Shearwater	*Puffinus baroli boydi*	b	b		B	?	B	B	B	B		?	●
White-faced Storm-petrel	*Pelagodroma marina eadesi*		B	B			B			B			
Cape Verde Purple Heron	*Ardea purpurea bournei*				B								
Cape Verde Kite	*Milvus fasciicauda*		?		b		b	b				b	b
Cape Verde Buzzard	*Buteo buteo bannermani*		b		B	?	●	b				●	B
Neglected Kestrel	*Falco tinnunculus neglectus*							B	●	●	B	B	B
Alexander's Kestrel	*Falco tinnunculus alexandri*	B	B	B	B	B	B						
Cape Verde Peregrine	*Falco peregrinus madens*	●	●		B	B	b	B	●	●		●	B
Cape Verde Barn Owl	*Tyto alba detorta*		B	B	B	?	B	B	B	B		●	B
Cape Verde Swift	*Apus alexandri*	●	●	●	B	B	B	B				B	B
Raso Lark	*Alauda razae*								B				
Cape Verde Warbler	*Acrocephalus brevipennis*				B			?		B			
Spectacled Warbler	*Sylvia conspicillata orbitalis*		B	B	B	B	B	B				B	B
Iago Sparrow	*Passer iagoensis*	B	B	B	B	?	B	B	B	?	B	B	B

BIBLIOGRAPHY

Alexander, B. 1898a. An ornithological expedition to the Cape Verde Islands. *Ibis* 4: 74–118.
Alexander, B. 1898b. Further notes on the ornithology of the Cape Verde Islands. Ibis 4: 277–285.
Bannerman, D. A. 1914. Distribution and nidification of the tubinares in the North Atlantic Islands. *Ibis* 10: 438–494.
Bannerman, D. A. 1919a. List of the Birds of the Canary Islands, with detailed reference to the migratory species and the accidental visitors. Part 1. *Ibis* 11: 84–131.
Bannerman, D. A. 1919b. List of the Birds of the Canary Islands, with detailed reference to the migratory species and the accidental visitors. Part 2. *Ibis* 11: 291–321.
Bannerman, D. A. 1919c. List of the Birds of the Canary Islands, with detailed reference to the migratory species and the accidental visitors. Part 3. *Ibis* 11: 457–495.
Bannerman, D. A. 1919d. List of the Birds of the Canary Islands, with detailed reference to the migratory species and the accidental visitors. Part 4. *Ibis* 11: 708–764.
Bannerman, D. A. 1920a. List of the Birds of the Canary Islands, with detailed reference to the migratory species and the accidental visitors. Part 5. *Ibis* 11: 97–132.
Bannerman, D. A. 1920b. List of the Birds of the Canary Islands, with detailed reference to the migratory species and the accidental visitors. Part 6. *Ibis* 11: 323–360.
Bannerman, D. A. 1920c. List of the Birds of the Canary Islands, with detailed reference to the migratory species and the accidental visitors. Part 7. *Ibis* 11: 519–569.
Bannerman, D. A. 1922. *The Canary Islands: Their History, Natural History and Scenery.* Gurney & Jackson, London.
Bannerman, D. A. 1963. *Birds of the Atlantic Islands. Volume 1: A History of the Birds of the Canary Islands and of the Salvages.* Oliver & Boyd, Edinburgh and London.
Bannerman, D. A. & Bannerman, W. M. 1965. *Birds of the Atlantic Islands. Volume 2: A History of the Birds of Madeira, the Desertas, and the Porto Santo Islands.* Oliver & Boyd, Edinburgh and London.
Bannerman, D. A. & Bannerman, W. M. 1966. *Birds of the Atlantic Islands. Volume 3: A History of the Birds of the Azores.* Oliver & Boyd, Edinburgh and London.
Bannerman, D. A. & Bannerman, W. M. 1968. *Birds of the Atlantic Islands. Volume 4: History of the Birds of the Cape Verde Islands.* Oliver & Boyd, Edinburgh and London.
Barone, R. 1997. Observaciones de aves migratorias en el archipiélago de Cabo Verde, Septiembre de 1997. *Revista de la Academia Canaria de Ciencias* IX: 87–96.
Beaman, M. & Madge, S. 1998. *The Handbook of Bird Identification for Europe and the Western Palearctic.* Christopher Helm, London.
Bergmann, H-H. & Schottler, B. H. 2001. Trends in systematics: Tenerife Robin – a species on its own? *Dutch Birding* 23: 140–146.
Bernström, J. 1951. Check-list of the Breeding Birds of the Archipelago of Madeira. *Boletim do Museu Municipal do Funcha* 5: 64–82.
Bocage, J. V. Barbosa du 1867. Aves das possessões portuguezas da Africa occidental que existem no Museu de Lisboa. *Jornal de Sciencias Mathematicas, Physicas e Naturaes, Academia Real das Sciencias de Lisboa* 1: 129–153.
Bolle, C. 1856. Die Vogelwelt auf den Inseln des grünen Vorgebirges. *Journal für Ornithologie* 4: 17–31.
Bolle, C. 1858. Der wilde Canarienvogel, eine Biographie. *Journal für Ornithologie* 6: 125–151.
Bolle, C. 1862a. *Anthus berthelotii*, eine neue Pieperart. *Journal für Ornithologie* 10: 357–360.
Bolle, C. 1862b. Sur l'Anthus des Canaries reconnu comme espèce nouvelle et nommé *Anthus berthelotii*. *Ibis* 4: 343–348.
Bourne, W. R. P. 1953. On the races of the Frigate Petrel, *Pelagodroma marina* (Latham) with a new race from the Cape Verde Islands. *Bulletin of the British Ornithologists' Club* 73: 79–82.
Bourne, W. R. P. 1955a. The birds of the Cape Verde Islands. Ibis 97: 508–556.
Bourne, W. R. P. 1955b. A new race of Kestrel from the Cape Verde Islands. *Bulletin of the British Ornithologists' Club* 75: 35–36.
Bourne, W. R. P. 1966. Further notes on the birds of the Cape Verde Islands. *Ibis* 108: 425–429.
Buxton, E. J. M. 1959. Notes on birds seen in Madeira: Winter 1958-59. *Bocagiana* 2: 1–2.
Buxton, E. J. M. 1960. Winter Notes from Madeira. *Ibis* 102: 127–129.
Cabrera, A. 1893. Catálogo de las aves del Archipiélago Canario. *Actas de la Real Sociedad Española de Historia Natural.* 22: 1–70.
Cámara, D. B. & Teixeira, A. M. 1980. Autumn occurrence of Palearctic migrants on Selvagem Grande Island (Madeira). *Bocagiana*, 50: 1-3.
Clarke, T. & Collins, D. (1996). *A Birdwatcher's Guide to the Canary Islands.* Prion, Perry, Cambridgeshire.
Concepción, D. 1992. *Avifauna del Parque Nacional de Timanfaya. Censo y Análisis. Red de Parques Nacionales.*

ICONA, Madrid.
Costa, H., Araujo, A., Farinha, J. C., Poças M. & Machado, A. 2000. *Nomes Portugueses das Aves do Paleárctico Ocidental*. Assírio and Alvim. Lisboa.
Cramp, S. (ed.). 1985. *The Birds of the Western Palearctic. Volume IV*. Oxford University Press, Oxford.
Cramp, S. (ed.). 1988. *The Birds of the Western Palearctic. Volume V.* Oxford University Press, Oxford.
Cramp, S. (ed.) 1992. *The Birds of the Western Palearctic. Volume VI*. Oxford University Press, Oxford.
Cramp, S. & Perrins, C. M. (eds) 1993. *The Birds of the Western Palearctic. Volume VII*. Oxford University Press, Oxford.
Cramp, S. & Perrins, C. M. (eds) 1994a. *The Birds of the Western Palearctic. Volume VIII*. Oxford University Press, Oxford.
Cramp, S. & Perrins, C. M. (eds) 1994b. *The Birds of the Western Palearctic. Volume IX*. Oxford University Press, Oxford.
Cramp, S. & Simmons, K. E. L. (eds) 1977. *The Birds of the Western Palearctic. Volume I*. Oxford University Press, Oxford.
Cramp, S. & Simmons, K. E. L. (eds) 1980. *The Birds of the Western Palearctic. Volume II*. Oxford University Press, Oxford.
Cramp, S. & Simmons, K. E. L. (eds) 1983. *The Birds of the Western Palearctic. Volume III*. Oxford University Press, Oxford.
Cullen, J. M., Guiton, P. E., Horridge, G. A. & Peirson, J. 1952. Birds on Palma and Gomera (Canary Islands). *Ibis* 94: 68–84.
Dalgleish, J. J. 1890. Letter on *Oestrelata mollis*. *Ibis*: p. 386.
Dalgleish, J. J. 1890-1892. Notes on the petrels of Madeira and the adjoining seas. *Proceedings of the Royal Physical Society of Edinburgh* 1891: 27–30.
Dapper. O. 1668. *Naukeurige beschrijvinge der Afrikaensche eylanden: als Madagaskar, of Sant Laurens, Sant Thomee, d'eilanden van Kanarien, Kaep de Verd, Malta, en andere, etc*. Van Meurs, Amsterdam.
de Chavigny, J. & Mayaud, N. 1932. Sur l'avifaune des Açores Généralités et Etude contributive. *Alauda* 4: 133–55, 304–48, 416–41.
de Chelmicki, J. C. C. & de Varnhagen, F. A. 1841. *Corografia cabo-verdiana ou descrição geográphico-histórica da provincial das ilhas de Cabo-Verde e Guiné*. Da Cunha, Lisbon.
de Noronha, A. & Schmitz, E. 1902. Aus dem vogelleben der Insel Porto Santo (Madeira). *Ornithologische Jahrbuch* 13: 130-135; 14: 119-137, 193-205; 15: 124-145.
Dohrn, H. 1871. Beiträge zur Ornithologie der Capverdischen. *Journal für Ornithologie* 19: 1–10.
Drouet, H. 1861. *Eléments de la Fauna Acoréenne*. Paris.
Emmerson, K. W. & Martín, A. 1985. Situación de la Avifauna de la Macaronesia (Azores, Madeira, Canarias y Cabo Verde). In M. Fernández-Cruz & J. Araujo, *Situación de la Avifauna de la Península Ibérica, Baleares y Macaronesia*. CODA-SEO, Madrid.
Emmerson, K. W., Martín, A., Bacallado, J. J. & Lorenzo, J. A. 1994. Catálogo y Bibliografía de la Avifauna Canaria. *Publicaciones Científicas del Cabildo de Tenerife, Museo de Ciencias Naturales, O.A.M.C.* 4.
Fea, L. 1898–1899. Dalle Isole del Capo Verde (5 Parts). *Bolletino della Societá Geografica Italiana* 11: 358–368, 537–552; 12: 7–26, 163–174, 302–312.
Folmer, O. & Ortvad, T. 1992. Observations of terrestrial birds on Selvagem Grande in September 1990. *Bocagiana* 160: 1–6.
Forster, J. G. A. 1777. *A Voyage round the World in His Britannic Majesty's Sloop 'Resolution', commanded by Capt. James Cook, during the years 1772, 3, 4 and 5*. London.
Fructuoso, G. 1591. *As Saudades da Terra*.
Fructuoso, G. 1873. *As Saudades da Terra, Vol. 2*. Funchal.
Gantlett, S. 1995 Identification forum: Field separation of Fea's, Zino's and Soft-plumaged Petrels. *Birding World* 8: 256–260.
Godman, F. du Cane 1866. On the birds of the Azores. *Ibis* 5: 88–109.
Godman, F. du Cane 1870. *Natural History of the Azores or Western Islands*. Van Voorst, London.
Godman, F. du Cane 1872. Notes on the resident and migratory birds of Madeira and the Canaries. *Ibis* 3: 158–177, 209–224.
Gould, J. 1841. *Birds. Part 3. The Zoology of the Voyage of H.M.S. Beagle, under the command of Captain Fitzroy, R.N., during the years 1832 to 1836*. Smith, Elder and Co., London.
Harcourt, E. V. 1851. *A Sketch of Madeira*. Edward Vernon, London.
Harcourt, E. V. 1855. Final list of residents and migrants combined. *Annals of Natural History* XV: 437–38.
Harris, H. E. 1901. *Essays and photographs: some birds of the Canary Islands and South Africa*. R.H. Porter, London.
Hartert, E. 1905. Eine neue Subspecies von Fringilla teydea. *Ornithologische Monatsberichte* 13: 164.
Hartert, E. and Ogilvie-Grant, W. R. 1905. On the Birds of the Azores. *Novitates Zoologicae* 12: 80–128.
Hartwig, W. 1886. Die Vögel Madeiras. *Journal für Ornithologie* 34: 452–85.
Hazevoet, C. J. 1993. *Aves de Cabo Verde*. Birdlife International, São Jorge dos Orgáos.

Hazevoet, C. J. 1995. *The Birds of the Cape Verde Islands. BOU Check-list no.13*. British Ornithologists' Union, Tring.

Hazevoet, C. J. 1997. Notes on distribution, conservation, and taxonomy of birds from the Cape Verde Islands, including records of six species new to the archipelago. *Bulletin Zoologisch Museum, University of Amsterdam* 15: 89–100.

Hazevoet, C. J. 1998. Third annual report on birds from the Cape Verde Islands, including records of seven taxa new to the archipelago. *Bulletin Zoologisch Museum, University of Amsterdam* 16: 65–72.

Hazevoet, C. J. 1999. Fourth report on birds from the Cape Verde Islands, including notes on conservation and records of 11 taxa new to the archipelago. *Bulletin Zoologisch Museum, University of Amsterdam* 17: 19–32.

Hazevoet, C. J. 1999. Notes on birds from the Cape Verde Islands in the collection of the Centro de Zoologia, Lisbon, with comments on taxonomy and distribution. *Bulletin of the British Ornithologists' Club* 119: 25–31.

Hazevoet, C. J. 2003. Fifth report on birds from the Cape Verde Islands, including records of 15 taxa new to the archipelago. *Arquivos do Museu Bocage* 3: 503–528.

Hazevoet, C. J., Fischer, S. & Deloison, G. 1996. Ornithological news from the Cape Verde Islands in 1995, including records of species new to the archipelago. *Bulletin Zoologisch Museum, University of Amsterdam* 15: 21–27.

Hazevoet, C. J., Monteiro, L. R. & Ratcliffe, N. 1999. Rediscovery of the Cape Verde Cane Warbler *Acrocephalus brevipennis* on São Nicolau in February 1998. *Bulletin of the British Ornithologists' Club* 119: 68–71.

Heineken, C. 1829. Notice of some of the birds of Madeira. *Edinburgh Journal of Science* 2: 229–33.

Holmes, P. F. 1939. Some oceanic records and notes on the winter distribution of the Phalaropes. *Ibis* 3: 329-342.

Howell, S. 1996. *Pterodroma* identification revisited. *Birding World* 9: 276–277.

Jardine, W. 1830. Observations on a collection of birds lately received from Madeira, with the description of some new species from that island. *Edinburgh Journal of Natural and Geographical Science* I: 241–245.

Jensen, A. 1981. Ornithological winter observations on Selvagem Grande. *Bocagiana* 62: 1–7.

Jepson, P. R. & Zonfrillo, B. 1988. Bird notes from Madeira, summer 1986. *Bocagiana* 117: 1–10.

Johnson, J. A., Watson, R. T. & Mindell, D. P. (2005). Prioritizing species conservation: does the Cape Verde kite exist? *Proceedings of the Royal Society B*. 272: 1365–1371 (published online).

Keulemans, J.G. 1866. Opmerkingen over de vogels van de Kaap-Verdische Eilanden en van Prins-Eiland (Ilha do Principe) in de bogt van Guinea gelegen. *Nederlandsch Tijdschrift voor De Dierkunde* 3: 363–401.

Koenig, A. 1890. Ornith. Forschungsergebnisse einer Reise nach Madeira und den Canarischen Inseln. *Journal für Ornithologie* 38: 257–488.

Lack, D. & Southern, H. N. 1949. Birds on Tenerife. *Ibis* 91: 607–626.

Le Grand, G. 1983. Bilan des observations sur les oiseaux d'origine Néarctique effectuées aux Açores (jusqu'en janvier 1983). *Arquipélago* 4: 73–83.

Le Grand, G. 1990. Catalogue des oiseaux observés aux Açores (jusqu'en 1990). Unpublished manuscript.

Ledrú, A. P. 1810. *Voyage aux iles de Ténériffe, la Trinité, Saint-Thomas, Sainte-Croix et Porto-Ricco*. Paris.

Lima, J. J., Lopes de 1844. *Ensaio sobre a statística das ilhas de Cabo-Verde no mar Atlântico e suas dependências na Guiné portugueza ao norte do equador. Ensaios sobre a statística das possessões portuguezas no ultramar, vol. 1*. Imprensa Nacional, Lisbon.

Lockley, R. M. 1942. *Shearwaters*. J. M. Dent, London.

Lockley, R. M. 1952. Notes on the Birds of the islands of Berlengas (Portugal), the Desertas and Baixo (Madeira) and the Salvages. *Ibis* 94: 144–157.

Lorenzo, J. A. and González, J. 1993. *Las aves de El Médano (Tenerife-Islas Canarias)*. Asociación Tinerfeña de Amigos de la Naturaleza, Santa Cruz de Tenerife.

Lowe, P. R. (undated). Check-list of birds collected during second cruise of SY *Zenaïda*, Oct. 13th 1906 to June 1907. Unpublished manuscript, Natural History Museum, London.

Macgillivray, J. 1852. The birds observed in Porto Grande and the neighbourhood. Unpublished journal notes, Hydrographic Department of the Admiralty, London.

Martín, A. 1987. Atlas de las aves nidificantes en la isla de Tenerife. *Instituto de Estudios Canarios, Monografia* 32.

Martín, A. & Lorenzo, J. A. 2001. *Aves del archipiélago Canario*. Francisco Lemus Editor, La Laguna.

Mathews, G. M. 1934. The Soft-plumaged Petrel, *Pterodroma mollis*, and its subspecies. *Bulletin of the British Ornithologists' Club* 54: 178–179.

Meade-Waldo, E. G. B. 1889a. Notes on some birds of the Canary Islands. *Ibis* 6: 1–13.

Meade-Waldo, E. G. B. 1889b. Further notes on the birds of the Canary Islands. *Ibis* 6: 503–520.

Meade-Waldo, E. G. B. 1889c. On a new species of tit. *Annals and Magazine of Natural History* 3: 490.

Meade-Waldo, E. G. B. 1890a. Further notes on the birds of the Canary Islands. *Ibis* 6: 429–438.

Meade-Waldo, E. G. B. 1890b. On a new species of tit. *Annals and Magazine of Natural History* 5: 103.

Meade-Waldo, E. G. B. 1893. List of birds observed in the Canary Islands. *Ibis* 6: 185–207.

Meinertzhagen, R. 1925. May in Madeira. *Ibis* 12: 600–621.

Morelet, A. 1860. *Notice sur l'Histoire Naturelle des Açores*. Paris.

Moreno, J. M. 1988. *Guia de las Aves de las Islas Canarias*. Edicions Interinsular Canaria, Santa Cruz de Tenerife.

Moreno, J. M. 2000. *Cantos y Reclamos de las Aves de Canarias.* Turquesa Ediciones, Santa Cruz de Tenerife.
Moseley, H. N. 1892. *Notes by a naturalist: An account of observations made during the voyage of* H.M.S. Challenger round the world in the years 1872–1876. John Murray, London.
Mougin, J-L., Roux, F., Zino, P. A., Jouanin, C., Stahl, J-C. & Despin, B. 1987. Les oiseaux visiteurs des Iles Selvagens. *Boletim do Museu Municipal do Funchal* 39: 5–24.
Murphy, R. C. 1924a. The marine ornithology of the Cape Verde Islands, with a list of all the birds of the archipelago. *Bulletin of the American Museum of Natural History* 50: 211–278.
Murphy, R. C. 1924b. On the avifauna of the Cape Verde Islands. *Science* 60: 94–95.
Murphy R. C. & Chapin, J. P. 1929. A collection of birds from the Azores. *American Museum Novitates* 384: 1–23.
Nicoll, M. J. 1904. Ornithological journal of a voyage round the world in the *Valhalla (November 1902 to August 1903).* Ibis 4: 32–67.
Noronha, A. & Schmitz, E. 1902. Aus dem Vogelleben der Insel Porto Santo. *Ornithologische Jahrbuch* 12: 130-135.
Noronha, A. & Schmitz, E. 1903. Aus dem Vogelleben der Insel Porto Santo. *Ornithologische Jahrbuch* 14: 119-137.
Noronha, A. & Schmitz, E. 1904. Aus dem Vogelleben der Insel Porto Santo. *Ornithologische Jahrbuch* 15: 124-125.
Ogilvie-Grant, W. R. 1890. Notes on some birds obtained at Madeira, Deserta Grande and Porto Santo. *Ibis* 1890: 438–45.
Oliveira, P. A. 1999. *Conservação e gestão das aves do arquipelago da Madeira.* Parque Natural Da Madeira, Funchal.
Pérez Padrón, F. 2003. *Las aves de Canarias. 4th Edicion.* Publicaciones Turquesa, Santa Cruz de Tenerife.
Polatzek, J. 1908a. Die Vögel der Canaren. *Ornithologische Jahrbuch* 19: 81–119.
Polatzek, J. 1908b. Die Vögel der Canaren. *Ornithologische Jahrbuch* 19: 161–197.
Polatzek, J. 1909a. Die Vögel der Canaren. *Ornithologische Jahrbuch* 20: 1–24.
Polatzek, J. 1909b. Die Vögel der Canaren. *Ornithologische Jahrbuch* 20: 117–134.
Polatzek, J. 1909c. Die Vögel der Canaren. *Ornithologische Jahrbuch* 20: 1–8
Polatzek, J. 1909d. Die Vögel der Canaren. *Ornithologische Jahrbuch* 20: 202–210.
Reid, S. G. 1887. Notes on the Birds of Tenerife. *Ibis* 6: 424–435.
Reid, S. G. 1888. Notes on the Birds of Tenerife. *Ibis* 5: 73–83.
Rufino, R. & Araujo A. 1981. *Migradores palearcticos nas ilhas Selvagens.* CEMPA, Lisboa.
Sacarrão, G. F. & Soares, A. A. 1979. Nomes Portugueses para as Aves da Europa, com Anotações. *Arquivos do Museo Bocage* 6: 395–480.
Salzburger, S., Martens J. & Sturmbauer, C. (2002). Paraphyly of the Blue Tit (*Parus caeruleus*) suggested from cytochrome b sequences. *Molecular Phylogenetics and Evolution* 24: 19–25.
Sánchez, T. 2002. *Aves de Canarias nidificantes.* Editorial Rueda, Madrid.
Sarmento, A. A. 1936. *As aves do arquipélago da Madeira; indígenas e de passagem.* Funchal.
Sarmento, A. A. 1948. *Vertebrados da Madeira.* Funchal.
Schmitz, E. 1893. Tagebuchnotizen von Madeira (1892). *Ornithologische Jahrbuch* 4: 30–32.
Schmitz, E. 1903. Tagebuchnotizen aus Madeira. *Ornithologische Jahrbuch* 14: 206–211.
Schmitz, E. 1909–1910. Brüten der Madeirataube (*Columba trocaz* Hein.) in Gefangenschaft. *Zeitschrift für Oologie* 19 (1909): 22–23; 20 (1910): 68–70.
Sloane, H. 1707. *A voyage to the islands of Madeira, Barbados, etc.* London.
Swash, A. R. H. 1986. Observations of birds in the Madeiran Archipelago, summer 1981. *Bocagiana* 94: 1–13.
Tove, M. 2001. Verification of ID Differences in Fea's and Zino's Petrels. *Birding World* 14: 283–289.
Tristram, H. B. 1889a. Ornithological notes on the island of Gran Canaria. *Ibis* 6: 13–32.
Tristram, H. B. 1889b. On a new species of Chaffinch. *Annals and Magazine of Natural History* 3: 489.
Tristram, H. B. 1890. Notes on the island of Palma in the Canary Group. *Ibis* 6: 67–76.
Trotter, W. D. C. 1970. Observations faunistiques sur L'Ile de Lanzarote (Canarias). *L'Oiseau et R.F.O* 40: 160–172.
Volsøe, H. 1949. A new Blackbird from the Canary Islands. *Dansk Ornitologisk Forenings Tidsskrift* 43: 81–84.
Volsøe, H. 1950. Spring observations on migrant birds in the Canary Islands. *Videnskabelige Meddelelser Dansk Naturhistorisk Forening* 112: 75–117.
Volsøe, H. 1951. The Breeding birds of the Canary Islands. I. Introduction and synopsis of the species. *Videnskabelige Meddelelser Dansk Naturhistorisk Forening* 113: 1–153.
Volsøe, H. 1955a. Origin and evolution of the Canarian Avifauna. *Acta XI Congressus Internationalis Ornitholigici, Basel, 1954.*
Volsøe, H. 1955b. The breeding birds of the Canary Islands. II. Origin and history of the Canarian avifauna. *Videnskabelige Meddelelser Dansk Naturhistorisk Forening* 117: 117–178.
Webb, P. B. & Berthelot, S. 1842. *Histoire Naturelle des Îles Canaries. Tome II.* Béthune, Paris.
Yate Johnson, J. 1885. *Madeira: Its Climate and Scenery. A Handbook.* Dulau & Co, London.
Zino, F., Biscoito, M. J. & Zino, P. A. 1995. Birds of the archipelago of Madeira and the Selvagens. New records and checklist. *Boletim do Museu Municipal do Funchal* 47: 63–100.
Zino, P. A. & Zino, F. 1986. Contribution to the study of the petrels of the genus *Pterodroma* in the archipelago of Madeira. *Boletim do Museu Municipal do Funchal* 38: 141–165.

INDEX

FIGURES IN PLAIN TEXT refer to page numbers, while those in **bold** refer to plate numbers. Species are indexed by their common (e.g. Pigeon, Bolle's) and scientific names; alternative names are also given.

Accentor, Alpine 55, **142**, 292
Accipiter gentilis 20, **72**, 211
　nisus 20, **72**, 212
Acridotheres tristis 65, **162**, 318
Acrocephalus arundinaceus 59, **150**, 303
　brevipennis 59, **150**, 302
　melanopogon 59, **150**, 302
　paludicola 59, **150**, 302
　palustris 59, **150**, 303
　schoenobaenus 59, **150**, 302
　scirpaceus 59, **150**, 303
Actitis hypoleucos 37, **106**, 246
　macularius 37, **106**, 246
Aix sponsa 13, **58**, 197
Ajaia ajaja 10, **52**, 193
Alaemon alaudipes 52, **136**, 283
Alauda arvensis 52, **136**, 285
　razae 52, **136**, 285
Alca torda 45, **122**, 267
Alcedo atthis 51, **134**, 280
Alectoris barbara 24, **80**, 219
　rufa 24, **80**, 218
Alle alle 45, **122**, 267
Ammomanes cinctura 52, **136**, 282
Anas acuta 13, **58**, 200
　americana 13, **58**, 197
　carolinensis 14, **60**, 199
　clypeata 14, **60**, 201
　crecca 14, **60**, 198
　discors 14, **60**, 201
　falcata 14, **60**, 198
　penelope 13, **58**, 197
　platyrhynchos 13, **58**, 199
　querquedula 14, **60**, 200
　rubripes 13, **58**, 199
　strepera 13, **58**, 198
Anser albifrons 12, **56**, 195
　anser 12, **56**, 195
　brachyrhynchus 12, **56**, 194
　caerulescens 12, **56**, 195
　fabalis 12, **56**, 194
Anthus berthelotii 54, **140**, 288
　campestris 54, **140**, 288
　cervinus 54, **140**, 289
　hodgsoni 54, **140**, 288
　petrosus 54, **140**, 290
　pratensis 54, **140**, 289
　richardi 54, **140**, 288
　spinoletta 54, **140**, 290
　trivialis 54, **140**, 289
Apus affinis 50, **132**, 279
　alexandri 50, **132**, 278
　apus 50, **132**, 278
　caffer 50, **132**, 279

　melba 50, **132**, 279
　pallidus 50, **132**, 278
　unicolor 50, **132**, 278
Aquila chrysaetos 21, **74**, 213
　fasciata 21, **74**, 214
　pennata 21, **74**, 214
Ardeirallus sturmii 7, **46**, 186
Ardea cinerea 9, **50**, 190
　herodias 9, **50**, 190
　purpurea 8, **48**, 191
Ardeola ralloides 7, **46**, 187
Arenaria interpres 37, **106**, 247
Asio capensis 49, **130**, 276
　flammeus 49, **130**, 276
　otus 49, **130**, 276
Auk, Little 45, **122**, 267
Avocet, Pied 28, **88**, 226
Aythya affinis 15, **62**, 204
　collaris 15, **62**, 202
　ferina 15, **62**, 202
　fuligula 15, **62**, 203
　marila 15, **62**, 203
　nyroca 15, **62**, 203
Bartramia longicauda 34, **100**, 242
Bee-eater, Blue-cheeked 51, **134**, 280
　European 51, **134**, 280
Bittern, American 7, **46**, 186
　Dwarf 7, **46**, 186
　Eurasian 7, **46**, 185
　Great 7, **46**, 185
　Least 7, **46**, 186
　Little 7, **46**, 186
Blackbird, Common 58, **148**, 299
Blackcap 60, **152**, 308
Bluethroat 56, **144**, 294
Bobolink 69, **170**, 330
Bonxie 38, **108**, 251
Booby, Brown 5, **42**, 183
　Red-footed 5, **42**, 183
Botaurus lentiginosus 7, **46**, 186
　stellaris 7, **46**, 185
Brambling 67, **166**, 322
Branta bernicla 11, **54**, 196
　canadensis 11, **54**, 195
　leucopsis 11, **54**, 196
Bubo bubo 275
　scandiacus 49, **130**, 275
Bubulcus ibis 7, **46**, 188
Bucanetes githagineus 67, **166**, 325
Bucephala albeola 17, **66**, 206
　clangula 17, **66**, 206
Budgerigar 272
Bufflehead 17, **66**, 206
Bullfinch, Azores 67, **166**, 325

Bulweria bulwerii 2, **36**, 177
Bunting, Cirl 69, **170**, 328
　Corn 69, **170**, 330
　Cretzschmar's 69, **170**, 329
　House 69, **170**, 329
　Lapland 327
　Little 69, **170**, 329
　Ortolan 69, **170**, 329
　Reed 69, **170**, 329
　Rock 69, **170**, 328
　Snow 69, **170**, 328
Burhinus oedicnemus 28, **88**, 227
Bustard, Houbara 27, **86**, 225
　Little 27, **86**, 225
Buteo buteo 20, **72**, 212
　lagopus 20, **72**, 213
　rufinus 20, **72**, 213
Butorides virescens 7, **46**, 187
Buzzard, Common 20, **72**, 212
　Long-legged 20, **72**, 213
　Rough-legged 20, **72**, 213
Calonectris diomedea 3, **38**, 177
　edwardsii 3, **38**, 178
Cahow 2, **36**, 175
Calandrella brachydactyla 52, **136**, 284
　rufescens 52, **136**, 284
Calcarius lapponicus 327
Calidris acuminata 32, **96**, 236
　alba 31, **94**, 233
　alpina 32, **96**, 237
　bairdii 32, **96**, 235
　canutus 31, **94**, 233
　ferruginea 32, **96**, 236
　fuscicollis 32, **96**, 235
　himantopus 37, **106**, 237
　maritima 32, **96**, 237
　mauri 31, **94**, 234
　melanotos 32, **96**, 236
　minuta 31, **94**, 234
　minutilla 32, **96**, 235
　pusilla 31, **94**, 233
　temminckii 31, **94**, 234
Canary, Atlantic 66, **164**, 323
Caprimulgus europaeus 48, **128**, 277
　ruficollis 48, **128**, 277
Carduelis cannabina 66, **164**, 324
　carduelis 66, **164**, 324
　chloris 66, **164**, 323
　spinus 66, **164**, 324
Catbird, Grey 55, **142**, 292
Catharus minimus 58, **148**, 298
Catoptrophorus semipalmatus 36, **104**, 247
Cecropis daurica 53, **138**, 287

363

Cercotrichas galactotes 56, 144, 293
Ceryle alcyon 51, 134, 280
Chaetura pelagica 50, 132, 277
Chaffinch, Blue 67, 166, 322
  Common 67, 166, 321
Charadrius alexandrinus 29, 90, 229
  dubius 29, 90, 228
  hiaticula 29, 90, 228
  mongolus 29, 90, 230
  morinellus 29, 90, 230
  semipalmatus 29, 90, 229
  undulata 27, 86, 225
  vociferus 29, 90, 229
Chat, Rufous-tailed Bush 56, 144, 293
Chersophilus duponti 52, 136, 283
Chiffchaff, Canary Islands 61, 154, 310
  Common 61, 154, 309
  Iberian 61, 154, 310
Chlidonias hybrida 44, 120, 265
  leucopterus 45, 122, 266
  niger 45, 122, 266
Chordeiles minor 48, 128, 277
Chough, Red-billed 64, 160, 315
Ciconia ciconia 9, 50, 191
  nigra 9, 50, 191
Cinclus cinclus 55, 142, 292
Circus aeruginosus 19, 70, 210
  cyaneus 19, 70, 210
  gallicus 19, 70, 210
  macrourus 19, 70, 211
  pygargus 19, 70, 211
Cisticola juncidis 59, 150, 301
Cisticola, Zitting 59, 150, 301
Clamator glandarius 48, 128, 273
Clangula hyemalis 16, 64, 205
Coccothraustes coccothraustes 67, 166, 326
Coccyzus erythrophthalmus 48, 128, 273
Columba bollii 47, 126, 269
  junoniae 47, 126, 270
  livia 47, 126, 268
  palumbus 47, 126, 268
  trocaz 47, 126, 269
Coot, American 24, 80, 224
  Common 24, 80, 223
  Crested 24, 80, 224
  Red-knobbed 24, 80, 224
Coracias garrulus 51, 134, 281
Cordon-bleu, Red-cheeked 66, 164, 321
Cormorant, Double-crested 6, 44, 184
  Great 6, 44, 183
Corvus corax 64, 160, 317
  corone 64, 160, 316
  frugilegus 64, 160, 316
  monedula 64, 160, 316
  ruficollis 64, 160, 316
Coturnix coturnix 24, 80, 219

Courser, Cream-coloured 28, 88, 227
Crake, African 221
  African Black 25, 82, 221
  Baillon's 25, 82, 221
  Black 25, 82, 221
  Corn 25, 82, 222
  Little 25, 82, 221
  Spotted 25, 82, 220
Crane, Common 27, 86, 224
  Sandhill 27, 86, 224
Crex crex 25, 82, 222
  egregia 221
Crossbill, Common 67, 166, 325
Crow, Carrion 64, 160, 316
Cuckoo, Black-billed 48, 128, 273
  Common 48, 128, 273
  Great Spotted 48, 128, 273
Cuculus canorus 48, 128, 273
Curlew, Eurasian 34, 100, 242
  Slender-billed 34, 100, 242
Cursorius cursor 28, 88, 227
Cyanistes ultramarinus 63, 158, 312
Cygnus olor 11, 54, 194
Dabchick 1, 34, 174
Delichon urbicum 53, 138, 287
Dendrocopos major 51, 134, 282
Dendrocygna bicolor 11, 54, 193
  viduata 11, 54, 194
Dendroica coronata 68, 168, 326
  magnolia 68, 168, 326
  petechia 68, 168, 326
  triata 68, 168, 326
Dipper, White-throated 55, 142, 292
Diver, Great Northern 1, 34, 173
Dolichonyx oryzivorus 69, 170, 330
Dotterel 29, 90, 230
Dove, African Collared 46, 124, 270
  Barbary 47, 126, 270
  Eurasian Collared 46, 124, 270
  European Turtle 46, 124, 271
  Laughing 46, 124, 271
  Namaqua 46, 124, 271
  Palm 46, 124, 271
  Rock 47, 126, 268
Dowitcher, Long-billed 34, 100, 240
  Short-billed 34, 100, 239
Duck, American Black 13, 58, 199
  Falcated 14, 60, 198
  Ferruginous 15, 62, 199
  Fulvous Whistling 11, 54, 193
  Long-tailed 16, 64, 205
  Marbled 14, 60, 201
  Ring-necked 15, 62, 202
  Tufted 15, 62, 203
  White-faced Whistling 11, 54, 194
  Wood 13, 58, 197
Dumetella carolinensis 55, 142, 292
Dunlin 32, 96, 237

Dunnock 55, 142, 292
Eagle, Bonelli's 21, 74, 214
  Booted 21, 74, 214
  Golden 21, 74, 213
  Short-toed Snake 19, 70, 210
  White-tailed Sea 21, 74, 209
Egret, Cattle 7, 46, 188
  Great White 8, 48, 190
  Intermediate 8, 48, 189
  Little 8, 48, 189
  Snowy 8, 48, 189
  Western Reef 8, 48, 188
Egretta alba 8, 48, 190
  garzetta 8, 48, 189
  gularis 8, 48, 188
  intermedia 8, 48, 189
  thula 8, 48, 189
Eider, Common 16, 64, 204
  King 16, 64, 204
Elanoides forficatus 18, 68, 207
Elanus caeruleus 18, 68, 207
Emberiza caesia 69, 170, 329
  calandra 69, 170, 330
  cia 69, 170, 328
  cirlus 69, 170, 328
  hortulana 69, 170, 329
  pusilla 69, 170, 329
  schoeniclus 69, 170, 329
  striolata 69, 170, 329
Eremopterix nigriceps 52, 136, 282
Erithacus rubecula 56, 144, 293
Estrilda astrild 66, 164, 321
  melpoda 321
Eurystomus glaucurus 51, 134, 281
Falco biarmicus 23, 78, 217
  columbarius 22, 76, 216
  eleonorae 23, 78, 217
  naumanni 22, 76, 215
  pelegrinoides 23, 78, 218
  peregrinus 23, 78, 218
  sparverius 22, 76, 216
  subbuteo 23, 78, 217
  tinnunculus 22, 76, 215
  vespertinus 22, 76, 216
Falcon, Barbary 23, 78, 218
  Eleonora's 23, 78, 217
  Lanner 23, 78, 217
  Peregrine 23, 78, 218
  Western Red-footed 22, 76, 216
Ficedula albicollis 312
  hypoleuca 62, 156, 312
  parva 62, 156, 312
Fieldfare 58, 148, 299
Finch, Trumpeter 67, 166, 325
Firecrest, Madeira 62, 156, 311
Flamingo, Greater 10, 52, 193
  Lesser 10, 52, 193
Flycatcher, Collared 312
  Pied 62, 156, 312
  Red-breasted 62, 156, 312
  Spotted 62, 156, 311

Fratercula arctica 45, 122, 267
Fregata magnificens 5, 42, 185
Fregetta grallaria 4, 40, 181
Frigatebird, Magnificent 5, 42, 185
Fringilla coelebs 67, 166, 321
   montifringilla 67, 166, 322
   teydea 67, 166, 322
Fulica americana 24, 80, 224
   atra 24, 80, 223
   cristata 24, 80, 224
Fulmar, Northern 2, 36, 175
Fulmarus glacialis 2, 36, 175
Gadwall 13, 58, 198
Galerida cristata 52, 136, 284
   theklae 52, 136, 285
Gallinago delicata 33, 98, 239
   gallinago 33, 98, 239
   media 33, 98, 239
Gallinula angulata 26, 84, 222
   chloropus 26, 84, 222
Gallinule, Allen's 26, 84, 223
   American Purple 26, 84, 223
Gannet, Northern 5, 42, 183
Garganey 14, 60, 200
Gavia immer 1, 34, 173
Gelochelidon nilotica 43, 118, 261
Geronticus eremita 9, 50, 192
Glareola pratincola 28, 88, 228
Godwit, Bar-tailed 34, 100, 241
   Black-tailed 34, 100, 240
Goldcrest 62, 156, 311
   Tenerife 62, 156, 311
Goldeneye, Common 17, 66, 206
Goldfinch, European 66, 164, 324
Goose, Barnacle 11, 54, 196
   Bean 12, 56, 194
   Brent 11, 54, 196
   Canada 11, 54, 195
   Greater White-fronted 12, 56, 195
   Greylag 12, 56, 195
   Pink-footed 12, 56, 194
   Snow 12, 56, 195
Goshawk, Northern 20, 72, 211
Grebe, Black-necked 1, 34, 174
   Eared 1, 34, 174
   Great Crested 1, 34, 174
   Horned 1, 34, 174
   Little 1, 34, 174
   Pied-billed 1, 34, 173
   Slavonian 1, 34, 174
Greenfinch, European 66, 164, 323
Greenshank, Common 35, 102, 244
Grus canadensis 27, 86, 224
   grus 27, 86, 224z
Guillemot, Brünnich's 45, 122, 267
   Common 45, 122, 266
Guineafowl, Helmeted 24, 80, 220
Gull, American Herring 41, 114, 257
   Audouin's 40, 112, 255

Black-headed 39, 110, 254
Bonaparte's 42, 116, 254
Common 40, 112, 256
Franklin's 39, 110, 253
Glaucous 42, 116, 259
Glaucous-winged 42, 116, 260
Great Black-backed 41, 114, 260
Great Black-headed 40, 112, 251
Herring 41, 114, 257
Iceland 42, 112, 259
Laughing 39, 110, 252
Lesser Black-backed 41, 114, 256
Little 39, 110, 253
Mediterranean 39, 110, 252
Mew 40, 256
Pallas's 40, 112, 251
Ring-billed 40, 112, 255
Sabine's 39, 110, 253
Slender-billed 40, 112, 254
Yellow-legged 41, 114, 258
Haematopus meadewaldoi 28, 88, 226
   ostralegus 28, 88, 225
Halcyon leucocephala 51, 134, 279
Haliaeetus albicilla 21, 74, 209
Harrier, Hen 19, 70, 210
   Marsh 19, 70, 210
   Montagu's 19, 70, 211
   Pallid 19, 70, 211
   Western Marsh 19, 70, 210
Hawfinch 67, 166, 326
Heron, Black 8, 48, 189
   Black-crowned Night 7, 46, 187
   Great Blue 9, 50, 190
   Green 7, 46, 187
   Green-backed 7, 46, 187
   Grey 9, 50, 190
   Little Blue 8, 48, 188
   Night 7, 46, 187
   Purple 8, 48, 191
   Squacco 7, 46, 187
   Tricoloured 8, 48, 188
Himantopus himantopus 28, 88, 226
Hippolais icterina 61, 154, 304
   opaca 61, 154, 304
   pallida 303
   polyglotta 61, 154, 304
   rustica 53, 138, 286
Hobby, Eurasian 23, 78, 217
Honey-buzzard, Western 18, 68, 207
Hoopoe 51, 134, 281
Hydrobates pelagicus 4, 40, 181
Hydranassa ardesiaca 8, 48, 189
   caerulea 8, 48, 188
   tricolor 8, 48, 188
Hydroprogne caspia 43, 118, 261

Hylocichla mustelina 58, 148, 298
Ibis, Bald 9, 50, 192
   Glossy 9, 50, 192
   Sacred 9, 50, 192
Ixobrychus exilis 7, 46, 186
   minutus 7, 46, 186
Jackdaw, Western 64, 160, 316
Jynx torquilla 51, 134, 281
Kestrel, American 22, 76, 216
   Common 22, 76, 215
   Lesser 22, 76, 215
Killdeer 29, 90, 229
Kingfisher, Belted 51, 134, 280
   Common 51, 134, 280
   Grey-headed 51, 134, 279
Kinglet, Tenerife 62, 156, 311
Kite, American Swallow-tailed 18, 68, 209
   Black 18, 68, 208
   Black-shouldered 18, 68, 207
   Cape Verde 18, 68, 208
   Red 18, 68, 209
Kittiwake, Black-legged 42, 116, 261
Knot, Red 31, 94, 233
Lanius collurio 63, 158, 314
   isabellinus 63, 158, 314
   meridionalis 63, 158, 315
   senator 63, 158, 315
Lapwing, Northern 30, 92, 232
   Sociable 30, 92, 232
   Spur-winged 30, 92, 232
   White-tailed 30, 92, 232
Lark, Bar-tailed Desert 52, 136, 282
   Calandra 52, 136, 283
   Crested 52, 136, 284
   Dupont's 52, 136, 283
   Greater Hoopoe 52, 136, 283
   Greater Short-toed 52, 136, 284
   Lesser Short-toed 52, 136, 284
   Raso 52, 136, 285
   Sky 52, 136, 285
   Thekla 52, 136, 285
Larus argentatus 41, 114, 257
   atricilla 39, 110, 252
   audouinii 40, 112, 255
   canus 40, 112, 256
   delawarensis 40, 112, 255
   fuscus 41, 114, 256
   genei 40, 112, 254
   glaucescens 42, 116, 260
   glaucoides 42, 116, 259
   hyperboreus 42, 116, 259
   ichthyaetus 40, 112, 251
   marinus 41, 114, 260
   melanocephalus 39, 110, 252
   michahellis 41, 114, 258
   minutus 39, 110, 253
   philadelphia 42, 116, 254
   pipixcan 39, 110, 253

ridibundus 39, 110, 254
sabini 39, 110, 253
smithsonianus 41, 114, 257
Limnocorax flavirostra 25, 82, 221
Limnodromus griseus 34, 100, 239
  scolopaceus 34, 100, 240
Limosa lapponica 34, 100, 241
  limosa 34, 100, 240
Linnet, Common 66, 164, 324
Locustella fluviatilis 59, 150, 301
  luscinioides 59, 150, 301
  naevia 59, 150, 301
Loxia curvirostra 67, 166, 325
Luscinia luscinia 56, 144, 293
  megarhynchos 56, 144, 294
  svecica 56, 144, 294
Lymnocryptes minimus 33, 98, 238
Mallard 13, 58, 199
Marmaronetta angustirostris 14, 60, 201
Martin, Brown-throated Sand 53, 138, 286
  Common House 53, 138, 287
  Eurasian Crag 53, 138, 286
  Plain 53, 138, 286
  Purple 53, 138, 285
  Sand 53, 138, 286
Melanitta nigra 16, 64, 205
  perspicillata 16, 64, 205
Melanocorypha calandra 52, 136, 283
Melopsittacus undulatus 272
Merganser, Hooded 17, 66, 206
  Red-breasted 17, 66, 207
Mergellus albellus 17, 66, 206
Mergus cucullatus 17, 66, 206
  serrator 17, 66, 207
Merlin 22, 76, 216
Merops apiaster 51, 134, 280
  persicus 51, 134, 280
Milvus fasciicauda 18, 68, 208
  migrans 18, 68, 208
  milvus 18, 68, 209
Monticola saxatilis 58, 148, 298
  solitarius 58, 148, 298
Montifringilla nivalis 67, 166, 320
Moorhen, Common 26, 84, 222
  Lesser 26, 84, 222
Morus bassanus 5, 42, 183
Motacilla alba 55, 142, 291
  cinerea 55, 142, 291
  citreola 55, 142, 291
  flava 55, 142, 290
Muscicapa striata 62, 156, 311
Myiopsitta monachus 48, 128, 272
Myna, Common 65, 162, 318
Neophron percnopterus 19, 70, 209
Netta rufina 14, 60, 202
Nighthawk, Common 48, 128, 277
Nightingale, Common 56, 144, 294

Thrush 56, 144, 294
Nightjar, European 48, 128, 277
  Red-necked 48, 128, 277
Nucifraga caryocatactes 64, 160, 315
Numenius arquata 34, 100, 242
  phaeopus 34, 100, 241
  tenuirostris 34, 100, 242
Numida meleagris 24, 80, 220
Nutcracker, Spotted 64, 160, 315
Nuthatch, Eurasian 62, 156, 313
Nycticorax nycticorax 7, 46, 187
Oceanites oceanicus 4, 40, 180
Oceanodroma castro 4, 40, 182
  leucorhoa 4, 40, 181
  monorhis 4, 40, 181
Oena capensis 46, 124, 271
Oenanthe deserti 57, 146, 297
  hispanica 57, 146, 297
  isabellina 57, 146, 296
  leucopyga 297
  lugens 57, 146, 297
  oenanthe 57, 146, 296
Onychoprion anaethetus 44, 120, 265
  fuscata 44, 120, 265
Oriole, Golden 63, 158, 314
Oriolus oriolus 63, 158, 314
Osprey 21, 74, 214
Otus scops 49, 130, 274
Ouzel, Ring 58, 148, 298
Owl, Barn 49, 130, 274
  Eurasian Eagle 275
  Eurasian Scops 49, 130, 274
  Long-eared 49, 130, 276
  Marsh 49, 130, 276
  Northern Hawk 49, 130, 275
  Short-eared 49, 130, 276
  Snowy 49, 130, 275
  Tawny 49, 130, 275
Oystercatcher, Canary Islands 28, 88, 226
  Canary Islands Black 28, 88, 226
  Eurasian 28, 88, 225
Pandion haliaetus 21, 74, 214
Parakeet, Monk 48, 128, 272
  Ring-necked 48, 128, 272
  Rose-ringed 48, 128, 272
Parrot, Senegal 272
Partridge, Barbary 24, 80, 219
  Red-legged 24, 80, 218
Parus major 313
Passer domesticus 65, 162, 318
  hispaniolensis 65, 162, 319
  iagoensis 65, 162, 319
  montanus 65, 162, 319
Passerculus sandwichensis 68, 168, 327
Pelagodroma marina 4, 40, 180
Pelecanus onocrotalus 6, 44, 184
  rufescens 6, 44, 185

Pelican, Great White 6, 44, 184
  Pink-backed 6, 44, 185
  White 6, 44, 184
Pernis apivorus 18, 68, 207
Petrel, Bermuda 2, 36, 175
  Bulwer's 2, 36, 177
  Cape Verde 2, 36, 176
  Fea's 2, 36, 176
  Madeira 2, 36, 176
  Soft-plumaged 2, 36, 176
  Trindade 2, 36, 175
  Zino's 2, 36, 176
Petrochelidon pyrrhonota 287
Petronia petronia 65, 162, 319
Phaethon aethereus 5, 42, 182
  lepturus 5, 42, 182
Phalacrocorax aristotelis 6, 44, 184
  auritus 6, 44, 184
  carbo 6, 44, 183
Phalarope, Grey 38, 108, 248
  Red 38, 108, 248
  Red-necked 38, 108, 248
  Wilson's 38, 108, 247
Phalaropus fulicarius 38, 108, 248
  lobatus 38, 108, 248
  tricolor 38, 108, 247
Philomachus pugnax 33, 98, 238
Phoenicopterus minor 10, 52, 193
  roseus 10, 52, 193
Phoenicurus ochruros 56, 144, 294
  phoenicurus 56, 144, 295
Phylloscopus bonelli 61, 154, 309
  canariensis 61, 154, 310
  collybita 61, 154, 309
  humei 61, 154, 309
  ibericus 61, 154, 310
  inornatus 61, 154, 308
  sibilatrix 61, 154, 309
  trochilus 61, 154, 310
Pigeon, Bolle's 47, 126, 269
  Feral 47, 126, 268
  Laurel 47, 126, 270
  Long-toed 47, 126, 269
  Trocaz 47, 126, 269
  White-tailed Laurel 47, 126, 270
Pintail, Northern 13, 58, 200
Pipit, Berthelot's 54, 140, 288
  Meadow 54, 140, 2897
  Olive-backed 54, 140, 288
  Red-throated 54, 140, 289
  Richard's 54, 140, 288
  Rock 54, 140, 290
  Tawny 54, 140, 288
  Tree 54, 140, 289
  Water 54, 140, 290
Platalea leucorodia 10, 52, 192
Plectrophenax nivalis 69, 170, 328
Plegadis falcinellus 9, 50, 192
Ploceus cucullatus 320
Plover, American Golden 30, 92, 230

INDEX

Black-bellied 30, 92, 231
Egyptian 28, 88, 227
European Golden 30, 92, 231
Grey 30, 92, 231
Kentish 29, 90, 229
Lesser Sand 29, 90, 230
Little Ringed 29, 90, 228
Mongolian 29, 90, 230
Pacific Golden 30, 92, 231
Ringed 29, 90, 228
Semipalmated 29, 90, 229
Sociable 30, 92, 232
Spur-winged 30, 92, 232
White-tailed 30, 92, 232
Pluvialis apricaria 30, 92, 231
 dominica 30, 92, 230
 fulva 30, 92, 231
 squatarola 30, 92, 231
Pluvianus aegyptius 28, 88, 227
Pochard, Common 15, 62, 202
 Red-crested 14, 60, 202
Podiceps auritus 1, 34, 174
 cristatus 1, 34, 174
 nigricollis 1, 34, 174
Podilymbus podiceps 1, 34, 173
Poicephalus senegalus 272
Porphyrula alleni 26, 84, 223
 martinica 26, 84, 223
Porzana carolina 25, 82, 220
 parva 25, 82, 221
 porzana 25, 82, 220
 pusilla 25, 82, 221
Pratincole, Collared 28, 88, 228
Progne subis 53, 138, 285
Prunella collaris 55, 142, 292
 modularis 55, 142, 292
Psittacula krameri 48, 128, 272
Pterocles alchata 46, 124, 268
 orientalis 46, 124, 268
Pterodroma arminjoniana 2, 36, 175
 cahow 2, 36, 175
 feae 2, 36, 176
 madeira 2, 36, 176
 mollis 2, 36, 176
Ptyonoprogne rupestris 53, 138, 286
Puffin, Atlantic 45, 122, 267
Puffinus baroli 4, 40, 179
 gravis 3, 38, 178
 griseus 3, 38, 178
 mauretanicus 3, 38, 179
 puffinus 3, 38, 179
Pyrrhocorax pyrrhocorax 64, 160, 315
Pyrrhula murina 67, 166, 325
Quail, Common 24, 80, 219
Quelea quelea 66, 164, 320
Quelea, Red-billed 66, 164, 320
Rail, Water 25, 82, 220
Rallus aquaticus 25, 82, 220
Raven, Common 64, 160, 317
 Brown-necked 64, 160, 316
Razorbill 45, 122, 267
Recurvirostra avosetta 28, 88, 226

Redshank, Common 35, 102, 243
 Spotted 35, 102, 243
Redstart, American 68, 168, 327
 Black 56, 144, 294
 Common 56, 144, 295
Redwing 58, 148, 300
Regulus madeirensis 62, 156, 311
 regulus 62, 156, 311
 teneriffae 62, 156, 311
Remiz pendulinus 62, 156, 314
Riparia paludicola 53, 138, 286
 riparia 53, 138, 286
Rissa tridactyla 42, 116, 261
Robin, European 56, 144, 293
 Rufous Bush 56, 144, 293
Roller, Broad-billed 51, 134, 281
 European 51, 134, 281
Rook 64, 160, 316
Ruff 33, 98, 238
Sanderling 31, 94, 233
Sandgrouse, Black-bellied 46, 124, 268
 Pin-tailed 46, 124, 268
Sandpiper, Baird's 32, 96, 235
 Buff-breasted 33, 98, 237
 Common 37, 106, 246
 Curlew 32, 96, 236
 Green 36, 104, 245
 Least 32, 96, 235
 Marsh 35, 102, 243
 Pectoral 32, 96, 236
 Purple 33, 98, 237
 Semipalmated 31, 94, 233
 Sharp-tailed 32, 96, 236
 Solitary 245
 Spotted 37, 106, 246
 Stilt 37, 106, 237
 Terek 36, 104, 246
 Upland 30, 92, 242
 Western 31, 94, 234
 White-rumped 32, 96, 235
 Wood 36, 104, 245
Saxicola dacotiae 57, 146, 295
 rubetra 57, 146, 295
 torquatus 57, 146, 296
Scaup, Greater 15, 62, 203
 Lesser 15, 62, 204
Scolopax rusticola 33, 98, 240
Scoter, Common 16, 64, 205
 Surf 16, 64, 205
Seiurus motacilla 68, 168, 327
 noveboracensis 68, 168, 327
Serin, European 66, 164, 323
Serinus canaria 66, 164, 323
 serinus 66, 164, 323
Setophaga ruticilla 68, 168, 327
Shag 6, 44, 184
Shearwater, Balearic 3, 38, 179
 Cape Verde 3, 38, 178
 Cory's 3, 38, 177
 Great 3, 38, 178
 Little 4, 40, 179

Macaronesian 4, 40, 179
 Manx 3, 38, 179
 Sooty 3, 38, 178
Shelduck, Common 12, 56, 196
 Ruddy 12, 56, 196
Shoveler 14, 60, 201
Shrike, Isabelline 63, 158, 314
 Red-backed 63, 158, 314
 Southern Grey 63, 158, 315
 Woodchat 63, 158, 315
Siskin, European 66, 164, 324
Sitta europaea 62, 156, 313
Skua, Arctic 38, 108, 249
 Great 38, 108, 251
 Long-tailed 38, 108, 250
 Pomarine 38, 108, 248
 South Polar 38, 108, 251
Skylark 52, 136, 285
Smew 17, 66, 206
Snipe, Common 33, 98, 239
 Great 33, 98, 239
 Jack 33, 98, 238
 Wilson's 33, 98, 239
Snowfinch, White-winged 67, 166, 320
Somateria mollissima 16, 64, 204
 spectabilis 16, 64, 204
Sora 25, 82, 220
Sparrow, Cape Verde 65, 162, 319
 Hedge 55, 142, 292
 House 65, 162, 318
 Iago 65, 162, 319
 Rock 65, 162, 319
 Savannah 68, 168, 327
 Spanish 65, 162, 319
 Tree 65, 162, 319
Sparrowhawk, Eurasian 20, 72, 212
Sparrow-lark, Black-crowned 52, 136, 282
Spoonbill, Eurasian 10, 52, 192
 Roseate 10, 52, 193
Sprosser 56, 144, 293
Starling, Common 65, 162, 317
 Rose-coloured 65, 162, 318
 Spotless 65, 162, 318
Stercorarius longicaudus 38, 108, 250
 maccormicki 38, 108, 251
 parasiticus 38, 108, 249
 pomarinus 38, 108, 248
 skua 38, 108, 251
Sterna bengalensis 43, 118, 261
 dougallii 43, 118, 263
 forsteri 44, 120, 264
 hirundo 44, 120, 263
 maxima 43, 118, 262
 paradisaea 44, 120, 263
 repressa 44, 120, 264
 sandvicensis 43, 118, 262
Sternula albifrons 43, 118, 264
Stilt, Black-winged 28, 88, 226
Stint, Little 31, 94, 234

367

Temminck's 31, 94, 234
Stonechat, Canary Islands 57, 146, 295
  Common 57, 146, 296
Stone-curlew 28, 88, 227
Stork, Black 9, 50, 191
  White 9, 50, 191
Storm-petrel, European 4, 40, 181
  Leach's 4, 40, 181
  Madeiran 4, 40, 182
  Swinhoe's 4, 40, 181
  White-bellied 4, 40, 181
  White-faced 4, 40, 180
  Wilson's 4, 40, 180
Streptopelia decaocto 46, 124, 270
  roseogrisea 46, 124, 270
  senegalensis 46, 124, 271
  turtur 46, 124, 271
Strix aluco 49, 130, 275
Sturnus roseus 65, 162, 318
  unicolor 65, 162, 318
  vulgaris 65, 162, 317
Sula leucogaster 5, 42, 183
  sula 5, 42, 183
Surnia ulula 49, 130, 275
Swallow, American Cliff 53, 138, 287
  Barn 53, 138, 286
  Red-rumped 53, 138, 287
Swan, Mute 11, 54, 194
Swift, Alpine 50, 132, 279
  Cape Verde 50, 132, 278
  Chimney 50, 132, 277
  Common 50, 132, 278
  Little 50, 132, 279
  Pallid 50, 132, 278
  Plain 50, 132, 278
  White-rumped 50, 132, 279
Sylvia atricapilla 60, 152, 308
  borin 60, 152, 308
  cantillans 60, 152, 306
  communis 60, 152, 307
  conspicillata 60, 152, 305
  curruca 60, 152, 307
  deserti 60, 152, 307
  deserticola 60, 152, 305
  hortensis 60, 152, 307
  melanocephala 60, 152, 306
  mystacea 60, 152, 306
Tachybaptus ruficollis 1, 34, 174
Tadorna ferruginea 12, 56, 196
  tadorna 12, 56, 196
Teal, Blue-winged 14, 60, 201
  Eurasian 14, 60, 198
  Green-winged 14, 60, 199
  Marbled 14, 60, 201
Tern, Arctic 44, 120, 263
  Black 45, 122, 266
  Bridled 44, 120, 265
  Caspian 43, 118, 261
  Common 44, 120, 263
  Forster's 44, 120, 264
  Gull-billed 43, 118, 261

Lesser Crested 43, 118, 262
Little 43, 118, 264
Roseate 43, 118, 263
Royal 43, 118, 262
Sandwich 43, 118, 262
Sooty 44, 120, 265
Whiskered 44, 120, 265
White-cheeked 44, 120, 264
White-winged Black 45, 122, 266
Tetrax tetrax 27, 86, 225
Threskiornis aethiopicus 9, 50, 192
Thrush, Blue Rock 58, 148, 298
  Dark-throated 58, 148, 299
  Grey-cheeked 58, 148, 298
  Mistle 58, 148, 210
  Rufous-tailed Rock 58, 148, 298
  Song 58, 148, 300
  Wood 58, 148, 298
Tit, African Blue 63, 158, 312
  Great 313
  Penduline 62, 156, 314
Tringa erythropus 35, 102, 243
  flavipes 35, 102, 244
  glareola 36, 104, 245
  melanoleuca 35, 102, 244
  nebularia 35, 102, 244
  ochropus 36, 104, 245
  solitaria 245
  stagnatilis 35, 102, 243
  totanus 35, 102, 243
Troglodytes troglodytes 55, 142, 292
Tropicbird, Red-billed 5, 42, 182
  White-tailed 5, 42, 182
Tryngites subruficollis 33, 98, 237
Turdus iliacus 58, 148, 300
  merula 58, 148, 299
  philomelos 58, 148, 300
  pilaris 58, 148, 299
  ruficollis 58, 148, 299
  torquatus 58, 148, 298
  viscivorus 58, 148, 300
Turnstone, Ruddy 37, 106, 247
Tyto alba 49, 130, 274
Upupa epops 51, 134, 281
Uraeginthus bengalus 66, 164, 321
Uria aalge 45, 122, 266
  lomvia 45, 122, 267
Vanellus gregarius 30, 92, 232
  leucurus 30, 92, 232
  spinosus 30, 92, 232
  vanellus 30, 92, 232
Vulture, Egyptian 19, 70, 209
Wagtail, Citrine 55, 142, 291
  Grey 55, 142, 291
  White 55, 142, 291
  Yellow 55, 142, 290
Waldrapp 9, 50, 192
Warbler, African Desert 60, 152, 307
  Aquatic 59, 150, 302
  Blackpoll 68, 168, 326
  Cape Verde 59, 150, 302

Cape Verde Swamp 59, 150, 302
Eastern Olivaceous 303
Fan-tailed 59, 150, 301
Garden 60, 152, 308
Common Grasshopper 59, 150, 301
Great Reed 59, 150, 303
Hume's Yellow-browed 61, 154, 309
Icterine 61, 154, 304
Magnolia 68, 168, 326
Marsh 59, 150, 303
Melodious 61, 154, 304
Ménétries's 60, 152, 306
Moustached 59, 150, 302
Myrtle 68, 168, 326
Reed 59, 150, 303
River 59, 150, 301
Sardinian 60, 152, 306
Savi's 59, 150, 301
Sedge 59, 150, 302
Spectacled 60, 152, 305
Subalpine 60, 152, 306
Tristram's 60, 152, 305
Western Bonelli's 61, 154, 309
Western Olivaceous 61, 154, 304
Western Orphean 60, 152, 307
Willow 61, 154, 310
Wood 61, 154, 309
Yellow 68, 168, 326
Yellow-browed 61, 154, 308
Yellow-rumped 68, 168, 326
Waterthrush, Louisiana 68, 168, 327
  Northern 68, 168, 327
Waxbill, Common 66, 164, 312
  Orange-cheeked 321
Weaver, Village 320
Wheatear, Black-eared 57, 146, 297
  Desert 57, 146, 297
  Isabelline 57, 146, 296
  Mourning 57, 146, 297
  Northern 57, 146, 296
  White-crowned Black 297
Whimbrel 34, 100, 241
Whinchat 57, 146, 295
Whitethroat, Common 60, 152, 307
  Lesser 60, 152, 307
Wigeon, American 13, 58, 197
  Eurasian 13, 58, 197
Willet 36, 104, 247
Woodcock 33, 98, 240
Woodpecker, Great Spotted 51, 134, 282
Woodpigeon, Common 47, 126, 268
Wren, Winter 56, 144, 292
Wryneck, Eurasian 51, 134, 281
Xenus cinereus 36, 104, 246
Yellowlegs, Greater 35, 102, 244
  Lesser 35, 102, 244